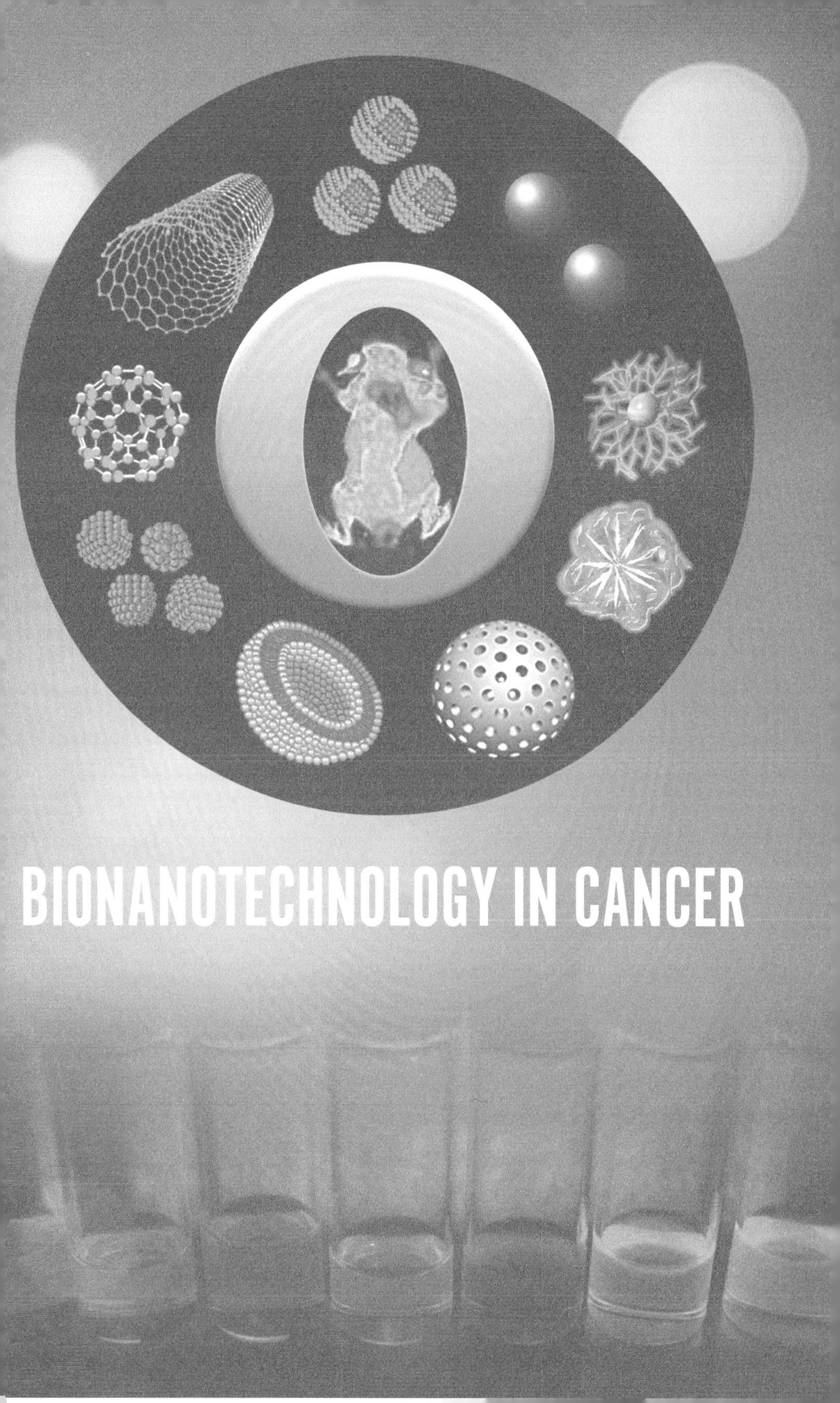

BIONANOTECHNOLOGY IN CANCER

BIONANOTECHNOLOGY IN CANCER
Diagnosis and Therapy

edited by
D. Sakthi Kumar
Aswathy Ravindran Girija

JENNY STANFORD
PUBLISHING

Published by

Jenny Stanford Publishing Pte. Ltd.
101 Thomson Road
#06-01, United Square
Singapore 307591

Email: editorial@jennystanford.com
Web: www.jennystanford.com

British Library Cataloguing-in-Publication Data
A catalogue record for this book is available from the British Library.

Bionanotechnology in Cancer: Diagnosis and Therapy

ISBN 978-981-4800-30-3 (Hardcover)
ISBN 978-0-429-42291-1 (eBook)

Contents

Preface xv
Foreword xxi

1. Introduction to Cancer, Conventional Therapies, and Bionano-Based Advanced Anticancer Strategies **1**

Sindhu C. Pillai and Jayasree Nair

1.1 Cancer 1
1.2 Understanding Cancer and Its Occurrence 3
1.2.1 Genes Activities Related to Cancer 5
1.3 Different Theories of Carcinogenesis 7
1.4 Factors Influencing Cancer Development 8
1.4.1 Intrinsic/Biological Factors 8
1.4.2 External Factors 10
1.5 Microenvironment of Cancer 13
1.5.1 Tumor Angiogenesis 14
1.5.2 Physical and Chemical Cues 14
1.5.3 Acidosis 15
1.5.4 High Interstitial Fluid Pressure 15
1.5.5 Enhanced Permeability Retention 16
1.5.6 Hypoxia 16
1.6 Cellular and Non-Cellular Components That Support Tumor Growth 17
1.6.1 Cellular Components 17
1.6.2 Non-Cellular Components 19
1.7 Difference between Normal Cells and Cancer Cells 21
1.8 Conventional Anti-Cancer Therapies 22
1.9 Cell Death and Cancer 26
1.10 Advanced Nanotechnology for Cancer Diagnosis and Therapy 29
1.10.1 Key Features of Nanomaterials for Therapy and Diagnostic Applications 30

1.10.1.1 Nanoparticle size 30
1.10.1.2 Nanoparticle surface properties 30
1.10.1.3 NPs in diagnosis 30
1.10.1.4 NPs in imaging 31
1.10.1.5 Targeted nanoparticles 31
1.10.1.6 Nanomaterials to target and regulate tumor and its microenvironment 32
1.10.2 Challenges Faced by NPs in Combatting Tumor 32
1.10.2.1 Challenges faced during circulation 32
1.10.2.2 Challenges after reaching tumor 32
1.11 Conclusion and Future Perspective 33

2. Understanding the Interaction of Nanoparticles at the Cellular Interface 45

Baiju G. Nair

2.1 Introduction 45
2.2 Nanotechnology 46
2.3 NPs and Cell Internalization 47
2.4 Transmembrane Penetration by Designed Nanomaterials 49
2.5 Engineered Protein Corona: A Better Drug Delivery System 50
2.6 Subcellular Interaction of NPs 53
2.7 Endosomal/Lysosomal Escape of NPs 55
2.7.1 Formation of Membrane Pores in Lysosomes/Endosomes 57
2.7.2 Proton Sponge Effect 58
2.7.3 Fusion with Endosomal Membrane 58
2.7.4 Photochemical Rupture of Endosomal Membrane 59

2.8 Tumor and Nanomedicine 59
2.9 Smart Material as Drug Delivery Nanoparticle 60
2.9.1 Polymeric Drug Nanocarrier: Act as a Smart Material 61
2.10 Conclusion and Future Perspective 63

NANODIAGNOSTICS

3. Radiolabeled Nanoparticles for Cancer Diagnosis 75

R. G. Aswathy and D. Sakthi Kumar

3.1 Introduction 75
3.2 SPECT 77
3.3 Positron Emission Tomography 79
3.4 Radiolabeled Nanoparticles 83
3.4.1 Polymer-Based RNPs 84
3.4.2 Liposome-Based RNPs 86
3.4.3 Carbon Nanotube-Based RNPs 90
3.4.4 Inorganic NPs as RNPs 92
3.5 Multimodal Imaging of PET and SPECT with MRI, CT, and NIRF 94
3.6 Conclusion and Future Prospects 97

4. Magnetic Nanoparticles for Cancer Diagnosis 109

R. G. Aswathy and D. Sakthi Kumar

4.1 Introduction 109
4.2 Principle of MRI 112
4.3 MNPs as MRI Contrast Agents 114
4.4 Dual-Mode (T_1 and T_2) MRI Contrast Agents 118
4.5 Multimodal Imaging: MRI with PET, SPECT, Optical, and Ultrasound 120
4.6 Novel Application of MNPs for Imaging 124
4.6.1 Magnetic Particle Imaging–MRI 124
4.6.2 MMUS–US Dual-Modal Imaging 126
4.6.3 MPA–US Dual-Modal Imaging 127
4.7 Conclusion and Future Prospects 128

5. Acoustically Reflective Nanoparticles for Tumor Diagnosis **143**

R. G. Aswathy and D. Sakthi Kumar

5.1 Introduction 143
5.2 Ultrasound Imaging 143
5.2.1 Microbubble as Contrast Agent 145
5.2.2 Different Nanobubbles as Contrast Agents 148
5.2.2.1 PFC emulsion nanodroplets 149
5.2.2.2 Echogenic Liposomes 150
5.2.2.3 Polymer-based nanobubbles 151
5.2.2.4 Solid NPs 151
5.3 Multimodal Imaging: US Imaging with MRI, PET, SPECT, and PA 154
5.4 Conclusion and Future Prospects 159

6. X-Ray Computed Tomography and Nanomaterials as Contrast Agents for Tumor Diagnosis **171**

R. G. Aswathy and D. Sakthi Kumar

6.1 Introduction 171
6.2 Iodinated and Gold-Based Nano Contrast Agents 173
6.3 Radiopaque Polymeric NPs 178
6.4 Inorganic Nanomaterials 179
6.5 Multi-Modal Imaging of X-Ray CT with MRI, PET, and FI 181
6.6 Conclusion and Future Prospects 186

7. Optical Nanoprobes for Diagnosis **195**

R. G. Aswathy and D. Sakthi Kumar

7.1 Introduction 195
7.2 Optical Imaging with Nanoparticles 196
7.2.1 Semiconductor QDs 200
7.2.2 Metallic NPs 212
7.2.3 Carbon Nanotubes 218

7.2.4 Polymeric Nanoparticles 226
7.2.5 UCNPs 234
7.2.6 Ceramic NPs 239
7.3 Conclusion and Future Prospects 243

NANOTHERAPY

8. Nanomaterials in Chemotherapy 271

P. K. Hashim and Anjaneyulu Dirisala

8.1 Introduction 271
8.1.1 DDS 272
8.1.2 DDS Evolution 273
8.1.3 Nanomedicine and DDS 274
8.2 DDS Behavior in Bloodstream 276
8.3 Drug Release Mechanisms 279
8.3.1 Diffusion-Controlled Release 281
8.3.2 Solvent-Controlled Release 282
8.3.3 Chemically Programmed Release 283
8.4 Lipid-Based DDSs 284
8.4.1 Composition and Preparation 284
8.4.2 Drug Loading into Liposomes 286
8.4.3 Targeting and Drug Release 287
8.4.4 Lipid-Coated Polymeric Nanoparticles 291
8.4.5 Hybrid Liposomes 292
8.5 Polymer-Based DDSs 293
8.5.1 Polymeric Micelles 294
8.5.1.1 Drug loading and release in polymeric micelles 296
8.5.2 Polymer Vesicles 298
8.5.3 Polymer–Drug Conjugates 298
8.5.4 Dendritic Polymers 300
8.5.5 Hyperbranched Polymers 301
8.6 Inorganic Nanomaterials 302
8.7 Carbon Nanostructures 303

8.8 Nanoscale Metal-Organic Frameworks 306
8.9 Clinically Approved Cancer Nano-Chemotherapeutics 308
8.10 Conclusion and Future Prospects 308

9. Magnetic Nanoparticles for Hyperthermia against Cancer 337

Sivakumar Balasubramanian and Allison J. Cowin

9.1 Introduction 337
9.2 Biology of Hyperthermia and Cell Death 340
9.3 Heat Generating Sources for Hyperthermia 341
9.4 Concepts of Nanotechnology and Hyperthermia: Nanothermotherapy 342
9.5 Mechanism of Heat Generation 344
9.5.1 Metal Nanoparticle Heating 344
9.5.2 MNP Heating 344
9.6 Factors Influencing Design of MNPs for Hyperthermia 346
9.7 Magnetic Nanomaterials for Hyperthermia 346
9.8 Significance of SPIONs in Hyperthermia 348
9.9 Engineered Smart Nanosystems: Drug Delivery via Hyperthermia 350
9.10 Examples of MNP Hyperthermia in Biology 352
9.11 Conclusion and Future Prospects 352

10. Phototherapy Using Nanomaterials 373

Resmi A. N., Resmi V. Nair, Rekha C. R., Lakshmi V. Nair, Shaiju S. Nazeer, and Ramapurath S. Jayasree

10.1 Introduction 373
10.2 Photothermal Therapy 374
10.2.1 Metal-Based Nanomaterial 377
10.2.1.1 Gold nanoshells 379
10.2.1.2 Gold nanorods 382
10.2.1.3 Hollow gold nanoshells 385
10.2.1.4 Gold nanocages 388
10.2.1.5 Gold nanostars 389

10.2.2 Carbon Nanomaterials 390
10.2.2.1 CNTs 392
10.2.2.2 Graphene 393
10.2.2.3 Fullerenes 395
10.3 Photodynamic Therapy – Introduction 395
10.3.1 Challenges to Clinical Adoption of PDT 396
10.3.1.1 Photosensitizers 397
10.3.1.2 Light wavelength 400
10.3.1.3 Selective drug delivery 401
10.4 Nanoparticles in PDT 403
10.4.1 Passive Nanoparticles 408
10.4.1.1 Biodegradable nanoparticle carriers 408
10.4.1.2 Non-biodegradable nanoparticle carriers 412
10.4.2 Active Nanoparticles 414
10.4.2.1 Photosensitizer nanoparticles 415
10.4.2.2 Self-lighting nanoparticles 419
10.4.2.3 Upconversion nanoparticles 420
10.5 Conclusion and Future Prospects 422

11. Nanotechnology-Mediated Radiation Therapy 445

Ankita Borah and D. Sakthi Kumar

11.1 Introduction 445
11.2 Radiotherapy: Principles and Examples in Various Cancers 446
11.2.1 Breast Cancer 448
11.2.2 Liver Cancer 448
11.2.3 Ovarian Cancer 448
11.2.4 Head and Neck Cancer 449
11.2.5 Prostate Cancer 449
11.3 Techniques of Radiation Therapy 449
11.3.1 Intensity Modulated Radiation Therapy 449
11.3.2 Image-Guided Radiation Therapy 450
11.3.3 Particle Therapy 452

11.3.4 3D Conformal Radiotherapy and Stereotactic Body Radiation Therapy 453
11.3.5 Internal Radiation Therapy 455
11.4 Radiation Induced Cell Death Mechanisms 456
11.4.1 Mitotic Cell Death 456
11.4.2 Apoptosis and Necrosis 458
11.4.3 Autophagy 459
11.4.4 Senescence 460
11.5 Risks Associated with Conventional Radiotherapy 461
11.6 Nanotechnology-Mediated Radiotherapy Treatments 463
11.6.1 Gold Nanoparticles 465
11.6.2 Platinum- and Silver-Based NPs 468
11.6.3 Gadolinium-Based NPs 470
11.6.4 Hafnium-Based NPs 472
11.6.5 Superparamagnetic Iron Oxide Nanoparticles 472
11.7 Conclusion and Future Perspectives 475

12. Role of Nanoparticles in Cancer Immunotherapy 495

Madhu Malinee and Alok Kumar

12.1 Introduction 495
12.1.1 Innate and Adaptive Immunity 496
12.1.2 Cancer and Treatment Methods 497
12.2 Tumor Immune Surveillance and Immunoediting 498
12.3 Cancer Immunotherapy 498
12.3.1 Enhancement Immunotherapy 501
12.3.1.1 Passive immunotherapy 501
12.3.1.2 Active immunotherapy 502
12.3.2 Normalization Immunotherapy 502
12.4 Immune Evasion Strategies by Tumor 507
12.4.1 Downregulating MHC Class I Expression 507
12.4.2 Developing Resistance to CTL-Mediated Killing Mechanisms 507

12.4.3 Turning Off Activated T-Cells via Direct Contact 508
12.4.4 Releasing Soluble Factors to Inhibit Immune Cells 509
12.4.5 Inhibiting T Cells through Bystander Effect 509
12.5 Tumor Immunotherapy: Advantages, Drawbacks, and Need of Combination Approaches 510
12.6 Nano Immunotherapy 511
12.6.1 Delivery of TAA and Adjuvants to APCs 512
12.6.2 Role of Artificial APCs 514
12.6.3 Direct Activation of TAA-Specific T Cells 515
12.6.4 Role of NPs in Targeting Immunosuppressive TME 516
12.7 Conclusion and Future Perspectives 517

13. Nanoparticle-Mediated Small RNA Deliveries for Molecular Therapies 527

Ramasamy Paulmurugan, Uday Kumar Sukumar, and Tarik F. Massoud

13.1 Introduction 527
13.1.1 Introduction to Small RNA Deliveries 528
13.2 Lipid-Based Nanovectors for Small RNA Deliveries 531
13.2.1 Liposomes/Lipoplexes 532
13.2.2 Stable Nucleic Acid Lipid Particles 534
13.3 Structured Nanoparticles for Small RNA Deliveries 535
13.3.1 Inorganic Nanoparticles for Small RNAs Deliveries 537
13.3.1.1 Carbon nanoparticles for small RNA deliveries 537
13.3.1.2 QD for small RNAs deliveries 539
13.3.1.3 Gold nanoparticles/nanorods/ nanostars for small RNAs deliveries 539

13.3.1.4 Other inorganic nanoparticles in small RNA deliveries 541
13.3.2 Organic Nanoparticles for Small RNA Deliveries 542
13.3.2.1 Polymeric nanoparticles 542
13.4 Natural Polymers 555
13.5 Small RNA Deliveries in Clinical Trials 556
13.6 Conclusion and Future Perspectives 559

14. Theranostics: A New Holistic Approach in Nanomedicine 573

Ankit Rochani and Sreejith Raveendran

14.1 Introduction 573
14.2 Bioconjugation Technology for Theranostic Materials 577
14.3 Polymeric Nano and Microstructures 579
14.4 Radio-Isotopic Nanomaterials 584
14.5 Nano Carbon Structures 586
14.5.1 Graphene as Theranostic Agent 587
14.5.2 Nanodiamonds as Theranostic Agents 590
14.5.3 Fullerene and Carbon Nanotubes 594
14.6 Quantum Dots 600
14.7 Gold Nanostructures 602
14.8 Magnetic Nanoparticles 605
14.9 Clinical Translations of Theranostic Materials 608
14.10 Conclusion and Future Perspectives 611

Glossary 625
Index 639

Preface

Nanotechnology is one of the latest scientific research fields which is trying to enhance human well-being, be it in electronics, communications, automobiles, home appliances, or different medical fields. During the developmental stages itself, nanotechnology found very interesting common and exciting application areas with many other scientific fields. As a result, many interdisciplinary fields sprang up. Among these interdisciplinary fields, the fusion of biotechnology and nanotechnology, or in other words, bionanoscience, was shown to be very promising and considered the most important and exciting field.

Although we boast about the achievements of most modern advancements in the field of medical sciences, the menace of cancer has been increasingly claiming more and more lives across the world every year. Though the current therapeutic options such as surgery, radiation, and chemotherapies benefited patients to an extent, a complete cure for cancer is still in question and these current treatment options only extend the life of patients by compromising the immunological system and the quality of life by causing pain, discomfort, mental trauma, etc. Additionally, the treatment cost is very high and there are various side effects of treatments.

At this juncture, bionanoscience is destined to provide many solutions to cancer-related problems for which the current medical field has not been able to find satisfactory answers so far. With the emergence of bionanoscience, the concept of "Magic Bullet" to kill cancer cells selectively, which was coined by German Scientist "Paul Ehrlich" decades ago, started giving good results. By finding the increased possibility of targeting the imaging moieties as well as drugs to the cancer cells selectively, "nano diagnosis and drug delivery" against cancer cells became a hot field among bionanoscience researchers. Nanodrug delivery provided many promising results, and the medical field is anxiously waiting to hear the ultimate good news from the scientists working in this field that, "Yes, we can cure cancer."

During our more than a decade's research in "nanoimaging and drug delivery" in cancer, although we came across several interesting papers and reviews, we could not see a highly useful book that would be concise and still discuss all fields related to the nano cancer research in diagnosis and drug delivery. To fill this vacuum, we bring forward the book *Bionanotechnology in Cancer: Diagnosis and Therapy*. This book, comprising 14 chapters, has been mainly divided into two parts primarily to accommodate the latest bionano research in cancer: nanodiagnostics and nanotherapy. We believe this book will be a valuable resource for students, scientists, and teachers who are involved in bionano research and will inspire scientists who wish to start their career in this fascinating research field.

The first two chapters have been designed as the gateway to enter the bionano research field in nanodiagnostics and nanotherapy and provide general and useful information about cancer and the interactions of nanoparticles at the cellular level. Chapters 3 to 7 fall under nanodiagnostics and Chapters 8 to 14 under nanotherapy.

In Chapter 1, "Introduction to Cancer, Conventional Therapies and Bionano Based Advanced Anticancer Strategies," we provide an introduction to cancer biology and solid tumor physiology for a better understanding of this fatal disease, the current therapies and their shortcomings, and advantages of nanotechnology in fusion with biotechnology that could bring the cure for this ailment.

Chapter 2 details the interaction of nanoparticles at the cellular interface. A good understanding of this phenomenon is highly needed for designing nanoparticles for diagnosis or for therapy to obtain fruitful results.

Chapters 3–7, under the "Nanodiagnostics" section, discuss different latest nanodiagnostic methods in cancer.

Chapter 3 describes radiolabeled nanoparticles for cancer diagnosis. For imaging the cancer cells with the help of radio-labeled nanoparticles, we use radioisotopes, which can provide information about the physiology and metabolism of the cells, leading to the understanding of the functional abnormalities in the cancer cells compared to healthy cells. This will help in the early detection of the disease. Radiolabeled nanoparticles can be designed and can be attached to the cancer-specific ligands so that these radioisotopes will be delivered to cancer cells, which limit the side effect–based off-targeting of the radioisotopes and drifting into normal cells. This chapter provides a lot of information about different successful strategies adopted in imaging by using "radiolabeled nanoparticles."

Chapter 4 deals with the magnetic nanoparticles for cancer diagnosis. Magnetic nanoparticles can be used as contrast agents in magnetic resonance imaging (MRI). Magnetic nanoparticles can be tagged with targeting moieties such as peptides, proteins, aptamers and several small molecules that can actively target the tumor via specific molecular recognition, which is an useful advance for tumor detection at the initial stage. This chapter deals with many nano magnetic contrast agents that can be used for the diagnostic purpose.

Chapter 5 introduces acoustically reflective nanoparticles. Ultrasound imaging is one of the operative, diagnostic methodologies to detect the tumor in different stages. Lots of research work has been carried out in this field, and many new nanotechnological methods have been developed to make the imaging very precise and sharp. Development of different types of nanobubbles, nanoparticles, and nanodroplets has made this imaging technique one of the essential imaging techniques for detecting tumors.

Chapter 6 details X-ray computed tomography by using nanomaterials as contrast agents. Currently, this technique is considered as the best and efficient method to diagnosis the tumor. Research in this field enabled 2D computed tomography uplifted to 3D tomography. Nanomaterials developed for supporting this imaging technique improved the resolution, and hence we are now in a position to visualize and understand the tumor mass in greater detail. This chapter gives in-depth knowledge of the nanotechnological advancement in the field of X-ray computed tomography.

Chapter 7 deals with different optical nanoprobes. Optical imaging has been transformed as an indispensable tool for preclinical experimental diagnosis and in medical diagnosis. This field is continuously evolving with new materials that added to its bioimaging armory. Regarding the materials used for imaging in this field, nanomaterials are providing the lion's share. Development in nanotechnology enhanced the impact of optical imaging in diagnostic field hugely. While going through this chapter, the readers will be familiarized with the different techniques as well as various nanomaterials used in this arena.

In Chapters 8–14, under the "Nanotherapy" section, we deal with different cancer therapy methods supported by nanomaterials.

Chapter 8 discusses chemotherapy using nanoparticulate materials. Nowadays by developing many nanodrug delivery vehicles, many drugs that were not much efficient in cancer therapy earlier have been made highly competent against tumors. Cleverly designing the nanodrug vehicles and entrapping the drugs inside, scientists developed newer ways to treat different cancers. While conventional chemotherapeutics are usually distributed throughout the body, which results in several side effects, nanodrug delivery ensures the delivery of the toxic drugs only to the diseased cells and in that way limiting the distribution of the drug throughout the body and nullifies or minimizes the side effects. This chapter provides information about various nanodrug delivery vehicles and their chemotherapeutic actions.

Chapter 9 contains the information about the nanomagnets that are useful in magnetic hyperthermia. By using external magnetic fields, we can develop magnetic hyperthermia after treating the cancer cells with superparamagnetic nanoparticles. By using hyperthermia, we can increase the temperature beyond the tolerance level of the cancer cells, and in that way, we can kill the cancer cells. A lot of research work has been conducted in this field and recently many biodegradable magnetic nanoparticles were also developed for the use in this arena. This chapter enhances the knowledge in the field of magnetic hyperthermia.

Chapter 10 provides the information about the photothermal ablation using nanoparticles. A lot of work has been carried out to develop the nanoparticles that can be used for photothermal treatments. Some materials generate heat when illuminated by light. Materials that produce heat under the near infra-red light region are generally used in this category to kill the cancer cells by increasing the temperature above the tolerance level of the cells. This chapter introduces many nanomaterials and their use for photothermal ablation of cancer cells. This chapter also deals the photodynamic therapy of tumors by using nanomaterials. Materials that can be active in the biological window in the near infra-red region (about 650–900 nm) have received considerable attention in conducting this treatment. The biological window will ensure the deep tissue penetration for enabling this therapy. NIR lasers and photodynamic nanomaterials that are active in the NIR region have become a hot subject in the development of this treatment. This chapter provides valuable information about this treatment method and the advantage of this technique over other methods.

Chapter 11 deals with nanotechnology-mediated radiation therapy. In recent years, technological advancements in nanotechnology have elevated the current conventional procedures in radiation therapy, which brought enhanced therapeutic outcomes in cancer treatment. This chapter gives a discernible understanding of the principles of radiation therapy, conventional approaches, and how the inclusivity of nanotechnology-mediated applications has revolutionized the current scenario of radiation-based treatments, thus providing an unconventional outlook in radiation therapy.

Chapter 12 is specially designed to introduce the latest research happening in the field of immunotherapy using nanoparticles against cancer. The immune system is responsible for checking and eliminating any foreign bodies in the human body and thus limiting the risk of damage caused by them. If we can develop a vaccine that can be used against all types of cancer, then it will be a giant leap in the medical field. This chapter discusses the nanomaterials that have been developed in this direction.

Chapter 13 describes bionanotechnologists' fight against cancer by using nanoparticle-mediated small RNA delivery. Once we came to know that cancer is mainly caused by mutation in a gene, we began searching for a method for gene therapy to get cured of this menace. Nanotechnology is providing its support to help us reach our goal. Recent developments in this field are very encouraging. This chapter gives a detailed account of our fight against cancer with the aid of gene therapy.

Chapter 14 details the latest method in the field of bionanoscience for the fight against cancer using a combination of diagnosis and therapy, termed theragnostics or theranostics. By combining various combinations of therapy and diagnosis mentioned in different chapters in this book under these categories, a new wave of development is going on to combine these two fields to increase the effectiveness and cure from cancer. The results that have emerged so far are very much encouraging and inspiring. This chapter provides more information about different techniques and nanomaterials used in theragnostics/theranostics.

We have paid utmost care and attention to cover almost all research fields coming under the application of bionanoscience to fight against cancer. Owing to the vastness of the field, in our book, we have selected the examples of the research in respective areas that can provide unambiguous information as well as great insight. Readers are highly recommended and requested to go through the references presented in the book and references therein for a profound and in-depth understanding of the field. We are confident that the book will act as the gateway for readers to enter the beautiful and elegant field of bionanoscience, which is considered as the last hope for the fight against cancer and will serve as a highly useful resourceful for students, researchers, and teachers working in this field or related fields.

D. Sakthi Kumar, PhD
Bio Nano Electronics Research Center
Graduate School of Interdisciplinary New Science
Toyo University, Kawagoe, Saitama, 350-8585
Japan
sakthi@toyo.jp
www.drsakthikumar.com

R. G. Aswathy, PhD
Future Industries Institute, Mawson Lakes Campus
University of South Australia
Adelaide, South Australia 5095
Australia
aswathy.ravindrangirija@unisa.edu.au, aswathyrg@gmail.com

Foreword

Bionanotechnology, a key integrative technology of the 21st century, aims to use the knowledge gathered from the natural construction of cellular systems for the advancement of science and engineering to target therapies directly and selectively to cancerous cells and neoplasms. Though microscale to nanoscale bionanotechnology research offers an excellent space for developing futuristic technologies, many challenges must be overcome. Due to the paucity of dedicated literature on bionanotechnology in cancer, to bridge the gap, this book has assembled 14 chapters in diverse areas of oncology focusing on nanoparticles, 2D nanomaterials in chemotherapy, imaging, diagnostics, nanoparticle-mediated small RNA deliveries, and theranostics in a single compendium, making it the only one in the burgeoning field of bionanotechnology.

Bionanotechnology combines the study of bio-nano interactions with the expanding field of nanotechnology. Proteins, RNA, and DNA are versatile biomolecules that provide the foundation for important biological functions such as molecule transport and DNA replication. The protein structures also make them suitable for forming functional biological materials as well as provides them the ability to self-assemble. The resulting developments in protein-based bionanotechnology include proteins being utilized as important elements in nanotechnology applications in the design of nanodevices. With these tools, clinicians can safely image and effectively deliver chemotherapy, radiotherapy, and the next generation of immuno- and gene therapies to the tumors. This book fills up the void in bionanotechnology education at the graduate level.

Prof. Venkatesan Renugopalakrishnan
Northeastern University, Boston Children's Hospital
MGB Center for COVID Innovation, Harvard Boston, Massachusetts
May 15, 2022
v.renugopalakrishnan@northeastern.edu
+1 508 983 4068, Skype: Renu190

Chapter 1

Introduction to Cancer, Conventional Therapies, and Bionano-Based Advanced Anticancer Strategies

Sindhu C. Pillai[a] **and Jayasree Nair**[b,c]

[a]*Bio Nano Electronics Research Center, Graduate School of Interdisciplinary New Science, Toyo University, Kawagoe, Saitama 350-8585, Japan*

[b]*Jennifer Goodman Linn Laboratory of New Drug Development, Department of Medicine, Memorial Sloan-Kettering Cancer Center, New York, USA*

[c]*Current Address: Bronx Charter School of Excellence, 1960 Benedict Ave, The Bronx, NY 1046, USA*

sindhu.09@gmail.com, jayasree_nair@hotmail.com

1.1 Cancer

Cancer is a set of diseases that can arise in almost any tissue or organ of the body. Simply, it is abnormal and uncontrollable cell growth [1]. The transformation of normal cells into cancerous cells is not an important event in cancer development; however, it is the inability of the body's immune cells to identify and kill these cancer cells, which is critical. Cancer risk is multiplied in persons who have a suppressed immune system [2]. The

Bionanotechnology in Cancer: Diagnosis and Therapy
Edited by D. Sakthi Kumar and Aswathy Ravindran Girija

ISBN 978-981-4800-30-3 (Hardcover), 978-0-429-42291-1 (eBook)
www.jennystanford.com

movement of tumor cells from a primary location to colonize distant organs, also known as tumor metastasis, is a great contributor to the deaths of most patients affected by cancer. There are various types of cancer and some of their names depend on the locations they are found, such as prostate, lung, colorectal, thyroid, liver, breast, and stomach cancer. Moreover, carcinoma is a type of cancer that affects organs or the skin, leukemia is a cancer of the blood, lymphoma is a cancer of the lymphatic system, and sarcoma is a cancer of the connective tissues [3].

Globally, cancer was considered the second leading cause of death in 2018. In men, the most common types of cancer are prostate, lung, colorectal, liver, and stomach cancer, while in women, the most common types of cancer are breast, lung, colorectal, thyroid, and cervical cancer. The transformation of normal body cells into cancer cells is a multistage process, resulting from the interaction between the genetic factors of individual and external agents, such as physical, chemical, and biological carcinogens [1]. The incidence of cancer has been increasing alarmingly, most probably due to the changing habits and lifestyle, and increasing life expectancy [2].

The burden of cancer continues to grow worldwide, exerting tremendous financial, physical, and emotional strain on affected individuals, their families, and health systems. Approximately 70% of cancer-related deaths occur in low- and middle-income countries, and about one-third of these deaths are due to the following dietary and behavioral risks: tobacco and alcohol use, low vegetable and fruit intake, lack of physical activity, and high body mass index. The use of tobacco is the key cancer risk factor, as it is responsible for around 22% of cancer-related deaths.

The impact of cancer on the economy is significantly increasing. In 2017, the total annual cost of cancer was estimated to be $161.2 billion. Early cancer detection plays is a major role in its effective treatment, as it is more likely to result in less morbidity, better probability of surviving, and less expensive treatment. Early diagnosis consists of easy-care access, awareness, clinical evaluation, diagnosis, and access to treatment. Due to lack of early diagnosis, cancer diagnosis takes place at the late stage, during which curative treatment options may no longer beneficial. It is essential to have a correct cancer diagnosis for effective and adequate treatment. A specific treatment regimen is required for every cancer type, encompassing one or more modalities, including chemotherapy, surgery, and radiotherapy. Generally, the treatment goal is to cure cancer, improve the quality of life of cancer patients, and to prolong survival [1].

1.2 Understanding Cancer and Its Occurrence

Carcinogenesis is the process by which the transformation of normal cells into cancer/tumor takes place. Tumor is an abnormal mass of tissue due to the uncontrolled growth of cells. Tumor can be benign (not cancer) or malignant (cancer). The roles of genes involved include the inhibition of cellular proliferation, which encompasses the function of tumor suppressor genes, and positive signaling of migration and proliferation, which contains oncogenes' function. In addition, other genes concerned involve these with roles in apoptosis control and DNA repair [4]. One of the most known models involved in carcinogenesis is the RAS gene activation. The activation mechanism of the RAS protooncogenes is normally via point mutations at either one or many activating positions within the sequence. The elevated expression of the RAS gene may lead to the transformation of normal cells, which induces the formation of tumors [5].

Mainly there are four main steps for carcinogenesis: (a) tumor initiation, (b) tumor promotion, (c) tumor progression, and (d) tumor metastasis [4], as shown in Figure 1.1.

(a) Tumor initiation

Tumor initiation is derived from the accumulation of mutations that arise in body cells, which can proliferate and survive the organism's lifetime. Genetic errors normally arise by a chemical carcinogen that modifies the DNA's molecular structure, subsequently leading to a potential mutation during DNA synthesis. A mutation mostly results through the formation of an adduct between the carcinogen and a DNA nucleotide [6]. Generally, the number of adducts between a chemical carcinogen and DNA and the number of cancers that develop are positively correlated.

Therefore, tissues that do not have carcinogen–DNA adducts rarely exhibit tumor formation. These adducts are necessary, but not sufficient, as prerequisites for the initiation of tumors. A tumor initiating event normally requires the DNA adduct to cause either protooncogene to be activated or tumor suppressor genes to be inactivated [7].

(b) Tumor promotion

Tumor promotion consists of selective expansion of initiated cells. This clonal expansion produces a large number of cells that have the risk to be further genetically mutated because the rate of accumulation of mutations is proportional to the cell division rate [8].

Tumor promoters are generally not carcinogenic alone, not mutagenic, and are often able to act without metabolic activation. These promoters are characterized by their capacity to decrease the latency period for cancer formation following the exposure of a tissue to tumor initiators or to increase the chance of tumor formation in that tissue. Furthermore, they promote the formation of cancer in conjunction with an initiator dose too low to cause cancer alone. Complete carcinogens are chemicals that are capable of both initiating a tumor and promoting it [9].

(c) Tumor progression

Tumor progression consists of the genetic expression that results in a malignant phenotype, in addition to the acquisition of malignant cells to more aggressive characteristics with time. The malignant phenotype has a prominent characteristic of uncontrolled growth and genetic instability. Moreover, tumor cells may secrete proteases during metastasis to allow invasion from the primary and immediate tumor location [10]. Further genetic changes are prone to occur during tumor progression. Again, these include the activation of oncogenes and the loss of function of tumor suppressor genes [11].

(d) Tumor metastasis

Metastasis is a complex multistep process that is responsible for around 90% of cancer-associated mortality. Sequentially, tumor metastasis includes local tissue invasion where the tumor has started, detachment from primary cancer, intravasation, survival in the circulatory system, and dissemination, adhesion, extravasation, proliferation, and formation of new blood vessels. Metastasis can take place either through the lymphatic system or through the blood circulatory system. Generally, detached cancer cells can directly go into the blood vessels or indirectly through the lymphatic system. Each type of cancer typically has a preferential metastatic organ. For instance, those of breast cancer are the liver, brain, bone, and lungs.

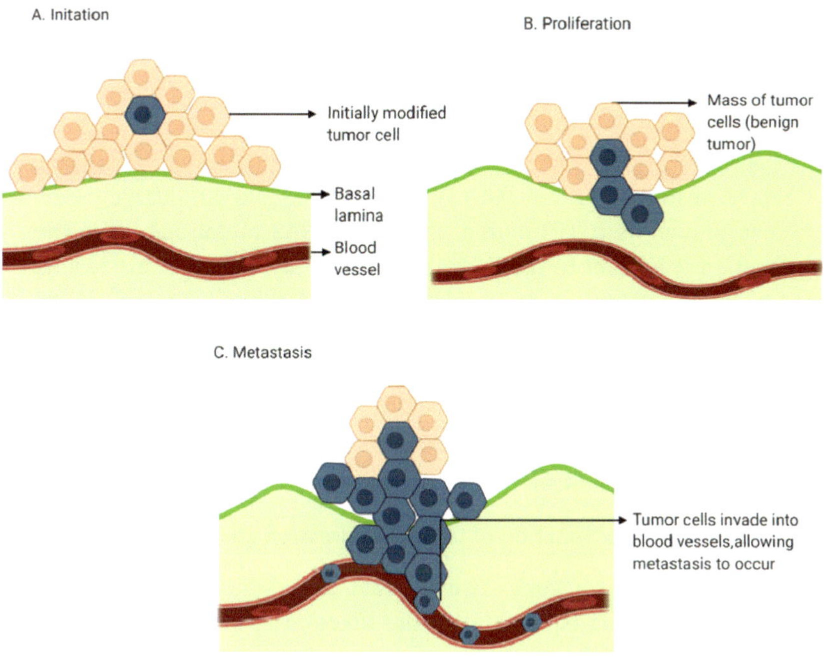

Figure 1.1 Tumor growth and metastasis: (A) Altered growth properties of a single cell. (B) Altered cells proliferate and form a mass of tumor cells called a benign tumor. (C) Tumor invasion of the basal lamina and becoming malignant, this malignant tumor exhibits metastasis and forms small clusters that dislodge from the tumor, invade the blood, or lymphatic vessels.

1.2.1 Genes Activities Related to Cancer

Cancer progression involves mainly two genetic events (a) activation of oncogenes, and (b) disruption of function of tumor suppressor genes [12].

(a) Activation of oncogenes

Oncogenes are homologs of normal genes that typically have a role in cell cycle regulation and cell growth pathways. A mutation in an oncogene may result in consecutive activation of that gene, leading to uncontrolled cellular proliferation [13]. Among the oncogenes, the RAS oncogene is implicated in cancer development. This oncogene exists in three variants on the cellular level, HRAS, KRAS, and NRAS. When mutated, all these three oncogenes are able to transform normal cells into cancer cells. However, in colorectal cancer, KRAS is the most frequently mutated [14, 15].

A family of tiny proteins that are responsible for regulating cellular signal transduction is encoded by the RAS oncogenes. These proteins act as a switch that transmits growth signals to the nucleus [16]. The proteins typically cycle between an inactive state, also known as (GDP)-bound state, and an active state, also known as (GTP)-bound phase. The RAS mutations leave the proteins resistant to the hydrolysis of GTP by GTPase. This results in constitutively active proteins that are GTP-bound, and subsequently in a continuous growth stimulus [17]. Another noteworthy gene that is commonly associated with several types of cancers, including prostate and breast, is the Bcl-2 gene, which is associated with the expression of several Bcl-2 proteins involved in apoptosis [18].

(b) Disruption of function of tumor suppression genes

Tumor suppressor genes usually have an inhibitory effect on the normal cell cycle equivalent to the function of breaks in vehicles. Once they are deleted, or their role is reduced due to mutation or so, the cells will act like vehicles without breaks and be directly involved in cancer development. In this case, cellular growth proceeds unchecked because standard control mechanisms are not operative anymore. At the cellular level, normal proteins are no longer functional only when both alleles of the tumor suppressor gene are inactivated. Retinoblastoma is a type of cancer that is caused by a mutation that inhibits the RB1 gene [19, 20]. The two forms of retinoblastoma that exist are: The inherited form of retinoblastoma involves a germline mutation that is either inherited or new, plus another somatic mutation that affects the remaining allele and occurs later during development [18]. The nonhereditary form of retinoblastoma involves two somatic mutations that arise spontaneously in one retinal cell [20].

1.3 Different Theories of Carcinogenesis

There are several theories that came into existence to explain carcinogenesis. These theories include (a) gene mutation theory, (b) aneuploidy theory, and (c) epigenetic theory.

(a) Gene mutation theory

The genetic theory of carcinogenesis is the most common theory. It states that normal cells become neoplastic due to mutations in the DNA [21]. The gene mutation theory predicts the spontaneous appearance of diploid polyclonal cancer cells following the accumulation of a sufficient number of transforming genes. The oncogenes of retroviruses are a class of genes that meet the predictions of the gene mutation theory because transformation takes place within one or limited cell generations, and the transformation depends on the continued role of viral proteins. Furthermore, the tumors seem to be diploid, or at least at the beginning [22].

(b) Aneuploidy theory

In humans, many cell divisions occur each minute, and appropriate segregation of chromosomes is required with every cell division to maintain a diploid karyotype. However, sometimes the segregation machinery of chromosomes may be imperfect, and a few spontaneous chromosome mis-segregations could occur during cell divisions. Even if the rate of mis-segregations is very low, selective pressure can allow rare aneuploidy cells to proliferate under certain conditions [23].

The aneuploidy theory states that carcinogens initiate cancer by creating aneuploidy, which unbalances several genes and therefore catalyzes chain reactions that lead to progressive aneuploidizations [24], and carcinogenesis may arise spontaneously. This irregular dosage of genes is created by the loss or gain of specific segments of chromosomes [21]. Studies have made it clear that aneuploidy may predispose to the development of tumors given to the strong evidence of high aneuploidy frequency in tumors [23]. In addition, studies have shown that malignancy is proportional to the degree of aneuploidy [21].

(c) Epigenetic theory

The epigenetic theory of carcinogenesis refers to mechanisms that modify the expression of genes leading to carcinogenesis without modifying the primary DNA sequence. Normally, epigenetic processes are inherited and can be reversed, and they include alterations in DNA methylation and histone modifications. Disruption of epigenetic mechanisms can result in modified gene function and malignant transformation of cells [25]. Deregulated epigenetic processes can initiate genetic instability that results in the acquisition of mutations in tumor suppressor genes and in oncogenes. Furthermore, the studies have shown that epigenetic modifications in tumors are mainly of a clonal nature, which indicates the occurrence of these epigenetic disruptions in early cell generations [26].

1.4 Factors Influencing Cancer Development

Certain individuals are affected with cancer throughout their lifetime, while others are not, and certain tissue types in humans develop tumors more often than others. Knowing the main risk factors that influence the development of cancer as well as their interactions may elicit ways of prevention. Generally, two types of risk factors contribute to the development of cancer: (a) intrinsic/biological factors and (b) external factors.

1.4.1 Intrinsic/Biological Factors

Intrinsic factors that contribute to the pathogenesis of cancer generally refer to unavoidable biological factors. These factors include typically (a) age, (b) hormone levels, (c) family history, and (d) genetic predisposition.

(a) Age

Cancer is an age-related disease since the incidence of most tumors increases with age [27]. Paradoxically, adults that have the longest longevity are usually less likely to develop tumors [28]. The cumulative risk for the different types of cancers increases with age, up until around the age of 70 years, and then decreases slightly [29]. For instance, studies have shown that patients who are younger than 65 years have a significantly higher incidence of esophageal cancer than in older patients [30]. Some cancers are not affected by the age of patients, such as lung cancer, for which the incidence was not found to be significant according to the age at diagnosis [29, 30]. Studies have also shown that after the age of 90 years, cancer becomes rare as a reason for disease or death [29].

(b) Hormone levels

Growth hormones have been shown to be linked to cancer progression. These hormones do not directly cause a malignant transformation; however, they can increase mutation risks by reducing the DNA repair time during the cell cycle's rapid progression [31]. In the breast cancer, growth hormones were shown to boost cancer cell survival, proliferation, vascularization, migration, and invasion, favoring metastatic growth, and inducing chemoresistance [32, 33]. Growth hormones were also shown to enhance the metastatic ability of prostate cancer cells through the stimulation of cellular motility and invasiveness [34]. In colorectal cancer, the expression of growth hormones is positively correlated with the size of the tumor and with lymph node metastasis [35].

(c) Family history

Family history is an essential risk factor for several types of cancers, such as breast cancer [36]. A family history of colorectal cancer has been found in almost 35% of patients with the disease [37]. A family history of bladder cancer increases the risk of bladder cancer by two-fold [38]. Also, studies have shown that a family history of gastric cancer and familial mucosal abnormalities in the stomach increased the risk of gastric cancer in individuals [39].

(d) Genetic predisposition

A genetic predisposition to cancer is the increased likelihood of developing cancer depends on a person's genes. A genetic predisposition to cancer comes from variations in genes, usually inherited from a parent, contributing to cancer development. Within the same family, some people that have the predisposing genetic variation will develop cancer, while others will not [40]. Some germline mutations within critical genes predispose to cancer. The names of these genes, such as BRCA1 and RB1, reflect the phenotypes that resulted in their identification [41].

The breast and ovarian cancer risks are increased in patients who carry a BRCA1 mutation, while BRCA2 carriers have an increased risk of prostate, pancreas, and skin cancers [42]. The identification of genetic predisposition to cancer permits screening processes to recognize patients who are at risk of developing certain types of cancer. This will subsequently help them modify potential cancer risk increasing behaviors [43]. Physicians recommend breast surveillance for BRCA heterozygotes, including self-examination, physician examination, and mammography [44].

1.4.2 External Factors

External factors that influence the development of cancer are usually modifiable and exogenous to the human body. These include, but are not limited to, (a) diet, (b) consumption of tobacco and alcohol, (c) exposure to radiations, (d) infections from viruses, and (e) free radicals.

(a) Diet

Overall dietary patterns, as well as specific components of the diet, have been studied to deduce their relation to cancer risk. A diet rich in animal fat has shown to be a risk factor for prostate cancer. Particularly, a large intake of alpha-linolenic acid and a low intake of linoleic acid, which is a common combination in red meat and certain dairy products, appear to increase the risk [45]. Other studies suggest that hyperinsulinemia may increase the risk of certain cancers [46]. Diabetic patients have more risk of cancers, such as pancreas, liver, and endometrium, throughout their lives [47]. Studies have also shown that unhealthy dietary patterns have been found to have an association with cancer development. In addition, ultra-processed foods are associated with an increased risk of breast cancer [48].

(b) Consumption of tobacco and alcohol

The use of tobacco accounts for 21% of total deaths related to cancer, worldwide. In the United States, smoking is responsible for around 30% of all deaths that are related to cancer [49]. It is considered to be the most potent risk factor for lung cancer, as it increases the risk by 10–20 folds [50]. Studies also suggest that it is associated as a causative factor for cancer in the oral and nasal cavity [51]. In addition, smoking is implicated with an increased incidence of prostate cancer as well as breast cancer [52, 53]. Quitting smoking results in significant health benefits, even for long-time smokers. These benefits can be noticed at all ages and can be measured almost directly after cessation. Furthermore, quitting smoking in patients who already have cancer may improve cancer treatment outcomes [54].

Consumption of alcohol is an important risk factor for multiple cancers. For instance, it was shown that even consumptions of small amounts of alcohol, between three and six glasses a week, has been associated with an increased risk for esophageal and oropharyngeal cancer, as well as breast cancer in women [55].

(c) Exposure to radiations

Another avoidable external contributor that increases the cancer incidence includes excessive exposure to solar radiations and artificial ultraviolet (UV) radiations. Each year, more than one million skin cancer cases are diagnosed, and the incidence continues to rise [56]. The primary cause of skin cancer that is melanomatous and non-melanomatous is solar radiation. UV radiation induces genetic mutations and affects the cutaneous immune system, thus limiting the ability of the body to reject abnormal cells. The risk of basal cell and squamous cell cancer seems to be correlated with total lifetime exposure to the sun. Studies have shown that cumulative exposure to the UV radiations from the sun may increase the risk of melanoma, but it is even more dangerous to have repeated intense exposures that lead to blistering burns, particularly at a young age [57]. In addition, UV radiation from tanning booths has been classified as a carcinogen to humans, with a 75% increase in the risk of melanoma in people who used tanning beds before the age of 35 [58].

(d) Infections from viruses

It is estimated that virus infections cause 17% of new cancers, globally. Viruses are microorganisms that increase the risk of cancer via cellular transformation, increased cell turnover rates, cell cycle control disruption, and immune suppression. Multiple links between viruses and cancer have been found. For example, human papillomavirus has been associated with cervical cancer and squamous cell cancers of the neck and the head. Hepatitis C and hepatitis B are associated with hepatocellular carcinoma. Human T-cell lymphotropic virus type 1 and adult T-cell leukemia have been found to have a link. Human herpes virus 8 is associated with primary effusion lymphoma and Kaposi sarcoma. Epstein–Barr virus has been associated with Burkitt lymphoma [59].

(e) Free radicals

Overproduced free radicals react with proteins and fatty acids of the cell membrane, impairing their function permanently. Furthermore, the abundance of free radicals can result in DNA damage and mutations that can predispose to cancer and old age-related disorders [60]. Tumor formation due to endogenous free radical reactions may be due to their initiation by ionizing radiation. Oxidative stress ensues following the body's inability to regulate overwhelming free radicals. Therefore, free radicals can adversely modify proteins, lipids, and DNA, and lead to the formation of a number of diseases. Applying external sources of antioxidants can aid in coping with oxidative stress [61].

1.5 Microenvironment of Cancer

Cancer is not just a mass of malignant cells, but a complex system to which various cells are recruited to the environment and can be transformed by cancer cells. The interplay between malignant cells and non-malignant cells create the microenvironment of cancer/tumor [62]. This microenvironment is created by cancer itself and influenced by cancer-induced interactions. Cells found in the cancer microenvironment generally include proliferating cancer cells, tumor stroma, inflammatory cells, and other associated tissue cells and blood vessels. The tumor microenvironment is a unique environment in which it emerges during the progression of the tumor because of the interaction of cancer with the host. In addition, its cellular events are orchestrated and dominated by the tumor itself [63].

Some of the cells found in the tumor microenvironment interact with the tumor to promote its growth. For instance, tumor-associated macrophages (TAMs) are found to be pro-tumorigenic as they are implicated in the migration, invasion, and metastasis of malignant cells. Studies have shown that an abundance of TAMs in the tumor microenvironment is linked to a poor prognosis. TAMs are also the main contributors to tumor angiogenesis. Furthermore, in hypoxic areas of tumors, TAMs are found to accumulate there due to the release of hypoxia-induced chemo-attractants. Other cells found in the tumor microenvironment and that are associated with tumor development are tumor-associated neutrophils (TANs). There is evidence that TANs contribute to tumor growth and metastasis. In addition, studies have shown that TANs have a pro-tumorigenic effect by suppressing the immune system and enhancing angiogenesis. Cancer-associated fibroblasts (CAFs) arise from residential fibroblasts that differentiate into myofibroblasts, during tissue injury. Studies have shown that CAFs increase the risk of development of malignant cells. In addition, the arrangement of CAFs in the tumor microenvironment enables them to restrict the ability of the chemotherapeutic agents from reaching malignant cells [62].

1.5.1 Tumor Angiogenesis

The growth of tumors that are solid in nature is normally accompanied by neovascularization. This growth of new capillaries is more continuous and vigorous than a capillary outgrowth that is observed in inflammation or fresh wounds [64]. The rapid development of a new vascular network is needed to sustain the high rate of proliferation of cancer cells. This is unlike the normal conditions of an adult body, in which blood vessels stay quiescent and rarely have newly formed branches. Tumors are unable to grow beyond 2 mm without a proper vascular supply. The formation of new vessels emerges from a preexistent vascular network. The phases of the angiogenic process are depicted in Figure 1.2 [65].

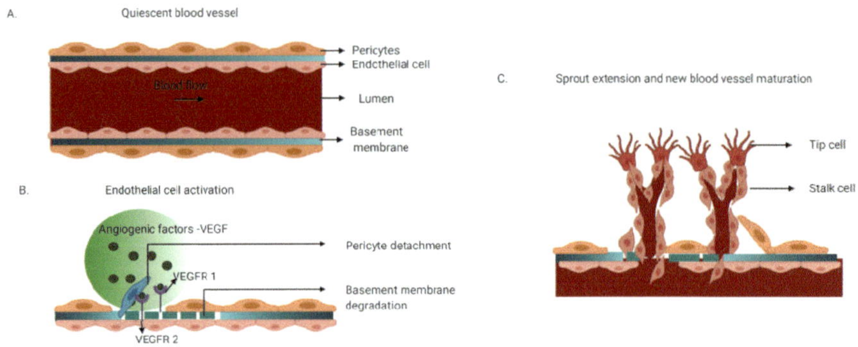

Figure 1.2 Phases of the angiogenic process: (A) Blood vessels remain quiescent under physiological conditions. (B) Blood vessels are destabilized. Pericyte dissociation from the basement membrane and a weakened extracellular matrix (ECM) are induced by angiogenic signals. (C) Endothelial cells sprout out in response to angiogenic signals; a leading cell called tip cell detects and migrates toward the gradient of growth factor, while stalk cells follow behind the tip cells so that blood flow can be initiated and form a new vascular network.

1.5.2 Physical and Chemical Cues

The integrated response of a tumor to the physical and chemical cues within its microenvironment enables tumor cell dissemination and malignant progression. These cues include pH gradients and hypoxia, and gradients of soluble cues, as well as alterations in the concentration, stiffness, and organization of ECM. Metastasis is shown to be enabled by efficient tumor cell invasion, which is fostered by the ECM and growth factor gradients that work together to promote the migration of cancer cells toward vessels. This cooperation also facilitates the extravasation of cancer cells from primary tumor location into the circulation and subsequently their intravasation into distant tissues. The physical attributes of tumors include a high tumor solid stress, which is promoted by the elevated interstitial pressure, the expanding tumor cell mass, and the resistance applied by the surrounding ECM stiffness. This stiffness reflects abundance in proteins, particularly collagens I and III. It also indicates the reorganization of these proteins and their posttranslational modifications. The ECM will hold distinct regions of stiffness, which lead to endogenous stiffness gradients that foster tumor cell migration [66].

1.5.3 Acidosis

Solid tumors have an acidic extracellular pH due to the combination of poor perfusion and production of lactic acid in high rates. Cells adapt to chronic acidosis through multiple pathways, all of which can be exploited for therapeutic goals [67]. For the "acid-mediated tumor invasion" hypothesis proposes, this acidic pH will cause normal cells to die and will promote the degradation of the ECM through the release of proteases [68]. Survival of cancer cells in a chronically acidic environment involves mechanism of adaptation, such as chronic autophagy. This mechanism allows cells that are constantly under stress to maintain homeostasis via self-digestion [69]. This autophagy can be a target for therapy for some cancer models [70]. Another adaptive mechanism is an increase in the turnover of lysosomes and redistribution. Lysosomes require an acidic environment for proper functioning. They are organelles that have enzymes that act on cell digestive processes, thus maintaining homeostasis of the macromolecular components of the cells [71].

1.5.4 High Interstitial Fluid Pressure

Interstitial fluid pressure (IFP) in solid tumors has been identified as an essential element in tumor growth and vascularization.

Studies have shown that oxygen extravasation is affected by IFP; a higher IFP distribution inhibits tumor growth with time by decreasing the transfer of oxygen to the tissue [72]. An unfavorable treatment response is usually seen when there is an increase in tumor cell hypoxia [73]. The outcome of advanced cervical carcinoma is influenced by the tumor IFP; a high IFP in the primary tumor seems to result in a poor cancer outcome [74]. The IFP can also be used as a prognostic factor for certain types of cancer, such as lung cancer, because its IFP showed association with several clinic-pathological factors. An enhanced recurrence-free survival was seen in patients exhibiting low IFP in the tumor than in patients having a high IFP [75].

1.5.5 Enhanced Permeability Retention

Most solid cancers have blood vessels that produce an extensive number of vascular permeability factors. Therefore, most solid tumors show enhanced vascular permeability that will enable a sufficient supply of oxygen and nutrients to the tumor for rapid growth. Angiogenesis results in high vascular density in these types of tumors, which exhibit extravasation and retention of drugs. This enhanced permeability retention (EPR) serves as a basis for anticancer therapy development. This is because the nature of blood vessels in tumors facilitates the transport of drugs into tumor tissues, unlike in normal tissue, in which this EPR effect driven the delivery of macromolecules does not occur. The EPR effect is dependent on the molecular weight of molecules; particles that are larger than 40 kDa seem to have a prolonged circulation time in the body, thus gradually permeating tumors in a selective manner. In addition, the accumulated macromolecules remain in the tumors for a relatively extended time. The EPR effect is also shown to be mediated by several vascular factors, hence modulating these factors seems critical in augmenting the EPR effect and, therefore, the anticancer drug effects [76].

1.5.6 Hypoxia

Hypoxic stress has several effects on cells, such as metabolic changes and altered cell growth [77]. A major driver of cancer angiogenesis is hypoxia [65]. Reducing hypoxia is a possible target that can help control cancer [77]. In response to low oxygenation, cells upregulate the hypoxia-inducible factor (HIF), which is a transcription factor that has a role in coordinating an adaptive response within the pathophysiological and physiological range [78]. Under conditions of hypoxia, HIF is able to activate the transcription of several genes that have a role in maintaining oxygen homeostasis. HIF influences the expression of target genes involved in cell metabolism, invasiveness, angiogenesis, and erythropoiesis. Studies have shown that HIF has been correlated with poor outcomes in patients affected by cancer [79].

1.6 Cellular and Non-Cellular Components That Support Tumor Growth

Tumor growth is affected by several cellular and non-cellular components. Cellular components include (a) endothelial cells and pericytes, (b) CAFs, and (c) inflammatory cells, while non-cellular components include (a) ECM, (b) matrix metalloproteinase (MMP), (c) soluble growth factors, and (d) tumor-specific receptors.

1.6.1 Cellular Components

(a) Endothelial cells and pericytes

Vascular physiology depends on the concerted interaction of several cells, including endothelial cells and pericytes. Pericytes are cells that are involved in support of vascular development and homeostasis. These cells also play a critical role in angiogenesis. This process begins with the formation of endothelial cells from pre-existing vessels, elongation of a tube, and introduction of perivascular cells [80]. Pericytes, which are found in the vascular basement membrane that is shared with endothelial cells, detach from vessel walls and promote the migration of endothelial cells into the surrounding ECM [81, 82].

Studies have shown that the migration of pericytes and endothelial cells is uncorrelated, which indicates that their contact takes place indirectly through the basal membrane. Thus, pericytes are able to move freely on vessels, and their movement is independent of the movement of endothelial cells. It was also shown that the recruitment of pericytes occurs on endothelial sprouts that lack any vascular function, further supporting the notion that pericytes are required for early vascular development [80].

(b) Cancer associated fibroblasts

Cancer associated fibroblasts (CAFs) have been shown to either promote the growth, spread, and survival of tumors through improvement in functionality or to retard tumorigenesis via unknown mechanisms [83]. At an early stage of tumor development, CAFs can mediate a tumor-enhancing inflammation derived by interleukin-1β [84].

CAFs were also shown to be involved in cancer immune evasion, as they induced an immune-suppressive environment to ease the escape of tumors from the body's immune surveillance [85]. The immune system deactivation by CAFs can happen directly by expressing the negative regulatory immune signals, such as programmed death ligands, or secreting immune-suppressive factors, including prostaglandin E2, to directly decrease the activation of natural killer and T cells [86, 87]. In addition, the peritumoral stroma, which is rich in CAFs, can physically block the effective interaction of the immune system with the tumor [88].

(c) Inflammatory cells

Recent studies have shown that inflammation is an important component of tumor progression. Several types of cancers can arise from sites of inflammation, and the microenvironment of tumors is greatly orchestrated by inflammatory cells [89]. During tissue injury, inflammation and proliferation subside after the repair is completed or the assaulting agent is taken out. On the contrary, proliferating cells that maintain mutagenic assault or DNA damage keep on proliferating in microenvironments that are rich in inflammatory cells. It is similar to tumors acting as wounds that cannot heal [90].

The repeated process of tissue damage followed by tissue regeneration, while oxygen species are released from inflammatory cells, induces the interaction with DNA in the epithelium, which results in permanent alterations of the genes. P53 mutations are observed at frequencies almost equal to those in cancers in chronic inflammatory diseases, including inflammatory bowel disease and rheumatoid arthritis [91].

1.6.2 Non-Cellular Components

(a) Extracellular matrix

The extracellular matrix (ECM) is very dynamic in nature because, during development, it is constantly being deposited, remodeled, and degraded to maintain tissue homeostasis. The composition and organization of the ECM are regulated to control cell differentiation and behavior. Any dysregulation of the ECM dynamics results in the development of various diseases, such as cancer [92]. The arrangement and orientation of the constituents of ECM play an important role in tumor progression [93]. The ECM is not only involved in remodeling, but also induces biophysical and biochemical cues to affect cell adhesion and migration [94]. Therefore, small alterations in the homeostasis of the microenvironment can have major effects on cancer cell proliferation. The primary functional characteristics of the matrix are dictated by collagen, which is the most abundant ECM component. Indeed, alterations in the degradation or deposition of collagen can result in the damage of ECM homeostasis [95].

As cancer cells proliferate, ECM surrounding the tumor undergoes significant modifications in an interplay between the cells and the microenvironment. These alterations, such as an enhanced secretion of collages I and III, illustrate that the tumor's progression requires continuous interaction between the tumor cells and the ECM [96].

(b) Matrix metalloproteinases

Matrix metalloproteinases (MMPs) are a family of ECM remodeling endopeptidases that act as the primary mediators of the modifications seen in the microenvironment during tumor progression [97]. MMPs are implicated in several physiological processes, such as uterine involution and organogenesis, and wound healing, as well as in various pathological conditions, such as vascular and auto-immune diseases, inflammatory disorders, and carcinogenesis. All stages of tumor progression, including tumor growth, invasion, metastasis, and angiogenesis, can be regulated by MMPs. The expression of MMPs depends on both cancer cells and neighboring stromal cells. MMPs facilitate angiogenesis by exerting their proteolytic activity to degrade physical barriers. In this manner, they also facilitate cancer cell invasion and metastasis. In addition, MMPs alter the interactions between cells and between cells and ECM, therefore both influencing cell phenotype and increasing cell migration [98]. MMPs are believed to be possible diagnostic and prognostic biomarkers for various types of cancer [99].

(c) Soluble growth factors and their receptors

Under normal conditions, growth factors are the signals received by cells from their surroundings to determine their fate. Although an alteration in homeostasis and the initiation of the tumor are instigated by mutations rather than by the involvement of growth factors, the latter significantly regulates all subsequent phases of tumor progression, such as clonal expansion, invasion, angiogenesis, and colonization [100]. The essential role of growth hormones for clonal expansion allows for the fixation of oncogenic mutations and the increase in the number of initiated cells that are susceptive to additional mutations. Several tumor cells acquire the capacity to synthesize growth factors to which these cells are responsive [101]. In addition, various mechanisms may result in a consecutive activation pathway in cancer. At the receptor level, this overexpression may lead tumor cells to become hyper-responsive to growth hormones [102]. Tumor progression takes place merely when the basement membrane secretory ducts are dissolved. Growth factors play an essential role in the disruption of the basement membrane, as well as the penetration by tumor cells [103].

(d) Tumor-specific receptors

Every tumor cell has an array of active surface molecules. Several of these molecules act as receptors for different ligands. These include cytokine receptors, human leukocyte antigens (HLAs), cell-adhesion molecules, and others. The expression of these receptors is a subject to modifications, typically to the advantage of tumors, supporting their growth and spread. Some receptors are completely new as they appear only on tumor cells, having no counterparts on normal cells. Tumor-cell receptors are critical in cancer research as they provide information that is important to create tools specific for tumor immunotherapy [104]. Drug delivery vehicle accumulation within targeted tissues can be enhanced via receptor-mediated targeting. Targeted drug delivery is essential for the enhanced tumor internalization. This occurs due to the interactions between ligand and receptor at the cell surface, rendering drug delivery vehicles to have a higher affinity for tissues affected with cancer and a lower affinity for the surrounding healthy tissue, as compared to non-targeted drugs. In this manner, receptor-mediated drug delivery can alter the bio-distribution of the drug by enhancing its accumulation in the tumor site [105].

An example of a tumor-specific receptor is the sigma receptor, which was first described as an opioid receptor, but later found to have the ability to bind to various drugs. Sigma receptors are notably upregulated in cancer cells and tumor tissues. Drugs that target sigma receptors, particularly sigma 2 receptors, showed to have anti-proliferative effects. One possibility is that drugs that target sigma receptors could have anti-cancerous effects through the modulation of ion channels, as well as other metastatic cellular behaviors could also be affected, such as motility, adhesion, and secretion [106].

1.7 Difference between Normal Cells and Cancer Cells

There are many ways in which cancer cells are different from normal cells that enable them to become invasive. One critical difference is that normal cells are more specialized than cancer cells. That is, unlike normal cells, cancer cells do not mature into distinct cell types that have specific functions. Cancer cells undergo uncontrolled cell division. Furthermore, the signals that induce normal cells to halt division or that induce apoptosis are ignored by cancer cells. In addition, cancer cells can influence the surrounding of normal cells and blood vessels to feed the tumor. For example, they can induce the normal cells to produce new blood vessels to supply the tumor with oxygen and nutrients for their growth. These newly formed blood vessels are also needed for the removal of waste products from tumors. The immune system can also remove damaged cells or normal cells that become abnormal in the body, unlike cancer cells, which can often evade the immune system [107].

1.8 Conventional Anti-Cancer Therapies

Several cancer treatment methods exist, and the utilization of each of them depends on the type of cancer and on its stage of advancement. Treatment options include (a) surgery, (b) radiation therapy (RT), (c) chemotherapy, (d) hormonal therapy, (e) immunotherapy, (f) photodynamic therapy (PDT), (g) targeted therapy, and others. In some cases, a combination of treatment methods may be used to maximize results. Each of these treatments has possible side effects to patients [108].

(a) Surgery

Surgery is a common procedure done for the treatment of non-hematological cancers. During surgery, a surgeon excises the cancer tissues, a procedure that can either result in a partial or a complete cure. The surgery is usually accompanied with side effects that depend on the cancer type and on the health condition of the patient. The complete removal of cancer from the body is not possible if there is metastasis of cancer to other body parts [109]. In general, tumor resection is typically done to localized cancers that are small in size. Some of these surgical procedures that remove tumors include prostatectomy for prostate cancer and mastectomy for breast cancer. Surgical procedures to resect a tumor do not ensure the complete removal of cancer cells. A single cancer cell that is left out can regrow into a tumor and metastasize to other body parts. Other cancer treatment procedures can be performed before or after tumor resection by surgery to remove the tissues that are affected [110, 111].

(b) Radiation therapy

Radiation therapy (RT) is a method of cancer treatment that uses high doses of radiation to destroy cancer cells. This treatment is typically used in conjunction with surgical methods to remove tumors or reduce their sizes. One of the adverse effects of RT is that it can damage normal cells due to the ionizing radiations used [112]. Most types of cancers are treated by RT, such as cancers of the breast, cervix, liver, larynx, lung, brain, prostate, stomach, skin, etc. Sometimes RT can also be used for the treatment of lymphoma and leukemia [113]. Specialized equipment is used to deliver calculated radiation doses to the cancer cells. The radiation dose chosen depends upon several factors, such as the type and location of the tumor, age of the person, and possible side effects on the target or nearby tissues. The RT kills cancer cells by destroying their DNA. The DNA damage happens either directly or indirectly by creating intracellular free radicals that will, in turn, destroy the DNA [108].

(c) Chemotherapy

Chemotherapy is the use of anticancer drugs to destroy cancer cells. The drugs work by killing cancer cells or interfering with tumor growth. Chemotherapy is usually considered one of the most effective cancer treatment methods; however, this type of therapy can induce severe side effects, as it can also destroy healthy cells. The adverse effects depend upon the type of cancer and the type of drugs used to treat it. Generally, the side effects are not associated with the treatment effectiveness, and once the treatment process is over, the side effects may stop. Normally, chemotherapy agents are prescribed to a patient in measured dosages and in specific intervals of time. Sometimes a combination chemotherapy is used, during which two or more drug agents are used at the same time [114].

(d) Hormonal therapy

Certain types of cancers depend on hormones to grow and propagate. Thus, hormonal therapy works by altering the amount of certain hormones in the body to fight cancers that are dependent on the chemicals. This method of treatment is used to treat cancers such as cancer of the prostate, breast, and reproductive systems. For instance, all patients who have hormone receptor-positive breast cancer are mandated to have hormonal therapy. In this case, hormone therapy aims to inhibit the interaction between a hormone, such as estrogen, and its dependent pathways that stimulate neoplastic cells. The side effects that arise from hormonal treatment depending on the age and sex of the patient, type of cancer, and the drug type used [115].

(e) Immunotherapy

Immunotherapy is a cancer treatment method that acts by boosting the immune system's ability to fight cancer. This cancer therapy works by stimulating the body's own disease-fighting mechanisms. A large number of studies have been conducted to treat different types of cancer by immunotherapy. For instance, researchers have used monoclonal antibodies that inhibit the function of specific proteins by binding to tumor cells, which, in turn, train the body's immune system to identify and attack cancer cells. No major side effects have been reported with immunotherapy. Despite the success of immunotherapy, it only works in certain types of cancers and in less number of cancer patients respond to immunotherapy [116].

Other mechanisms of action of immunotherapy include the use of vaccinations and checkpoint inhibitors. For instance, in 1990, the FDA approved the utilization of the Bacillus Calmette–Guerin (BCG) vaccine to treat superficial bladder cancer. The FDA also approved another vaccine, sipuleucel-T, to extend the overall survival rate of castrate-resistant prostate cancer patients. Generally, immunotherapies are divided into two types: active and passive immunotherapies. The use of active immunotherapy implies the direct stimulation of immune memory, immune response, and lasting response. The use of passive immunotherapy may lead to the production of specific but short-lived responses, which, therefore, require a consistent treatment administration [117].

(f) Photodynamic therapy

Photodynamic therapy (PDT) is a non-invasive method of anticancer therapy. This method is based on the systemic or local application of a photosensitizer, a photosensitive compound that accumulates in pathological tissues. This compound absorbs light of an appropriate wavelength, which induces the initiation of the activation processes that lead to the selective damage of the cancer cells. The selective destruction is enabled due to the occurrence of photo-cytotoxic reactions only in the area of compound distribution within the pathological tissues. The mechanism of action of PDT is dependent on the oxygen molecules within the cells. After entering the cell, a photosensitizer is irradiated with light having a specific wavelength and is converted into an excited energy state due to the absorption of a photon. A portion of the energy will be radiated, while the remaining portion directs the compound into another excited state, which can generate reactive oxygen species (ROS) that can kill cancer cells by damaging its DNA (Figure 1.3). Generally, patients well tolerate PDT because of its selective action. This type of therapy is usually painless and very simply applicable, thus allowing for outpatient use [117, 118].

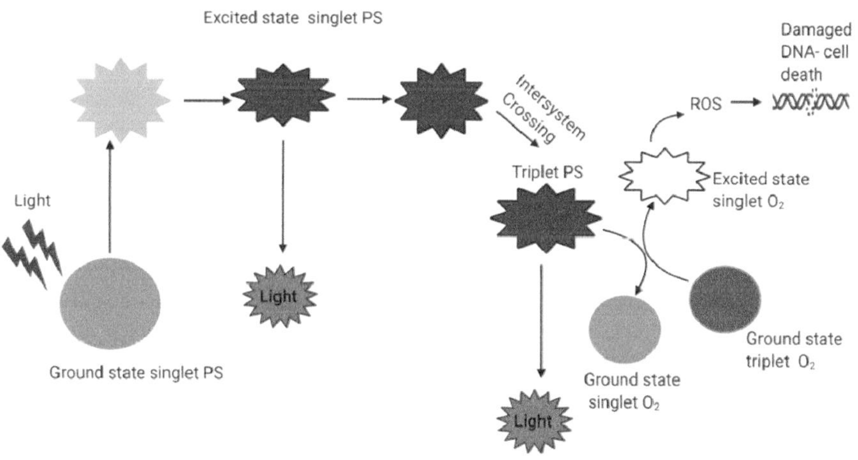

Figure 1.3 PDT to treat cancer. A photosensitizer enters the cell and is irradiated with a specific wavelength of light. This allows the conversion of the photosensitizer into an excited energy state. A part of the energy will be radiated, while the other part brings the photosensitizer into another excited state, which can generate ROS that can kill cancer cells by damaging their DNA.

(g) Targeted therapy

Targeted cancer therapies target specific characteristics in cancer cells compared to normal cells that give the cancer cells the ability to divide, proliferate, and spread. These specific alterations that happen on the genetic level of the cell can be utilized as a molecular target to develop target cancer therapies. Specific agents are used in targeted treatment for proteins that are deregulated in cancer cells. The agents used in this method of cancer treatment are generally drugs that act as inhibitors of enzymatic domains found on the mutated and overexpressed proteins in cancer cells. The targeted agents can include cell-cycle proteins, signaling molecules, growth factors, molecules that promote angiogenesis, and modulators of apoptosis, among many others. The mechanism of action of targeted therapies can be seen in Figure 1.4 [119, 120].

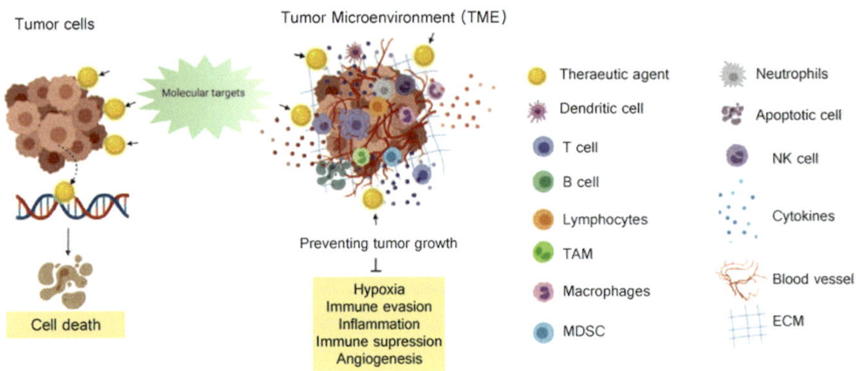

Figure 1.4 Molecular targeted therapy. Illustration of targeted therapy development that targets specific characteristics of cancer cells that give them the ability to divide, proliferate, and metastasize compared to normal cells.

1.9 Cell Death and Cancer

Cell death plays an important and vital part in the development of cells in multicellular organisms. During the cell death phase the excess cells that are produced by mitosis, in order to control the cell number. In the development and morphogenesis, a strategy of defense takes place where the infected, mutated and damaged cells are removed. The normal cells deaths are suicides in which they activate their own intracellular death program and kill themselves in a controlled way, and we call it as programmed cell death.

According to the nomenclature community on cell death (NCCD), apoptosis was termed as programmed cell death type 1, autophagy was termed as programmed cell death type 2. When a cell attains a state of permanent cell cycle arrest, then this phenomenon is called as senescence. Let us now discuss in detail about cell deaths according to their morphological appearance, biochemical feature, functional appearance, and immunological behavior. Advanced studies and research in the past decades had helped to classify further the death types according to their corresponding feature and appearance as mentioned above.

(a) Apoptosis

In apoptosis, (originated from Greek word meaning "falling off" as leaves from tree) cell death occurs by releasing intracellular materials, chromatin condensation, nuclear breakdown, loss of plasma membrane symmetry, formation of small membrane blebs, and shedding these blebs as apoptotic bodies. These apoptotic bodies soon will be engulfed by phagocytosis by neighboring cells and/or macrophages before they can spill out contents from the apoptotic cell for lysosomal degradation. In this way, the cell dies neatly and rapidly, cleared without causing damage or any inflammatory response. Though we already discussed about morphological changes, and in addition to the same, there are biochemical changes that also occur, which include the activation of cysteine-dependent aspartate-directed proteases known as caspase, the leakage of cytochrome from mitochondria into the cytosol, the activation of endonucleus that depends on Ca^{2+} and Mg^{2+}, which leaves inter-nucleosomal DNA, the degradation of poly-ADP-ribose polymerase (PARP), and the exposure of phosphatidylserine to the exterior lipid bilayer, which allows phagocytosis to detect and engulf the apoptotic cell for degradation and clearance.

(b) Autophagy

It is a self-digestion process, which is also another source of cell death. It is the process where it removes and degrades the damaged cellular constituents by which a small portion of cellular components are isolated from the rest of the cell to form a double layered vesicle called autophagosome. Autophagosome further fuses with lysosome having low pH (acidic) and later endosome to form autolysosome. When autophagy and lysosome degradation is disturbed, Cathespin B is released from lysosome to trigger mitochondrial permeabalization and caspase activation. Cathespin D is translocated from lysosome to cytosol, which is responsible for Bax activation and apoptotic induced factor release. In addition, caspase cleave autophagy relative proteins, such as Beclin1, Bcl-2, P53, and UV-irradiation resistance gene (UVRAG), are several tumor proteins that regulate autophagy.

(c) Senescence

Cellular stress that activates P53 induces apoptosis. During the cell cycle arrest, the weak and sustained activation can promote cell death known as senescence. Cellular senescence is pathophysiological condition by which the cells permanently lose their proliferative capacity. Senescence cell shows specific-morphological features, including flattening, nuclear enlargement, enriched with vacuoles, and altered chromatin structure. In the biochemical level, cellular senescence is characterized by increased β-galactocidase activity that increases senescence markers, such as P15, P16, P21, P53, and ARF.

(d) Necrosis

Necrosis is a death mechanism, in which cells die accidently in response to an acute insult such as snake biting, trauma or lack of blood supply, etc. In the case of necrosis, the cells undergo plasma membrane permeabilization, swelling, and rupture, and by that the necrotic cells spill out the cellular contents over their neighbors, which invite the damage associated molecular patterns (DAMPs) and may initiate inflammation. Necrosis is also termed as necroptosis and it is mediated by death receptor

TNFR1 (tumor necrosis factor receptor). On TNF stimulation, TNFR1 goes through a confirmation change and recruits TNFR-associated death domain protein (TRADD), TRARF2, cellular inhibitor of apoptosis protein 1 (cIAP1), cIAP2, and receptor interacting protein kinase 1 (RIPK1), to form complex 1. RIPK1 is polyubiquinated by cIPA1 and c1PA2 activates NFkB. Caspase 8 cleavage induces apoptosis, in certain condition, and caspase 8 inhibition will induce necroptosis. Necroptosis involves several factors such as ROS production, lysosomal permeabalization, AIF release, and PARP activation. When caspase activation is not involved, necroptosis is associated with the formation of autophagic vesicles.

1.10 Advanced Nanotechnology for Cancer Diagnosis and Therapy

Nanoscale agents are materials that have the ability to carry various drugs and imaging agents [119]. Solid lipid nanoparticles (SLNs) are mainly comprised lipids that self-assemble. SLNs are perfect carriers for nucleic acids and hydrophobic drugs. Studies have shown that SLNs increased the uptake and delivery of drugs within the liver. Other types of nanoparticles (NPs) include those that have an inorganic base system. For instance, gold NPs are inorganic drug delivery vehicles that can be synthesized in various shapes and sizes. These are easily altered with ligands to target drugs or cells, making them ideal for chemotherapeutics. Similarly, iron oxide NPs are other drug delivery vehicles that have proven to enable ligand and drug conjugation and to improve biocompatibility. Selection of the most appropriate vehicle for targeted drug delivery is a critical decision. During the formulation of NPs, it is necessary to encapsulate the drug efficiently, release it to the desired site, and achieve the desired function. In addition, the choice of vehicle can be dependent on the route of administration, target location, and desired delivery timeline. Typically, targeted drug delivery vehicles are taken up quickly by the targeted cells; thus, these vehicles are greatly dependent on having a strong adhesion with cancer cell receptors [105].

1.10.1 Key Features of Nanomaterials for Therapy and Diagnostic Applications

1.10.1.1 Nanoparticle size

NPs have a typical size of 1–100 nm, at least in one dimension among its three dimensions. These particles are able to achieve a high ligand density for targeting purposes due to their high surface area to volume ratio. This desirable nanoparticle size will enable them to penetrate into tumors and is advisable to avoid renal clearance. Therefore, precise control of the size is necessary for pharmacokinetics, distribution, tumor penetration, and accumulation of the nanoparticle therapeutic system [120].

1.10.1.2 Nanoparticle surface properties

Nanoparticle drug delivery systems have various structures, and these include, but are not limited to, polymeric micelles, dendrimers, and liposomes. The basis of the characteristics of micelles is that they are amphiphilic block copolymers that gather to nano-size structures in aqueous media. Polymeric micelles act as an interplay between hydrophilic regions and hydrophobic core regions, in which hydrophobic drugs are found for stabilization, therefore, making the entire entity water soluble. The transportation of drugs in polymeric micelles can take place via either physical encapsulation or chemical adhesion [121]. The surface of dendrimers, which are polymeric macromolecules, exhibits many radial branched monomers. Targeted delivery via dendrimers is enabled by their distinctive characteristics, such as their adjustable surface functionality, water solubility, water valency, internal space for drugs, and mono-disperse size [122]. On the other hand, liposomes are closed colloidal structures that consist of lipid layers. They exhibit a spherical structure in which the center is an aqueous space that is surrounded by a lipid bilayer as the outer surface [123].

1.10.1.3 NPs in diagnosis

Several NPs are used for cancer diagnosis. For instance, those of cadmium selenide can glow if exposed to UV light. They go within cancer tumors when injected into the human body, and surgeons can detect the glowing tumor and utilize these NPs for more accurate tumor resection. Furthermore, sensor test chips, which contain thousands of nanowires, are able to detect remnant biomarkers left by cancer cells. This may enable the diagnosis of cancer in its early stages from just a few drops of the patient's blood [124].

1.10.1.4 NPs in imaging

A number of nanoparticle applications have been developed for angiogenesis imaging, tracking of immune cells, cancer staging, and for cellular or even molecular targeting. With the advent of NPs, a more accurate detection, staging, and therapeutic planning of cancer are enabled, due to the combination of nanoscale devices with new nuclear imaging techniques. Research has found that gold NPs may serve as bright imaging agents. By linking gold NPs to antibodies that bind to tumors, researchers have found that the NPs will align themselves on the cancer cells' surfaces in an ordered manner, therefore providing a distinctive optical signature for cancer cells [124].

1.10.1.5 Targeted nanoparticles

Nanoparticles (NPs) are considered as a treatment option against cancer that can help improve therapeutic index accompanied with decreased side effects [125]. This is because the use of targeted NPs allows for an enhanced drug bioavailability and a reduced dosage and frequency are needed [126]. Applications of NPs include targeted anti-cancer drug delivery, radiation sensitizers, lymph node mapping, investigation of apoptosis, and imaging [127, 128]. With the advent of NPs, they are used for the delivery of anti-cancer drugs into unhealthy cells and subsequently avoiding the toxicity of normal cells, targeted drug delivery into a specific position, increment of carrier capacity and drug stability, adjusting the release of drugs to target sites, enabling efficient permeation of endothelial and epithelial barriers, and co-delivering of several drugs for transporting non-water-soluble drugs, which are mediated by water-soluble substance became possible [129].

1.10.1.6 Nanomaterials to target and regulate tumor and its microenvironment

Nanomaterials are artificially engineered materials that greatly enhance the therapeutic efficacy of anti-neoplastic drugs, while decreasing nonspecific toxicity. Therefore, this allows the achievement of safer and more effective cancer treatment. Nanomaterials can interact with biomolecules at particular intracellular compartments, thus allowing for possible manipulation of specific signal pathways that are involved in the survival and proliferation of tumors. Nanomaterials also offer possibilities to circumvent the biological defense mechanisms and the vascular barriers of the body [130]. NPs used in targeted drug delivery systems are derived from nanomaterials and organometallic combinations [129].

1.10.2 Challenges Faced by NPs in Combatting Tumor

1.10.2.1 Challenges faced during circulation

Several challenges are faced by NPs before reaching cancer cells, such as the physiological barriers that are intrinsic to the tumor microenvironment. For instance, these include the dense matrix composed of proteins and collagen fibers and the high IFP induced by the lack of functional lymphatics in tumor tissue. All of these have the impact of reducing the transport of NPs across the wall of vessels into the interstitial space. The diffusion of NPs is further limited by their nonspecific uptake by other cells. Another limitation faced by NPs during circulation is the high tumor inter-capillary distances, rendering it challenging for NPs to reach tumor cells that are far from vessels [131]. In addition, the prolonged residence time of NPs in the systemic circulation is considered a major challenge due to their rapid sequestration by macrophages [132].

1.10.2.2 Challenges after reaching tumor

NPs are efficient to deliver anti-cancer drugs to their target sites. Even though the objective of using NPs is to allow more drugs to reach the target cells and potentially reduce side effects;

however, toxicity of some NPs has been clearly shown [133]. Several studies have shown that potentially toxic NPs, such as gold NPs, are practicable tools for short-term use only. However, due to their harmful effects on the genetic material and to the accumulation of gold deposits in several organs, it is strongly recommended to question their use in the chronic treatment of cancer [134].

1.11 Conclusion and Future Perspective

Current treatments against cancer are chemotherapeutic drugs, surgical intervention, and radiation. Generally, side effects are an indispensable part of these treatments. Thus, the focus of current studies is directed toward finding new strategies of therapy without significant side effects. In recent years, better comprehension of the molecular basis of tumors and the advent of de novo diagnostic technologies have decreased the mortality rate. Therefore, researchers continue to research and find strategies to enhance survival chances and to improve the quality of life of cancer patients. Nowadays, NPs used in the treatment of cancer have been receiving surging attention in the pharmaceutical research field, due to the increasing importance of nanotechnology. These NPs are providing a safe alternative to typical chemotherapeutic agents, due to their increased solubility and stability of drugs, thus drastically reducing drug-related toxicities through the decrease of non-specific bio-distribution. A range of nanoscale systems for drug delivery have been developed. These NPs exhibit prolonger circulation, greater cellular uptake, and better efficiency in accessing the targeted tumor site, as compared with the typical anticancer therapy. Furthermore, they are able to carry a significant number of therapeutic drugs, due to their large surface area to volume ratio, therefore, penetrating into the tumor vessels and subsequently delivering the drugs to tumor tissues. The use of NPs that carry anti-cancer drugs is a promising alternative for formulating drug products; however, more and more research efforts need to be carried out to get a 100% satisfactory result in treating cancer.

References

1. World Health Organization. Cancer. **2020**.
2. Roy PS, Saikia BJ. Cancer and cure: A critical analysis. *Indian J Cancer.* 53, **2016**, 441–442.
3. Patricia S. Targeting metastasis. 16, **2016**, 201–218.
4. Ainsley W, Curtis CH. Multistage carcinogenesis. *Holland-Frei Cancer Medicine.* 6th edition. Hamilton (ON): BC Decker; 2003. https://www.ncbi.nlm.nih.gov/books/NBK13982/.
5. Balmain A, Brown K. Oncogene activation in chemical carcinogenesis. *Adv Cancer Res.* 51, **1988**, 147–182.
6. Yuspa SH, Poirier MC. Chemical carcinogenesis: From animal models to molecular models in one decade. *Adv Cancer Res.* 50, **1988**, 25–70.
7. Neumann HG. Role of extent and persistence of DNA modifications in chemical carcinogenesis by aromatic amines. *Recent Results Cancer Res.* 84, **1983**, 77–89.
8. Cairns J. Mutation selection and the natural history of cancer. *Nature.* 255, **1975**, 197–200.
9. Yuspa SH, Dlugosz AA, Denning MF, Glick AB. Multistage carcinogenesis in the skin. *J Investig Dermatol Symp Proc.* 1, **1996**, 147–150.
10. Lengauer C, Kinzler KW, Vogelstein B. Genetic instabilities in human cancers. *Nature.* 396, **1998**, 643–649.
11. Bingmei MF. Tumor metastasis in the microcirculation. *Molecular, Cellular, and Tissue Engineering of the Vascular System.* 1097, **2018**, 201–218.
12. Eva YHPL, William JM. Oncogenes and tumor suppressor genes. *Cold Spring Harb Perspect Biol.* 2, **2010**, a003236.
13. Sherr CJ. Cancer cell cycles. *Science.* 274, **1996**, 1672–1677.
14. Takayama T, Ohi M, Hayashi T, Miyanishi K, Nobuoka A, Nakajima T, Satoh T, Takimoto R, Kato J, Sakamaki S, Niitsu Y. Analysis of K-ras, APC, and beta-catenin in aberrant crypt foci in sporadic adenoma, cancer, and familial adenomatous polyposis. *Gastroenterology.* 121, **2001**, 599–611.
15. Shibata D, Schaeffer J, Li ZH, Capella G, Perucho M. Genetic heterogeneity of the c-K-ras locus in colorectal adenomas but not in adenocarcinomas. *J Natl Cancer Inst.* 85, **1993**, 1058–1063.

16. Bourne HR, Sanders DA, McCormick F. The GTPase superfamily: Conserved structure and molecular mechanism. *Nature*. 349, **1991**, 117–127.

17. Byoung-San M, Woo-Jeong J, Jieun P, Tae IIK, Do SM, Kang-Yell C. Role of oncogenic K-Ras in cancer stem cell activation by aberrant Wnt/β-catenin signaling. *J Natl Cancer Inst*. 106, **2014**, 373.

18. Vasanti S, Vikas S, Manikanta M. Current overview on the clinical update of Bcl-2 anti-apoptotic inhibitors for cancer therapy. *European Journal of Pharmacology*. 862, **2019**, 172655.

19. Knudson Jr AG. Mutation and cancer: Statistical study of retinoblastoma. *Proc Natl Acad Sci USA*. 68, **1971**, 820–823.

20. Knudson Jr AG. Hereditary cancer, oncogenes, and antioncogenes. *Cancer Res*. 45, **1985**, 1437–1443.

21. Peter D, Ruhong L, David R. Aneuploidy approaching a perfect score in predicting and preventing cancer. *Cell Cycle*. 3, **2004**, 823–828.

22. Ruhong L, Arvinf S, Reinhard S, David R, Peter D. Aneuploidy vs. gene mutation hypothesis of cancer: Recent study claims mutation but is found to support aneuploidy. *PNAS*. 97, **2000**, 3236–3241.

23. D David JG, Benjamin R, David P. Causes and consequences of aneuploidy in cancer. *Nature Reviews Genetics*. 13, **2012**, 189–203.

24. Ankit H, Mathew B, Peter D. Speciation theory of carcinogenesis explains karyotypic individuality and long latencies of cancers. *Genes (Basel)*. 9, **2018**, 402.

25. Kanwal R, Gupta S. Epigenetic modifications in cancer. *Clin Genet*. 81, **2012**, 303–311.

26. Leigh E, Peter WA, Ricky WJ. Epigenetics in cancer: Targeting chromatin modifications. *Mol Cancer Ther*. 8, **2009**, 1409–1420.

27. Centers for Disease Control and Prevention. US cancer statistics. **2013**.

28. Nicholas P, Giorgio S, Riccardo AA. Cancer prevalence and mortality in centenarians: A systematic review. *Crit Rev Oncol Hematol*. 83, **2012**, 145–152.

29. Mary CW, Dawn MH, Jennifer EB, Lucy AP, Melissa G, S Jane H. Age and cancer risk. *Am J Prev Med*. 46, **2014**, S7–15.

30. Taro I, Ryu I, Toshitaka M, et al. Impact of age at diagnosis of head and neck cancer on incidence of metachronous cancer. *BMC Cancer*. 19, **2019**, 3.

31. Andrej P, Marta NVA, Krysta Y, Viktorija P, Eszter N, Tripti G, Richard AM, William ES, Anna C, Zoltan U. The GH/IGF-1 axis in a critical period early in life determines cellular DNA repair capacity by altering transcriptional regulation of DNA repair-related genes: Implications for the developmental origins of cancer. *GeroScience*. 39, **2017**, 147–160.

32. Ramadevi S, Sushmita BN, Diego P, Rajkumar L. Role of growth hormone in breast cancer. *Endocrinology*. 158, **2017**, 147–160.

33. Yi-Jun C, Xiao Z, Zheng-Sheng W, Jing-Jing W, Amy YCL, Tao Z, Peter EL. Autocrine human growth hormone stimulates the tumor initiating capacity and metastasis of estrogen receptor-negative mammary carcinoma cells. *Cancer Lett*. 365, **2015**, 182–189.

34. Alona ON, Brian MS. Growth hormone enhances LNCaP prostate cancer cell motility. *Endocr Res*. 40, **2015**, 97–105.

35. Jing-Jing W, Qing-Yun C, Xin-Bao S, Ming-Liang Y, Vijay P, Yi-Jun C, Qui-Shi Z, Dong-Xu L, Lan M, Zgheng-Sheng W, Tao Z, Peter EL. Autocrine hGH stimulates oncogenicity, epithelial-mesenchymal transition and cancer stem cell-like behavior in human colorectal carcinoma. *Oncotarget*. 8, **2017**, 103900–103918.

36. Hannah RB, Michael EJ, Minouk JS, Alan A, Anthony JS. Family history and risk of breast cancer: An analysis accounting for family structure. *Breast Cancer Res Treat*. 165, **2017**, 193–200.

37. Fay K, N. Jewel S, Randall WB. Use of family history and genetic testing to determine risk of colorectal cancer. *Gastroenterology*. 158, **2020**, 389–403.

38. Federica T, Cristina B, Jerry P, Diego S, Maurizio M, Massimo L, Gaetano F, Monica F, Alessandra T, Carlo LV, Eva N. Family history of cancer and the risk of bladder cancer: A case-control study from Italy. *Cancer Epidemiol*. 48, **2017**, 29–35.

39. Huan S, Isabella GE, Alexander P, Jan E, Olof N, Weimin Y. Family history of gastric mucosal abnormality and the risk of gastric cancer: A population-based observational study. *Int J Epidemiol*. 47, **2017**, 440–449.

40. NIH U.S. National Library of Medicine. What does it mean to have a genetic predisposition to disease? **2020**.

41. Nazneen R. Realising the promise of cancer predisposition genes. *Nature*. 505, **2014**, 302–308.

42. Easton D, Thompson D, McGuffog L, et al. Cancer risks in BRCA2 mutation carriers. *J Natl Cancer Inst*. 91, **1999**, 1310–1316.

43. Sarah EJ, John DC. Personalized cancer medicine. *International Journal of Cancer*. 137(2), **2014**, 262–266.

44. Burke W, Daly M, Garber J, Botkin J, Kahn MJ, Lynch P, McTiernan A, Offit K, Perlman J, Petersen G, Thomson E, Varrichhio C. Recommendations for follow-up care of individuals with an inherited predisposition to cancer. II. BRCA1 and BRCA2. *Cancer Genetics Studies Consortium. JAMA*. 277, **1997**, 997–1003.

45. Franceschi S, Maso LD, Augustin L, Negri E, Parpinel E, Boyle P, Jenkins DJ, Vecchia CL. Dietary glycemic load and colorectal cancer risk. *Ann Oncol*. 12, **2001**, 173–178.

46. Giovannucci E. Insulin and colon cancer. *Cancer Causes Control*. 6, **1995**, 164–179.

47. Paolo V, Fransceco F, Laura S, Giuseppe P, Riccardo V. Diabetes and cancer. *Endocr Relat Cancer*. 16, **2009**, 1103–1123.

48. Thibault F, Bernard S, Laury S, Emmanuelle KG, Benjamin A, Caroline M, Melanie D, Philippine F, Paule LM, Marie B, Serge H, Celine L, Carlos AM, Chantal J, Mathilde T. Consumption of ultra-processed foods and cancer risk: Results from NutriNet-Santé prospective cohort. *BMJ*. 360, **2018**, 322.

49. Jane SH, Cheryll CT, Saida RS, Behnoosh M, Greta MM, Deborah MW, Brian SA, Lisa CR. Vital signs: Disparities in tobacco-related cancer incidence and mortality in United States, 2004–2013. *JSTOR*. 65(44), **2016**, 1212–1218.

50. Curry SJ, Byers T, Hewitt M. *Fulfilling the Potential of Cancer Prevention and Early Detection*. Washington (DC): National Academies Press (US); 2003. https://www.ncbi.nlm.nih.gov/books/NBK223926/ doi: 10.17226/10263.

51. Sasco AJ, Secretan MB, Straif K. Tobacco smoking and cancer: A brief review of recent epidemiological evidence. *Lung Cancer*. 45, **2004**, S3–9.

52. Adam BM, Folasade A, Yaw AN, Mignonne CG, Iman KM, Courtney MPH, Kelly W, Rick AK, Chiledum A. Smoking and prostate cancer in a multi-ethnic cohort. *Prostate*. 73, **2013**, 1518–1528.

53. Alina M, Philippe A, Mathieu B, Peter B. Active and passive smoking and risk of breast cancer: A meta-analysis. *Breast Cancer Res Treat* 154, **2015**, 213–224.

54. Vineis P, Alavanja M, Buffler P, Fontham E, Franceschi S, Gao YT, Gupta PC, Hackshaw A, Matos E, Samer J, Sitas F, Smith J, Stayner L,

Straif K, Thun MJ, Wichmann HE, Wu AH, Zaridze D, Peto R, Doll R. Tobacco and cancer: Recent epidemiological evidence. *J Natl Cancer Inst*, 96, **2004**, 99–106.

55. Alcohol use and burden for 195 countries and territories, 1990–2016: A systematic analysis for the global burden of disease study 2016. *Lancet*. 392(10152), **2018**, 1015–1035.
56. Rebecca LS, Kimberly DM, Ahmedin J. Cancer statistics, 2020. *CA Cancer J Clin*. 70, **2020**, 7–30.
57. Whiteman DC, Whiteman CA, Green AC. Childhood sun exposure as a risk factor for melanoma: A systematic review of epidemiologic studies. *Cancer Causes Control*. 12, **2001**, 69–82.
58. International Agency for Research on Cancer Working Group on artificial ultraviolet (UV) light and skin cancer. The association of use of sunbeds with cutaneous malignant melanoma and other skin cancers: A systematic review. *Int J Cancer*. 120(5), 1116–1122. doi: 10.1002/ijc.22453. Erratum in: *Int J Cancer*. **2007**, 120(11), 2526.
59. Graham AC. Cancer Prevention. UpToDate. **2020**.
60. Nemat K, Yadollah S, Mahdi M. Chronic inflammation and oxidative stress as a major cause of age-related diseases and cancer. *Recent Pat Inlamm Allergy Drug Discov*. 3, **2009**, 73–80.
61. Lobo V, Patil A, Phatak A, Chandra N. Free radicals, antioxidants and functional food: Impact on human health. *Pharmacogn Rev*. 4, **2010**, 118–126.
62. Frances RB, Melania C, Thorsten H. The tumor microenvironment at a glance. *Journal of Cell Science*. 125, **2012**, 5591–5596.
63. Whiteside TL. The tumor microenvironment and its role in promoting tumor growth. *Oncogene*. 27, **2008**, 5904–5912.
64. Judah F. Tumor angiogenesis: Therapeutic implications. *N Engl J Med*. 285, **1971**, 1182–1186.
65. Claire V, Bruno L. Tumor angiogenesis and vascular normalization: Alternative therapeutic targets. *Angiogenesis*. 20, **2017**, 409–426.
66. Madeleine JO, Valerie MW. Physical and Chemical Gradients in the Tumor Microenvironment Regulate Tumor Cell Invasion, Migration, and Metastasis. Cold Spring Laboratory Press. **2016**.
67. Arig IH, Veronica E. Acidosis and cancer: From mechanism to neutralization. *Cancer Metastasis Rev*. 38, **2019**, 149–155.

68. Robert AG, Edward TG, Arthur FG, Brant K, Robert JG. Acid-mediated tumor invasion: A multidisciplinary study. *Cancer Res.* 66, **2006**, 5216–5223.

69. Noboru M, Daniel JK. Protein turnover via autophagy: Amplications for metabolism. *Annu Rev Nutr.* 27, **2007**, 19–40.

70. Paola P, Angela S, Chiara Z, Maria HO, Maria B, Stig L, Angelo DM. Acidic extracellular pH neutralizes the autophagy-inhibiting activity of chloroquine: Implications for cancer therapies. *Autophagy.* 10, **2014**, 562–571.

71. Danielle EJ, Philip O, Valentin J, Sergio G. The position of lysosomes within the cell determines their luminal pH. *J Cell Biol.* 212, **2016**, 677–692.

72. Min W, Hermann BF, Steven RM, Mark AJC, Vittorio C, John L. The effect of interstitial pressure on tumor growth: Coupling with the blood and lymphatic vascular systems. *J Theor Biol.* 320, **2013**, 131–151.

73. Rakesh KJ. Normalization of tumor vasculature: An emerging concept in antiangiogenic therapy. *Science.* 307, **2005**, 58–62.

74. Trude GS, Kjersti VL, Tord H, Gunnar BK, Einar KR. DCE-MRI-derived measures of tumor hypoxia and interstitial fluid pressure predict outcomes in cervical carcinoma. *Radiation Oncology.* 102, **2018**, 1193–1201.

75. Takeshi M, Takamasa K, Hidekatsu S, Koei I, Kenji S, Makoto S, Ken-ichi I. Interstitial fluid pressure correlates clinicopathological factors of lung cancer. *Ann Thorac Cardiovasc Surg.* 21, **2015**, 201–208.

76. Jun F, Hideaki N, Hiroshi M. The EPR effect: Unique features of tumor blood vessels for drug delivery, factors involved, and limitation and augmentation of the effect. *Advanced Drug Delivery Reviews.* 63, **2011**, 136–151.

77. Keith IB, Charlotte G, Leroy L, et al. Designing a broad-spectrum integrative approach for cancer prevention and treatment. *Semin Cancer Biol.* 35, **2015**, S276–304.

78. Philips SM, James MA, Christopher WP, Atsushi Y. Hypoxia and HIF pathway in cancer and the placenta. *Placenta.* 56, **2017**, 8–13.

79. Omar K, Kate S, Camile SF. The role of hypoxia in oral cancer and potentially malignant disorders: A review. *Journal of Oral Pathology & Medicine.* 46(4), **2016**, 246–252.

80. Giulia C, Laura DB, Valentina M, Massimo A, Miriam P, Barbara P, Marianela VM, Alberto P, Luca P. Dynamic interplay between pericytes and endothelial cell during sprouting angiogenesis. *Cells*. 8, **2019**, 1109.

81. Annika A, Guillem G, Christer B. Pericytes: Developmental, physiological, and pathological perspectives, problems, and promises. *Dev Cell*. 21, **2011**, 193–215.

82. Peter C, Rakesh KJ. Molecular mechanisms and clinical applications of angiogenesis. *Nature*. 473, **2011**, 298–307.

83. Krystyna AG, Lisa MB, Daniel LW, Susan LW. Cancer-associated fibroblasts: Heroes or villains? *BR J Cancer*. 121, **2019**, 293–302.

84. Neta E, Morgan T, Peter O, Sarah TA, Douglas H. Cancer-associated fibroblasts are activated in incipient neoplasia to orchestrate tumor-promoting inflammation in an ND-kappaB-dependent manner. *Cancer Cell*. 17, **2009**, 135–147.

85. Matthew K, Paul JB, James NA, Edward WR, Lukasz M, James OJ, Aarthi G, David AT, Douglas TF. Suppression of antitumor immunity by stromal cells expressing fibroblast activation protein-alpha. *Science*. 330, **2010**, 827–830.

86. Raghu K. The biology and function of fibroblasts in cancer. *Nat Rev Cancer*. 16, **2016**, 582–598.

87. Tuanjie L, Yang Y, Xuefeng H, Guoying W, Wei L, Changchang J, Yan T, Qi Z, Guihua C. Hepatocellular carcinoma-associated fibroblasts trigger NK cell dysfunction via PGE2 and IDO. *Cancer Lett*. 318, **2012**, 154–161.

88. Sanjeev M, Shannon JT, Dorothee N, et al. TGFβ attenuates tumour response to PD-L1 blockade by contributing to exclusion of T cells. *Nature*. 554, **2018**, 544–548.

89. Lisa MC, Zena W. Inflammation and cancer. *Nature*. 420, **2002**, 860–867.

90. Dvorak HF. Tumors: Wounds that do not heal. Similarities between tumor stroma generation and wound healing. *N Engl J Med*. 315, **1986**, 1650–1659.

91. Yuji Y, David LB, Sanna R, Douglas RG, Nathan JZ, Gary SF. Regional analysis of p53 mutations in rheumatoid arthritis synovium. *Proc Natl Acad Sci USA*. 99, **2002**, 10025–10030.

92. Cameron W, Elijah M, Armando RH. Role of extracellular matrix in development ad cancer progression. *Int J Mol Sci*. 19, **2018**, 3028.

93. Thomas RC, Janine TE. Remodeling and homeostasis of the extra-cellular matrix: Implications for fibrotic diseases and cancer. *Dis Model Mech.* 4, **2011**, 165–178.
94. Benjamin G, Kenneth MY. Molecular architecture and function of matrix adhesions. *Cold Spring Harb Perspect Biol.* 3, **2011**, a005033.
95. Paolo PP, Kevin WE, Jay MC, David RI, John GW, Patricia JK. Collagen reorganization at the tumor-stromal interface facilitates local invasion. *BMC Med.* 4(1), **2006**, 38.
96. Ruchi M, Peter IL, Edna C. Biomechanical and biochemical remodeling of stromal extracellular matrix in cancer. *Trends Biotechnol.* 33, **2015**, 230–236.
97. Kai K, Vicki P, Zena W. Matrix metalloproteinases: Regulators of the tumor microenvironment. *Cell.* 141, **2010**, 52–67.
98. Chrisostomi G, Achilleas DT, Nikos KK. Roles of matrix metalloproteinases in cancer progression and their pharmacological targeting. *FEBS J.* 278(1), **2011**, 16–27.
99. Roopali Rm Jiang Y, Marsha AM. Matrix metalloproteinases as novel biomarkers and potential therapeutic targets in human cancer. *J Clin Oncol.* 27, **2009**, 5287–5297.
100. Esther W, Michael S, Yosef Y. Roles for growth factors in cancer progression. *Physiology (Bethesda).* 25, **2010**, 85–101.
101. Sporn MB, Todaro GJ. Autocrine secretion and malignant transformation of cells. *N Engl J Med.* 303, **1980**, 878–880.
102. Huang HS, Nagane M, Klingbeil CK, Lin H, Nishikawa R, Ji XD, Huand, CM, Gill GN, Wiley HS, Cavenee WK. The enhanced tumorigenic activity of a mutant epidermal growth factor receptor common in human cancers is mediated by threshold levels of constitutive tyrosine phosphorylation and unattenuated signaling. *J Biol Chem.* 272, **1997**, 2927–2935.
103. Mamut Y, Gerhard C. EMT, the cytoskeleton, and cancer cell invasion. *Cancer Metastasis Rev.* 28, **2009**, 15–33.
104. Jan Z. Significance of tumor-cell receptors in human cancer. *Arch Immunol Ther Exp.* 50, **2002**, 105–110.
105. Danielle EL, Jonathan RS, Jacob H, Debra TA. Advances in receptor-mediated, tumor-targeted drug delivery. *Adv Ther.* 2(1), **2018**, 1800091.
106. Ebru A, Christopher PP, Mustafa BD. Sigma Receptors and Cancer. American Association for Cancer Research. **2004**.

107. National Cancer Institute. What Is Cancer? **2015**.

108. Wang J-J, Lei K-F, Han F. Tumor microenvironment: Recent advances in various cancer treatments. *Arch Immunol Ther Exp*. 22, **2018**, 3855–3864.

109. Subotic S, Weiss H, Wyler S, Rentsch CA, Rassweiler J, Bachmann A, Teber D. Surgical treatment of localised renal cancer. *World J Urol*. 31, **2013**, 689–695.

110. Mieog JS, van der Hage JA, van de Velde CJ. Neoadjuvant chemotherapy for operable breast cancer. *Br J Surg*. 94. **2007**, 1189–1200.

111. Jafri, SH, Mills, G. Neoadjuvant chemotherapy in lung cancer. *Therapy*. 8, **2011**, 23–31.

112. D'Amico AV, Whittington R, Malkowicz SB, Schultz D, Blank K, Broderick GA, Tomaszewski JE, Renshaw AA, Kaplan I, Beard CJ, Wein A. Biochemical outcome after radical prostatectomy, external beam radiation therapy, or interstitial radiation therapy for clinically localized prostate cancer. *JAMA*. 280, **1998**, 969–974.

113. Evans AE. Central nervous system involvement in children with leukemia. *Cancer*. 17, **1964**, 256–258.

114. Takimoto C, Calvo E. Principles of Oncologic Pharmacotherapy (2008). Ch 3 Appendix 3, in Pazdur R, Wagman LD, Camphausen KA, Hoskins WJ (Eds) *Cancer Management: A Multidisciplinary Approach*, 11th edition. Available: http://www.cancernetwork.com/cancermanagement-11.

115. Madalina D, Codruta C. Hormone therapy in breast cancer. *Chirurgia (Bucur)*. 112, **2017**, 413–417.

116. Yiping Y. Cancer immunotherapy: Harnessing the immune system to battle cancer. *J Clin Invest*. 125, **2015**, 3335–3337.

117. Maura A, Yelena U. The history and background of immunotherapy. *Semin Oncol Nurs*. 35(5), **2019**, 150923.

118. Stanislaw K, Bartosz K, Dawid P, Jolanta S, Ewa K, Karolina KC, Jolanta K, Olga M, Krzysztof K, Julita K. Photodynamic therapy: Mechanisms, photosensitizers and combinations. *Biomedicine & Pharmacotherapy*. 106, **2018**, 1098–1107.

119. Yeuan TL, Yi JR, Chern EO. Molecular targeted therapy: Treating cancer with specificity. *Eur J Pharmacol*. 834, **2018**, 188–196.

120. Zaimy MA, Saffarzadeh N, Mohammadi A, Pourghadamyari H, Izadi P, Sarli A, Moghaddam LK, Paschepari SR, Azizi H, Torkamandi

S, Tavakkoly-Bassaz J. New methods in the diagnosis of cancer and gene therapy of cancer based on nanoparticles. *Cancer Gene Ther.* 24, **2017**, 233–243.

121. Horacio C, Kazunori K. Progress of drug-loaded polymeric micelles into clinical studies. *J Control Release.* 190, **2014**, 465–476.

122. Jeipin Y, Qiang Z, Hong C, Yiyun C. Surface-engineered dendrimers in gene delivery. *Chem Rev.* 115, **2015**, 5274–5300.

123. Templeton NS, Lasic DD, Frederik PM, Strey HH, Roberts DD, Pavlakis GN. Improved DNA: Liposome complexes for increased systemic delivery and gene expression. *Nat Biotechnol.* 15, **1997**, 647–652.

124. Hirendra NB, Mukesh V. Application of nanotechnology in cancer. *Technology in Cancer Research and Treatment.* 7, **2008**, 149–154.

125. Ranjita S, Rainer HM. Nanocrystal: Industrially feasible multifunctional formulation technology for poorly soluble actives. 399, **2010**, 129–139.

126. Jain KK. Nanomedicine: Application of nanobiotechnology in medical practice. *Med Princ Pract.* 399, **2008**, 89–101.

127. Arap W, Pasqualini R, Ruoslahti E. Cancer treatment by targeted drug delivery to tumor vasculature in a mouse model. *Science.* 279, **1998**, 377–380.

128. Astrid JS, Robbert JK, Henk EM, Erwin JB, Johannes JP, Dirk KFM, Lou FMHL, Grietje M. Targeting of RGD-modified protein to tumor vasculature: A pharmacokinetic and cellular distribution study. *Int J Cancer.* 102, **2002**, 469–475.

129. Sok BL, Amrita B, Hayat O. Improvement of drug safety by the use of lipid-based nanocarriers. *J Control Release.* 163, **2012**, 34–45.

130. Ki HB, Hyun JC, Tae GP. Nanomaterials for cancer therapy and imaging. *Mol Cells.* 31, **2011**, 295–302.

131. Jinjun S, Philip WK, Richard W, Omid CF. Cancer nanomedicine: Progress, challenges and opportunities. *Nat Rev Cancer.* 17, **2017**, 20–37.

132. Geetha M, Chun-Wai M, Mohd Z, Christine D, Doryn MYT, Ju-Yen F. Challenges and opportunities of nanotechnology as delivery platform for tocotrienols in cancer therapy. *Front Pharmacol.* 9, **2018**.

133. Yan-Peng J, Bu-Yun M, Xia-Wei W, Zhi-Yong Q. The in vitro and in vivo toxicity of gold nanoparticles. *Chinese Chemical Letters.* 28, **2017**, 691–702.

134. Carlos LC, Juan SA, Maria MB, Jorg B, Juan L, Cristina SG. Gold nanoparticles: Distribution bioaccumulation and toxicity. In vitro and in vivo studies. *Nanomedicine: Nanotechnology, Biology and Medicine*. 14, **2018**, 1–12.

Chapter 2

Understanding the Interaction of Nanoparticles at the Cellular Interface

Baiju G. Nair

School of Biotechnology, NIT Calicut, Kerala, India

bgnair@nitc.ac.in

2.1 Introduction

Researchers and healthcare professionals are looking out for more innovative and combined therapeutic strategies for cancer treatment, apart from conventional methods [1]. Chemotherapy is one of the oldest forms of the prescribed procedures and often faces the limitations of therapeutic agents' water solubility and their side effects [2, 3]. The evolving resistance to drugs is an emerging challenge faced by most of the successful treatment of cancer till now [4, 5]. Current chemotherapy has been flooded with various drugs with diverse actions on cancer cells [6]. Surgery is one of the known invasive methods to combat cancer. However, triggering a metastasis tumor environment and challenges in the tumor demarcation are significant problems in surgical procedures [7]. To answer these hurdles, nanotechnology

Bionanotechnology in Cancer: Diagnosis and Therapy
Edited by D. Sakthi Kumar and Aswathy Ravindran Girija

ISBN 978-981-4800-30-3 (Hardcover), 978-0-429-42291-1 (eBook)
www.jennystanford.com

has been evolved into various platforms, primarily to address the multiple problems of cancer-related treatments [8]. As a nanoformulation, the materials can have their unique properties at their smallest dimensions. Based on these new developments, cancer nanotechnology has emerged as an interdisciplinary branch of science for extensive cancer treatments from diagnosis to treatments [9]. Engineered nanoparticles (NPs), compared to traditional therapeutics, provide site-specific targeting, solubility, bioavailability, and compatibility, evolved as targeted drug delivery that became an advantage for the drug delivery against cancer [10]. Hence, it is highly imperative to understand nanomaterials' interactions at the bio-nano interface and various cell internalization routes and its activities that will be used to kill the cancer cells effectively.

2.2 Nanotechnology

The term nano came from a Greek word termed for 'dwarf'. The idea of nanotechnology emerged into the current stage after the famous talk by Richard Feynman, "there is plenty of room at the bottom." Later, many nanomaterials have been developed to treat various diseases in which cancer has obtained significant attention. Remote-controlled NPs and stimuli-responsive polymer NPs are some of the novel strategies employed so far in the realm of cancer therapies. The use of NPs in cancer therapy is going from stealth to smart and intelligent nanosystems for various therapeutic applications. Nanotechnology always provide us with significant varieties of novel materials for disease diagnosis, drug design, delivery, and implants [11]. Apart from any other technologies, nanotechnology can generate significant surface-modified nanomaterials due to its high surface-to-volume ratio that can detect cancer cells even in its minute concentration.

Similarly, the combination of material science and molecular biology has augmented cancer therapy into more personalized nanomedicines. However, to establish a successful application of nanomaterial as a nanomedicine or in a diagnostic platform, certain factors have to be considered in advance. Nanosystems can be used as nanodrugs, as they should bypass the biological cell barrier and successfully enhance their cellular uptake to deliver the active drugs they carry to cellular components, which can be used to kill the cancer cells selectively and effectively.

2.3 NPs and Cell Internalization

The transportation of NPs across the cell membrane is crucial for cell internalization and therapeutic cargos delivery. Generally, particles' transportation through diffusion into cells or tissues depends on their size, charge, and physicochemical properties of the extracellular matrix [12]. Mostly, lipophilic molecules and polar molecules can cross cell membrane freely by passive transport mechanisms in accord with Fick's first law of diffusion [13, 14].

Fick's law states that the flux or rate at which a molecule diffuses across the plasma membrane is proportional to the concentration gradient, the membrane surface area, and the molecule's permeability coefficient [15]. Many molecules have a low permeability coefficient, and they tend to diffuse slowly. However, these molecules are rapidly transported across the membrane by specific transmembrane proteins as a facilitated diffusion, which is a far part of the Fick's law of diffusion. Small NPs, less than 1 nm, are internalized by diffusion. The entry of engineered NPs into the cell may fall into different categories, as shown in Figure 2.1, based on the size and surface functionalities [16]. Most of the larger particles are internalized either by phagocytosis or micropinocytosis pathways, which are independent of clathrin and caveolin pathways. Phagocytosis or micropinocytosis pathways start membrane protrusions due to actin polymerization, followed by the engulfing of substance to be internalized into the cell. In receptor-mediated endocytosis, there is always a highly specific uptake of NPs with the help of functionalized ligands on the surface of NPs. It is observed that NPs rotate during the internalization process with the diffusion of receptors as a prolog to the wrapping mechanism exhibited by membranes during receptor-mediated endocytosis [17]. Clathrin-mediated endocytosis has been identified as a cellular entry mechanism to internalize specific molecules into the cells [18]. Clathrin is a particular protein coated at the intracellular face of the plasma membrane, which further forms a pit for internalizing specific molecules or engineered NPs in the presence of adaptor and accessory proteins [19]. There is a cascade of signals that are activated during the contact of NPs on the cell surface. Epsin, one of the known adaptor proteins, induces membrane curvature on the cell. That will be followed by the deep invagination by an accessory protein called dynamin, which will help NPs enter the cell. Generally, viruses are considered as a naturally made nanoparticle, and they follow a variety of endocytic pathways, based on their structure, for the entry into a mammalian cell [20]. Several chemical inhibitors can be used as a tool for direct probing of endocytic events in live cells [21].

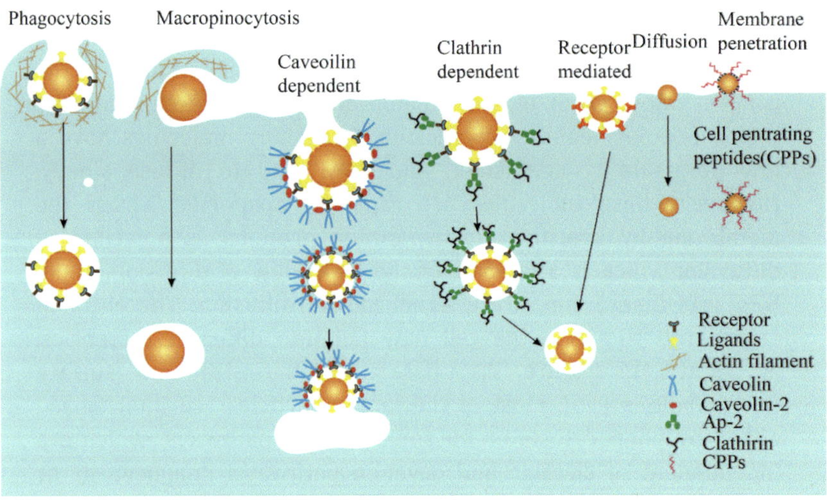

Figure 2.1 Nanoparticle-cell interactions and different internalization pathways.

There are several engineered NPs with targeting ligands, which could be internalized explicitly by Clathrin-mediated endocytosis [22]. Mammalian cells, such as adipocytes, endothelial cells, and muscle cells, show an incredibly high density of flask-shaped caveolin vesicles, which can increase the surface area of the plasma membrane two-fold [23, 24]. Caveola-mediated nanoparticle delivery has been successfully carried out by using several engineered NPs [25]. According to the reports, there are many factors that impact the internalization of NPs into the cells, such as the shape of NPs, surface ligands, tissue environment, and disease conditions, as shown in Figure 2.2 [16]. According to mathematical modelling, initially a cationic nanoparticle is essential for targeting the tumors and changing the charge to neutral after exiting the blood vessels [12]. Recent trends are in the line of bypassing endocytic pathways through a transmembrane penetration with the help of engineered NPs.

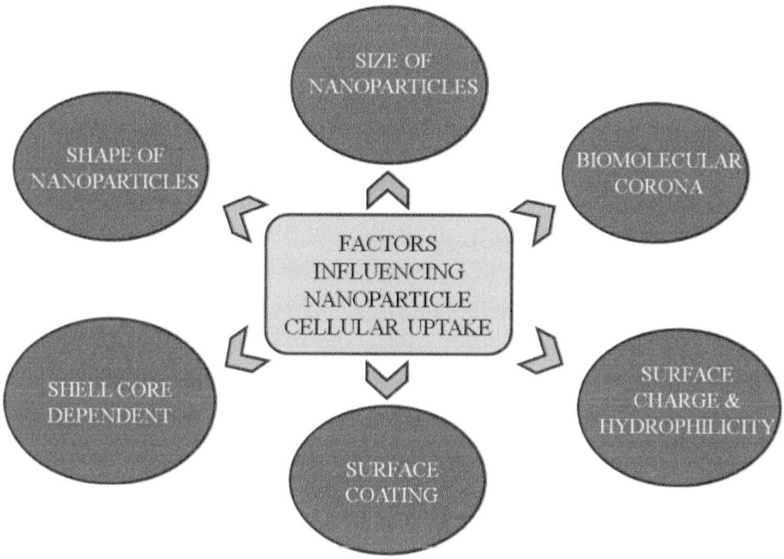

Figure 2.2 Factors influencing nanoparticle uptake by cell.

2.4 Transmembrane Penetration by Designed Nanomaterials

Passive penetration of the cell membrane is an alternative strategy for the intracellular delivery of NPs to bypass the endocytic pathways [26]. A proper selection of NPs and surface chemistry is inevitable for the particle's significant translocation across the membrane [27]. Cell-penetrating peptides (CPPs) [28] are one of the most widely used ligands for direct penetration of membrane by NPs for various therapeutic applications [29]. Highly cationic arginine content is a major responsible factor for membrane translocation by CPPs [30]. Among the first CPP used, transactivator of transcription (TAT) peptide derived from the TAT of human immunodeficiency virus [31] is a best-chosen candidate for most drug delivery applications. One of the previous works on 3D electron tomography has identified the potential of TAT conjugated superparamagnetic iron oxide NPs to breach cell membrane and escape endocytosis [32]. Due to the increasing demand for new targeting moieties for cargo transportation, many CPPs have been identified and analyzed from various sources. Comparative studies on the internalization of CPPs tagged green fluorescent probe into different cell lines have been revealed the high cell-penetrating potential of cyclic CPPs compared to their linear ones [33]. Functionalization is highly recommended for the redirection of NPs to specific targeting of cell organelle and plays a significant role in interfering with physiological intracellular trafficking. Interestingly, targeting peptides has further triggered a particular interest in the organelle-specific targeting using various cell organelle-specific peptide-like mitochondrial penetrating peptides (MPPs) [34].

2.5 Engineered Protein Corona: A Better Drug Delivery System

Any application of nanoparticle in a biological fluid, such as blood, results in the adherence and aggregation of proteins on the surface of NPs; and forms an entirely new entity on the surfaces [33]. This new entity in an in vivo condition is called protein corona (PC) and is divided into two categories: soft and hard corona [35]. Initially, the soft corona is formed when NPs come in contact with biological fluids by the proteins with low-binding affinity. Later on, these loosely bound proteins are replaced with a hard corona composed of less abundant proteins with high-binding affinity [36]. Processing of protein adsorption and corona formation is controlled by a dynamic phenomenon known as the Vroman effect, as provided in Figure 2.3.

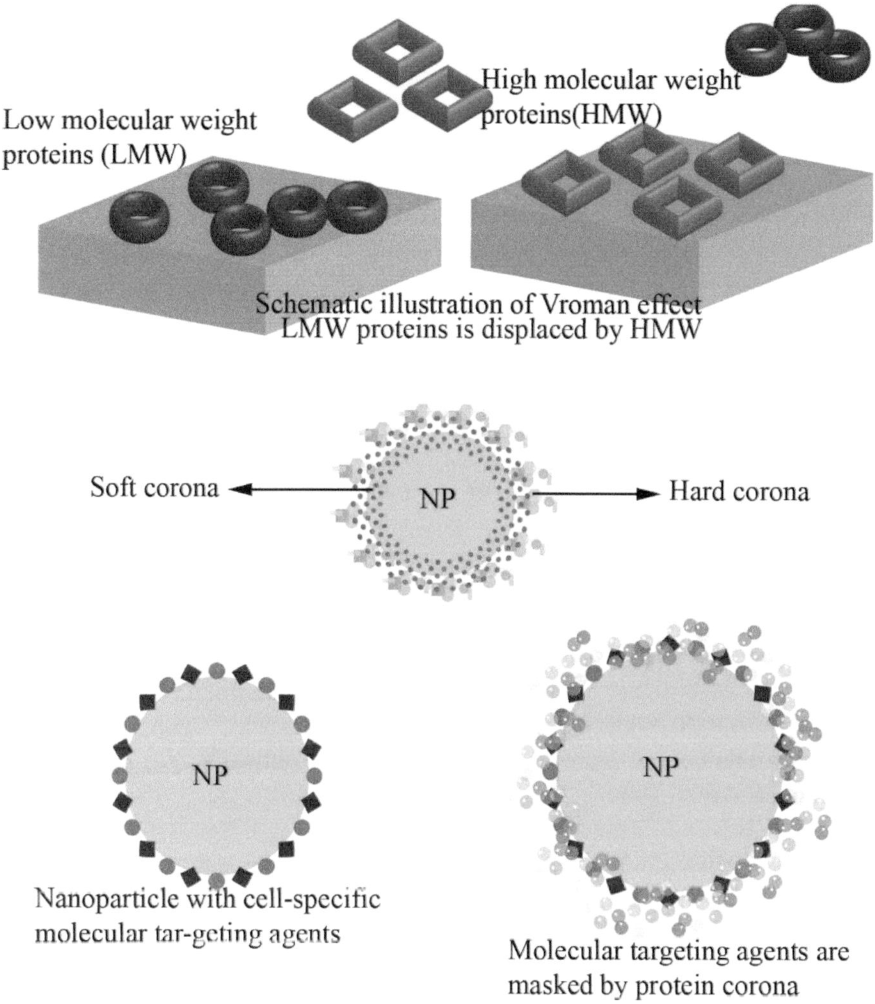

Figure 2.3 Schematic illustration of Vroman effect and PC formation on a conventional nanoparticle.

According to Leo Vroman, there is a competitive protein exchange where one protein may get replaced by another protein with a strong binding affinity and cause changes in corona's composition [37]. Later on, Cedervall and colleagues introduced the effect of PC on a nanoparticle; and they found that, when they move from one protein component to another, there is a significant difference in protein binding on NPs, based on the abundance, affinities, and incubation time [38].

It is interesting to observe the physicochemical properties of NPs (size, shape, surface chemistry, curvature, porosity, crystallinity, heterogeneity, roughness, hydrophobicity/hydrophilicity, and roughness) have a significant influence on the formation of the corona [39, 40]. As we know, the high surface free energy of NPs drives the adsorption of proteins in the in vivo conditions with the help of various binding forces such as van der Waals interactions, hydrogen bonds, hydrophobic interactions, electrostatic interactions, and π–π stacking [41].

Any change in the size, shape, surface charge, hydrophobicity, and surface topography is found to alter PC's properties on NPs [42]. There are several methods for identifying the PC component and mechanism behind the formation of PC on NPs [43]. The fate of conventional nanocarriers is intensely regulated by the adsorbed serum proteins and immediately cleared from the body by macrophages as an immune regulatory mechanism. Consequently, nanoparticle drug carriers with a site-specific targeting ligand lose their targeting ability in a biological environment [44]. Several methods have been identified to reduce the protein adsorption by modifying the surface with stealth polymers, such as polyethylene glycol (PEG), polyvinylpyrrolidone, or dextran. Though these polymers can provide a stealth effect for NPs and increase the half-life for the circulation in the blood, the ability to conjugate the surface targeting ligands on the surface on NPs may be suppressed. It is interesting to note that PC is shielded by an impressive array of NPs formed from the supramolecular assembly of a genetically engineered fusion protein (HER2-binding affibody (Afb) and glutathione-S-transferase (GST)) into the pores of mesoporous silica NPs. The in vivo model assays have also recorded the highest fluorescent signals from this PC shield NPs [45]. We can observe rapid progress in overcoming the issues associated with PC in the future of cancer nanotechnology [46]. Understanding and engineering the PC are highly recommended while designing and implementing any drug delivery systems. Moreover, animal serum proteins may vary from species to species, and PCs they develop and, hence, a proper understanding of PCs is inevitable before human trials.

2.6 Subcellular Interaction of NPs

A cell is a fundamental unit of life bestowed with vivid cell organelles for controlling unique functions. For instance, their roles in cell repair, waste removal, and growth give them the privilege to tasks as a mini-organ. In these contexts, intracellular interaction of NPs, after a successful internalization process, plays a pivotal role in various therapeutic strategies against multiple diseases such as cancer. Tuning and taming these organelles' extraordinary properties using nanomaterials can revolutionize the treatment strategy against most of the conditions of the prevailing disease by providing sufficient dosage, controlled release, and address the limitation of drug resistance. So far, the mitochondria, Golgi apparatus, and nucleus are the most explored cell organelle for the direct nanoparticle interactions to check the cytotoxicity activities (Figure 2.4).

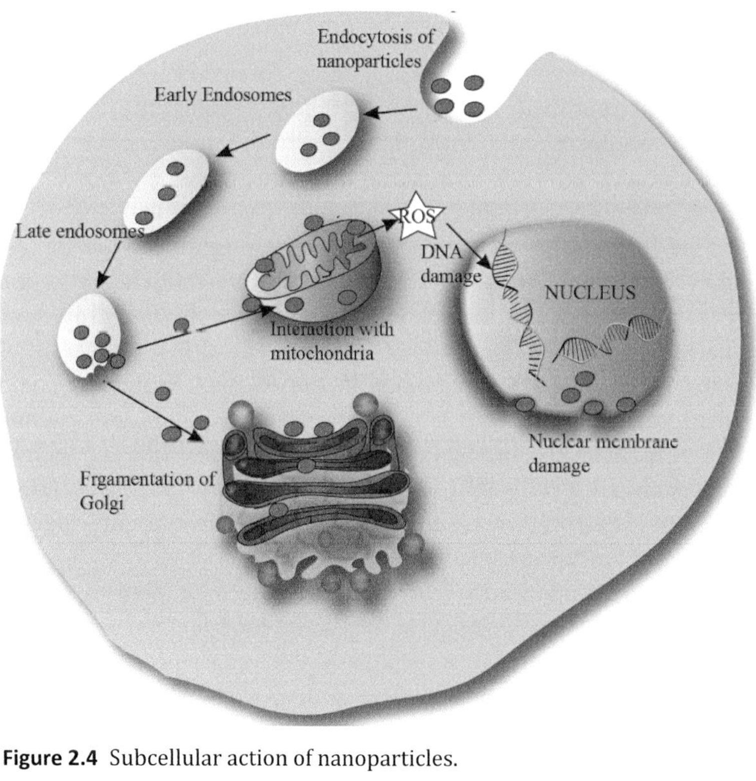

Figure 2.4 Subcellular action of nanoparticles.

The antimicrobial properties of silver have been famous from ancient times, and considerable variations in size, composition, and structure have been utilized in the biomedical field until today. In a study conducted on the periodontal implant, silver NPs were present in the perinuclear region, but were absent inside the mitochondria [47]. Regarding mitochondria, if we disturb the mitochondrial functionality, it can significantly disrupt the respiratory chain, leading to oxidative stress. Generally, oxidative stress in a cell can lead to the massive release of reactive oxygen species (ROS). Another investigation using silver NPs on liver cells revealed the presence of aggregates of these NPs inside the mitochondria [48]. The accumulation of silver NPs inside mitochondria increases ROS in cells, which leads to apoptosis or necrosis of the cells. It is interesting to note that most of the NPs can interact with the mitochondrial membrane, though some of them provided ROS and some of them not. In another experiment, gold NPs were able to come in contact with the mitochondria without any toxic effects on cells [49].

As mentioned earlier, the surface chemistry of NPs has a significant role in designing NPs for both cellular internalization and subcellular internalization. Versatile nanoparticle-based formulations with several surface modifications have been so useful in guiding them to the site of specific intracellular organelles for performing particular actions. According to Gallud et al., ammonium-modified gold NPs can enter the mitochondria and revealed significant mitochondrial dysfunction by proteomic analysis and RNA sequencing [50]. There was down regulation of many genes, codes for subunits, in the electron transport chain when cells treated with ammonium modified gold NPs. Along this line, various mitochondriotropic moieties like triphenyl-phosphonium (TPP) [51], dequalinium (DQA) [52], and MPPs [53] are also useful in modifying the nanoparticle for the specific targeting of mitochondria.

The Golgi apparatus is a vital organelle as it plays a crucial role in protein processing. The organelle functions as an integral unit in a cell because of its shape. Any anomaly in the condition of the Golgi apparatus is known to cause abnormalities in the cell. Once inside the cell, NPs interact with various cell organelles, one of which is the Golgi apparatus. Ma et al. have conducted a study to identify the effects of gold NPs on the Golgi apparatus [54]. Interestingly, gold NPs (50 nm) in normal rat kidney epithelial cells (NRK Line) have disrupted the calcium level inside the cell leading to the endoplasmic reticulum (ER) stress. The calcium ion concentration in cell organelles plays a crucial role in maintaining organelle function and structure. Similarly, the Golgi apparatus's configuration changed from the familiar ribbon-like design into mini stacks, which were not linked or aligned. To study the Golgi apparatus's functioning, mannosidase II and Galactosyltransferase were selected, and cells treated with gold NPs showed a significant decrease in their production. Cluster analysis of genes expressed in control and section treated with gold NPs showed substantial expression patterns. There is a change in the regulation of integrin-mediated cellular adhesion, which plays a crucial role in cell-extracellular matrix adhesion.

Moreover, there is a consistent change in cellular behavior (adhesion), seen from the KEGG and GO pathways. Although the gold nanoparticle treated kidneys showed changes like early renal fibrosis, the organ morphology remains intact. Overall, the use of gold NPs on NEK cells leads to the accumulation of 50 nm gold NPs in cells, disruption of homeostasis of calcium, and fragmentation of the organelle that is visible with poor cellular adhesion. Even though 90% of the cells go unaffected, such small changes also need to be studied when we look into cellular toxicity due to NPs.

2.7 Endosomal/Lysosomal Escape of NPs

The dynamics of the interaction of NPs with cells and their organelle explains the process of how the NPs end-up inside lysosomes and endosomes. These play a significant role in regulating the intracellular transport and degradation of foreign bodies. Utilizing NPs as carriers of biomolecules or drugs has found a new dimension into medical requirements with improved technical expertise. NPs, after endocytosis, are transported into early endosomes where most of the sorting occurs. Parallelly, the transport of some NPs moves to recycle endosomes, where finally they are excreted out from the cells. NPs inside the early endosomes move into the cell interior along the microtubules. The transport of cargos from late endosomes to lysosomes is a unidirectional process where late endosomes fuse with lysosomal bodies [55]. There are approximately 50–1000 lysosomes of sizes ranging from as small as 0.1 μm to as large as 1.5 μm in a cell. The size and number of lysosomes vary widely based on the environmental cue, type of cell, and disease conditions [56]. The primary function of lysosome is to hydrolyze any biomolecule that comes into it, implying the presence of hydrolases inside the lumen of the organelle. Its lumen also contains various enzyme activators, transport proteins (NPC2), and other protective factors with an acidic pH of 4.5–5.0 maintained with the help of vacuolar ATPase [57].

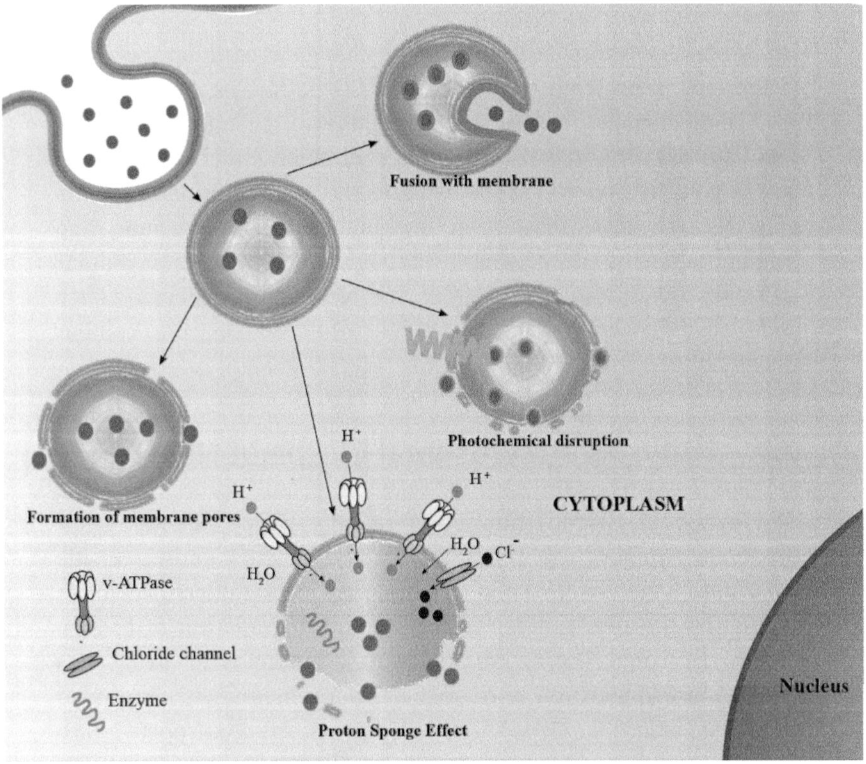

Figure 2.5 Escape of nanoparticles from endosome/lysosome vesicles.

As far as the fate of NPs in the cytoplasm is concerned, for any specific function or targeting of another organelle, a complete escape from the endosomes/lysosomes is mandatory. As mentioned earlier, overcoming the cells' biological barrier remains a significant challenge in therapeutic delivery against specific diseases. Several approaches address these issues, and nanotechnology has secured a particular interest in this aspect. The mechanism for the escape of NPs from endosomes is an exciting area where progress is attaining every day [58]. New means and methods are being developed for the NPs to escape the natural system unscathed to deliver their payload effectively into desired organelles. A better understanding of several mechanisms by which NPs can escape from the endosomal/lysosomal pathways is required [59]. Formation of membrane pores in the endosomes or lysosomes, proton sponge effect, fusion with the membrane, photochemical disruption of endosomal/lysosomal membranes (Figure 2.5) are significant mechanisms identified until now.

2.7.1 Formation of Membrane Pores in Lysosomes/Endosomes

In general, there is membrane tension in the cell membrane, which enlarges the pore, and there is line tension that closes the pore. The binding of cationic amphiphilic peptides to the membrane causes internal membrane tension, creating pores in the membrane [60]. Some proteins bind to the pore's border, which reduces line tension, decreasing the peptide numbers leading to internal membrane tension, which keeps the pore open. Two of the pores' models are (a) the barrel-stave model and (b) the toroidal model. In the barrel-stave model, the peptides orient to form staves that shape into a barrel-like cluster perpendicular to the membrane, thus creating a pore. In the toroidal model, peptide aggregates become part of the membrane perpendicularly, and then the membrane curves inward, and a pore is formed with the peptides bordering it [11–13]. It has been found that individual engineered polymeric pH-sensitive

NPs can effectively escape the endosomal vesicles at high (pH 7.2) or low (pH 4.9) pH level by dissembling themselves [61]. Similarly, pH-induced pore formation can be seen by gold NPs functionalized with the pore-forming Listeria monocytogenes Listeriolysin O (LLO) [62]. LLO belongs to a cholesterol-dependent cytolysins family, which can form pores of 300 Å diameter in the membrane by the mechanism of self-assembly [63].

2.7.2 Proton Sponge Effect

Cationic polymers' ability to form polyplexes by incorporating nucleic acids by electrostatic interactions has been a significant area of research from the past years. It is possible to have high transfection rates using cationic polymers with a buffering capacity under the physiological pH. A hypothesis by Boussif et al. stated that any change in osmolarity would lead to swelling and disruption of endosomes [64]. Later, it was identified as the process called the proton sponge effect. In this effect, vacuolar ATPases pump proton into the endosome, leading to protons' accumulation. Parallelly, protonation leads to the influx of chloride ions and water results in the increased ionic concentration inside the endosome. This osmotic swelling can disrupt endosomes [65]. There is an evidence that highly branched polymers, such as glycogen or polyethyleneimine (PEI), can induce the proton sponge effect if they are immobilized on NPs [66]. Though the proton sponge mechanism is highly debated, it is one of the most widely accepted phenomena for endosomal escape [67]. Charge reversal by polymeric NPs on exposure to pH is another way to escape endosomes using the proton sponge effect [68].

2.7.3 Fusion with Endosomal Membrane

The development of NPs encapsulated in combination of lipids and proteins can fuse with the endosomal membrane helping its escape. The use of this method for the endosomal flight of NPs involves studying all attributes of both the shell (liposomes, biomimetic materials, or emulsions) and the core (NPs). Intracellular concentration, surface reactivity, and exposure time are playing a vital role in this process's success. He et al. developed lipid-based liquid crystalline NPs, which can transform the intracellular acidic environment structure and perform gene silencing against cancer [69]. These lipid NPs are also known as nano transformers, and they possess satisfactory payload (e.g., siRNA) efficiency and low cytotoxicity. Moreover, they can undergo a conformational transition of lipids to fuse with the endosomal membrane, further promoting siRNA release into the cytoplasm.

2.7.4 Photochemical Rupture of Endosomal Membrane

Photochemical internalization (PCI) is the photochemically releasing of biological or non-biological compounds into the cytosol from the endosomal pathway. Photosensitizer-mediated chemiluminescence occurs only on proper stimulation. The simulation results from singlet oxygen production indicating an increase in ROS levels, disrupting the endosomal or lysosomal membrane. The singlet oxygen has a short lifetime, thus destroying the membrane while leaving other contents of the organelle unharmed. Some of the photosensitizers used are TPPS4, A1PcS2a, TPPS2a, etc. [69, 70].

This dual-mode mechanism is suitable for the targeted delivery of NPs and the selective destruction of cancer cells. Similarly, the singlet oxygen species formed due to this mechanism can also destroy the nuclear membrane if the nanoparticle is inside the nucleus. Besides, disruption of the lysosomal membrane triggered the release of cathepsins due to lysosome membrane permeabilization induced by ROS surge. It activates the endogenous apoptotic stimuli [71], suitable for the selective destruction of various cells [72, 73].

2.8 Tumor and Nanomedicine

Cancer cell is quite different from normal cells in many ways; therefore, the interaction of NPs with cancer cells would encourage a new area of research. The selective accumulation of therapeutic nanomaterials of a particular size at a tumor site is based on the controversial enhanced permeability and retention effect (EPR). The leaky vasculature and tumor microenvironment, production of an abnormal amount of vascular growth factors, and vascular permeability enhancing elements (such as bradykinin, nitric oxide, and prostaglandins) are significant factors for controlling the EPR effect [74]. It has been found that less than a 2-fold increase of nanomedicine in tumor site compared to normal organ is sufficient to cure cancer completely [75]. There are numerous methods for increasing the concentration of NPs at a tumor site. Remote controlled and stimuli-responsive smart NPs are good candidates for therapeutic strategies.

2.9 Smart Material as Drug Delivery Nanoparticle

Over the years, biomaterials have been employed in many diverse controlled releases of drug materials, through the interdisciplinary combination of chemistry, biology, physics, and pharmacology, with inputs from clinicians. These smart biomaterials are tailored to various points of care treatment modalities for cancer. A material gets identified as smart when it can get manipulated from various resources. Generally, intelligent materials may use internal or external stimuli as a triggering mechanism to initiate a proposed application. Enzymatic changes and variation of microenvironment pH are the major internal stimuli for most of the smart biomaterials. External stimuli are mainly composed of electromagnetic, light, and acoustic energy [76]. Stimuli-responsive nanocarriers offer tremendous advantage for drug and gene delivery as various therapeutic strategies. Among all these, pH-responsive nanocarriers have a unique gift for releasing drugs at the targeted sites by responding to different pH locations. Generally, the extracellular pH of normal living cells is at a slightly basic side (pH 7.4), while tumor cells keep an acidic side (pH 5–6). As a result, when nanocarrier enters a lower pH level, its components respond to pH change and release the drug cargos into the cytoplasmic areas [77].

The use of Zwitterionic materials, which have both anionic and cationic groups, such as O-carboxymethyl-chitosan (CMC), can respond to pH stimuli and increase circulation [78]. Zwitterionic polymeric materials have great potential to act as a drug delivery vehicle with many improved properties [79, 80]. These materials can respond to pH and undergo a charge reversal mechanism by the protonation of polymers or proton sponge effect. Moreover, the charge reversal can modify the properties of NPs for various activities, such as enhanced cellular uptake, improved endosomal escape, resistance to nonspecific protein adsorption, and controlled release. Zwitterionic nanocarriers possess several advantages over PEG derivatives as a biomimetic strategy [81].

A light source such as UV is among the most studied external stimuli for self-assembly and routing of nanocarriers to enable controlled release of therapeutics [82]. At present, many light sources (UV, NIR, visible, and various lasers) are being investigated for guiding the nanocarriers [83]. The use of magnetic nanocarriers has further augmented the wide range of external magnetic stimulation as a guiding platform for controlled drug delivery, monitoring, distribution, and release kinetics [84]. There are several advantages to magnetic NPs as a drug carrier. An external magnetic field can improve not only the endocytic rate, but also maneuver the internalizing pathways [85]. Nanosurgeon is the right person for excising tumors from the brain, where surgery is a significant challenge, with the help of an external 3D rotational magnetic field, which could create a strangling force on super paramagnetic iron oxide nanoparticles (SPIONs) to remove tumors [86]. It has been observed that the unique ability of polymer-coated SPIONs with a site-specific ligand can manipulate cells for various biomedical applications. Interestingly, polymeric nanocarriers are the best-studied smart NPs that can respond to multiple stimuli to improve drug delivery applications.

2.9.1 Polymeric Drug Nanocarrier: Act as a Smart Material

Polymers are large molecules composed of repeating small units interlinked to each other by physical or chemical forces. Generally, polymers are obtained from natural resources and synthetic methods with immense potential for drug delivery applications with properties showing bio-compatibility, biodegradability, and least toxicity. Polymer-based nanotechnology is an emerging field toward the treatment of various diseases through targeted drug delivery applications [87]. As we know, certain drugs or biomolecules are susceptible to intracellular environments, such as enzymes and defensive proteins. Therefore, the therapeutic load may be ineffective at the time of medication due to the insufficient quantity required for a particular disease. Polymeric nanocarriers can protect the drug molecules and proteins from the various stringent microenvironments, owing to their high biocompatibility. Moreover, polymeric nanocarriers can improve the pharmacokinetics of drugs and augment the accumulation of drugs at the tumor sites [88]. Consequently, polymeric NPs are being evolved into a state of new generation stealth treatment modalities with comprehensive options for personalized medicines, as shown in Figure 2.6 [89].

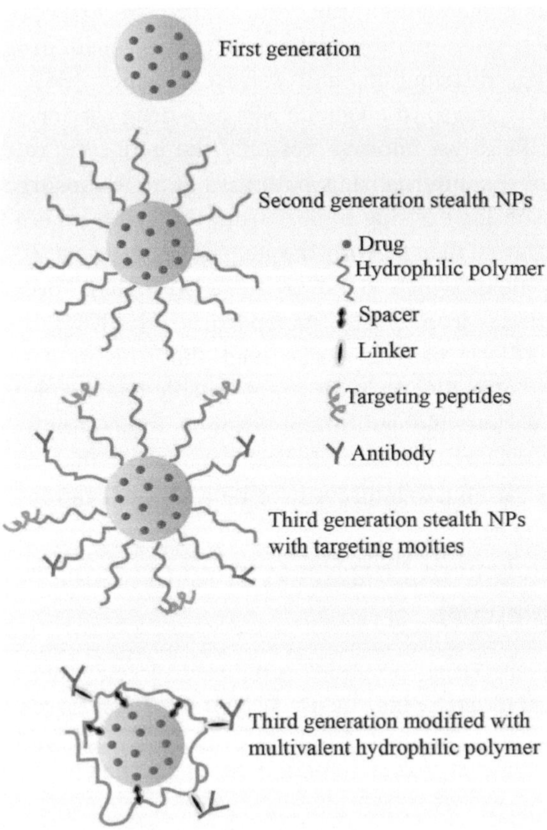

Figure 2.6 Evolution of various generations of polymeric nanocarriers.

2.10 Conclusion and Future Perspective

Nanotechnology has been progressing from almost all science fields into a multitude of options for various futuristic applications. Among all, nanotechnology for biomedical application stands out as the complex human system. A better understanding of the nanoparticle's surface engineering is essential for analyzing its interaction at the cellular interface. This part is highly crucial for both in vitro and in vivo applications. From conventional NPs to most modern stimuli-responsive engineered NPs follow the specific internalization pathways depending on the targeting moieties. Precise subcellular targeting is possible through various peptides by escaping the endosomal/lysosomal vesicles. A smart, intelligent stimulus-responsive polymeric nanocarrier will make many changes in conventional drug delivery, particularly against cancer.

References

1. Charmsaz, S.; Collins, D. M.; Perry, A. S.; Prencipe, M., Novel strategies for cancer treatment: Highlights from the 55th IACR Annual Conference. *Cancers (Basel)* **2019,** *11* (8).
2. Nurgali, K.; Jagoe, R. T.; Abalo, R., Adverse effects of cancer chemotherapy: Anything new to improve tolerance and reduce sequelae, *Front Pharmacol* **2018,** *9*, 245.
3. Partridge, A. H.; Burstein, H. J.; Winer, E. P., Side effects of chemotherapy and combined chemohormonal therapy in women with early-stage breast cancer. *J Natl Cancer Inst Monogr* **2001,** *30*, 135–142.
4. Nikolaou, M.; Pavlopoulou, A.; Georgakilas, A. G.; Kyrodimos, E., The challenge of drug resistance in cancer treatment: A current overview. *Clin Exp Metastasis* **2018,** *35* (4), 309–318.
5. Rueff, J.; Rodrigues, A. S., Cancer drug resistance: A brief overview from a genetic viewpoint. *Methods Mol Biol* **2016,** *1395*, 1–18.
6. Huang, C.-Y.; Ju, D.-T.; Chang, C.-F.; Muralidhar Reddy, P.; Velmurugan, B. K., A review on the effects of current chemotherapy drugs and natural agents in treating non-small cell lung cancer. *Biomedicine (Taipei)* **2017,** *7* (4), 23.

7. Tohme, S.; Simmons, R. L.; Tsung, A., Surgery for cancer: A trigger for metastases. *Cancer Res* **2017,** *77* (7), 1548–1552.
8. Charmsaz, S.; Prencipe, M.; Kiely, M.; Pidgeon, G. P.; Collins, D. M., Innovative technologies changing cancer treatment. *Cancers (Basel)* **2018,** *10* (6).
9. Grobmyer, S. R.; Iwakuma, N.; Sharma, P.; Moudgil, B. M., What is cancer nanotechnology? *Methods Mol Biol* **2010,** *624*, 1–9.
10. Chaturvedi, V. K.; Singh, A.; Singh, V. K.; Singh, M. P., Cancer nanotechnology: A new revolution for cancer diagnosis and therapy. *Curr Drug Metab* **2019,** *20* (6), 416–429.
11. Ramos, A. P.; Cruz, M. A. E.; Tovani, C. B.; Ciancaglini, P., Biomedical applications of nanotechnology. *Biophys Rev* **2017,** *9* (2), 79–89.
12. Stylianopoulos, T.; Poh, M. Z.; Insin, N.; Bawendi, M. G.; Fukumura, D.; Munn, L. L.; Jain, R. K., Diffusion of particles in the extracellular matrix: The effect of repulsive electrostatic interactions. *Biophys J* **2010,** *99* (5), 1342–1349.
13. Baynes, R. E.; Riviere, J. E., Chapter 3: Absorption. Pesticide Biotransformation and Disposition. Hodgson, E. (ed.), Academic Press: Boston, **2012**; 25–52.
14. Panariti, A.; Miserocchi, G.; Rivolta, I., The effect of nanoparticle uptake on cellular behavior: Disrupting or enabling functions? *Nanotechnol Sci Appl* **2012,** *5*, 87–100.
15. Fick, A., On liquid diffusion. *The London, Edinburgh, and Dublin Philosophical Magazine and Journal of Science* **1855,** *10* (63), 30–39.
16. Zhang, S.; Gao, H.; Bao, G., Physical principles of nanoparticle cellular endocytosis. *ACS Nano* **2015,** *9* (9), 8655–8671.
17. Tang, H.; Zhang, H.; Ye, H.; Zheng, Y., Receptor-mediated endocytosis of nanoparticles: Roles of shapes, orientations, and rotations of nanoparticles. *The Journal of Physical Chemistry B* **2018,** *122* (1), 171–180.
18. Kaksonen, M.; Roux, A., Mechanisms of clathrin-mediated endocytosis. *Nat Rev Mol Cell Biol* **2018,** *19* (5), 313–326.
19. Akinc, A.; Battaglia, G., Exploiting endocytosis for nanomedicines. *Cold Spring Harb Perspect Biol* **2013,** *5* (11), a016980.
20. Cossart, P.; Helenius, A., Endocytosis of viruses and bacteria. *Cold Spring Harb Perspect Biol* **2014,** *6* (8).
21. Zhao, J. C.; Stenzel, M. H., Entry of nanoparticles into cells: The importance of nanoparticle properties. *Polym Chem-Uk* **2018,** *9* (3), 259–272.

22. Phuc, L. T. M.; Taniguchi, A., Epidermal growth factor enhances cellular uptake of polystyrene nanoparticles by clathrin-mediated endocytosis. *Int J Mol Sci* **2017,** *18* (6).

23. Parton, R. G., Caveolae: Structure, function, and relationship to disease. *Annual Review of Cell and Developmental Biology* **2018,** *34* (1), 111–136.

24. Wang, Z., Caveolae-mediated delivery of therapeutic nanoparticles across blood-endothelial barrier. *Austin J Anal Pharm Chem* **2014,** *1* (4).

25. Hao, X.; Wu, J.; Shan, Y.; Cai, M.; Shang, X.; Jiang, J.; Wang, H., Caveolae-mediated endocytosis of biocompatible gold nanoparticles in living Hela cells. *J Phys Condens Matter* **2012,** *24* (16), 164207.

26. Zhang, H.; Ji, Q.; Huang, C.; Zhang, S.; Yuan, B.; Yang, K.; Ma, Y. Q., Cooperative transmembrane penetration of nanoparticles. *Sci Rep* **2015,** *5*, 10525.

27. Wang, W.; Yang, R.; Zhang, F.; Yuan, B.; Yang, K.; Ma, Y., Partner-facilitating transmembrane penetration of nanoparticles: A biological test in silico. *Nanoscale* **2018,** *10* (24), 11670–11678.

28. Bechara, C.; Sagan, S., Cell-penetrating peptides: 20 years later, where do we stand? *FEBS Lett* **2013,** *587* (12), 1693–1702.

29. Silva, S.; Almeida, A. J.; Vale, N., Combination of cell-penetrating peptides with nanoparticles for therapeutic application: A review. *Biomolecules* **2019,** *9* (1).

30. Allolio, C.; Magarkar, A.; Jurkiewicz, P.; Baxova, K.; Javanainen, M.; Mason, P. E.; Sachl, R.; Cebecauer, M.; Hof, M.; Horinek, D.; Heinz, V.; Rachel, R.; Ziegler, C. M.; Schrofel, A.; Jungwirth, P., Arginine-rich cell-penetrating peptides induce membrane multilamellarity and subsequently enter via formation of a fusion pore. *Proc Natl Acad Sci USA* **2018,** *115* (47), 11923–11928.

31. Chauhan, A.; Tikoo, A.; Kapur, A. K.; Singh, M., The taming of the cell penetrating domain of the HIV Tat: Myths and realities. *J Control Release* **2007,** *117* (2), 148–162.

32. Nair, B. G.; Fukuda, T.; Mizuki, T.; Hanajiri, T.; Maekawa, T., Intracellular trafficking of superparamagnetic iron oxide nanoparticles conjugated with TAT peptide: 3-dimensional electron tomography analysis. *Biochemical and Biophysical Research Communications* **2012,** *421* (4), 763–767.

33. Patel, S. G.; Sayers, E. J.; He, L.; Narayan, R.; Williams, T. L.; Mills, E. M.; Allemann, R. K.; Luk, L. Y. P.; Jones, A. T.; Tsai, Y. H., Cell-

penetrating peptide sequence and modification dependent uptake and subcellular distribution of green florescent protein in different cell lines. *Sci Rep* **2019,** *9* (1), 6298.

34. Zhao, T.; Singh, S.; Zhang, Y.; Belfield, K. D., Novel mitochondria penetrating peptide for live-cell long-term tracking of mitochondria. **2019**.
35. Docter, D.; Westmeier, D.; Markiewicz, M.; Stolte, S.; Knauer, S. K.; Stauber, R. H., The nanoparticle biomolecule corona: Lessons learned – challenge accepted. *Chemical Society Reviews* **2015,** *44* (17), 6094–6121.
36. Foroozandeh, P.; Aziz, A. A., Merging worlds of nanomaterials and biological environment: Factors governing protein corona formation on nanoparticles and its biological consequences. *Nanoscale Research Letters* **2015,** *10* (1), 221.
37. Leonard, E. F.; Vroman, L., Is the Vroman effect of importance in the interaction of blood with artificial materials? *J Biomater Sci Polym Ed* **1991,** *3* (1), 95–107.
38. Cedervall, T.; Lynch, I.; Lindman, S.; Berggård, T.; Thulin, E.; Nilsson, H.; Dawson, K. A.; Linse, S., Understanding the nanoparticle-protein corona using methods to quantify exchange rates and affinities of proteins for nanoparticles. *Proc Natl Acad Sci USA* **2007,** *104* (7), 2050–2055.
39. Bertrand, N.; Grenier, P.; Mahmoudi, M.; Lima, E. M.; Appel, E. A.; Dormont, F.; Lim, J. M.; Karnik, R.; Langer, R.; Farokhzad, O. C., Mechanistic understanding of in vivo protein corona formation on polymeric nanoparticles and impact on pharmacokinetics. *Nat Commun* **2017,** *8* (1), 777.
40. Capjak, I.; Goreta, S. Š.; Jurašin, D. D.; Vrček, I. V., How protein coronas determine the fate of engineered nanoparticles in biological environment. **2017,** *68* (4), 245.
41. Yang, S.-T.; Liu, Y.; Wang, Y.-W.; Cao, A., Biosafety and bioapplication of nanomaterials by designing protein–nanoparticle interactions. *Small* **2013,** *9* (9 & 10), 1635–1653.
42. Nguyen, V. H.; Lee, B. J., Protein corona: A new approach for nanomedicine design. *Int J Nanomedicine* **2017,** *12,* 3137–3151.
43. Lundqvist, M.; Augustsson, C.; Lilja, M.; Lundkvist, K.; Dahlback, B.; Linse, S.; Cedervall, T., The nanoparticle protein corona formed in human blood or human blood fractions. *PLoS One* **2017,** *12* (4), e0175871.

44. Han, G. S.; Domaille, D. W., Protein corona, with a twist. *ACS Cent Sci* **2020,** *6* (1), 14–15.

45. Oh, J. Y.; Kim, H. S.; Palanikumar, L.; Go, E. M.; Jana, B.; Park, S. A.; Kim, H. Y.; Kim, K.; Seo, J. K.; Kwak, S. K.; Kim, C.; Kang, S.; Ryu, J. H., Cloaking nanoparticles with protein corona shield for targeted drug delivery. *Nat Commun* **2018,** *9* (1), 4548.

46. Barui, A. K.; Oh, J. Y.; Jana, B.; Kim, C.; Ryu, J.-H., Cancer-targeted nanomedicine: Overcoming the barrier of the protein corona. *Advanced Therapeutics* **2020,** *3* (1), 1900124.

47. Bressan, E.; Ferroni, L.; Gardin, C.; Rigo, C.; Stocchero, M.; Vindigni, V.; Cairns, W.; Zavan, B., Silver nanoparticles and mitochondrial interaction. *Int J Dent* **2013,** *2013,* 312747.

48. Arora, S.; Jain, J.; Rajwade, J. M.; Paknikar, K. M., Interactions of silver nanoparticles with primary mouse fibroblasts and liver cells. *Toxicol Appl Pharmacol* **2009,** *236* (3), 310–8.

49. Karatas, O. F.; Sezgin, E.; Aydin, O.; Culha, M., Interaction of gold nanoparticles with mitochondria. *Colloids Surf B Biointerfaces* **2009,** *71* (2), 315–318.

50. Gallud, A.; Kloditz, K.; Ytterberg, J.; Ostberg, N.; Katayama, S.; Skoog, T.; Gogvadze, V.; Chen, Y. Z.; Xue, D.; Moya, S.; Ruiz, J.; Astruc, D.; Zubarev, R.; Kere, J.; Fadeel, B., Cationic gold nanoparticles elicit mitochondrial dysfunction: A multi-omics study. *Sci Rep* **2019,** *9* (1), 4366.

51. Li, W.-Q.; Wu, J.-Y.; Xiang, D.-X.; Luo, S.-L.; Hu, X.-B.; Tang, T.-T.; Sun, T.-L.; Liu, X.-Y., Micelles loaded with puerarin and modified with triphenylphosphonium cation possess mitochondrial targeting and demonstrate enhanced protective effect against isoprenaline-induced H9c2 cells apoptosis. *Int J Nanomedicine* **2019,** *14,* 8345–8360.

52. Ju, R. J.; Mu, L. M.; Li, X. T.; Li, C. Q.; Cheng, Z. J.; Lu, W. L., Development of functional docetaxel nanomicelles for treatment of brain glioma. *Artif Cell Nanomed B* **2018,** *46,* S1180–S1190.

53. Wang, Z.; Guo, W.; Kuang, X.; Hou, S.; Liu, H., Nanopreparations for mitochondria targeting drug delivery system: Current strategies and future prospective. *Asian Journal of Pharmaceutical Sciences* **2017,** *12* (6), 498–508.

54. Ma, X.; Sun, J.; Zhong, L.; Wang, Y.; Huang, Q.; Liu, X.; Jin, S.; Zhang, J.; Liang, X. J., Correction to evaluation of turning-sized gold nanoparticles on cellular adhesion by golgi disruption in vitro and in vivo. *Nano Lett* **2020,** *20* (1), 799.

55. Hesketh, G. G.; Wartosch, L.; Davis, L. J.; Bright, N. A.; Luzio, J. P., The lysosome and intracellular signalling. *Prog Mol Subcell Biol* **2018,** *57,* 151–180.

56. de Araujo, M. E. G.; Liebscher, G.; Hess, M. W.; Huber, L. A., Lysosomal size matters. *Traffic* **2020,** *21* (1), 60–75.

57. Johnson, D. E.; Ostrowski, P.; Jaumouillé, V.; Grinstein, S., The position of lysosomes within the cell determines their luminal pH. *Journal of Cell Biology* **2016,** *212* (6), 677–692.

58. Selby, L. I.; Cortez-Jugo, C. M.; Such, G. K.; Johnston, A. P. R., Nanoescapology: Progress toward understanding the endosomal escape of polymeric nanoparticles. *Wires Nanomed Nanobi* **2017,** *9* (5).

59. Liu, C. G.; Han, Y. H.; Kankala, R. K.; Wang, S. B.; Chen, A. Z., Subcellular performance of nanoparticles in cancer therapy. *Int J Nanomedicine* **2020,** *15,* 675–704.

60. Yang, J. R.; Luo, Y. C.; Shibu, M. A.; Toth, I.; Skwarczynski, M., Cell-penetrating peptides: Efficient vectors for vaccine delivery. *Curr Drug Deliv* **2019,** *16* (5), 430–443.

61. Kongkatigumjorn, N.; Smith, S. A.; Chen, M.; Fang, K. T.; Yang, S. L.; Gillies, E. R.; Johnston, A. P. R.; Such, G. K., Controlling endosomal escape using pH-responsive nanoparticles with tunable disassembly. *Acs Appl Nano Mater* **2018,** *1* (7), 3164–3173.

62. Plaza-Ga, I.; Manzaneda-Gonzalez, V.; Kisovec, M.; Almendro-Vedia, V.; Munoz-Ubeda, M.; Anderluh, G.; Guerrero-Martinez, A.; Natale, P.; Montero, I. L., pH-triggered endosomal escape of pore-forming Listeriolysin O toxin-coated gold nanoparticles. *J Nanobiotechnol* **2019,** *17* (1).

63. Tweten, R. K.; Hotze, E. M.; Wade, K. R., The unique molecular choreography of giant pore formation by the cholesterol-dependent cytolysins of gram-positive bacteria. *Annu Rev Microbiol* **2015,** *69,* 323–340.

64. Boussif, O.; Lezoualch, F.; Zanta, M. A.; Mergny, M. D.; Scherman, D.; Demeneix, B.; Behr, J. P., A versatile vector for gene and oligonucleotide transfer into cells in culture and in-vivo polyethylenimine. *Proc Natl Acad Sci USA* **1995,** *92* (16), 7297–7301.

65. Vermeulen, L. M. P.; De Smedt, S. C.; Remaut, K.; Braeckmans, K., The proton sponge hypothesis: Fable or fact? *Eur J Pharm Biopharm* **2018,** *129,* 184–190.

66. Wojnilowicz, M.; Glab, A.; Bertucci, A.; Caruso, F.; Cavalieri, F., Super-resolution imaging of proton sponge-triggered rupture of endosomes and cytosolic release of small interfering RNA. *ACS Nano* **2019,** *13* (1), 187–202.

67. Benjaminsen, R. V.; Mattebjerg, M. A.; Henriksen, J. R.; Moghimi, S. M.; Andresen, T. L., The possible "proton sponge" effect of polyethylenimine (PEI) does not include change in lysosomal pH. *Mol Ther* **2013,** *21* (1), 149–57.

68. Deng, X.; Wang, Y.; Zhang, F.; Yin, Z.; Hu, Q.; Xiao, X.; Zhou, Z.; Wu, Y.; Sheng, W.; Zeng, Y., Acidic pH-induced charge-reversal nanoparticles for accelerated endosomal escape and enhanced microRNA modulation in cancer cells. *Chemical Communications* **2016,** *52* (15), 3243–3246.

69. He, S. F.; Fan, W. W.; Wu, N.; Zhu, J. J.; Miao, Y. Q.; Miao, X. R.; Li, F. F.; Zhang, X. X.; Gan, Y., Lipid-based liquid crystalline nanoparticles facilitate cytosolic delivery of sirna via structural transformation. *Nano Lett* **2018,** *18* (4), 2411–2419.

70. Jerjes, W.; Theodossiou, T. A.; Hirschberg, H.; Hogset, A.; Weyergang, A.; Selbo, P. K.; Hamdoon, Z.; Hopper, C.; Berg, K., Photochemical internalization for intracellular drug delivery from basic mechanisms to clinical research. *J Clin Med* **2020,** *9* (2).

71. Perez-Hernandez, M.; del Pino, P.; Mitchell, S. G.; Moros, M.; Stepien, G.; Pelaz, B.; Parak, W. J.; Galvez, E. M.; Pardo, J.; de la Fuente, J. M., Dissecting the molecular mechanism of apoptosis during photothermal therapy using gold nanoprisms. *ACS Nano* **2015,** *9* (1), 52–61.

72. Sosic, L.; Selbo, P. K.; Kotkowska, Z. K.; Kundig, T. M.; Hogset, A.; Johansen, P., Photochemical internalization: Light paves way for new cancer chemotherapies and vaccines. *Cancers (Basel)* **2020,** *12* (1).

73. Martinez-Jothar, L.; Beztsinna, N.; van Nostrum, C. F.; Hennink, W. E.; Oliveira, S., Selective cytotoxicity to HER2 positive breast cancer cells by saporin-loaded nanobody-targeted polymeric nanoparticles in combination with photochemical internalization. *Mol Pharmaceut* **2019,** *16* (4), 1633–1647.

74. Greish, K., Enhanced permeability and retention (EPR) effect for anticancer nanomedicine drug targeting. *Methods Mol Biol* **2010,** *624*, 25–37.

75. Nakamura, Y.; Mochida, A.; Choyke, P. L.; Kobayashi, H., Nanodrug delivery: Is the enhanced permeability and retention effect sufficient for curing cancer? *Bioconjugate Chemistry* **2016,** *27* (10), 2225–2238.

76. Wells, C. M.; Harris, M.; Choi, L.; Murali, V. P.; Guerra, F. D.; Jennings, J. A., Stimuli-responsive drug release from smart polymers. *J Funct Biomater* **2019,** *10* (3).

77. Yuan, T.; Dong, J.; Han, G.; Wang, G., Polymer nanoparticles self-assembled from photo-, pH- and thermo-responsive azobenzene-functionalized PDMAEMA. *RSC Advances* **2016,** *6* (13), 10904–10911.

78. Xie, Y.; Qiao, H.; Su, Z.; Chen, M.; Ping, Q.; Sun, M., PEGylated carboxymethyl chitosan/calcium phosphate hybrid anionic nano-particles mediated hTERT siRNA delivery for anticancer therapy. *Biomaterials* **2014,** *35* (27), 7978–91.

79. Hao, L.; Lin, L.; Zhou, J., pH-responsive zwitterionic copolymer DHA–PBLG–PCB for targeted drug delivery: A computer simulation study. *Langmuir* **2019,** *35* (5), 1944–1953.

80. Zhou, L. Y.; Zhu, Y. H.; Wang, X. Y.; Shen, C.; Wei, X. W.; Xu, T.; He, Z. Y., Novel zwitterionic vectors: Multi-functional delivery systems for therapeutic genes and drugs. *Comput Struct Biotechnol J* **2020,** *18,* 1980–1999.

81. Jin, Q.; Chen, Y.; Wang, Y.; Ji, J., Zwitterionic drug nanocarriers: A biomimetic strategy for drug delivery. *Colloids and Surfaces B: Biointerfaces* **2014,** *124,* 80–86.

82. Chen, Y.; Wang, Z.; He, Y.; Yoon, Y. J.; Jung, J.; Zhang, G.; Lin, Z., Light-enabled reversible self-assembly and tunable optical properties of stable hairy nanoparticles. *Proceedings of the National Academy of Sciences* **2018,** *115* (7), E1391–E1400.

83. Wells, C. M.; Harris, M.; Choi, L.; Murali, V. P.; Guerra, F. D.; Jennings, J. A., Stimuli-responsive drug release from smart polymers. *J Funct Biomater* **2019,** *10* (3).

84. Price, P. M.; Mahmoud, W. E.; Al-Ghamdi, A. A.; Bronstein, L. M., Magnetic drug delivery: Where the field is going? *Frontiers in Chemistry* **2018,** *6* (619).

85. Li, J.; Sun, J.; Chen, T.; Li, X., Towards a better understanding of the effects of the magnetic nanoparticles size and magnetic field on cellular endocytosis. *Journal of Physics D: Applied Physics* **2020,** *53* (17), 175401.

86. Nair, B. G.; Nagaoka, Y.; Morimoto, H.; Yoshida, Y.; Maekawa, T.; Kumar, D. S., Aptamer conjugated magnetic nanoparticles as nanosurgeons. *Nanotechnology* **2010**, *21* (45), 0957–4484.

87. Akash, M. S.; Rehman, K.; Chen, S., Polymeric-based particulate systems for delivery of therapeutic proteins. *Pharm Dev Technol* **2016**, *21* (3), 367–378.

88. Li, B.; Li, Q.; Mo, J.; Dai, H., Drug-loaded polymeric nanoparticles for cancer stem cell targeting. *Front Pharmacol* **2017**, *8* (51).

89. Ulbrich, K.; Holá, K.; Šubr, V.; Bakandritsos, A.; Tuček, J.; Zbořil, R., Targeted drug delivery with polymers and magnetic nanoparticles: Covalent and noncovalent approaches, release control, and clinical studies. *Chemical Reviews* **2016,** *116* (9), 5338–5431.

NANODIAGNOSTICS

Chapter 3

Radiolabeled Nanoparticles for Cancer Diagnosis

R. G. Aswathy [a] and D. Sakthi Kumar [b]

[a]*Future Industries Institute, Mawson Lakes Campus, University of South Australia, Adelaide, South Australia 5095, Australia*
[b]*Bio Nano Electronics Research Center, Graduate School of Interdisciplinary New Science, Toyo University, Kawagoe, Saitama 350-8585, Japan*

aswathy.ravindrangirija@unisa.edu.au, aswathyrg@gmail.com, sakthi@toyo.jp

3.1 Introduction

Nuclear imaging utilizes radioisotopes for the investigation and examination of the physiological and metabolic effects of the body. In nuclear imaging, radiopharmaceuticals are non-invasively administrated to the patients and the radiation emitted is recorded. The data acquired is used for diagnosis specifically for detecting functional abnormalities and early detection of tumors. The most common types of nuclear imaging methods are based on single photon emission computed tomography (SPECT) and positron emission tomography (PET) (Figure 3.1).

Bionanotechnology in Cancer: Diagnosis and Therapy
Edited by D. Sakthi Kumar and Aswathy Ravindran Girija

ISBN 978-981-4800-30-3 (Hardcover), 978-0-429-42291-1 (eBook)
www.jennystanford.com

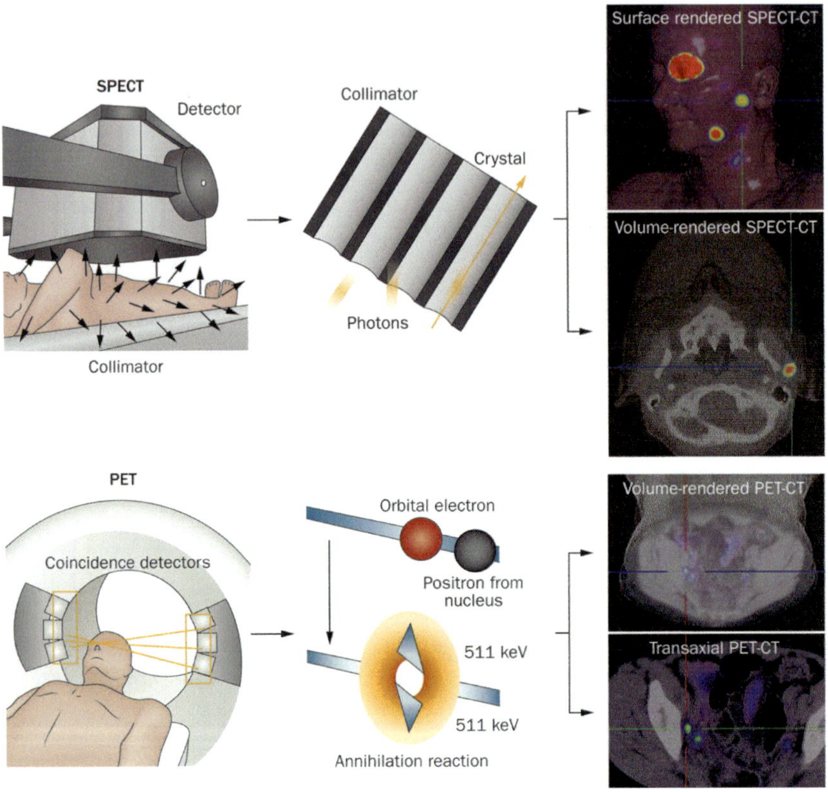

Figure 3.1 Schematic diagram on the pinciple and application of SPECT and PET. Reprinted with permission from Ref. [1]. Copyright © 2012, Springer Nature.

The approaches in molecular imaging with radioisotopes include the design and development of molecules that target precise receptors, and conjugation of radioisotopes emitting photons for SPECT or positrons for PET without interrupting the inherent biological effects of target molecules for in vivo imaging, by high resolution and high sensitivity techniques. The administration of contrast agents permits functional information that can be acquired by imaging modalities including computerized tomography (CT), magnetic resonance imaging (MRI), ultrasound (US), optical imaging, and several combined imaging modalities. Hence, molecular imaging is not constrained exclusively to SPECT and PET. These imaging modalities are all of pronounced clinical significance in cancer for the early detection and the degree and extent of malignancy based on the biochemical changes associated with malignancy that normally precedes anatomical changes.

3.2 SPECT

Imaging with SPECT is based on the radioactive decay of a nuclide with emission of a γ photon where the energy is depended on the nature of the nuclide. The radioisotopes for SPECT imaging emit gamma rays with the energy ranging between ~25–250 keV. The emitted photons are detected after passing through a single or a set of collimated radiation detectors. The camera rotates around the animal's or patient's body and detects the areas of emission (Figure 3.2). The projection data acquired in SPECT is from different angles around the animal/patient. The detectors used in SPECT are mostly based on a single or multiple sodium iodide NaI (TI) scintillation detectors. SPECT scan acquires multiple 2D images/projections from several angles. The computer algorithm then develops tomographic reconstruction of the 2D projections thereby generating 3D image [2–4].

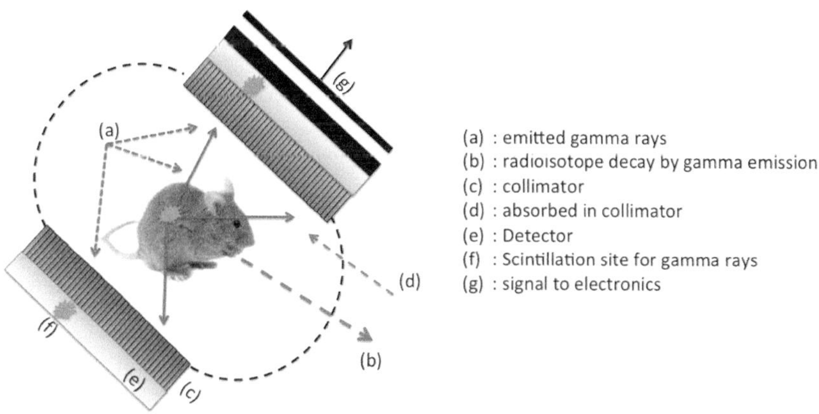

Figure 3.2 Schematic diagram of SPECT.

The most commonly employed radioisotopes for SPECT imaging include technetium (^{99}Tc, $t_{1/2}$: 6.0 h), indium (^{111}In, $t_{1/2}$: 2.8 days), and radioiodine (^{131}I, $t_{1/2}$: 8.0 days) and other radioisotopes as given in Table 3.1. Technetium-99m (^{99m}Tc, is an isotope of technetium, short-lived metastable radionuclide) is the most commonly employed radionuclide for nuclear imaging. Owing to the remarkable physical properties of ^{99m}Tc including the short half-life (6 h) and gamma photon emission (140 keV), this material is highly beneficial for efficient imaging as well as good for patient's safety. Additionally, ^{99m}Tc holds latent chemical properties, which allow this material to be used in kits of numerous types for labeling for multipurpose diagnostic applications. SPECT is often used for imaging ligands, including antibodies, peptides, hormones, and selectins, which are labeled with ^{99m}Tc or with other radioisotopes. These molecules slowly diffuse into tissue and exhibit slow clearance from blood that extends for several hours to even days. Some SPECT isotopes with long half-life, such as thallium-201 (^{201}Tl), tin-117m (^{117m}Sn), and iodine-125 (^{125}I), are used for imaging of slow biological processes, including cell division, inflammatory process, and effect of therapeutic radiopharmaceuticals.

Table 3.1 Important radionuclides and their half-life time for SPECT [4–6]

Radionuclide	Main emission energy	Half-life ($t_{½}$)
^{99m}Tc	140 keV	6.02 h
^{123}I	150 keV	13.3 h
^{111}In	171, 245 keV	2.8 days
^{201}Tl	135, 167 keV	3.0 days
^{67}Ga	93, 185 keV	3.3 days
^{131}I	362 keV	8.2 days

The reported spatial resolution of clinical SPECT in the range of 8–12 mm and that of preclinical SPECT is ≤1 mm [3]. SPECT possesses the distinctive capability to analyze different molecular pathways at the same time by employing isotopes with multiple emission energies. Thus, it is possible to monitor various organs or functions simultaneously that have great implication for the determination of treatment potency in tumors, neurodegenerative disorders, psychiatric abnormalities, gene expression studies, etc. The habitation of dopamine receptors (D2) by a competing drug has been measured in brain by ultra-high-resolution SPECT [7, 8]. Radiopharmaceuticals that are used for imaging inflammation are also available. Based on their specificity, they are broadly categorized into specific radiopharmaceuticals (radiolabeled monoclonal antibodies, in vitro labeled white blood cells (leucocytes), and small receptor proteins and peptides), and non–specific radiopharmaceuticals (radiolabeled macromolecules including human immunoglobulin, human serum albumin, various small molecules, and nanoparticles (NPs) such as liposomes and nanocolloids) [9].

3.3 Positron Emission Tomography

Positron Emission Tomography is a radio imaging modality that detects gamma rays emitted by a positron-emitting radionuclide administrated into the body mostly by intravenous administration. The radioisotope that experiences positron emission decay discharges a positron which encounters with an electron. The process results in the generation of a pair of gamma photons, which move in opposite directions and are detected by the photomultiplier tubes or silicon photodiodes placed at opposite direction (180°) (Figure 3.3) [10].

Usually ^{11}C, ^{13}N, ^{15}O, ^{18}F, ^{64}Cu, ^{62}Cu, ^{124}I, ^{76}Br, ^{82}Rb, ^{68}Ga, and ^{18}F are used for PET, however, ^{11}C, ^{13}N, ^{15}O, ^{18}F are most commonly used (Table 3.2). PET provides considerably superior resolution than SPECT and permits tomographic dynamic studies. The spatial resolution of clinical PET is in the range of 4–6 mm and of preclinical PET is 1–2 mm [3]. However, the short-life isotopes result in the overall increased expense and limit the applications. These are the major drawback of PET when compared with the SPECT imaging.

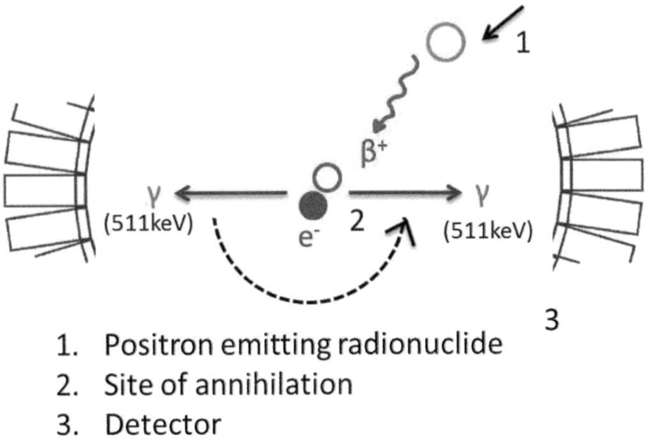

Figure 3.3 Schematic diagram of PET.

Table 3.2 Important radionuclides and their half-life time for PET [4–6]

Radionuclide	Mean β^+ energy	Half-life ($t_{1/2}$)
^{15}O	735 keV	2 min
^{11}C	386 keV	20 min
^{68}Ga	830 keV	68 min
^{18}F	250 keV	110 min
^{64}Cu	278 keV	12.7 h
^{76}Br	1180 keV	16.3 h
^{89}Zr	396 keV	3.27 days
^{124}I	820 keV	4.17 days

Though initially PET imaging was restricted for the imaging of the brain and the heart, a glucose analog, fluorine-18-fluorodeoxyglucose (FDG) has been introduced that enabled the diagnosis and staging of various malignant tumors [11–13]. Positron emitting isotopes are similar to biological compounds, such as ^{11}C, ^{13}N, ^{15}O, ^{18}F, and several other tracers, including ^{64}Cu, ^{68}Ga, and ^{76}Br, are being used as tracers for cancer diagnosis. 3-Deoxy-3-[^{18}F]-fluorothymidine (FLT) is another significant tracer, which is important for controlling response to therapy, owing to the fact that cancer drugs affect cell proliferation significantly than metabolism of glucose [14–16]. The effective collaboration of molecular imaging with molecular biology has enabled the development of techniques to quantify reporter gene expression under in vivo conditions along with PET imaging. Expression of two genes (Herpes simplex virus 1 thymidine kinase (HSV-Tk) gene and dopamine type 2 receptor (D_2R) gene) in a single tumor induced mice has been imaged successfully by PET [17]. The mouse was injected with two reporter probes specific for one of the gene reporter systems, 3-(2′-[^{18}F]fluoroethyl)-spiperone (FESP) and 8-[^{18}F]fluoropenciclovir (FPCV). The image demonstrated the selective potential of the radiolabeled probes in different tumors in the same mouse. Thus, gene therapy protocols could be monitored in vivo, both in animal models and, eventually, in humans by PET.

PET has also been reported for cell apoptosis imaging [18–36]. For molecular imaging of apoptosis, an ideal probe should incorporate distinctive features as selective and specific detection of apoptotic cells, adequate performance in in vivo conditions (high stability, rapid distribution and clearance, sufficient dosimetry, and lack of immunogenic reactivity), and compatibility with clinical practice. The small-molecule PET probes for apoptosis can be categorized into probes for different functions such as probes for detecting caspase activation, disruption of mitochondrial membrane potential (MMP), and detecting apoptotic membrane signature comprising irremediable damage of plasma membrane potential and perpetual acidification inside cytosol. The commonly used PET probes for investigation of cell apoptosis under these categories include ^{11}C, ^{18}F, ^{18}F-ML-10, ^{18}F-ICMT-11, ^{18}F-caspase binding radio-ligand, ^{11}C-WC-98, ^{18}F-WC-IV-3, [^{18}F-radiolabeled organic cations such as ^{18}F-fluorobenzyl triphenylphosphonium cation (^{18}F-FBnTP)], and Aposense compounds (a patented technology; Aposense Ltd.) such as DDC, ML-10, ML-9, NST-732, and NST-729 [36]. A comparison of major characteristics of SPECT and PET is outlined in Table 3.3.

Table 3.3 Comparison of the major characteristics of SPECT and PET [37]

	SPECT	PET
Principle	Detection of single photon Traces emit photons of energy range of ~70–~300 keV	Detection of photon pair Traces emit positrons of various energies that annihilate with free electron and result in generation of two antiparallel photons of energy 511 keV
Spatial resolution	8–12 mm in clinical applications ≤1 mm in preclinical studies	4–6 mm in clinical applications 1–2 mm in preclinical studies
Sensitivity	~0.01%	Up to 10%
Types of acquisition	Planar image with static heads	Planar coincidence image with static opposite heads
Acquisition time	Dynamic studies (static scintigraphic studies tomographic image with rotating heads): Few seconds to 30 min For entire SPECT (for 2D & 3D image reconstruction, depending on number of heads on the imaging system): Few seconds per projections up to 30 min	Dynamic studies: Few seconds to few minutes for static coincidence studies Tomographic imaging without heads: Few minutes or even few seconds Tomographic imaging fitted with rotary heads, ring scanner: Few minutes Whole body clinical scanning, multiple bed positions: up to 30 min
Clinical applications	Oncology, myocardial ischemia, HIV study, thyroid imaging, angiogenesis study, reticuloendothelial system (RES)	Brain imaging, oncology, cardiovascular diseases

	SPECT	PET
Advantages	Possibility of monitoring slow body process by using isotopes with long half-life Simultaneous imaging with one or more tracers for observation of one or more processes Easy available/accessible radiopharmaceuticals	High sensitive, whole body imaging, prospect of dynamic tomographic studies
Disadvantages	Limited spatial resolution Low sensitivity Possibility of planar dynamic studies	Isotopes with short half-life limit the monitoring of slow processes High costs of equipment and radiopharmaceuticals

3.4 Radiolabeled Nanoparticles (RNPs)

NPs are being designed for targeting cancer cells and are used in application of molecular imaging. The fundamental objective of the development of multifunctional NPs is the specific non-invasive diagnosis of cancer. The advantage of targeted NPs over non-targeted ones for non-invasive diagnosis is the tumor cell specific uptake and the visualization of tumor without interfering normal healthy cells. RNPs for the imaging of cancerous tissue generally comprise three major components: core, radionuclide, and targeting moiety (Figure 3.4).

The radionuclide is often conjugated to core component surface via different labeling methods, such as direct (nucleophilic or electrophilic) or indirect labeling (coordination chemistry, nucleophilic halogen exchange chemistry) [38–42]. Indirect attachment to the NPs core is possible via linker molecules, such as a peptide, hydrocarbon chain, or polyethylene glycol (PEG) unit. NPs are generally classified into organic and inorganic NPs based on the chemical composition of core. Organic NPs are mainly polymers, dendrimers, liposomes, carbon

were developed [67]. The protein used for the development of nanodiscs was 11 α-helix MSP1E3D1 and 1-palmitoyl-2-oleoyl-sn-glycero-3-phosphatidylcholine lipids. The reaction was performed by carefully conserving the amphipathic protein fold and labeling of the protein portion of a lipid-protein particle, and biodistribution of nanodiscs was evaluated in vivo by PET. Nanodiscs of approximately 13 nm in size were uptaken by cancer cells in tumor bearing mouse, which makes the nanodiscs as good candidates as advanced theragnostic (Therapy+Diagnosis) agents [67].

Angiogenesis, the formation of new blood vessels can be studied with RNPs. Integrins are expressed on tumor cells as well as actively proliferating endothelial cells that promote tumor cell invasion, angiogenesis that resulting in metastasis [68–72]. Radiolabeled tracers such as ^{18}F-galacto-RGD, ^{111}In-RP747, and ^{99m}Tc-NC100692 for SPECT imaging have been reported for detecting αvβ3 integrin expression related to angiogenesis in tumor [73–76]. Integrin αvβ3-targeted ^{111}In-labeled perfluorocarbon (PFC) NPs were used to detect angiogenesis in tumor tissue in New Zealand white rabbits bearing tumors and at 18 h after injection, and it was observed the mean tumor radioactivity of integrin αvβ3-targeted PFC NP was four-times than those with non-targeted group (control group) [77].

3.4.3 Carbon Nanotube-Based RNPs

The biodistribution of PEGylated radiolabeled single wall carbon nanotubes (SWCNTs) [^{64}Cu ($t_{1/2}$: 12.7 h)-labeled SWCNTs] in mice by PET imaging and ex vivo biodistribution, and Raman spectroscopy was reported [78]. PEGylated SWCNTs presented relatively longer circulation half-life (up to 2 h) and lower uptake by the RES. Also, targeting the nanocomposites with RGD peptide resulted in the targeting of integrin αvβ3-positive tumor in mice. The inherent Raman signatures from targeted SWCNTs were also employed to investigate SWCNTs in mice tissues and confirmed the PET imaging results (Figure 3.8).

Carbon-based NPs for biomedical imaging comprise fullerenes, CNTs, graphene oxide NPs, and perfluorocarbon nanoemulsions. For PET imaging, positron emitting radionuclides can either be conjugated or implanted into CNTs. ^{86}Y labeled CNTs was generated from amine-functionalized water-soluble CNTs by covalent conjugation of many DOTA chelates and radiolabeling yttrium-86 [79]. Whole-body PET images in mice signified the clearance of ^{86}Y-CNTs from blood within 3 h and mostly distributed in kidneys, liver, spleen, and bone.

Grapheneoxide (GO) is gaining considerable attention, owing to its excellent optical, mechanical, thermal, and electronic properties. ^{66}Ga-labeled nanographene tagged with TRC105 was developed for tumor vasculature imaging. MicroPET imaging of 4T1 tumor-bearing mice demonstrated accumulation of ^{66}Ga-NOTA-GO-TRC105 in tumors and uptake of probe in tumor remained stable upon treatment [80]. ^{66}Ga-labeled nanographene displayed excellent tumor specificity. However, significant radioactivity observed in liver limit its clinic application.

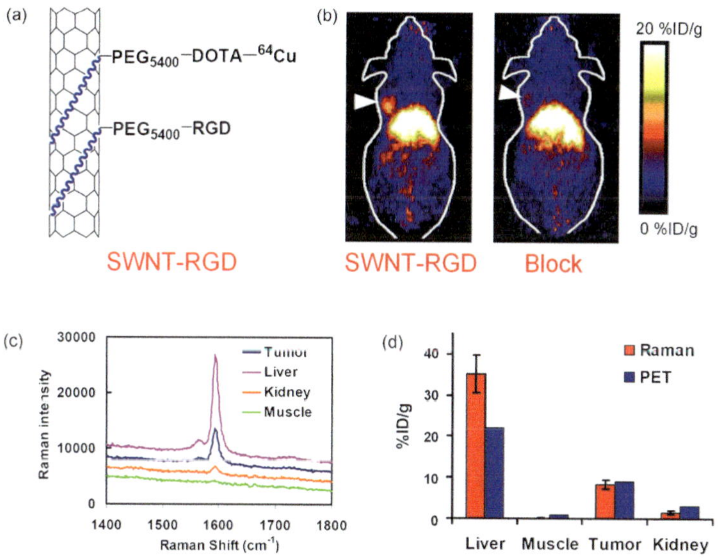

Figure 3.8 ^{64}Cu-labeled SWCNTs for tumor integrin αvβ3 targeting. (a) Schematic of functionalized SWCNTs. (b) 2D projection of PET images of U87MG tumor-bearing mice at 8 h post-administration of ^{64}Cu-labeled, RGD-conjugated SWCNTs with (block) or without co-injection of RGD peptides. Arrowheads specify tumors. (c) Raman spectra of tissue homogenate exhibiting presence of SWCNT in the tumor. (d) Biodistribution studies by PET and ex vivo Raman measurements. Reprinted with permission from Ref. [78].

3.4.4 Inorganic NPs as RNPs

Inorganic NPs include metal NPs, iron oxide NPs, metal sulfide NPs, quantumdots, upconversion NPs, silica NPs, and others. Excellent radio imaging for various disorders can be attained by the efficient radiolabeling of these inorganic NPs.

Iron oxide NPs are ideal nanoplatforms, which have been widely explored as potential MRI contrast agents [81–83]. The radiolabeling of these probes permits the dual imaging capability (PET/MR) for studying the biodistribution of NPs, tumor regression, and PEGylated, and cRGD peptide tagged ^{64}Cu labeled polyaspartic acid-coated iron oxide NPs were synthesized as probes for PET/MR imaging [84]. The radiolabeled NPs were injected to mice bearing U87MG tumor and in vivo PET imaging displayed integrin specific uptake of the nanoprobe and correlated with MR images. In another study, drug delivery was coupled with PET/MR imaging [85]. ^{64}Cu labeled iron oxide NPs tagged with cRGD and conjugated with doxorubicin was used as theranostic probe for as therapeutics and diagnostic agents. Dextran-coated iron oxide NPs labeled with ^{64}Cu based on microwave synthesis was demonstrated for PET/MR imaging [86].

Researches demonstrated tumor uptake, pharmacokinetics, and therapeutic effectiveness of an ^{111}Indium labeled chimeric L6 (ChL6) monoclonal antibody linked to carboxylated PEG on dextran-coated iron oxide (IO) NPs in athymic mice with human breast cancer tumors [87]. The time for the NPs to remain in the circulation was enough to leave from the blood vessels and to reach malignant cells. Also, the antibody (mAb)-linked IO NPs offered the prospect to improve magnetic hyperthermia for metastatic cancer in the presence of external applied alternating magnetic field (AMF).

Recently, there has been great interest in the application of metal sulfide NPs, especially CuS NPs, owing to their steady near infra-red (NIR) absorption and ease of synthesis for imparting photo thermal ablation in cancer cells. The radiolabeling of Cu component of the NP with ^{64}Cu offers remarkable potential as a theranostic agent. Intrinsically radiolabeled PEGylated ^{64}CuS NPs was developed for effective PET imaging and photo thermal ablation in U87MG tumors in mice model [88]. The same nanoconstruct was used for mapping sentinel lymph node in various tumor models [89–91]. The targeting capability of the nanoconstruct was demonstrated with folic acid and folic acid conjugated ^{64}CuSPEG NPs was used for PET imaging and photo thermal ablation in folate receptor positive tumors models. Ultra-small CuS NPs within ferritin nanocages were developed based on biomineralization [92]. The radiolabeled NPs were studied for multimodal imaging and photo thermal therapy in mice with U87MG tumor xenograft.

Radiolabeled upconversion NPs have established substantial attention, owing to their superior in vivo imaging potential. Most of the nanoconstructs based on upconversion NPs are focused on multimodal imaging and are discussed in the multimodal imaging section.

Silica NPs are ideal nanocarriers, owing to their cytocompatibility, flexible surface modification strategies for delivering drugs, and integrating contrast agents. Hollow mesoporous silica NPs were used as tumor targeted drug delivery nanocarriers by conjugating TRC105 [93] and RGD peptide [94] radiolabeled with ^{64}Cu and encapsulating chemotherapeutic drugs. In a recent study, p-isothiocyanatobenzyldesferrioxamine (DFO)-based chelator was used for radiolabeling ^{89}Zr to mesoporous silica NPs and functioned as a potential PET probe both in in vitro and in vivo studies [95]. Chelator-free technique was reported for radiolabeling ^{68}Ga [96] and ^{64}Cu [97] in mesoporous silica NPs for in vivo PET guided drug delivery applications. Recently, radiolabeled biodegradable dendritic mesoporous silica NPs for delivering several cargos and self-destruction of nanoprobe after the delivery of payloads was developed [98]. Chelator-free approach for labeling ^{89}Zr was employed and the probe was PEGylated and tagged with TRC105. In vivo PET imaging demonstrated rapid uptake in tumor at 4 h post-injection and constantly afterwards up to 48 h.

3.5 Multimodal Imaging of PET and SPECT with MRI, CT, and NIRF

It is hard to precisely quantify fluorescence signal in living subjects, particularly the deep tissue imaging. The radionuclide-based imaging techniques are highly sensitive, but they have comparatively poor resolution. In the case of MRI, it has high resolution, but lacks with low sensitivity. Therefore, the integration of different imaging modalities can benefit synergistic returns compared with imaging based on single modality. The advantage of NPs is that they have huge surface areas and therefore several functionalities can be integrated for the progress of multimodal NPs for multimodal imaging.

The application of radiolabeled IO NPs can be used for PET as well as for MRI. The benefits of MRI over radionuclide imaging include improved spatial resolution (in the range of sub-millimeter), absence of radiation, and excellent contrast in soft tissue. The inherent low sensitivity is considered as the key disadvantage of MRI, which can be solved by PET. IO NPs with crystalline IO core encapsulated by a polymer (e.g., dextran or PEG) are being used as NP-based MR contrast agents. The synergistic application PET/MR can be used for imaging neurological disorders, several types of cancer, and also in stem cell therapy.

Monocrystalline iron-oxide nanoparticle (MION) labeled with $^{Cu}64$ was developed for imaging macrophages in inflammatory atherosclerosis [99]. apoE−/− mice were scanned twice, initially 1 h post-injection of ^{18}FDG. The mouse was injected with ^{64}Cu trireporter NP (^{64}Cu-TNP) on the next day and images were acquired 24 h after injection. Strong PET signal was observed from the aortic root and arch. The anatomical information obtained from PET/CT and the injection of an iodinated CT contrast agent permitted the identification of those structures in functional images. ^{64}Cu-TNP offered 50% greater uptake compared to the normal ^{18}FDG and offered trimodal imaging with complementary MRI, PET, and CT.

An IO NP-based nanoprobe for PET/MRI scanning of tumor specific integrin $\alpha v\beta 3$ expression was studied [100]. The poly (aspartic acid)-coated IO NPs (PASP-IO) were conjugated to cyclic RGD peptides and DOTA chelators as PET tracers after radiolabeling with ^{64}Cu. The microPET imaging contributed by radioisotope and T2-weighted MRI contributed by IO confirmed integrin $\alpha v\beta 3$-specific delivery of RGD-PASP-IO nanoconjugates to the U87MG human glioblastoma tumor. ^{124}I ($t_{1/2}$: 4.2 days)-labeled IO NP was studied as dual-modal PET/MR probe for imaging the lymph node in rats. The study demonstrated precise localization of lymph nodes that was crucial for cancer staging as the lymphatic system is providing significant track for cancer metastasis [101].

Radiolabeled quantum dots (QDs) were used for studying vascular targeting. The interaction of the NPs with the RES, the uptake and clearance of NPs in the body were demonstrated [102]. CdTe NPs were synthesized and radiolabeled QD developed by a mixture of ^{125m}Te/^{124}Te. The RNPs were attached to a monoclonal antibody to mice lung thrombomodulin (mAb 201B) expressed in the lumen of lung blood vessels or a control antibody (mAb 33). Whole-body MicroSPECT/CT image was performed on dual-modality platform. After the administration of the NPs, mice were imaged. SPECT images demonstrated a prevalence of ^{125m}Te in the lungs, whereas the NPs linked to mAb 33 concentrated in the liver and spleen of the mouse.

The development of a dual-modal imaging agent comprising a PET isotope and a NIR QD results in very sensitive, accurate tumor-targeting efficacy, and significantly assist clinical applications. QD-based nanoprobe for synergistic NIR fluorescence (NIRF) and PET imaging has been studied [103]. Nanoconjugates have been developed with cyclic RGD peptides and DOTA chelators conjugated QD (maximum emission 705 nm) for targeted dual-modality PET/NIRF imaging after radiolabeling with ^{64}Cu. The quantitative evaluation of tumor-targeting efficacy was attained using dual-modal PET/NIRF probe and it was observed that the majority of nanoconjugate was localized within the tumor blood vessel. The PET/NIRF probe offered appropriate contrast demonstrable by PET at a lower concentration than necessary for in vivo NIRF imaging, thus considerably decreasing the probable cytotoxicity induced by cadmium-based QDs.

QD-based dual-modal PET/NIRF nanoconstructs could be predominantly suitable for an initial whole-body PET scan to identify the exact location of tumor(s), and subsequently using optical imaging to guide the resection of tumor. The excellent combination of PET and NIRF imaging permits quantitative in vivo targeted deep tissue imaging, vital for potential image-guided surgery through highly specific, very sensitive, and real-time intra-operative imagining in normal and diseased condition.

An UC-based nanoconstruct potential for radioimaging, MRI, and upconversion luminescent imaging, was developed by labeling ^{18}F to $Gd^{3+}/Yb^{3+}/Er^{3+}$ co-doped $NaYF_4$ nanophosphors NPs [104]. In vivo PET imaging and T1-weighted MRI in mice and upconversion luminescence imaging of live cells were demonstated. Dual imaging (PET/upconversion luminescence imaging) was demonstrated with ^{18}F-labeled Y_2O_3, $NaYF_4$, and upconversion NPs. Micelle-encapsulated radiolabeled upconversion NPs were used for in vivo PET imaging and upconversion luminescence imaging [105]. The study signified the clearance of NPs primarily through hepatobiliary excretion that was further confirmed by TEM imaging.

Development of $Fe_3O_4@NaYF_4$ core/shell NPs with different cation dopants either in the shell or core, to generate $Co_{0.16}Fe_{2.84}O_4@NaYF_4$ (Yb, Er) and $Fe_3O_4@NaYF_4$ (Yb, Tm) NPs, was reported [106]. The radiolabeled bisphosphonate PEGylated nanoconstruct presented trimodal (MRI, PET/SPECT, and optical) imaging of sentinel lymph nodes in mice model. The biodistribution study by PET/MR revealed negatively charged $CO_{0.16}Fe_{2.84}O_4@NaYF_4$ (Yb, Er)-PEGylated NPs slowly cleared from the blood pool than positively charged $Fe_3O_4@NaYF_4$ (Yb, Tm)-PEGylated NPs.

$_{64}Cu$ labeled porphyrin phospholipid coated core-shell UC NPs of $NaYbF_4$: Tm-$NaYF_4$ capable for hexa-model imaging was developed. The nanoconstruct was functional for conventional fluorescence, upconversion luminescence, photoacoustic, PET, computed tomography, and Cerenkov luminescence [107].

Multimodal imaging probes with radionuclide-based highly sensitive and quantitative imaging (SPECT or PET), and non-radionuclide-based strategies, based on optical imaging or MRI offering high resolution and anatomical information, might be promising for future bioimaging applications [108–115]. The application of targeted RNPs offers several benefits over traditional molecular imaging probes in terms of signal amplification and specificity attained by active targeting. Single as well as a multiple imaging labels of various imaging modalities can be integrated to a single NP, which can result in significant contrast enhancement. Effective targeting with potentially different targeting ligands can be attached on the NPs to offer augmented receptor-binding specificity. The two major advantages of RNPs compared with other NPs are their ability for quantitative imaging and excellent sensitivity.

3.6 Conclusion and Future Prospects

RNPs have immense potential to be employed for disease diagnosis. The prospect of labeling targeting moiety for precise imaging offers superior receptor binding affinity and also the potential to bypass biological barriers (e.g., blood brain barrier). Extreme sensitivity and precise quantitative imaging are the major benefits of RNPs when compared with other imaging modalities. The major challenges in the development of RNPs include the efficient entrapment or integration of radionuclide to the NPs, conjugation of targeting ligands, and safety profile along with regulatory issues. Although, there is development in the multimodal imaging, it is still in its infancy. Extensive research is essential to gain the comprehensive benefits of multimodal imaging that could offer imaging regime with noninvasive imaging, targeting potential, quantitation capability, high sensitivity, anatomical information, and also the prospect of integrating treatment along with imaging for future biomedical applications.

References

1. Hicks RJ, Hofman MS. Is there still a role for SPECT–CT in oncology in the PET–CT era? *Nat. Rev. Clin. Oncol.* 12, **2012**, 712–720.
2. Accorsi R. Brain single-photon emission CT physics principles. *Am. J. Neuroradiol.* 29, **2008**, 1247–1256.

3. Khalil MM, Tremoleda JL, Bayomy TB, Gsell W. Molecular SPECT imaging: An overview. *Int. J. Mol. Imaging.* 2011, **2011**, 1–15.
4. Xing Y, Zhao J, Conti PS, Chen K. Radiolabeled nanoparticles for multimodality tumor imaging. *Theranostics.* 4, **2014**, 290–306.
5. Liu S, Edwards DS. ^{99m}Tc-labeled small peptides as diagnostic radiopharmaceuticals. *Chem. Rev.* 99, **1999**, 2235–2268.
6. Wadas TJ, Wong EH, Weisman GR, Anderson CJ. Coordinating radiometals of copper, gallium, indium, yttrium, and zirconium for PET and SPECT imaging of disease. *Chem. Rev.* 110, **2010**, 2858–2902.
7. Acton PD, Hou C, Kung MP, Plössl K, Keeney CL, Kung HF. Occupancy of dopamine D2 receptors in the mouse brain measured using ultra-high-resolution single-photon emission tomography and [123]IBF. *Eur J Nucl Mol Imaging.* 29, **2002**, 1507–1515.
8. Brücke T, Wenger S, Asenbaum S, Fertl E, Pfafflmeyer N, Müller C, Podreka I, Angelberger P. Dopamine D2 receptor imaging and measurement with SPECT. *Adv Neurol.* 60, **1993**, 494–500.
9. Ting G, Chang CH, Wang HE. Cancer Nanotargeted radio-pharmaceuticals for tumor imaging and therapy. *Anticancer Res.* 29, **2009**, 4107–4118.
10. Ziegler SI, Positron emission tomography: Principles, technology, and recent developments. *Nucl. Phys. A.* 752, **2005**, 679c–687c.
11. Brown RK, Bohnen NI, Wong KK, Minoshima S, Frey KA. Brain PET in suspected dementia: Patterns of altered FDG metabolism. *RadioGraphics.* 34, **2014**, 684–701.
12. Almuhaideb A, Papathanasiou N, Bomanji J. 8F-FDG PET/CT imaging in oncology, *Ann Saudi Med.* 31, **2011**, 3–13.
13. Seam P, Juweid ME, Cheson BD. The role of FDG-PET scans in patients with lymphoma. *Blood.* 110, **2007**, 3507–3516.
14. Wang Z, Wang Y, Sui X, Zhang W, Shi R, Zhang Y, Dang Y, Qiao Z, Zhang B, Song W, Jiang J. Performance of FLT-PET for pulmonary lesion diagnosis compared with traditional FDG-PET: A meta-analysis. *Eur. J. Radiol.* 84, **2015**, 1371–1377.
15. Dittmann H, Dohmen BM, Paulsen F, Eichhorn K, Eschmann SM, Horger M, Wehrmann M, Machulla HJ, Bares R. [^{18}F]FLT PET for diagnosis and staging of thoracic tumours. *Eur. J. Nucl. Med. Mol. Imaging.* 30, **2003**, 1407–1412.
16. Chen W, Cloughesy T, Kamdar N, Satyamurthy N, Bergsneider M, Liau L, Mischel P, Czernin J, Phelps ME, Silverman DH. Imaging

proliferation in brain tumors with ^{18}F-FLT PET: Comparison with ^{18}F-FDG. *J. Nucl. Med.* 46, **2005**, 945–952.

17. Iyer M, Barrio JR, Namavari M, Bauer E, Satyamurthy N, Nguyen K, Toyokuni T, Phelps ME, Herschman HR, Gambhir SS. 8-[^{18}F] Fluoropenciclovir: An improved reporter probe for imaging HSV1-tk reporter gene expression in vivo using PET. *J. Nucl. Med.* 42, **2001**, 96–105.
18. De Saint-Hubert M, Prinsen K, Mortelmans L, Verbruggen A, Mottaghy FM. Molecular imaging of cell death. *Methods.* 48, **2009**, 178–187.
19. Reshef A, Shirvan A, Akselrod-Ballin A, Wall A, Ziv I. Small-molecule biomarkers for clinical PET imaging of apoptosis. *J. Nucl. Med.* 51, **2010**, 837–840.
20. Lee D, Long SA, Murray JH, Adams JL, Nuttall ME, Nadeau DP, Kikly K, Winkler JD, Sung CM, Ryan MD, Levy MA, Keller PM, DeWolf WE Jr. Potent and selective nonpeptide inhibitors of caspases 3 and 7. *J. Med. Chem.* 44, **2001**, 2015–2026.
21. Kopka K, Faust A, Keul P, Wagner S, Breyholz HJ, Höltke C, Schober O, Schäfers M, Levkau B. 5-pyrrolidinylsulfonyl isatins as a potential tool for the molecular imaging of caspases in apoptosis. *J. Med. Chem.* 49, **2006**, 6704–6715.
22. Zhou D, Chu W, Rothfuss J, Zeng C, Xu J, Jones L, Welch MJ, Mach RH. Synthesis, radiolabeling, and in vivo evaluation of an ^{18}F-labeled isatin analog for imaging caspase-3 activation in apoptosis. *Bioorg. Med. Chem. Lett.* 16, **2006**, 5041–5046.
23. Podichetty AK, Wagner S, Schröer S, Faust A, Schäfers M, Schober O, Kopka K, Haufe G. Fluorinated isatin derivatives. Part 2. New N-substituted 5pyrrolidinyl sulfonyl isatins as potential tools for molecular imaging of caspases in apoptosis. *J. Med. Chem.* 52, **2009**, 3484–3495.
24. Chen DL, Zhou D, Chu W, Herrbrich PE, Jones LA, Rothfuss JM, Engle JT, Geraci M, Welch MJ, Mach RH. Comparison of radiolabeled isatin analogs for imaging apoptosis with positron emission tomography. *Nucl. Med. Biol.* 36, **2009**, 651–658.
25. Nguyen QD, Smith G, Glaser M, Perumal M, Arstad E, Aboagye EO. Positron emission tomography imaging of drug-induced tumor apoptosis with a caspase-3/7 specific [^{18}F]-labeled isatin sulfonamide. *Proc. Natl. Acad. Sci. USA.* 106, **2009**, 16375–16380.
26. Zhou D, Chu W, Chen DL, Wang Q, Reichert DE, Rothfuss J, D'Avignon A, Welch MJ, Mach RH. [^{18}F]- and [^{11}C]-labeled N-benzyl-isatin sulfonamide analogues as PET tracers for apoptosis: Synthesis,

radiolabeling mechanism, and in vivo imaging study of apoptosis in Fas-treated mice using [^{11}C] WC-98. *Org. Biomol. Chem.* 7, **2009**, 1337–1348.

27. Madar I, Huang Y, Ravert H, Dalrymple SL, Davidson NE, Isaacs JT, Dannals RF, Frost JJ. Detection and quantification of the evolution dynamics of apoptosis using the PET voltage sensor ^{18}F-fluorobenzyl triphenyl phosphonium. *J. Nucl. Med.* 50, **2009**, 774–780.

28. Madar I, Ravert H, Nelkin B, Abro M, Pomper M, Dannals R, Frost JJ. Characterization of membrane potential-dependent uptake of the novel PET tracer ^{18}F-fluorobenzyl triphenylphosphonium cation. *Eur. J. Nucl. Med. Mol. Imaging.* 34, **2007**, 2057–2065.

29. Aloya R, Shirvan A, Grimberg H, Reshef A, Levin G, Kidron D, Cohen A, Ziv I. Molecular imaging of cell death in vivo by a novel small molecule probe. *Apoptosis.* 11, **2006**, 2089–2101.

30. Damianovich M, Ziv I, Heyman SN, Rosen S, Shina A, Kidron D, Aloya T, Grimberg H, Levin G, Reshef A, Bentolila A, Cohen A, Shirvan A. ApoSense: A novel technology for functional molecular imaging of cell death in models of acute renal tubular necrosis. *Eur. J. Nucl. Med. Mol. Imaging.* 33, **2006**, 281–291.

31. Cohen A, Ziv I, Aloya T, Levin G, Kidron D, Grimberg H, Reshef A, Shirvan A. Monitoring of chemotherapy-induced cell death in melanoma tumors by N,N9′ Didansyl-L-cystine. *Technol. Cancer. Res. Treat.* 6, **2007**, 221–234.

32. Reshef A, Shirvan A, Waterhouse RN, Grimberg H, Levin G, Cohen A, Ulysse LG, Friedman G, Antoni G, Ziv I. Molecular imaging of neurovascular cell death in experimental cerebral stroke by PET. *J. Nucl. Med.* 49, **2008**, 1520–1528.

33. Zeng W, Yao ML, Townsend D, Kabalka G, Wall J, Le Puil M, Biggerstaff J, Miao W. Synthesis, biological evaluation and radiochemical labeling of a dansylhydrazone derivative as a potential imaging agent for apoptosis. *Bioorg. Med. Chem. Lett.* 18, **2008**, 3573–3577.

34. Cohen A, Shirvan A, Levin G, Grimberg H, Reshef A, Ziv I. From the Gla domain to a novel small-molecule detector of apoptosis. *Cell Res.* 19, **2009**, 625–637.

35. Shirvan A, Reshef A, Yogev-Falach M, Ziv I. Molecular imaging of neurodegeneration by a novel cross-disease biomarker. *Exp. Neurol.* 219, **2009**, 274–283.

36. Shirvan A, Reshef A, Yogev-Falach M, Ziv I. Monitoring of tumor response to chemotherapy in vivo by a novel small-molecule detector of apoptosis. *Apoptosis.* 14, **2009**, 257–267.

37. Loudos G, Kagadis GC, Psimadas D. Current status and future perspectives of in vivo small animal imaging using radiolabeled nanoparticles. *Eur. J. Radiol.* 78, **2011**, 287–295.

38. Liu S, Edwards DS. ^{99}mTc-labeled small peptides as diagnostic radiopharmaceuticals. *Chem. Rev.* 99, **1999**, 2235–2268.

39. Holland JP, Williamson MJ, Lewis JS. Unconventional nuclides for radiopharmaceuticals. *Mol. Imaging.* 9, **2010**, 1–20.

40. Wadas TJ, Wong EH, Weisman GR, Anderson CJ. Coordinating radiometals of copper, gallium, indium, yttrium, and zirconium for PET and SPECT imaging of disease. *Chem. Rev.* 110, **2010**, 2858–2902.

41. Miller PW, Long NJ, Vilar R, Gee AD. Synthesis of ^{11}C, ^{18}F, ^{15}O, and ^{13}N radiolabels for positron emission tomography. *Angew. Chem. Int. Ed. Engl.* 47, **2008**, 8998–9033.

42. Shokeen M, Anderson CJ. Molecular imaging of cancer with copper-64 radiopharmaceuticals and positron emission tomography (PET). *Acc. Chem. Res.* 42, **2009**, 832–841.

43. Lee DE, Koo H, Sun IC, Ryu JH, Kim K, Kwon IC. Multifunctional nanoparticles for multimodal imaging and theragnosis. *Chem. Soc. Rev.* 41, **2012**, 2656–2672.

44. Cheki M, Moslehi M, Assadi M. Marvelous applications of quantum dots. *Eur. Rev. Med. Pharmaco.* 17, **2013**, 1141–1148.

45. Hardman R. A toxicologic review of quantum dots: Toxicity depends on physicochemical and environmental factors. *Environ. Health. Perspect.* 114, **2006**, 165–172.

46. Ling Y, Wei K, Luo Y, Gao X, Zhong S. Dual docetaxel/superparamagnetic iron oxide loaded nanoparticles for both targeting magnetic resonance imaging and cancer therapy. *Biomaterials.* 32, **2011**, 7139–7150.

47. Zhou J, Sun Y, Du X, Xiong L, Hu H, Li F. Dual-modality in vivo imaging using rare-earth nanocrystals with near-infrared to near-infrared (NIR-to-NIR) upconversion luminescence and magnetic resonance properties. *Biomaterials.* 31, **2010**, 3287–3295.

48. Nyk M, Kumar R, Ohulchanskyy TY, Bergey EJ, Prasad PN. High contrast in vitro and in vivo photoluminescence bioimaging using near infrared to near infrared up-conversion in Tm^{3+} and Yb^{3+} doped fluoride nanophosphors. *Nano. Lett.* 8, **2008**, 3834–3838.

49. Wang F, Banerjee D, Liu Y, Chen X, Liu X. Upconversion nanoparticles in biological labeling, imaging, and therapy. *Analyst.* 135, **2010**, 1839–1854.

50. Longmire M, Choyke PL, Kobayashi H. Dendrimer-based contrast agents for molecular imaging. *Curr. Top. Med. Chem.* 8, **2008**, 1180–1186.

51. Santra S. Fluorescent silica nanoparticles for cancer imaging. *Methods Mol. Biol.* 624, **2010**, 151–162.

52. Torchilin VP. Structure and design of polymeric surfactant-based drug delivery systems. *J. Control Release.* 73, **2001**, 137–172.

53. Torchilin VP. Micellar nanocarriers: Pharmaceutical perspectives. *Pharm Res.* 24, **2007**, 1–16.

54. Elbayoumi TA, Torchilin VP. Current trends in liposome research. *Methods Mol. Biol.* 605, **2010**, 1–27.

55. Ghanem GE, Joubran C, Arnould R, Lejeune F, Fruhling J. Labelled polycyanoacrylate nanoparticles for human in vivo use. *Appl. Radiat. Isot.* 44, **1993**, 1219–1224.

56. Videira MA, Botelho MF, Santos AC, Gouveia LF, de Lima JJ, Almeida AJ. Lymphatic uptake of pulmonary delivered radiolabelled solid lipid nanoparticles. *J Drug Target.* 10, **2002**, 607–613.

57. Reddy LH, Sharma RK, Chuttani K, Mishra AK, Murthy RR. Etoposide-incorporated Tripalmitin nanoparticles with different surface charge: Formulation, characterization, radiolabeling, and biodistribution studies. *AAPS J.* 6, **2004**, e23 (or) 55–64.

58. Cai W, Chen K, He L, Cao Q, Koong A, Chen X. Quantitative PET of EGFR expression in xenograft-bearing mice using ^{64}Cu-labeled cetuximab, a chimeric anti-EGFR monoclonal antibody. *Eur. J. Nucl. Mol. Imaging.* 34, **2007**, 850–858.

59. Smith-Jones PM, Solit D, Afroze F, Rosen N, Larson SM. Early tumor response to Hsp90 therapy using HER2 PET: Comparison with ^{18}F-FDG PET. *J. Nucl. Med.* 47, **2006**, 793–796.

60. Chakrabarti A, Zhang K, Aruva MR, Cardi CA, Opitz AW, Wagner NJ, Thakur ML, Wickstrom E. Radiohybridization PET imaging of KRAS G12D mRNA expression in human pancreas cancer xenografts with [(64)Cu]DO3A-peptide nucleic acid-peptide nanoparticles. *Cancer Biol. Ther.* 6, **2007**, 948–956.

61. Amirkhanov NV, Zhang K, Aruva MR, Thakur ML, Wickstrom E. Imaging human pancreatic cancer xenografts by targeting mutant KRAS2 mRNA with [111In]DOTAn-Poly(diamidopropanoyl)m-KRAS2 PNA-d (Cys-Ser-Lys-Cys) nanoparticles. *Bioconju. Chem.* 21, **2010**, 731–740.

62. Bartlett DW, Su H, Hildebrandt IJ, Weber WA, Davis ME. Impact of tumor-specific targeting on the biodistribution and efficacy of siRNA nanoparticles measured by multimodality in vivo imaging. *Proc. Natl. Acad. Sci. USA*. 104, **2007**, 15549–15554.

63. Ogihara-Umeda I, Sasaki T, Kojima S, Nishigori H. Optimal radiolabeled liposomes for tumor imaging. *J. Nucl. Med*. 37, **1996**, 326–332.

64. Andersen TL, Peterson AL, Hansen AE, Henriksen JR. Contribution of nuclear medicine to cancer nanotheragnostics. Chapter 7 in Nanotherganostic for personalized medicine. Simona Mura, Patrick Couvre (eds.). World Scientific publishing. Co. Pte. Ltd., Singapore, **2016**.

65. Harrington KJ, Rowlinson-Busza G, Syrigos KN, Uster PS, Abra RM, Stewart JSW. Biodistribution and pharmacokinetics of^{111}In-DTPA-labelled pegylated liposomes in a human tumour xenograft model: Implications for novel targeting strategies. *Br. J. Cancer*. 83, **2000**, 232–238.

66. Petersen AL, Binderup T, Rasmussen P, Henriksen JR, Elema DR, Kjær A, Andresen TL. ^{64}Cu loaded liposomes as positron emission tomography imaging agents. *Biomaterials*. 32, **2011**, 2334–2341.

67. Huda P, Binderup T, Pedersen MC, Midtgaard SR, Elema DR, Kjær A, Jensen M, Arleth L. PET/CT based in vivo evaluation of ^{64}Cu labelled nanodiscs in tumor bearing mice. *PLoS One*. 10, **2015**, e0129310.

68. Folkman J. Role of angiogenesis in tumor growth and metastasis. *Semin. Oncol*. 29, **2002**, 15–18.

69. Stromblad S, Cheresh DA. Integrins, angiogenesis and vascular cell survival. *Chem. Biol*. 3, **1996**, 881–885.

70. Hood JD, Cheresh DA. Role of integrins in cell invasion and migration. *Nat. Rev. Cancer*. 2, **2002**, 91–100.

71. Cai W, Chen X. Anti-angiogenic cancer therapy based on integrin $\alpha_v\beta_3$ antagonism. *Anti-Cancer Agents Med. Chem*. 6, **2006**, 407–428.

72. Cai W, Niu G, Chen X. Imaging of integrins as biomarkers for tumor angiogenesis. *Curr. Pharm. Des*.14, **2008**, 2943–2973.

73. Haubner R, Wester HJ, Reuning U, Senekowitsch-Schmidtke R, Diefenbach B, Kessler H, Stöcklin G, Schwaiger M. Radiolabeled $\alpha_v\beta_3$ integrin antagonists: A new class of tracers for tumor targeting. *J. Nucl. Med*. 40, **1999**, 1061–1071.

74. Coleman PJ, Brashear KM, Askew BC, Hutchinson JH, McVean CA, Duong LT, Feuston BP, Fernandez-Metzler C, Gentile MA, Hartman GD, Kimmel DB, Leu CT, Lipfert L, Merkle K, Pennypacker

B, Prueksaritanont T, Rodan GA, Wesolowski GA, Rodan SB, Duggan ME. Nonpeptide $\alpha_v\beta_3$ antagonists. Part 11. Discovery and preclinical evaluation of potent $\alpha_v\beta_3$ antagonists for the prevention and treatment of osteoporosis. *J. Med. Chem.* 47, **2004**, 4829–4837.

75. Onthank DC, Liu S, Silva PJ, Barrett JA, Harris TD, Robinson SP, Edwards DS. ^{90}Y and ^{111}In complexes of a DOTA-conjugated integrin $\alpha_v\beta_3$ receptor antagonist: Different but biologically equivalent. *Bioconjug. Chem.* 15, **2004**, 235–241.

76. Chen X, Park R, Hou Y, Khankaldyyan V, Gonzales-Gomez I, Tohme M, Bading JR, Laug WE, Conti PS. MicroPET imaging of brain tumor angiogenesis with ^{18}F-labeled PEGylated RGD peptide. *Eur. J. Nucl. Med. Mol. Imaging.* 31, **2004**, 1081–1089.

77. Hu G, Lijowski M, Zhang H, Partlow KC, Caruthers SD, Kiefer G, Gulyas G, Athey P, Scott MJ, Wickline SA, Lanza GM. Imaging of Vx-2 rabbit tumors with $\alpha_v\beta_3$-integrin-targeted ^{111}In nanoparticles. *Int. J. Cancer.* 120, **2007**, 1951–1957.

78. Liu Z, Cai W, He L, Nakayama N, Chen K, Sun X, Chen X, Dai H. In vivo biodistribution and highly efficient tumour targeting of carbon nanotubes in mice. *Nat. Nanotechnol.* 2, **2007**, 47–52.

79. McDevitt MR, Chattopadhyay D, Jaggi JS, Finn RD, Zanzonico PB, Villa C, Rey D, Mendenhall J, Batt CA, Njardarson JT, Scheinberg DA. PET imaging of soluble yttrium-86-labeled carbon nanotubes in mice. *PLoS One.* 3, **2007**, e907.

80. Hong H, Zhang Y, Engle JW, Nayak TR, Theuer CP, Nickles RJ, Barnhart TE, Cai W. In vivo targeting and positron emission tomography imaging of tumor vasculature with (66)Ga-labeled nano-graphene. *Biomaterials.* 33, **2012**, 4147–4156.

81. Ittrich H, Peldschus K, Raabe N, Kaul M, Adam G. Superparamagnetic iron oxide nanoparticles in biomedicine: Applications and developments in diagnostics and therapy. *Rofo.* 185, **2013**, 1149–1166.

82. Amstad E, Textor M, Reimhult E. Stabilization and functionalization of iron oxide nanoparticles for biomedical applications. *Nanoscale.* 3, **2011**, 2819–2843.

83. Iv M, Telischak N, Feng D, Holdsworth SJ, Yeom KW, Daldrup-Link HE. Clinical applications of iron oxide nanoparticles for magnetic resonance imaging of brain tumors. *Nanomedicine (Lond).* 10, **2015**, 993–1018.

84. Lee HY, Li Z, Chen K, Hsu AR, Xu C, Xie J, et al. PET/MRI dual-modality tumor imaging using arginine-glycine-aspartic (RGD)-conjugated

radiolabeled iron oxide nanoparticles. *J Nucl Med.* 49, **2008**, 1371–1379.

85. Yang X, Hong H, Grailer JJ, Rowland IJ, Javadi A, Hurley SA, Xiao Y, Yang Y, Zhang Y, Nickles RJ, Cai W, Steeber DA, Gong S. cRGD-functionalized, dox-conjugated, and ^{64}Cu-labeled superparamagnetic iron oxide nanoparticles for targeted anticancer drug delivery and PET/MR imaging. *Biomaterials.* 32, **2011**, 4151–4160.

86. Wong RM, Gilbert DA, Liu K, Louie AY. Rapid size-controlled synthesis of dextran-coated, ^{64}Cu -doped iron oxide nanoparticles. *ACS Nano.* 6, **2012**, 3461–3467.

87. DeNardo SJ, DeNardo GL, Miers LA, Natarajan A, Foreman AR, Gruettner C, Adamson GN, Ivkov R. Development of tumor targeting bioprobes (^{111}In-chimeric L6 monoclonal antibody nanoparticles) for alternating magnetic field cancer therapy. *Clin. Cancer. Res.* 11, **2005**, 7087s–7092s.

88. Zhou M, Zhang R, Huang M, Lu W, Song S, Melancon MP, Tian M, Liang D, Li C. A chelator-free multifunctional [64Cu]CuS nanoparticle platform for simultaneous micro-PET/CT imaging and photothermal ablation therapy. *J Am Chem Soc.* 132, **2010**, 15351–15358.

89. Zhou M, Zhao J, Tian M, Song S, Zhang R, Gupta S, Tan D, Shen H, Ferrari M, Li C. Radiophotothermal therapy mediated by a single compartment nanoplatform depletes tumor initiating cells and reduces lung metastasis in the orthotopic 4T1 breast tumor model. *Nanoscale.* 7, **2015**, 19438–19447.

90. Zhou M, Chen Y, Adachi M, Wen X, Erwin B, Mawlawi O, Lai SY, Li C. Single agent nanoparticle for radiotherapy and radio-photothermal therapy in anaplastic thyroid cancer. *Biomaterials.* 57, **2015**, 41–49.

91. Zhou M, Melancon M, Stafford RJ, Li J, Nick AM, Tian M, Sood AK, Li C. Precision nanomedicine using dual PET and MR temperature imaging guided photothermal therapy. *J Nucl Med.* 57, **2016**, 1778–1783.

92. Wang Z, Huang P, Jacobson O, Wang Z, Liu Y, Lin L, Lin J, Lu N, Zhang H, Tian R, Niu G, Liu G, Chen X. Biomineralization-inspired synthesis of copper sulfide-ferritin nanocages as cancer theranostics. *ACS Nano.* 10, **2016**, 3453–3460.

93. Chen F, Hong H, Shi S, Goel S, Valdovinos HF, Hernandez R, Theuer CP, Barnhaart TE, Cai W. Engineering of hollow mesoporous silica nanoparticles for remarkably enhanced tumor active targeting efficacy. *Sci Rep.* 4, **2014**, 5080.

94. Chakravarty R, Goel S, Hong H, Chen F, Valdovinos HF, Hernandez R, Barnhaart TE, Cai W. Hollow mesoporous silica nanoparticles for tumor vasculature targeting and pet image-guided drug delivery. *Nanomedicine (Lond)*. 10, **2015**, 1233–1246.

95. Miller L, Winter G, Baur B, Witulla B, Solbach C, Reske S, Lindén M. Synthesis, characterization, and biodistribution of multiple ^{89}Zr labeled pore-expanded mesoporous silica nanoparticles for PET. *Nanoscale*. 6, **2014**, 4928–4935.

96. Shaffer TM, Wall MA, Harmsen S, Longo VA, Drain CM, Kircher MF, Grimm J. Silica nanoparticles as substrates for chelator-free labelling of oxophilic radioisotopes. *Nano Lett.* 15, **2015**, 864–868.

97. Shaffer TM, Harmsen S, Khwaja E, Kircher MF, Drain CM, Grimm J. Stable radiolabeling of sulfur-functionalized silica nanoparticles with copper-64. *Nano Lett.* 16, **2016**, 5601–5604.

98. Goel S, Chen F, Luan S, Valdovinos HF, Shi S, Graves SA, Ai F, Barnhaart TE, Theuer CP, Cai W. Engineering intrinsically Zirconium-89 radiolabeled self-destructing mesoporous silica nanostructures for in vivo biodistribution and tumor targeting studies. *Adv Sci (Weinh)*. 3, **2016**, 1600122.

99. Nahrendorf M, Zhang H, Hembrador S, Panizzi P, Sosnovik DE, Aikawa E, Libby P, Swirski FK, Weissleder R. Nanoparticle PET-CT imaging of macrophages in inflammatory atherosclerosis. *Circulation*. 117, **2008**, 379–387.

100. Lee HY, Li Z, Chen K, Hsu AR, Xu C, Xie J, Sun S, Chen X. PET/MRI dual-modality tumor imaging using arginine-glycine-aspartic (RGD)-conjugated radiolabeled iron oxide nanoparticles. *J. Nucl. Med.* 49, **2008**, 1371–1379.

101. Wehrl HF, Judenhofer MS, Wiehr S, Pichler BJ. Preclinical PET/MR: technological advances and new perspectives in biomedical research. *Eur. J. Nucl. Med. Mol. Imaging*. 36, **2009**, S56–S68.

102. Kennel SJ, Woodward JD, Rondinone AJ, Wall J, Huang Y, Mirzadeh S. The fate of MAb-targeted Cd(125m)Te/ZnS nanoparticles in vivo. *Nucl. Med. Biol.* 35, **2008**, 501–514.

103. Cai W, Chen K, Li ZB, Gambhir SS, Chen X. Dual-function probe for PET and near-infrared fluorescence imaging of tumor vasculature. *J. Nucl. Med.* 48, **2007**, 1862–1870.

104. Zhou J, Yu M, Sun Y, Zhang X, Zhu X, Wu Z, Wu Z, Wu D, Li F. Fluorine-18-labeled $Gd^{3+}/Yb^{3+}/Er^{3+}$ co-doped $NaYF_4$ nanophosphors for multimodality PET/MR/UCL imaging. *Biomaterials*. 32, **2011**, 1148–1156.

105. Seo HJ, Nam SH, Im HJ, Park JY, Lee JY, Yoo B, Lee YS, Jeong JM, Hyeon T, Kim JW, Lee JS, Jang IJ, Cho JY,Hwang DW, Suh YD, Lee DS. Rapid hepatobiliary excretion of micelle-encapsulated/radiolabeled upconverting nanoparticles as an integrated form. *Sci Rep.* 5, **2015**, 15685.

106. Cui X, Mathe D, Kovacs N, Horvath I, Jauregui-Osoro M, Torres Martin De Rosales R, et al. Synthesis, characterization, and application of core-shell $Co_{0.16}Fe_{2.84}O_4$@$NaYF_4$(Yb, Er) and Fe_3O_4@ $NaYF_4$(Yb, Tm) nanoparticle as trimodal (MRI, PET/SPECT, and Optical) imaging agents. *Bioconjug Chem.* 27, **2016**, 319–328.

107. Rieffel J, Chen F, Kim J, Chen G, Shao W, Shao S,Chitqupi U, Hernandez R, Graves SA, Nickles RJ, Prasad PN, Kim C, Cai W, Lovell JF. Hexamodal imaging with porphyrin-phospholipid-coated upconversion nanoparticles. *Adv Mater.* 27, **2015**, 1785–1790.

108. Forte E, Fiorenza D, Torino E, Costagliola di Polidoro A, Cavaliere C, Netti PA, Salvatore M, Aiello M. Radiolabeled PET/MRI nanoparticles for tumor imaging. *J. Clin. Med.* 9, **2020**, 89.

109. Yang CT, Ghosh KK, Padmanabhan P, Langer O, Liu J, Eng DNC, Halldin C, Gulyás B. PET-MR and SPECT-MR multimodality probes: Development and challenges. *Theranostics.* 8, **2018**, 6210–6232.

110. Thakare V, Tran VL, Natuzzi M, Thomas E, Moreau M, Romieu A, Collin B, Courteau A, Vrigneaud JM, Louis C, Roux S, Boschetti F, Tillement O, Lux F, Denat F. Functionalization of theranostic AGuIX® nanoparticles for PET/MRI/optical imaging. *RSC Adv.* 9, **2019**, 24811–24815.

111. Bouziotis P, Stellas D, Thomas E, Truillet C, Tsoukalas C, Lux F, Tsotakos T, Xanthopoulos S, Paravatou-Petsotas M, Gaitanis A, Moulopoulos LA, Koutoulidis V, Anagnostopoulos CD, Tillement O. ^{68}Ga-radiolabeled AGuIX nanoparticles as dual-modality imaging agents for PET/MRI-guided radiation therapy. *Nanomedicine, Future Medicine,* 12, **2017**, 1561–1574.

112. Zhao L, Wen S, Zhu M, Li D, Xing Y, Shen M, Shi X, Zhao J. ^{99m}Tc-labelled multifunctional polyethylenimine-entrapped gold nanoparticles for dual mode SPECT and CT imaging. *Artificial Cells, Nanomedicine, and Biotechnology.* 46, **2018**, 488–498.

113. Farzin L, Sheibani S, Moassesi ME, Shamsipur M. An overview of nanoscale radionuclides and radiolabeled nanomaterials commonly used for nuclear molecular imaging and therapeutic functions. *J Biomed Mater Res Part A.* 107A, **2019**, 251–285.

114. Thomas G, Boudon J, Maurizi L, Moreau M, Walker P, Severin I, Oudot A, Goze C, Poty S, Vrigneaud JM, Demoisson F, Denat F, Brunotte F, Millot N. Innovative magnetic nanoparticles for PET/MRI bimodal imaging. *ACS Omega*. 4, **2019**, 2637–2648.

115. Sun H, Zhang B, Jiang X, Liu H, Deng S, Li Z, Shi H. Radiolabeled ultra-small Fe_3O_4 nanoprobes for tumor-targeted multimodal imaging. *Nanomedicine*. 14, **2019**, 5–17.

Chapter 4

Magnetic Nanoparticles for Cancer Diagnosis

R. G. Aswathy[a] and D. Sakthi Kumar[b]

[a]*Future Industries Institute, Mawson Lakes Campus,*
University of South Australia,
Adelaide, South Australia 5095, Australia
[b]*Bio Nano Electronics Research Center,*
Graduate School of Interdisciplinary New Science,
Toyo University, Kawagoe, Saitama 350-8585, Japan

aswathy.ravindrangirija@unisa.edu.au, aswathyrg@gmail.com, sakthi@toyo.jp

4.1 Introduction

Magnetic nanoparticles (MNPs) have immense potential in diverse scientific disciplines, including magnetic fluids, magnetic resonance imaging (MRI), biotechnology/biomedicine, catalysis, data storage, environmental remediation, etc. [1–11]. Due to their exclusive physical properties, MNPs are a major group of nanoparticles (NPs) with the potential to revolutionize existing clinical diagnostic and therapeutic systems. With a wide range of applications in the detection, diagnosis, and treatment of several illnesses, such as cancer, cardiovascular disease, and neurological disease, MNPs have evidenced its noteworthy role in the healthcare sector [12, 13].

Bionanotechnology in Cancer: Diagnosis and Therapy
Edited by D. Sakthi Kumar and Aswathy Ravindran Girija

ISBN 978-981-4800-30-3 (Hardcover), 978-0-429-42291-1 (eBook)
www.jennystanford.com

MNPs with different chemical structures and compositions have been developed for bio applications to study and analyze magnetic potentials in nano regime, such as superparamagnetic behavior and improved magnetic moments. Various structural and chemical factors, such as size, morphology, and surface chemistry, are modified to enhance magnetic properties and performance of NPs for in vivo applications [14, 15]. Thus, for biomedical applications, MNPs that are generally used are composed of an inorganic NP as core with surface coating to enhance biocompatibity to stipulate equilibrium under physiological conditions.

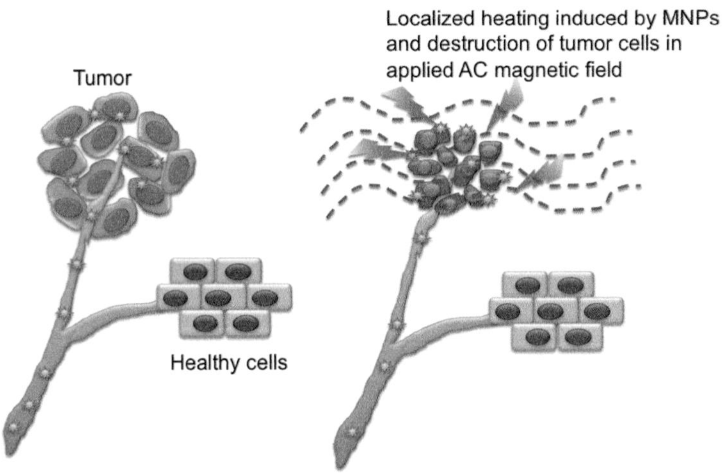

Figure 4.1 Effect of magnetic hyperthermia by using MNPs in tumor.

The introduction of appropriate surface chemistry to the NPs permits the incorporation of functional ligands (targeting ligands) [16–18] for offering a well-regulated mode of targeting. Since the nanomaterials exhibit magnetic behavior, they can be influenced by an external magnetic field gradient. The inherent penetrability of magnetic fields into human tissue can be regulated via 'action at a distance' that unlocks several applications, such as immobilization or transport of MNPs, or magnetically labeled biological stuffs including cells or ligands. The MNPs can be customized to function resonating to a time-dependent magnetic field, which results in the energy transfer from the excited field to the NPs. Also the MNPs can be used to heat up, as magnetic hyperthermia agents, distributing thermal energy to targeted bodies. For example, in the case of tumors along with chemotherapy and radiotherapy enhancement agents, application of MNPs induces a reasonable degree of tissue warming that results in more efficient malignant cell extermination (Figure 4.1). All these exceptional properties enable MNPs to execute multiple functions synergistically, such as drug delivery, multimodal imaging, real-time monitoring, hyperthermia, and combined therapeutic methods.

Due to water-rich system in in vivo conditions, the hydrodynamic size turns to be significant than the size of iron oxide (IO) NP in the dry conditions. Therefore, colloidal IO NPs having superior biocompatibility are classified based on their hydrodynamic size, as small particles of iron oxide (SPIO) and ultra-small particles of iron oxide (USPIO). MNPs with hydrodynamic size greater than 40 nm are termed as SPIO contrast agents, and hydrodynamic size lesser than 40 nm are known as USPIO contrast agents. The uptake of SPIOs is rapid due to reticuloendothelial system (RES) and accumulates in the spleen or liver. Therefore, SPIOs are commonly used for detection of lesions and tumors in the liver. On the other hand, USPIO, due to their small size, does not build up in the RES and shows longer blood circulating time. USPIOs are useful for the detection of metastases in the lymph nodes. Hypointense signals were observed in lymph nodes when USPIOs of less than 10 nm were administrated intravenously [19–27]. Metal-doped IO NPs are another category of MNPs that are used for biomedical applications, owing to superior magnetic properties. The metal ferrites are of MFe_2O_4, where M is +2 cations of metals from Mn, Fe, Co, or Ni, which are doped to specifically tune their magnetic properties [28]. The metallic MNPs, composed of iron, cobalt, or nickel, are used for biomedical applications, owing to their exceptional magnetic properties. Metallic MNPs are normally secured by coatings, such as PLGA/PEG/silica, to form a core/shell structure because they readily form oxides in the presence of oxygen and water. Some metallic NPs, like iron NPs with high magnetization, could sustain superparamagnetism at bigger particle size when compared to the oxides. Therefore, research is in progress on metallic NPs with magnetic properties to the exclusive advantages that some of them can be used for biological applications especially in diagnosis of cancer cells.

4.2 Principle of MRI

MRI is one among the potential non-invasive imaging modalities employed in clinical medical applications [29–31]. It offers high spatial resolution, superior soft tissue contrast, tomographic capabilities, and real-time monitoring approach with non-radiation hazard. However, reduced sensitivity for locating anatomical changes and precise early detection of tumor are some of the limitations associated with MRI. MNP as a contrast agent for MRI could address the limitations of conventional MRI.

MRI is the diagnostic modality based on principle of nuclear magnetic resonance (NMR). It quantifies the signal from the hydrogen nuclei of water, which will be changed by the adjacent chemical environment. NMR offers data on chemical shifts and coupling constants, in that way MRI offers spatial distribution of the intensity of the water proton signal in the volume of the body. It envisages contrasts in images based on examining the response of water protons to external magnetic field. Mainly the signal intensity depends on three factors: (a) the density of proton spins in a given volume, (b) the longitudinal and transverse relaxation times T_1 and T_2 of these proton spins, and (c) it can be weighted with respect to T_1 or T_2 using different radio frequency (RF) pulse sequences. MRI exploits the exceptional property of atomic nuclei rotating in a strong magnetic field. These nuclei possess special 'resonance' frequency that depends on the magnetic field. The energy of the nucleus can be augmented by absorbing radio waves of the same frequency. As they return to the lower energy state, radio waves are re-emitted by the nuclei and the time taken for the radio wave to do the same known as the 'relaxation time'. Different relaxation times result in varying bright and dark spots on the image.

In clinical diagnosis, when a strong magnetic field (1.5 T or 3 T) is applied, the magnetic moments of protons present in the sample align parallel with the direction of external magnetic field, thus inducing longitudinal magnetization. When RF pulse of resonating frequency (5–100 MHz) is applied, protons in the sample rotate the magnetic moments away from the longitudinal axis, excite resonantly and align antiparallel with external magnetic field, causing transverse magnetization and reducing the longitudinal magnetization. When RF pulse is removed, the excited protons relax to equilibrium (ground state) and align parallel with external magnetic field by discharging the energy attained from the RF pulse [32]. As mentioned earlier, the time that magnetic moments require to return back to the initial orientation in the presence of external magnetic field is termed as the relaxation time, and it is dependent on the type of tissue for imaging. The relaxation occurs through two different independent processes: (i) T_1 longitudinal relaxation, and (ii) T_2 transverse relaxation.

T_1 longitudinal relaxation (spin-lattice relaxation) signifies the speed of magnetization that is parallel to static magnetic field recovery after inducing a disruption. Rapidly relaxing protons (short T_1) recover complete magnetization along the longitudinal axis and yield high intensity signal. Complete magnetization on the longitudinal axis is not redeemed for protons that relax slowly (long T_1) before successive RF pulses, and generally produce a low intensity signals [32]. T_2 transverse relaxation (spin-spin relaxation) signifies how fast magnetization in the plane perpendicular to static magnetic field losses the phase coherence. When RF pulse is applied, proton nuclei spin in phase and after the pulse, the magnetic fields of all the nuclei interact and the energy is exchanged. The nuclei lose their phase coherence and spin in random manner [33].

In MRI, the relaxation process is fast and repeated in rapid succession of RF pulses. The longitudinal and transverse relaxation routes are accomplished independently. MRI records the relaxation events and recreates into gray scale images. MRI images are classified into two imaging modes, T_1-weighted images and T_2-weighted images. The T_1-weighted MRI images with fast T_1 relaxation rate (R_1) contribute bright contrast, and T_2-weighted MRI images with fast T_2 relaxation rate (R_2) produces dark contrast. In soft tissues, the endogenous MRI contrast results from local alterations of proton density originating from the concentration of water in the tissue that subsequently differs in the values of T_1 and T_2. The sensitivity can be improved by increasing the magnetic field or by using exogenous contrast agents [34].

4.3 MNPs as MRI Contrast Agents

Depending on relaxivity coefficients (r_1 or r_2), MRI contrast agents are being classified into T_1 or T_2. The important parameter to determine the efficacy of contrast agent is the relaxivity. The relaxivity of MRI contrast agent is the increase in the relaxation rate of the solvent (water) induced by active ion of the contrast agent and dependent on the magnetic field applied and temperature. The contrast agents in MRI reduce the relaxation time of water protons in the region of interest (ROI) and offer enhanced contrast. Contrast is enhanced depending on the tissues; some tissues may have higher affinity for contrast agents. Uptake of contrast agents is different in diseased cells and healthy cells, owing to their cellular metabolism. For fine structural image for anatomy information, T_1-weighed images are preferred. The T_1-weighed images are often preferred for pathological studies. The r_2/r_1 ratio is another parameter to specify the efficiency of contrast. When the ratio is high, the efficiency of a T_2 contrast agent is high and produces hypointense signal where organs are dark in the image. The lower r_2/r_1 describes to T_1 contrast agents, where images appear to be brighter and clearer (Figure 4.2). Owing to the inherent magnetic properties and possibility for tuning the magnetism to optimize r_2 and r_1 values, MNPs function as either T_2 or T_1 MRI contrast agents. Currently available T_1 contrast agents fall under paramagnetic category and most of the T_2 contrast agents are superparamagnetic iron oxides [35].

MRI contrast effects are greatly influenced by the size and composition of MNPs. T_1 contrast effect is enhanced by the reduction of the sizes of the NPs. The T_1 contrast effect (r_1) of 3 nm IO NPs was 4.77 $mM^{-1}s^{-1}$, which was two times greater than the 12 nm NPs with r_1 of 2.37 $mM^{-1}s^{-1}$. The extremely small-sized IO NPs exhibited sensitivity to acquire MRI images of blood vessels with small diameter of 0.2 mm, signifying their application in blood pool MR angiography [36]. In contrast, T_2 relaxivity is greatly influenced by the increase in size of NPs. R_2 value increases as the size of NPs increases. IO NPs of different sizes (4–6, 9, and 12 nm) were investigated to study size effects on the T_2 contrast enhancement [37]. The saturation magnetization and T_2 relaxivity coefficient (r_2) increased as the sizes of NPs increased. The saturation magnetization values were 25–43, 80, and 102 emu for 4–6, 9, and 12 nm sized MNPs.

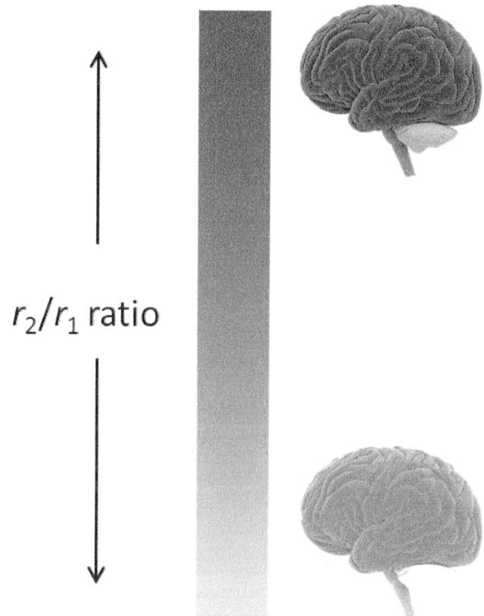

Figure 4.2 Effect of r_2/r_1 relation on contrast agents. High r_2/r_1 values of T_2 contrast agents (generally superparamagnetic iron oxides) produce hypointense signals (dark images) and low r_2/r_1 values of T_1 contrast agents (generally paramagnetic iron oxides) produce bright images.

In addition to size, the composition of MNPs is another parameter that affects MRI contrast. Transition metal dopants are often introduced to tune the magnetic properties. T_1 contrast effect is enhanced by doping rare-earth metals (e.g., Gadolinium (Gd)) into ferrite NPs. Gadolinium doping influence the spin orientation by affecting surface spins and disrupting interior spins resulting in a low r_2/r_1 value, thereby enhancing T_1 contrast effects. Gadolinium iron oxide (GdIO) MNPs of 5 nm exhibit higher T_1 contrast effects (r_1 = 7.85 $mM^{-1}s^{-1}$ (Fe + Gd)) than IO NPs (r_1 = 6.14 $mM^{-1}s^{-1}$) of the similar size [38]. The effect of metal doping on T_2 contrast effects was studied by doping different transitional metals such as Ni^{2+}, Co^{2+}, and Mn^{2+} by replacing $Fe^{2+}$of 12 nm NPs. The saturation magnetization values for MFe_2O_4, where M = Ni, Co, Fe, and Mn, were 85, 99, 101, and 110 emu g^{-1} (Fe +M) and r_2 values of 152, 172, 218, and 358 $mM^{-1}s^{-1}$, respectively. Mn-doped NPs exhibited the highest r_2 value and was used for imaging small tumor xenograft. The saturation magnetization values can be influenced by introducing non-magnetic dopant to metal ferrite NPs. When Zn^{2+} was used as dopant, the saturation magnetization and r_2 values were 161 emu g^{-1} (Fe +Zn) and 687 $mM^{-1}s^{-1}$, respectively [39].

Presently employed MRI contrast agents are mostly paramagnetic complexes or MNPs [40, 41]. Gadolinium (Gd^{3+}) or manganese (Mn^{2+}) chelates are the commonly used paramagnetic complexes that hasten longitudinal (T_1) relaxation of water protons to offer bright contrast in areas where they localize. For example, gadolinium diethylenetriamine pentaacetate (Gd-DTPA) has been commonly used paramagnetic complexes that are mainly engaged for the detection of damage in blood–brain barrier (BBB) and variations in flow dynamics, vascularity, and perfusion.

Gadolinium chelate-based contrast agents are currently used and they are limited, owing to edema that surrounds the tumor tissue and diffuse from the tumor vasculature. On the other hand, contrast agents based on MNPs present extended demarcation of tumor margins owing to improved cellular internalization and take comparatively long time for clearance from the tumor site [42, 43]. Blood pool imaging with MNPs necessitates contrast agents to stay in blood for adequate time for acquisition of data during the radio imaging process. The NPs below 5 nm are usually fast equilibrated between the NPs administered intravenously and the extracellular space, owing to the average 5 nm pore size of mammalian blood vessels. Particles of larger size upon proper surface modification could impart extended blood half-life and reduced renal clearance; however, transport across the endothelium is remarkably slow [44].

IO NPs are the widely studied MNP-based contrast agents for MRI. The major drawbacks of MNPs are the necessity of lots of particles with high saturation magnetization values to attain adequate contrast. The efficacy of IO NPs as MRI contrast agent depends mostly on the physicochemical features, including their size, surface chemistry, etc. The spin-spin relaxation time (T_2) is dependent on particle size, and particles of large size demonstrate large T_2 effect. The intrinsic magnetization of the NPs is significant and alloys or metallic NPs are better than IO while considering the saturation magnetization of the latter.

IO NPs are employed as T_2 contrast agents, owing to their T_2-relaxation effects, which lead to reduction in the signal on T_2-weighted images. Both passive (non-targeted) and active (targeted) mode imaging are being practiced for in vivo applications. MRI detects phagocytes in inflammatory and degenerative diseases, including atherosclerosis, stroke, kidney disorders, oscteoarticular infection, and multiple sclerosis, and rejection of cardiac graft, by employing RES directed USPIOs. Another use of USPIO MNPs is in improving the detection of brain tumor boundaries and computing tumor volumes [42, 45–48]. However, their application requires further improvement in functionalization and biostabilization to reduce their inherent toxicity. The negative contrast and magnetic susceptibility of IO NPs are also considered as substantial limitations. Efficient capping of the NPs with non-antigenic, nonimmunogenic, and protein-resistant coatings of biocompatible materials, such as polyethylene glycol (PEG), could help greatly in averting aggregation, improving biocompatibility, blood retention, enabling cellular uptake of NPs, and providing binding sites for targeting ligands. For the potential diagnostic applications of MRI and to overcome the above-mentioned shortcomings of T_1 and T_2 contrast agents, the demand for an alternate class of contrast agent that fulfills the criteria, such as (a) positive (T_1) contrast ability, (b) effective intracellular uptake and accumulation for monitoring cellular biodistribution, (c) surface modification and labeling with targeting ligands for imaging (molecular and cellular), and (d) promising pharmacokinetics, biodistribution, and harmless clearance from patients with negligible side effects, is required.

MRI-based active targeting for detecting disorders such as early tumors has garnered much attention in research over the past several years. The target-specific delivery of NPs in active imaging can make use of targeting molecules, such as protein, peptide, and folic acid, which are mostly covalently bonded to IO particles [49–66]. Monoclonal antibodies are covalently conjugated to IO NPs to form magnetic IO particle-antibody probes to provide in vivo tumor detection. Fe_3O_4-herceptin probe was widely used for the in vivo imaging of breast cancer [67–69].

Although the application of antibody for tumor targeting is considered to be appropriate, the antibody NPs conjugate suffers from several disadvantages, such as large hydrodynamic size and poor diffusion and uptake by RES. Therefore, small targeting moieties are gaining more attention. Arginine-glycine-aspartic acid (RGD) [70–73] peptides are often coupled with IO NPs for the detection of various cancers, including breast cancer, squamous cell carcinomas, and malignant melanomas. Owing to the multivalent binding property, binding efficacy of IO particles to the RGD peptide can be improved. Thus, more numbers of RGD peptides have been attached to the IO particles, due to their very small size and multivalent binding property [74]. Another functional peptide employed was Chlorotoxin (36 amino acid peptide), concurrently used for both molecular imaging by MRI and tumor therapeutics [75]. Adhesion molecule vascular cell adhesion protein 1 (VCAM-1) was recognized as a target for macrophage and endothelial cells that may lead to atherosclerosis. VCAM-1 targeted peptide was conjugated with MNPs and MRI contrast enhancement of early lesions in juvenile mice and resected human carotid artery plaques was observed [76]. Studies of in vitro cellular targeting experiments suggest that Fe_3O_4-folic acid (FA) nanoconjugate binds specifically to tumor cells [66]. Tumor detection studies with SPIO-PEG-FA as MRI contrast agent demonstrated signal intensity difference in positive in human nasopharyngeal epidermal carcinoma cells (KB cells) tumors (around 20–25%), which was considerably less than the negative HT-1080 cell tumor from pre-contrast to post-contrast image of the tumor in in vivo MR imaging.

4.4 Dual-Mode (T_1 and T_2) MRI Contrast Agents

Conventional NPs functioning as MRI contrast agents are commonly effective in a single imaging mode either generating bright (T_1)

or dark (T_2) signal enhancement. The signal artifacts such as dark or bright signals that arise from endogenous factors, including hemorrhage, clots, and fat, in the MR images often lead to ambiguities in diagnostics [77]. The combination of T_1 and T_2 contrast effects into a single contrast agent that could offer accurate MRI through self-confirmation is effective for superior differentiation of healthy and diseased tissues, thus eliminating the potential ambiguity associated with single-mode contrast agent with in vivo artifacts [78].

The design and development of single nanostructure with T_1–T_2 dual-mode contrast agents are often challenging. The integration of T_1 (Gd- or Mn-based chelates) and T_2 components (metal ferrite MNPs) by preserving the inherent imaging potential is one of the reported methods. It can be attained either by the covalent conjugation strategy or by embedding T_1 material to T_2 NPs (Figure 4.3a). Gd–DTPA (Magnevists), a T_1 MRI contrast agent, was covalently attached to dopamine-coated IO NPs covalent conjugation chemistry [79]. The resultant Gd-labeled magnetite NPs exhibited T_1 and T_2 contrasts, with an r_1 value of 11.17 $mM^{-1}s^{-1}$ (Gd) and an r_2 value of 30.32 $mM^{-1}s^{-1}$ (Fe), and were used for in vivo MRI. In the second approach, in conjugated system, T_1 paramagnetic elements were embedded into T_2 MNPs (Figure 4.3b) [80, 81]. Mn was embedded into IO NPs and the resultant nanoconjugate exhibited r_1 and r_2 values of 18 $mM^{-1}s^{-1}$ and 45.9 $mM^{-1}s^{-1}$ (Fe+Mn), respectively [81]. In another study, Gd_2O_3-embedded IO NPs exhibited r_1 and r_2 values of 69.5 of $mM^{-1}s^{-1}$ (Gd) and 146.5 $mM^{-1}s^{-1}$ (Fe), respectively, and were used for T_1–T_2 dual-modal MRI of hepatic tumor in mice [80].

Another strategy for the advancement of T_1–T_2 contrast agent is magnetic decoupling. During the synthetic procedure, a non-magnetic separating layer such as SiO_2 can be introduced between the T_2 contrast MNPs in the core and the T_1 contrast material in the shell (Figure 4.3c). An example is the development of core shell T_1–T_2 dual-contrast agent, where T_1 contrast material was $Gd_2O(CO_3)_2$ of 1.5 nm positioned on the shell for the direct contact with water molecules to offer enhanced T_1 contrast. Superparamagnetic T_2 contrast component $MnFe_2O_4$ of 15 nm was used as the core structure for inducing magnetic field for the relaxation of water molecules at a longer-range. The distance between the core and shell was controlled by a SiO_2

layer. The thickness of silica layer was tuned and a thickness of 16 nm offered both high T_1 and T_2 contrast that was better than contrast generated by single mode [82].

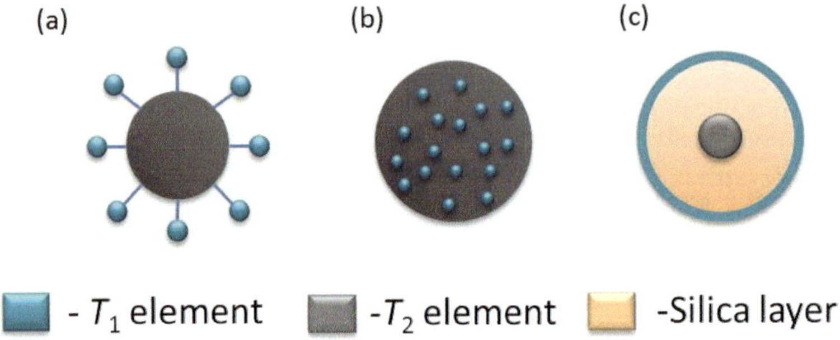

Figure 4.3 Schematic representation of different types of T_1–T_2 dual-mode contrast agents. (a) T_1 element labeled T_2 nanoparticle via covalent conjugation, (b) T_1 element embedded T_2 nanoparticle, and (c) magnetically decoupled core shell T_1–T_2 dual-mode contrast agents (DMCA) separated with silica layer.

4.5 Multimodal Imaging: MRI with PET, SPECT, Optical, and Ultrasound

Integration of different imaging modalities is the appropriate strategy to complement the drawbacks associated with single imaging modality. The low sensitivity associated with MRI can be improved by integrating other imaging (PET, SPECT, optical, and US) modalities to generate superior contrast agents for imaging. The combination of MRI- and radioisotope-based imaging offers highly sensitive and tomographic images with superior resolution [83–87]. Commonly used radionuclide probes as radiotracers or therapeutic agents are ^{64}Cu, ^{18}F, ^{67}Ga, ^{68}Ga, ^{124}I, ^{111}In, ^{177}Lu, ^{99m}Tc, ^{90}Y, and ^{89}Zr [88–92]. In general, gamma-rays emitted from ^{64}Cu, ^{18}F, ^{68}Ga, and ^{124}I are used for PET and ^{131}I, ^{111}In, and ^{99m}Tc are employed for SPECT imaging. The combination of these traces along with MNPs offers dual-modal imaging with either PET or SPECT along with MRI. The dual-modal imaging offers precise quantitative biological information, including in vivo biodistribution, deep tissue imaging, pharmacokinetics, etc.

MNPs are often integrated with radioisotopes for PET/SPECT via chemical conjugation strategies. Serum albumin coated on Mn-doped ferrite NPs are conjugated to ^{124}I through amino acid residues of the coated protein and dual MRI–PET imaging probe exhibited signals in both MRI and PET images [85]. The integrated MRI/PET images were capable of visualizing small sentinel lymph nodes. Chelating agents, such as DOTA (1,4,7,10-tetraazacyclododecane-N, N0, N00, N000-tetraacetic acid) and bifunctional chelators bisphosphonate (BP) and dithiocarbamate (DTC) that chelate the radioisotope and bind to the surfaces of MNPs, are employed for the design of dual-modal imaging agents, and the resultant imaging agents offer superior high-resolution imaging capabilities of tumor, lymph nodes, liver, etc. [84, 87].

Encapsulation technique is also another technique adapted for the development of integrated imaging agents. PET/MRI dual-modality imaging probe with IO NPs and ^{68}Ga using specially designed functional amphiphiles was developed by encapsulation method for targeted lymph nodes imaging [93]. Superparamagnetic iron oxide nanoparticle (SPION) with negative MRI contrast was radiolabeled with ^{67}Ga (SPECT radioisotope) or with ^{64}Cu (PET radioisotope) at high temperature via ligand-free radiolabeling. Stable biocompatible multimodal imaging probe for PET/SPECT-MRI was developed by using sterically stabilized SPIONs with diblock copolymers with methoxypolyethylene glycol (MPEG) or primary amine NH_2 end groups [94]. Design of physiologically stable radiolabeled MNPs with a tunable functional end group on the SPIONs facilitates multimodal PET/SPECT-MRI agent on T_2 contrast MRI probe.

The combination of optical imaging probes (UV/visible–absorbance and/or fluorescence) with MRI is one of the most explored dual-modality imaging combination. Optical imaging (OI) and MRI have unique associations, owing to their pro and cons that compensate each other. Remarkable advantages of OI include high sensitivity, superior temporal resolution, and possibility of multichannel imaging by integrating various fluorescent probes. However, image data is surface-weighted quantitative with any quantitative information. A combination of the benefits of MRI and OI and the compensation of their drawbacks facilitate the development of novel multimodal imaging system. NPs with inherent optical properties, such as quantum dots (QDs), upconversion NPs, gold, etc., or NPs loaded/labeled with fluorophore are coupled either chemically or physically [95–99] with MRI contrast agent, allowing MRI and optical fluorescent signals for imaging applications [100–106]. IO MNPs, owing to the inherent magnetic properties, are the commonly used T_2-weighted contrast agents for in vitro and in vivo studies [107–110].

MRI-optical probes are designed by conjugating fluorophores to polymer-coated MNPs [98, 99]. Although these integrated nanoprobes demonstrated superior nanomolar level sensitivity in vivo, the fluorescence quenching that arises from energy transfer between the fluorophore and MNPs limits its clinical application. The distance between the fluorophores and the NPs plays a crucial role in fluorescent quenching. However, the distance can be tuned by introducing silica layer or long chemical linkers [111].

Hybrid 'core–satellite' nanostructures of dye-doped silica NPs (core) decorated with multiple IO (Fe_3O_4) NPs (satellite) are developed [95]. The core–satellite dual imaging nano probe exhibited a 3.4 times enhancement in the T_2 MRI signal and 1.7 times increase in fluorescence signal when compared with dye-conjugated Fe_3O_4 NPs.

Organosilica capable to functionalize molecules, such as fluorescent dye and magnetic nanomaterials, has potential application in in vitro and in vivo imaging, and single cell analysis [112–119]. Organosilica NPs incorporating MNPs and rhodamine B demonstrated shortened T_2 relaxation time, originating from MNPs and longer excitation wavelength of the fluorophore, rhodamine B [119]. Hepatic tumor tissue using multimodal imaging with integrated fluorescent imaging (FI) and T_1- and T_2-weighted MRI using thiol modified organosilica NPs incorporated with MNPs and rhodamine (thiol-OS-MNP/Rho) was demonstrated [120]. The in vivo MRI exposed tumors in the liver and semi-quantitatively, owing to a labeling of macrophages in liver. The FI displayed irregular fluorescent patterns of the liver at the mesoscopic level, which was intermediate between macroscopic and microscopic levels. Mesoscopic imaging of ex vivo pathological liver on MRI and FI demonstrated "differential" multimodal imaging from bright field imaging and provided information on the peripheral distribution of labeled cells in invisible pathological area from MRI. However, histological FI demonstrated distribution of labeled cells in line with pathological results with mesoscopic FI. Thus, imaging using thiol-OS-MNP/Rho offered continuous and sequential "identical" multimodal imaging of pathological liver tissue. The study demonstrated the difference between the bright field imaging and that of MRI, and emphasized the both "identical" and "differential" multimodal imaging, which are very useful to evaluate pathological regions.

Apart from fluorescent dyes, optically active components, such as QDs, upconversion NPs (UPCN), are integrated with MNPs to offer OI along with MRI. Owing to large extinction coefficients and tunable absorption properties, QDs offer high fluorescence signal with excellent photostability. Near infrared (NIR) emitting QDs or UPCNs are preferred because they can be used in the biological diagnostic window (about 650–950 nm) to offer deep tissue imaging without interfering the absorbance from endogenous factors, including blood and tissue fluids. Integration of MNPs with UCNPs of lanthanide-doped rare-earth elements with enhanced photostability, narrow emission peak, high detection sensitivity, potential for deep tissue imaging, and low autofluorescence, has been reported for OI and MRI [121, 122]. Gd^{3+} doped UCNP for UCL/MRI is a promising nanoplatform for future theranostic (Therapy + diagnosis) applications. Doping Tm^{3+} ions in UCNP is efficient for imaging, owing to the NIR-to-NIR upconversion process that enables high contrast for deep tissue imaging [123–127]. However, more research directed toward the design and development of nanpocomposites based on UCNPs and MNPs is essential for superior MRI sensitivity, surface modification, and for efficient in vivo targeted imaging.

The design of nanoprobes for multimodal bio imaging with gold-IO (plasmonic-magnetic) NPs, owing to the exceptional features of IO and gold nanomaterials, has been reported [128]. The NPs performed as contrast agents for diverse imaging modalities including X-ray computed tomography (CT), T_2-weighted nuclear MRI, photoacoustic imaging (PA), dark-field and bright-field optical microscopy, and surface enhanced Raman spectroscopy (SERS). The complementary non-invasive imaging techniques permitted visualization of cellular uptake with artifact-free mode, multiplexing proficiencies, and with information on the fate particles inside the cells.

4.6 Novel Application of MNPs for Imaging

Owing to the inherent magnetic properties, MNPs have been explored in various emerging magnetic-based multi-modal imaging modalities. It includes (a) magnetic particle imaging (MPI), (b) magneto-motive ultrasound (MMUS) imaging, and (c) magneto-photoacoustic (MPA) imaging [129–131]. MPI is explored in combination with MRI, whereas MMUS and MPA are used in conjunction with ultrasound (US) imaging. In the above mentioned imaging modalities, the inherent magnetic behavior is explored and hence the integration of optical or radioactive component is completely avoided. Although these magnetic particles-based novel imaging modalities are in their infancy, comprehensive information on each of these modalities will be briefly discussed.

4.6.1 Magnetic Particle Imaging–MRI

MPI enables 3D imaging with superior sensitivity, spatial and temporal resolution without any hazardous ionizing radiation, nephrotoxic (iodine based), or nephrogenic systemic fibrosis (gadolinium based) contrast agents, and the used SPIONs are removed via the iron metabolism. In MPI, the signal intensity is directly proportional to the amount of MNPs and hence the imaging offers quantitative data. Real-time imaging is possible and no background signals are generated, owing to the diamagnetic tissues and organs. Fast and precise location of NPs is possible with the combination of MPI and MRI with high resolution anatomical image.

In general, when an external magnetic field (H) is present, the magnetic moment of the MNPs parallelly aligns toward the direction of the external magnetic field, leading to magnetization vector (M), which gets saturated, as external magnetic field increases [129, 132, 133]. The nonlinear relationship between

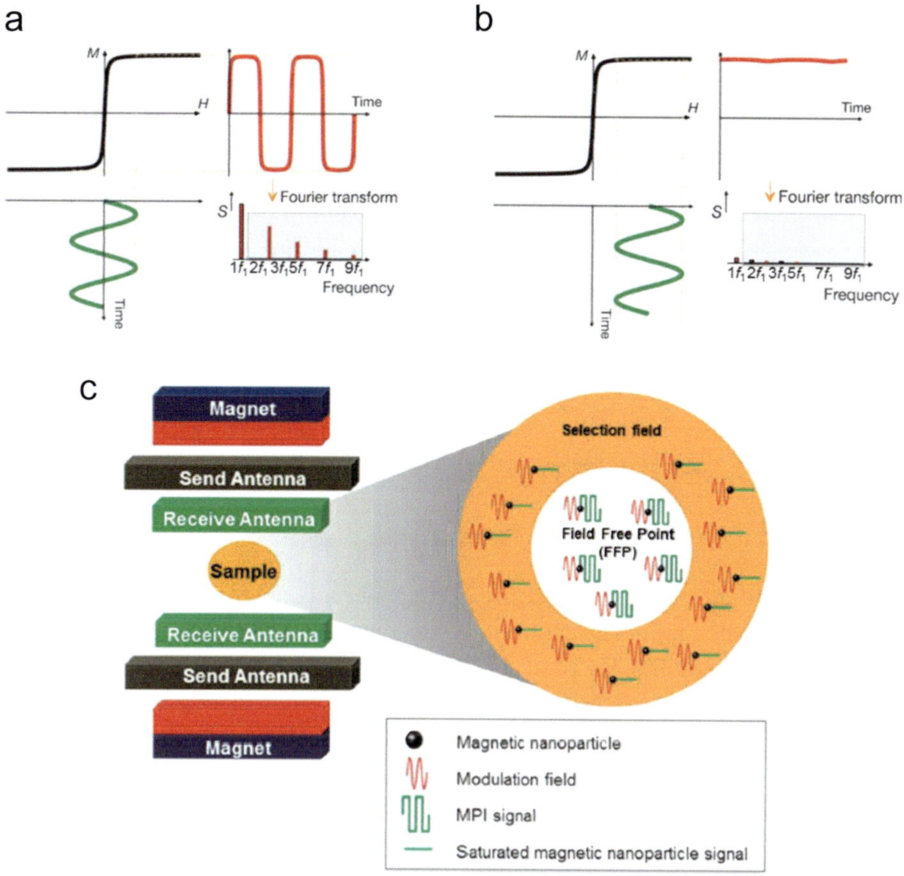

Figure 4.4 Basic concept of MPI. (a) Response of MNPs within the FFP with excitation frequency f and higher harmonics, (b) schematic of FFP where MNPs inside and near to the non-saturated areas respond to the excitation field, and (c) SPIONs present external to FFP are magnetically saturated and do not significantly respond to the excitation field. Reprinted with permission from Ref. [129, 133].

magnetic field and magnetization vector is exploited in MPI. A time-dependent magnetization M(t) is observed with a drive frequency (f_1) along with a sequence of harmonic frequencies (Figure 4.4a) [129]. When strong external magnetic field is applied, MNPs get magnetically saturated and result in the suppression of the generation of harmonic frequency (Figure 4.4b). Depending on this effect, the MPI signal can be acquired in a spatially selective mode (Figure 4.4c) [133]. A spatial selectivity was attained by an extra strong static magnetic field (or selection field) that could saturate the magnetization effect of MNPs. However, at the center of the selection field, magnetic field will be zero, which is the field-free point (FFP) and the MNPs that are present in FFP produce an MPI signal. FFP is subjected to raster scanning until all point in the field-of-view to generate MPI image [132]. The properties, such as resolution and sensitivity of MPI, are dependent on the physicochemical properties of the MNPs and their application as MPI tracer is in immature stage. The early studies demonstrated the effect of size of the MNPs, clustering effects, compositions, and surface chemistry on the quality of the MPI image and magnetic particle imaging–MRI (MPI–MRI) dual-modal imaging [132, 134–136].

4.6.2 MMUS–US Dual-Modal Imaging

In MMUS imaging modality, the inherent magnetic property of MNPs induced motion is examined using US imaging technique [130]. MNPs are subjected to a focused high intensity pulsed magnetic field and their induced motion or displacement is used to analyze the distribution of the MNPs. Owing to the diamagnetic behavior of biological components, they remain silent in the presence of an external magnetic field. However, MNPs respond to magnetic fields by moving toward the magnetic field and during the displacement they undergo tissue resistance force against magnetically generated magneto-motive force (Figure 4.5). These two forces induce vibrations, which are detected using US-based techniques. In dual MMUS and US imaging modalities, MNPs are visualized in MMUS and US and provide the anatomical details. Similar to MPI–MRI imaging, MMUS is in its initial phase of development, which has huge potential for dynamic imaging in the cellular and molecular levels, to monitor viscoelastic property of soft tissues to determine abnormality [137].

A recent study proposed MMUS imaging with NP-based contrast agent, which could be employed for pre-operative PET/MRI imaging [138]. Multifunctional nanoparticle (^{68}Ga-labeled SPION) was used as contrast agent PET and MRI for pre-operative imaging and the same nanoparticle was used for MMUS as an intra-operative guide, for visualization of sentinel lymph nodes. Intra-operative sentinel lymph nodes detection could be completed a number of days after the administration of NPs even when the radioactivity has diminished. In short, during multimodal imaging, PET/CT images can be obtained in an hour after the administration of NPs then the same can be used by MMUS imaging few days later (two or four days) and then MRI can also performed to confirm SPIONs.

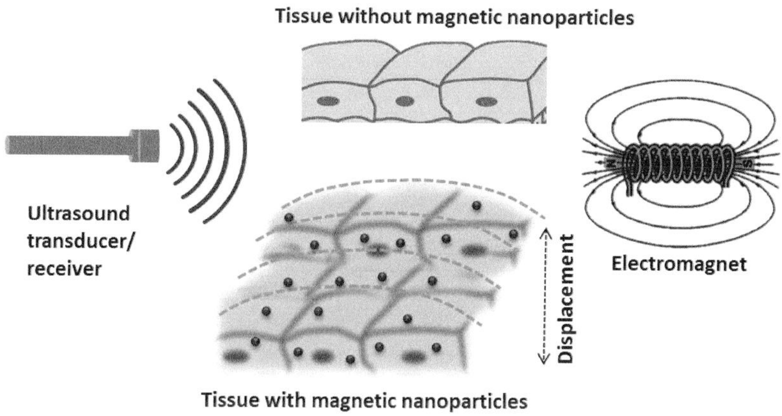

Figure 4.5 Schematic representation of MMUS–US dual-modal imaging depicting displacement of the tissue through magnetic induction with MNPs generating MMUS signal sensed by an US receiver.

4.6.3 MPA–US Dual-Modal Imaging

Although photoacoustic (PA) and US (PA–US) dual imaging are efficient imaging modalities, the artifacts associated with photoacoustic background signals from endogenous photo-absorbers present in tissues remain as a drawback (Figure 4.6). The combination of PA with MNPs and MPA, is a recently developed imaging modality to improve the superiority of PA–US dual imaging [131]. During the acquisition of PA image, when a pulsed external magnetic field was applied, MNPs vibrated while endogenous photo-absorbers (non-magnetic) did not respond. PA signals generated from the NPs could be differentiated from the background signal by detecting the motion of the NPs

(Figure 4.6) [131]. MNP-aided PA imaging was demonstrated to capture circulating tumor cells in the bloodstream of mice and further detected in PA imaging [139]. The capture of cells and PA signal enhancement capabilities of MNPs could be exploited for the detection of circulating tumor cells that function as key biomarkers in some cancer.

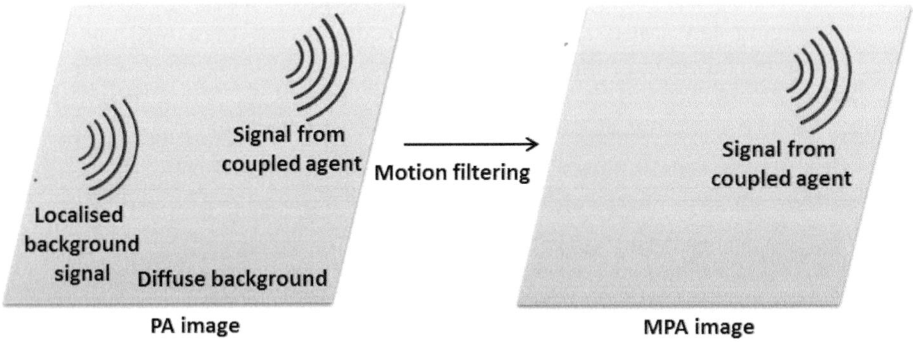

Figure 4.6 Schematic of elimination of background signal through motion filtering process where MNP-based contrast agent respond to pulsed magnetic field and is differentiated from background signals of photoacoustic absorbers.

4.7 Conclusion and Future Prospects

Magnetic IO NPs have been widely studied as MRI contrast agents, owing to their short T_2 relaxation times, reduced uptake, and accumulation in RES cells. IO particle-based MRI contrasts are classified as SPION and USPION based on their particle sizes. The synthesized IO NPs are subjected to various surface-modification methods, such as ligand exchange, encapsulation with biocompatible polymeric shells to render hydrophilicity and biocompatibility. MNPs used for targeted imaging are commonly functionalized with ligands, such as antibodies, peptides, and oligonucleotides, which helped the site-specific accumulation of NPs and targets organs/tumors. The development of theranostic NPs is garnering great attention, owing to the simultaneous multimodal imaging and therapy through the combinations of several nanomaterials [140–145]. Although the research in the development of MRI contrast agents is gaining considerable significance, most of them are still in the preliminary animal studies. Several issues associated with toxicity, stability, and pharmacokinetics of MRI contrast agents are to be addressed before being used extensively in laboratory for diagnosis. The development of dual T_1–T_2 MRI contrast agents and the emerging imaging technologies, including MPI, MMUS, and MPA, offer new platform for enhanced sensitive and precise imaging along with better understanding of biological systems. Also, the interdisciplinary research in the development of MRI contrast agents with active targeting can be ultimately used for the design of personalized diagnosis.

References

1. Chikazumi S, Taketomi S, Ukita M, Mizukami M, Miyajima H, Setogawa M, Kurihara Y. Physics of magnetic fluids. *J. Magn. Magn. Mater.* 65, **1987**, 245–251.
2. Li Z, Wei L, Gao MY, Lei H. One-pot reaction to synthesize biocompatible magnetite NPs. *Adv. Mater.* 17, **2005**, 1001–1005.
3. Gupta AK, Gupta M. Synthesis and surface engineering of iron oxide NPs for biomedical applications. *Biomaterials.* 26, **2005**, 3995–4021.
4. Lu AH, Schmidt W, Matoussevitch N, Bönnemann H, Spliethoff B, Tesche B, Bill E, Kiefer W, Schüth F. Nanoengineering of a magnetically separable hydrogenation catalyst. *Angew. Chem. Int. Ed.* 43, **2004**, 4303–4306.
5. Hyeon T. Chemical synthesis of magnetic NPs. *Chem. Commun.* 34, **2003**, 927–934.
6. Takafuji M, Ide S, Ihara H, Xu Z. Preparation of poly(1-vinylimidazole)-grafted magnetic NPs and their application for removal of metal ions. *Chem. Mater.* 16, **2004**, 1977–1983.
7. Ferrari M. Cancer nanotechnology: Opportunities and challenges. *Nat. Rev. Cancer.* 5, **2005**, 161–171.
8. Hwu JR, Lin YS, Josephrajan T, Hsu M, Cheng F, Yeh CS, Su WC, Shieh DB. Targeted paclitaxel by conjugation to iron oxide and gold NPs. *J. Am. Chem. Soc.* 131, **2009**, 66–68.
9. Zhang J, Rana S, Srivastava RS, Misra RDK. On the chemical synthesis and drug delivery response of folate receptor-activated, polyethylene glycol-functionalized magnetite NPs. *Acta Biomater.* 4, **2008**, 40–48.

10. Sonvico F, Mornet S, Vasseur S, Dubernet C, Jaillard D, Degrouard J, Hoebeke J, Duguet E, Colombo P, Couvreur P. Folate-conjugated iron oxide NPs for solid tumor targeting as potential specific magnetic hyperthermia mediators: Synthesis, physicochemical characterization, and in vitro experiments. *Bioconju. Chem.* 16, **2005**, 1181–1188.

11. Wu PC, Wang WS, Huang YT, Sheu HS, Lo YW, Tsai TL, Shieh DB, Yeh CS. Porous iron oxide-based nanorods developed as delivery nanocapsules. *Chem. Eur. J.* 13, **2007**, 3878–3885.

12. Wickline SA, Neubauer AM, Winter PM, Caruthers SD, Lanza GM. Molecular imaging and therapy of atherosclerosis with targeted NPs. *J. Magn. Reson. Imaging.* 25, **2007**, 667–680.

13. Corot C, Petry KG, Trivedi R, Saleh A, Jonkmanns C, Le Bas JF, Blezer E, Rausch M, Brochet B, Foster-Gareau P, Balériaux D, Gaillard S, Dousset V. Macrophage imaging in central nervous system and in carotid atherosclerotic plaque using ultrasmall superparamagnetic iron oxide in magnetic resonance imaging. *Invest. Radiol.* 39, **2004**, 619–625.

14. Tartaj P, Morales MD, Veintemillas-Verdaguer S, Gonzalez-Carreno T, Serna CJ. The preparation of magnetic NPs for applications in biomedicine. *J. Phys. D: Appl. Phys.* 36, **2003**, R182–R197.

15. Gupta AK, Gupta M. Synthesis and surface engineering of iron oxide nanoparticles for biomedical applications. *Biomaterials.* 26, **2005**, 3995–4021.

16. Sivakumar B, Aswathy RG, Nagaoka Y, Iwai S, Suzuki M, Venugopal K, Kato K, Yoshida Y, Maekawa T, Sakthikumar DN. Aptamer targeted theragnostic multifunctional magnetic NPs as nanoplatform for pancreatic cancer therapy. *RSC Adv.* 3, **2013**, 20579–20598.

17. Balasubramanian S, Girija AR, Nagaoka Y, Iwai S, Suzuki M, Kizhikkilot V, Yoshida Y, Maekawa T, Nair SD. Curcumin and 5-FU loaded, folate and transferrin decorated polymeric magnetic nanoformulation: A synergistic cancer therapeutic approach, accelerated by magnetic hyperthermia. *Int. J. Nanomedicine.* 9, **2014**, 437–459.

18. Sivakumar B, Aswathy RG, Nagaoka Y, Suzuki M, Fukuda T, Yoshida Y, Maekawa T, Sakthikumar DN. Multifunctional carboxymethyl cellulose-based magnetic nanovector as a theragnostic system for folate receptor targeted chemotherapy, imaging, and hyperthermia against cancer. *Langmuir.* 29, **2013**, 3453–3466.

19. Benderbous S, Corot C, Jacobs P, Bonnemain B. Superparamagnetic agents: Physicochemical characteristics and preclinical imaging evaluation. *Acad Radiol.* 3 (Suppl 2), **1996**, S292–S294.

20. Elias A, Tsourkas A. Imaging circulating cells and lymphoid tissues with iron oxide NPs. *Hematology Am Soc Hematol Educ Program*. 2009, **2009**, 720–726.

21. Huang P, Li Z, Lin J, Yang D, Gao G, Xu C, Bao L, Zhang C, Wang K, Song H Hu H, Cui D. Photosensitizer-conjugated magnetic NPs for in vivo simultaneous magnetofluorescent imaging and targeting therapy. *Biomaterials*. 32, **2011**, 3447–3458.

22. Wunderbaldinger P, Josephson L, Weissleder R. Crosslinked iron oxides (CLIO): A new platform for the development of targeted MR contrast agents. *Acad Radiol*. 9 (Suppl 2), **2002**, S304–S306.

23. Di Marco M, Sadun C, Port M, Guilbert I, Couvreur P, Dubernet C. Physicochemical characterization of ultrasmall superparamagnetic iron oxide particles (USPIO) for biomedical application as MRI contrast agents. *Int. J. Nanomedicine*. 2, **2007**, 609–622.

24. Mahmoudi M, Sahraian MA, Shokrgozar MA, Laurent S. Superparamagnetic iron oxide NPs: Promises for diagnosis and treatment of multiple sclerosis. *ACS Chem Neurosci*. 2, **2011**, 118–140.

25. Li M, Kim HS, Tian L, Yu MK, Jon S, Moon WK. Comparison of two ultra-small superparamagnetic iron oxides on cytotoxicity and MR imaging of tumors. *Theranostics*. 2, **2012**, 76–85.

26. Pultrum BB, van der Jagt EJ, van Westreenen HL, van Dullemen HM, Kappert P, Groen H, Sietsma J, Oudkerk M, Plukker JT, van Dam GM. Detection of lymph node metastases with ultrasmall superparamagnetic iron oxide (USPIO)-enhanced magnetic resonance imaging in oesophageal cancer: A feasibility study. *Cancer Imaging*. 9, **2009**, 19–28.

27. Wu L, Cao Y, Liao C, Huang J, Gao F. Diagnostic performance of USPIO-enhanced MRI for lymph-node metastases in different body regions: A meta-analysis. *Eur. J. Radiol*. 80, **2011**, 582–589.

28. Willard MA, Kurihara LK, Carpenter EE, Calvin S, Harris VG. Chemically prepared magnetic NPs. *Int. Mater. Rev*. 49, **2004**, 125–170.

29. Felton C, Karmakar A, Gartia Y, Ramidi P, Biris AS, Ghosh A. Magnetic NPs as contrast agents in biomedical imaging: recent advances in iron- and manganese-based magnetic NPs. *Drug Metab. Rev*. 46, **2014**, 142–154.

30. Chaughule RS, Purushotham S, Ramanujan RV. Magnetic NPs as contrast agents for magnetic resonance imaging. *Proc. Natl. Acad. Sci., India, Sect. A Phys. Sci*. 82, **2012**, 257–268.

31. Li L, Jiang W, Luo K, Song H, Lan F, Wu Y, Gu Z. Superparamagnetic iron oxide NPs as MRI contrast agents for non-invasive stem cell labeling and tracking. *Theranostics.* 3, **2013**, 595–615.

32. Stephen ZR, Kievit FM, Zhang, M. Magnetite NPs for medical MR imaging. *Mater Today.* 14, **2011**, 330–338.

33. Mornet S, Vasseur S, Grasset F, Duguet E. Magnetic nanoparticle design for medical diagnosis and therapy. *J Mater Chem.*14, **2004**, 2161–2175.

34. Lam T, Pouliot P, Avti PK, Lesage F, Kakkar AK. Superparamagnetic iron oxide based nanoprobes for imaging and theranostics. *Adv Colloid Interface Sci.* 199–200, **2013**, 95–113.

35. Laurent S, Vander Elst L, Roch A, Muller RN. Structure, synthesis and characterization of contrast agents for magnetic resonance molecular imaging. Carreta P, Lascialfari A (eds.). NMR–MRI, SR and Mössbauer Spectroscopies in Molecular Magnets. Milan, Italy, Springer, **2007**, 71–87.

36. Kim BH, Lee N, Kim H, An K, Park YI, Choi Y, Shin K, Lee Y, Kwon SG, Na HB, Park JG, Ahn TY, Kim YW, Moon WK, Choi SH, Hyeon T. Large-scale synthesis of uniform and extremely small-sized iron oxide NPs for high-resolution T1 magnetic resonance imaging contrast agents. *J. Am. Chem. Soc.* 133, **2011**, 12624–12631.

37. Jun YW, Huh YM, Choi JS, Lee JH, Song HT, Kim S, Yoon S, Kim KS, Shin JS, Suh JS, Cheon J. Nanoscale size effect of magnetic nanocrystals and their utilization for cancer diagnosis via magnetic resonance imaging. *J. Am. Chem. Soc.* 127, **2005**, 5732–5733.

38. Zhou Z, Wang L, Chi X, Bao J, Yang L, Zhao W, Chen Z, Wang X, Chen X, Gao J. Engineered iron-oxide-based NPs as enhanced T1 contrast agents for efficient tumor imaging. *ACS Nano.* 7, **2013**, 3287–3296.

39. Jang JT, Nah H, Lee JH, Moon SH, Kim MG, Cheon J. Critical enhancements of MRI contrast and hyperthermic effects by dopant-controlled magnetic NPs. *Angew. Chem., Int. Ed.* 48, **2009**, 1234–1238.

40. Caravan P, Ellison JJ, McMurry TJ, Lauffer RB. Gadolinium(III) chelates as MRI contrast agents: Structure, dynamics, and applications. *Chem. Rev.* 99, **1999**, 2293–2352.

41. Weinmann HJ, Ebert W, Misselwitz B, Schmitt-Willich H. Tissue-specific MR contrast agents. *Eur. J. Radiol.* 46, **2003**, 33–44.

42. Enochs WS, Harsh G, Hochberg F, Weissleder R. Improved delineation of human brain tumors on MR images using a long-circulating,

superparamagnetic iron oxide agent. *J. Magn. Reson. Imaging.* 9, **1999**, 228–232.

43. Varallyay P, Nesbit G, Muldoon LL, Nixon RR, Delashaw J, Cohen JI, Petrillo A, Rink D, Neuwelt EA. Comparison of two superparamagnetic viral-sized iron oxide particles ferumoxides and ferumoxtran-10 with a gadolinium chelate in imaging intracranial tumors. *AJNR. Am. J. Neuroradiol.* 23, **2002**, 510–519.
44. Bremerich J, Bilecen D, Reimer P. MR angiography with blood pool contrast agents. *Eur. Radiol.* 17, **2007**, 3017–3024.
45. Gibson, R.E. Molecular imaging: Principles and practice. *Mol. Imaging Biol.* 13, **2011**, 594.
46. Weinstein JS, Varallyay CG, Dosa E, Gahramanov S, Hamilton B, Rooney WD, Muldoon LL, Neuwelt EA. Superparamagnetic iron oxide NPs: Diagnostic magnetic resonance imaging and potential therapeutic applications in neurooncology and central nervous system inflammatory pathologies, a review. *J. Cereb. Blood Flow Metab.* 30, **2010**, 15–35.
47. Lee JH, Huh YM, Jun YW, Seo JW, Jang JT, Song HT, Kim S, Cho EJ, Yoon HG, Suh JS, Cheon JW. Artificially engineered magnetic NPs for ultra-sensitive molecular imaging. *Nat. Med.* 13, **2007**, 95–99.
48. Neuwelt EA, Várallyay P, Bagó AG, Muldoon LL, Nesbit G, Nixon R. Imaging of iron oxide NPs by MR and light microscopy in patients with malignant brain tumours. *Neuropathol. Appl. Neurobiol.* 30, **2004**, 456–471.
49. Josephson L, Tung C-H, Moore A, Weissleder R. High-efficiency intracellular magnetic labeling with novel superparamagnetic-tat peptide conjugates. *Bioconj. Chem.* 10, **1999**, 186–191.
50. Schellenberger E, Rudloff F, Warmuth C, Taupitz M, Hamm B, Schnorr J. Protease-specific nanosensors for magnetic resonance imaging. *Bioconj. Chem.* 19, **2008**, 2440–2445.
51. Xie J, Chen K, Lee H, Xu C, Hsu AR, Peng S, Chen X, Sun S. Ultrasmall c(RGDyK)-coated Fe_3O_4 NPs and their specific targeting to integrin αvβ3-rich tumor cells. *J. Am. Chem. Soc.* 130, **2008**, 7542–7543.
52. Puertas S, Batalla P, Moros M, Polo E, Pino P, Guisán JM, Grazú V, Fuente JM. Taking advantage of unspecific interactions to produce highly active magnetic nanoparticle-antibody conjugates. *ACS Nano.* 5, **2011**, 4521–4528.
53. Xu H, Aguilar ZP, Yang L, Kuang M, Duan H, Xiong Y, Wei H, Wang A. Antibody conjugated magnetic iron oxide NPs for cancer cell separation in fresh whole blood. *Biomaterials.* 32, **2011**, 9758–9765.

54. Veiseh O, Sun C, Fang C, Bhattarai N, Gunn J, Kievit F, Du K, Pullar B, Lee D, Ellenbogen RG, Olson J, Zhang M. Specific targeting of brain tumors with an optical/magnetic resonance imaging nanoprobe across the blood-brain barrier. *Cancer Res.* 69, **2009**, 6200–6207.

55. Montet X, Montet-Abou K, Reynolds F, Weissleder R, Josephson L. Nanoparticle imaging of integrins on tumor cells. *Neoplasia.* 8, **2006**, 214–222.

56. Guthi JS, Yang SG, Huang G, Li S, Khemtong C, Kessinger CW, Peyton M, Minna JD, Brown KC, Gao J. MRI-visible micellar nanomedicine for targeted drug delivery to lung cancer cells. *Mol. Pharm.* 7, **2010**, 32–40.

57. Simberg D, Duza T, Park JH, Essler M, Pilch J, Zhang L, Derfus AM, Yang M, Hoffman RM, Bhatia S, Sailor MJ, Ruoslahti E. Biomimetic amplification of nanoparticle homing to tumors. *Proc. Natl. Acad. Sci. USA.* 104, **2007**, 932–936.

58. Martin AL, Hickey JL, Ablack AL, Lewis JD, Luyt LG, Gillies ER. Synthesis of bombesin-functionalized iron oxide NPs and their specific uptake in prostate cancer cells. *J. Nanopart. Res.* 12, **2009**, 1599–1608.

59. Zhang Y, Yang M, Park JH, Singelyn J, Ma H, Sailor MJ, Ruoslahti E, Ozkan M, Ozkan C. A surface-charge study on cellular-uptake behavior of F3-peptide-conjugated iron oxide NPs. *Small.* 5, **2009**, 1990–1996.

60. Reddy GR, Bhojani MS, McConville P, Moody J, Moffat BA, Hall DE, Kim G, Koo YE, Woolliscroft MJ, Sugai JV, Johnson TD, Philbert MA, Kopelman R, Rehemtulla A, Ross BD. Vascular targeted NPs for imaging and treatment of brain tumors. *Clin. Cancer Res.* 12, **2006**, 6677–6686.

61. Jiang J, Gan Z, Yang, Y. Du B, Qian M, Zhang P. A novel magnetic fluid based on starch-coated magnetite NPs functionalized with homing peptide. *J. Nanopart. Res.* 11, **2009**, 1321–1330.

62. Meng J, Fan J, Galiana G, Branca RT, Clasen PL, Ma S, Zhou J, Leuschner C, Kumar CSSR, Hormes J, Otiti T, Beye AC, Harmer MP, Kiely CJ, Warren W, Haataja MP, Soboyejo WO. LHRH-functionalized superparamagnetic iron oxide NPs for breast cancer targeting and contrast enhancement in MRI. *Mater. Sci. Eng.* 29, **2009**, 1467–1479.

63. John R, Rezaeipoor R, Adie SG, Chaney EJ, Oldenburg AL, Marjanovic M, Haldar JP, Sutton BP, Boppart SA. In vivo magnetomotive optical molecular imaging using targeted magnetic nanoprobes. *Proc. Natl. Acad. Sci. USA.* 107, **2010**, 8085–8090.

64. Hadjipanayis CG, Machaidze R, Kaluzova M, Wang L, Schuette AJ, Chen H, Wu X, Mao H. EGFRvIII antibody-conjugated iron oxide NPs for magnetic resonance imaging-guided convection-enhanced delivery and targeted therapy of glioblastoma. *Cancer Res.* 70, **2010**, 6303–6312.

65. Cho YS, Yoon TJ, Jang ES, Soo Hong K, Young Lee S, Ran Kim O, Park C, Kim YJ, Yi GC, Chang K. Cetuximab-conjugated magneto-fluorescent silica NPs for in vivo colon cancer targeting and imaging. *Cancer Lett.* 299, **2010**, 63–71.

66. Chen TJ, Cheng TH, Hung YC, Lin KT, Liu GC, Wang YM. Targeted folic acid-PEG NPs for noninvasive imaging of folate receptor by MRI. *J. Biomed. Mater. Res A.* 87, **2008**, 165–175.

67. Huh Y, Jun Y, Song H, Kim S, Choi J, Lee J, Yoon S, Kim K, Shin J, Suh J, Cheon J. In vivo magnetic resonance detection of cancer by using multifunctional magnetic nanocrystals. *J. Am. Chem. Soc.* 127, **2005**, 12387–12391.

68. Gupta J, Mohapatra J, Bhargava P, Bahadur D. A pH-responsive folate conjugated magnetic nanoparticle for targeted chemo-thermal therapy and MRI diagnosis. *Dalton Trans.* 45, **2016**, 2454–2461.

69. Arosio P, Orsini F, Piras AM, Sandreschi S, Chiellini F, Corti M, Masa M, Múčková M, Schmidtová Ľ, Ravagli C, Baldi G, Nicolato E, Conti G, Marzola P, Lascialfari A. MR imaging and targeting of human breast cancer cells with folate decorated NPs. *RSC Adv.* 5, **2015**, 39760–39770.

70. Munshi N, De TK, Maitra A. Size modulation of polymeric NPs under controlled dynamics of microemulsion droplets. *J. Colloid Interface Sci.* 190, **1997**, 387–391.

71. Xie J, Chen K, Lee HY, Xu C, Hsu AR, Peng S, Chen X, Sun S. Ultrasmall c(RGDyK)-coated Fe_3O_4 NPs and their specific targeting to integrin alpha(v)beta3-rich tumor cells. *J. Am. Chem. Soc.* 130, **2008**, 7542–7543.

72. Xu J, Yang HB, Fu WY, Du K, Sui YM, Chen JJ, Zeng Y, Li MH, Zou G. Preparation and magnetic properties of magnetite NPs by sol–gel method. *J. Magn. Magn. Mater.* 309, **2007**, 307–311.

73. Sunderland CJ, Steiert M, Talmadge JE, Derfus AM, Barry SE. Targeted NPs for detecting and treating cancer. *Drug Dev. Res.* 67, **2006**, 70–93.

74. Montet X, Montet-Abou K, Reynolds F, Weissleder R, Josephson L. Nanoparticle imaging of integrins on tumor cells. *Neoplasia.* 8, **2006**, 214–222.

75. Sun C, Veiseh O, Gunn J, Fang C, Hansen S, Lee D, Sze R, Ellenbogen RG, Olson J, Zhang M. In vivo MRI detection of gliomas by chlorotoxin-conjugated superparamagnetic nanoprobes. *Small.* 4, **2008**, 372–379.

76. Nahrendorf M, Jaffer FA, Kelly KA, Sosnovik DE, Aikawa E, Libby P, Weissleder R. Noninvasive vascular cell adhesion molecule-1 imaging identifies inflammatory activation of cells in atherosclerosis. *Circulation*. 114, **2006**, 1504–1511.

77. Bulte WM, Kraitchman DL. Iron oxide MR contrast agents for molecular and cellular imaging. *NMR Biomed.* 17, **2004**, 484–499.

78. Yoo D, Lee JH, Shin TH, Cheon J. Theranostic magnetic NPs. *Acc Chem Res.* 44, **2011**, 863–874.

79. Bae KH, Kim YB, Lee Y, Hwang J, Park H, Park TG. Bioinspired synthesis and characterization of gadolinium-labeled magnetite NPs for dual contrast T1- and T2-weighted magnetic resonance imaging. *Bioconjugate Chem.* 21, **2010**, 505–512.

80. Zhou Z, Huang D, Bao J, Chen Q, Liu G, Chen Z, Chen X, Gao J. A synergistically enhanced T_1–T_2 dual-modal contrast agent. *Adv. Mater.* 24, **2012**, 6223–6228.

81. Huang G, Li H, Chen J, Zhao Z, Yang L, Chi X, Chen Z, Wang X, Gao J. Tunable T1 and T2 contrast abilities of manganese-engineered iron oxide NPs through size control. *Nanoscale.* 6, **2014**, 10404–10412.

82. Choi JS, Lee JH, Shin TH, Song HT, Kim EY, Cheon J. Self-confirming "AND" logic NPs for fault-free MRI. *J. Am. Chem. Soc.*132, **2010**, 11015–11017.

83. Sandiford L, Phinikaridou A, Protti A, Meszaros LK, Cui X, Yan Y, Frodsham G, Williamson PA, Gaddum N, Botnar RM, Blower PJ, Green MA de Rosales RT. Bisphosphonate-anchored PEGylation and radiolabeling of superparamagnetic iron oxide: Long-circulating NPs for in vivo multimodal (T1 MRI-SPECT) imaging. *ACS Nano.* 7, **2013**, 500–512.

84. Lee HY, Li Z, Chen K, Hsu AR, Xu C, Xie J, Sun S, Chen X. PET/MRI dual-modality tumor imaging using arginine-glycine-aspartic (RGD)-conjugated radiolabeled iron oxide NPs. *J. Nucl. Med.* 49, **2008**, 1371–1379.

85. Choi JS, Park JC, Nah H, Woo S, Oh J, Kim KM, Cheon GJ, Chang Y, Yoo J, Cheon J. A Hybrid nanoparticle probe for dual-modality positron emission tomography and magnetic resonance imaging. *Angew. Chem., Int. Ed.* 47, **2008**, 6259–6262.

86. Chen F, Ellison PA, Lewis CM, Hong H, Zhang Y, Shi S, Hernandez R, Meyerand ME, Barnhart TE, Cai W. Chelator-free synthesis of a dual-modality PET/MRI agent. *Angew. Chem., Int. Ed.* 52, **2013**, 13319–13323.

87. Torres Martin de Rosales R, Tavare´ R, Paul RL, Jauregui-Osoro M, Protti A, Glaria A, Varma G, Szanda I, Blower PJ. Synthesis of $^{64}Cu^{II}$–bis(dithiocarbamatebisphosphonate) and its conjugation with superparamagnetic iron oxide NPs: In vivo evaluation as dual-modality PET–MRI agent. *Angew. Chem., Int. Ed.* 50, **2011**, 5509–5513.

88. Torres Martin de Rosales R, Tavaré R, Glaria A, Varma G, Protti A, Blower PJ. ^{99m}Tc-bisphosphonate-iron oxide nanoparticle conjugates for dual-modality biomedical imaging. *Bioconjug Chem.* 22, **2011**, 455–465.

89. Wadas TJ, Wong EH, Weisman GR, Anderson CJ. Coordinating radiometals of copper, gallium, indium, yttrium, and zirconium for PET and SPECT imaging of disease. *Chem. Rev.* 110, **2010**, 2858–2902.

90. Werner RA, Bluemel C, Allen-Auerbach MS, Higuchi T, Herrmann K. (68)Gallium- and (90)yttrium-/(177)lutetium: "Theranostic twins" for diagnosis and treatment of NETs. *Ann Nucl Med.* 29, **2015**, 1–7.

91. Stelter L, Pinkernelle JG, Michel R, Schwartländer R, Raschzok N, Morgul MH, Koch M, Denecke T, Ruf J, Bäumler H, Jordan A, Hamm B, Sauer IM, Teichgräber U. Modification of aminosilanized superparamagnetic nanoparticles: Feasibility of multimodal detection using 3T MRI, small animal PET, and fluorescence imaging. *Mol Imaging Biol.* 12, **2010**, 25–34.

92. Thorek DL, Ulmert D, Diop NF, Lupu ME, Doran MG, Huang R, Abou DS, Larson SM, Grimm J. Non-invasive mapping of deep-tissue lymph nodes in live animals using a multimodal PET/MRI nanoparticle. *Nat Commun.* 5, **2014**, 3097.

93. Yang BY, Moon SH, Seelam SR, Jeon MJ, Lee YS, Lee DS, Chung JK, Kim YI, Jeong JM. Development of a multimodal imaging probe by encapsulating iron oxide NPs with functionalized amphiphiles for lymph node imaging. *Nanomedicine (Lond).* 10, **2015**, 1899–1910.

94. Pham TN, Lengkeek NA, Greguric I, Kim BJ, Pellegrini PA, Bickley SA, Tanudji MR, Jones SK, Hawkett BS, Pham BT. Tunable and noncytotoxic PET/SPECT-MRI multimodality imaging probes using colloidally stable ligand-free superparamagnetic iron oxide NPs. *Int. J. Nanomedicine.* 12, **2017**, 899–909.

95. Lee JH, Jun YW, Yeon SI, Shin JS, Cheon J. Dual-mode nanoparticle probes for high-performance magnetic resonance and fluorescence imaging of neuroblastoma. *Angew. Chem., Int. Ed.* 45, **2006**, 8160–8162.

96. Cheng L, Yang K, Li Y, Chen J, Wang C, Shao M, Lee ST, Liu Z. Facile preparation of multifunctional upconversion nanoprobes for multimodal imaging and dual-targeted photothermal therapy. *Angew. Chem., Int. Ed.* 50, **2011**, 7385–7390.

97. Shibu ES, Ono K, Sugino S, Nishioka A, Yasuda A, Shigeri Y, Wakida S, Sawada M, Biju V. Photouncaging NPs for MRI and fluorescence imaging in vitro and in vivo. *ACS Nano*. 7, **2013**, 9851–9859.

98. Lee H, Yu MK, Park S, Moon S, Min JJ, Jeong YY, Kang HW, Jon S. Thermally cross-linked superparamagnetic iron oxide NPs: Synthesis and application as a dual imaging probe for cancer in vivo. *J. Am. Chem. Soc.* 129, **2007**, 12739–12745.

99. Josephson L, Kircher MF, Mahmood U, Tang Y, Weissleder R. Near-infrared fluorescent NPs as combined MR/optical imaging probes. *Bioconjugate Chem*. 13, **2002**, 554–560.

100. Kircher MF, Mahmood U, King RS, Weissleder R, Josephson L. A multimodal nanoparticle for preoperative magnetic resonance imaging and intraoperative optical brain tumor delineation. *Cancer Res.* 63, **2003**, 8122–8125.

101. Pittet MJ, Swirski FK, Reynolds F, Josephson L, Weissleder R. Labeling of immune cells for in vivo imaging using magnetofluorescent NPs. *Nat Protoc.* 1, **2006**, 73–79.

102. Koo H, Huh MS, Sun IC, Yuk SH, Choi K, Kim K, Kwon IC. In vivo targeted delivery of NPs for theranosis. *Acc. Chem. Res.* 44, **2011**, 1018–1028.

103. Cormode DP, Skajaa T, van Schooneveld MM, Koole R, Jarzyna P, Lobatto ME, Calcagno C, Barazza A, Gordon RE, Zanzonico P, Fisher EA, Fayad ZA, Mulder WJ. Nanocrystal core high-density lipoproteins: a multimodality contrast agent platform. *Nano Lett.* 8, 2008, 3715–3723.

104. Devaraj NK, Keliher EJ, Thurber GM, Nahrendorf M, Weissleder R. ^{18}F labeled NPs for in vivo PET-CT imaging. *Bioconjug Chem.* 20, **2009**, 397–401.

105. Lin JJ, Chen JS, Huang SJ, Ko JH, Wang YM, Chen TL, Wang LF. Folic acid–pluronic F127 magnetic nanoparticle clusters for combined targeting, diagnosis, and therapy applications. *Biomaterials*. 30, **2009**, 5114–5124.

106. Ke JH, Lin JJ, Carey JR, Chen JS, Chen CY, Wang LF. A specific tumor-targeting magnetofluorescent nanoprobe for dual-modality molecular imaging. *Biomaterials.* 31, **2010**, 1707–1715.

107. Song HT, Jordan EK, Lewis BK, Liu W, Ganjei J, Klaunberg B, Despres D, Palmieri D, Frank JA. Rat model of metastatic breast cancer monitored by MRI at 3 tesla and bioluminescence imaging with histological correlation. *J. Transl. Med.* 7, **2009**, 88.

108. Eamegdool SS, Weible MW II, Pham BT, Hawkett BS, Grieve SM, Chan-ling T. Ultrasmall superparamagnetic iron oxide nanoparticle prelabelling of human neural precursor cells. *Biomaterials.* 35, **2014**, 5549–5564.

109. Pham BT, Jain N, Kuchel PW, Chapman BE, Bickley SA, Jones SK, Hawkett BS. The interaction of sterically stabilized magnetic NPs with fresh human red blood cells. *Int. J. Nanomedicine.* 10, **2015**, 6645–6655.

110. Lohrke J, Frenzel T, Endrikat J, Alves FC, Grist TM, Law M, Lee JM, Leiner T, Li KC, Nikolaou K, Prince MR, Schild HH, Weinreb JC, Yoshikawa K, Pietsch H. 25 years of contrast-enhanced MRI: Developments, current challenges and future perspectives. *Adv. Ther.* 33, **2016**, 1–28.

111. Ow H, Larson DR, Srivastava M, Baird BA, Webb WW, Wiesner U. Bright and stable core-shell fluorescent silica NPs. *Nano Lett.* 5, **2005**, 113–117.

112. (20–27) Nakamura M, Ozaki S, Abe M, Doi H, Matsumoto T, Ishimura K. Size-controlled synthesis, surface functionalization, and biological applications of thiol-organosilica particles. *Colloids Surf. B: Biointerfaces.* 79, **2010**, 19–26.

113. Nakamura M, Ozaki S, Abe M, Matsumoto T, Ishimura K. One-pot synthesis and characterization of dual fluorescent thiol-organosilica NPs as non-photoblinking quantum dots and their applications for biological imaging. *J. Mater. Chem.* 21, **2011**, 4689–4695.

114. Awaad A, Nakamura M, Ishimura K. Histochemical and biochemical analysis of the size-dependent nanoimmunoresponse in mouse peyer's patches using fluorescent organosilica particles. *Int. J. Nanomedicine.* 7, **2012**, 1423–1439.

115. Awaad A, Nakamura M, Ishimura K. Imaging of size dependent uptake and identification of novel pathway in mouse peyer's patches using fluorescent organosilica particles. *Nanomedicine.* 8, **2012**, 627–636.

116. Nakamura M, Awaad A, Hayashi K, Ochiai K, Ishimura K. Thiol-organosilica particles internally functionalized with propidium iodide as a multicolor fluorescence and X-ray computed tomography probe and application for non-invasive functional gastrointestinal tract imaging. *Chem. Mater.* 24, **2012**, 3772–3779.

117. Nakamura M, Miyamoto K, Hayashi K, Awaad A, Ochiai M, Ishimura K. Time-lapse fluorescence imaging and quantitative single cell and endosomal analysis of peritoneal macrophages using fluorescent organosilica NPs. *Nanomedicine.* 9, **2013**, 274–283.

118. Nakamura, M. Hayashi K, Nakano M, Kanadani T, Miyamoto K, Kori T, Horikawa K. Identification of polyethylene glycol-resistant macrophages on stealth imaging in vitro using fluorescent organosilica NPs. *ACS nano.* 9, **2015**, 1058–1071.

119. Nakamura M, Hayashi K, Kubo H, Kanadani T, Harada M, Yogo T. Relaxometric property of organosilica NPs internally functionalized with iron oxide and fluorescent dye for multimodal imaging. *J Colloid Interf. Sci.* 492, **2017**, 127–135.

120. Nakamura M, Hayashi K, Kubo H, Harada M, Izumi K. Mesoscopic multimodal imaging provides new insight to tumor tissue evaluation: An example of macrophage imaging of hepatic tumor using organosilica NPs. *Sci. Rep.* 7, **2017**, 3953.

121. Zhou J, Liu Z, Li F. Upconversion nanophosphors for small-animal imaging. *Chem Soc Rev.* 41, **2012**, 1323–1349.

122. Haase M, Schäfer H. Upconverting NPs. *Angew. Chem. Int. Ed.* 50, **2011**, 5808–5829.

123. Nyk M, Kumar R, Ohulchanskyy TY, Bergey EJ, Prasad PN. High contrast in vitro and in vivo photoluminescence bioimaging using near infrared to near infrared up-conversion in TM^{3+} and Yb^{3+} doped fluoride nanophosphors. *Nano Lett.* 8, **2008**, 3834–3838.

124. Xing H, Bu W, Zhang S, Zheng X, Li M, Chen F, He Q, Zhou L, Peng W, Hua Y, Shi J. Multifunctional nanoprobes for upconversion fluorescence, MR and CT trimodal imaging. *Biomaterials.* 33, **2012**, 1079–1089.

125. Hou Z, Li C, Ma P, Li G, Cheng Z, Peng C, Yang D, Yang P, Lin J. Electrospinning preparation and drug-delivery properties of an up-conversion luminescent porous $NaYF^4$:Yb^{3+}, Er^{3+}@silica fiber nanocomposite. *Adv. Funct. Mater.* 21, **2011**, 2356–2365.

126. Tian G, Gu Z, Zhou L, Yin W, Liu X, Yan L, Jin S, Ren W, Xing G, Li S, Zhao Y. Mn^{2+} dopant-controlled synthesis of $NaYF^4$:Yb/Er upconversion

nanoparticles for in vivo imaging and drug delivery. *Adv. Mater.* 24, **2012**, 1226–1231.

127. Wang J, Wang F, Wang C, Liu Z, Liu X. Single-band upconversion emission in lanthanide-doped KMnF(3) nanocrystals. *Angew. Chem. Int. Ed.* 50, **2011**, 10369–10372.

128. Reguera J, Jiménez de Aberasturi D, Henriksen-Lacey M, Langer J, Espinosa A, Szczupak B, Wilhelm C, Liz-Marzán LM. Janus plasmonic-magnetic gold-iron oxide NPs as contrast agents for multimodal imaging. *Nanoscale.* 9, **2017**, 9467–9480.

129. Gleich B, Weizenecker J. Tomographic imaging using the nonlinear response of magnetic particles. *Nature.* 435, **2005**, 1214–1217.

130. Mehrmohammadi M, Shin TH, Qu M, Kruizinga P, Truby RL, Lee JH, Cheon J, Emelianov SY. In vivo pulsed magneto-motive ultrasound imaging using high-performance magnetoactive contrast nanoagents. *Nanoscale.* 5, **2013**, 11179–11186.

131. Jin Y, Jia C, Huang SW, O'Donnell M, Gao X. Multifunctional NPs as coupled contrast agents. *Nat. Commun.* 1, **2010**, 41–48.

132. Goodwill PW, Saritas EU, Croft LR, Kim TN, Krishnan KM, Schaffer DV, Conolly SM. X-space MPI: Magnetic NPs for safe medical imaging. *Adv. Mater.* 24, **2012**, 3870–3877.

133. Pablico-Lansigan MH, Situ SF, Samia AC. Magnetic particle imaging: Advancements and perspectives for real-time in vivo monitoring and image-guided therapy. *Nanoscale.* 5, **2013**, 4040–4055.

134. Starmans LW, Burdinski D, Haex NP, Moonen RP, Strijkers GJ, Nicolay K, Grüll H. Iron oxide nanoparticle-micelles (ION-micelles) for sensitive (molecular) magnetic particle imaging and magnetic resonance imaging. *PLoS One.* 8, **2013**, e57335.

135. Ferguson RM, Khandhar AP, Krishnan KM. Tracer design for magnetic particle imaging. *J. Appl. Phys.* 111, **2012**, 07B318–5.

136. Eberbeck D, Wiekhorst F, Wagner S, Trahms L. How the size distribution of magnetic NPs determines their magnetic particle imaging performance. *Appl. Phys. Lett.* 98, **2011**, 182502.

137. Mehrmohammadi M, Oh J, Mallidi S, Emelianov SY. Pulsed magneto-motive ultrasound imaging using ultrasmall magnetic nanoprobes. *Mol. Imaging.* 10, **2011**, 102–110.

138. Evertsson M, Kjellman P, Cinthio M, Andersson R, Tran TA, In't Zandt R, Grafström G, Toftevall H, Fredriksson S, Ingvar C, Strand SE, Jansson T. Combined magnetomotive ultrasound, PET/CT, and MR imaging of ^{68}Ga-labelled superparamagnetic iron oxide NPs in rat sentinel lymph nodes in vivo. *Sci. Rep.* 7, **2017**, 4824.

139. Galanzha EI, Shashkov EV, Kelly T, Kim JW, Yang L, Zharov VP. In vivo magnetic enrichment and multiplex photoacoustic detection of circulating tumour cells. *Nat. Nanotechnol.* 4, **2009**, 855–860.

140. Forte E, Fiorenza D, Torino E, Costagliola di Polidoro A, Cavaliere C, Netti PA, Salvatore M, Aiello M. Radiolabeled PET/MRI nanoparticles for tumor imaging. *J. Clin. Med.* 9, **2020**, 89.

141. Thomas G, Boudon J, Maurizi L, Moreau M, Walker P, Severin I, Oudot A, Goze C, Poty S, Vrigneaud JM, Demoisson F, Denat F, Brunotte F, Millot N. Innovative magnetic nanoparticles for PET/MRI bimodal imaging. *ACS Omega.* 4, **2019**, 2637–2648.

142. Yang CT, Ghosh KK, Padmanabhan P, Langer O, Liu J, Eng DNC, Halldin C, Gulyás B. PET-MR and SPECT-MR multimodality probes: Development and challenges. *Theranostics.* 8, **2018**, 6210–6232.

143. Bouziotis P, Stellas D, Thomas E, Truillet C, Tsoukalas C, Lux F, Tsotakos T, Xanthopoulos S, Paravatou-Petsotas M, Gaitanis A, Moulopoulos LA, Koutoulidis V, Anagnostopoulos CD, Tillement O. ^{68}Ga-radiolabeled AGuIX nanoparticles as dual-modality imaging agents for PET/MRI-guided radiation therapy. *Nanomedicine, Future Medicine.* 12, **2017**, 1561–1574.

144. Sun H, Zhang B, Jiang X, Liu H, Deng S, Li Z, Shi H. Radiolabeled ultrasmall Fe_3O_4 nanoprobes for tumor-targeted multimodal imaging. *Nanomedicine.* 14, **2019**, 5–17.

145. Thakare V, Tran VL, Natuzzi M, Thomas E, Moreau M, Romieu A, Collin B, Courteau A, Vrigneaud JM, Louis C, Roux S, Boschetti F, Tillement O, Lux F, Denat F. Functionalization of theranostic AGuIX® nanoparticles for PET/MRI/optical imaging. *RSC Adv.* 9, **2019**, 24811–24815.

Chapter 5

Acoustically Reflective Nanoparticles for Tumor Diagnosis

R. G. Aswathy[a] and D. Sakthi Kumar[b]

[a]*Future Industries Institute, Mawson Lakes Campus,*
University of South Australia,
Adelaide, South Australia 5095, Australia
[b]*Bio Nano Electronics Research Center,*
Graduate School of Interdisciplinary New Science,
Toyo University, Kawagoe, Saitama 350-8585, Japan

aswathy.ravindrangirija@unisa.edu.au, aswathyrg@gmail.com, sakthi@toyo.jp

5.1 Introduction

In the medical field, medical practitioners as well as patients are always interested to use non-invasive nature of diagnosis methods to reduce complications caused by invasive natured ones. In the class of non-invasive diagnosis, ultrasound (US) imaging is the second popular clinical diagnostic modality after X-ray radiography-based imaging. Application of nanotechnology in US imaging brought many welcoming advanced level changes that would be highly beneficial in the medical field.

5.2 Ultrasound Imaging

Ultrasound is a well-established and the most commonly used topographic imaging methodology and widely accepted due to

Bionanotechnology in Cancer: Diagnosis and Therapy
Edited by D. Sakthi Kumar and Aswathy Ravindran Girija

ISBN 978-981-4800-30-3 (Hardcover), 978-0-429-42291-1 (eBook)
www.jennystanford.com

its non-invasive nature, absence of ionizing irradiation, broad diagnostic applicability, low cost, high spatial and temporal resolution (real-time examination), and easy in usage [1–4]. The visualization of flow of blood and blood vessels by power and color Doppler suggests the application of US imaging for vascular diagnosis, tumors, and other organs after transplantation [5, 6]. Also, US imaging has the potential to distinguish lesions in soft tissues, including liver, urogenital tract, etc. US waves are above the frequency heard by human ears. US imaging is based on the pulse–echo principle, with emitted and received wave's frequencies higher than 20 kHz [7]. US-based contrast is based on the reflection and refraction of sound waves through media, which depend on the sound itself, but also on the nature of the medium and density. These sound waves are reflected back by tissues leading to the creation of cross-sectional anatomical picture of what is inside the body. It is basically a soft tissue imaging methodology, because it cannot resolute interior of bone or regions of gas filled organs like lungs or bowel. The clinical use of US imaging comprises sound waves ranging from 2–3 MHz for pediatric and 5–12 MHz for adult imaging and offering spatial resolution ranging from 0.2 mm to 1 mm.

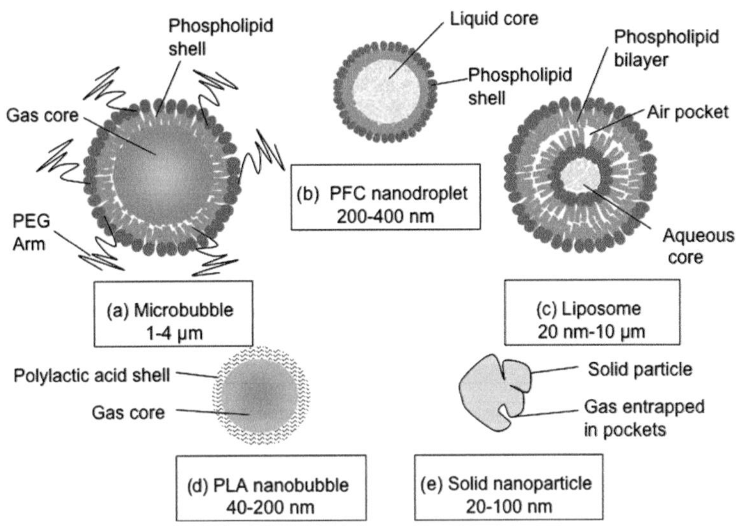

Figure 5.1 Various categories of US contrast agents: (a) MB, (b) perfluorocarbon (PFC) nanodroplets, (c) liposomes, (d) nanobubbles, (e) solid nanoparticles (NPs). Reproduced with permission from Ref. [10].

The mechanism of the contrast in US imaging is a mismatch in acoustic impedance. The contrast agents for US are elements that exhibit altered acoustic impedance relative to the surrounding biological backgrounds. Efficient application of US molecular imaging requires: (a) capability to image retained contrast with sensitivity; (b) removal of any unbound contrast from the imaged area; and (c) retention of the contrast agent by the molecular target [8]. US imaging uses contrast agents of two types: (a) microbubble (MB) and (b) non-MB-based contrast agents (Figure 5.1) [9, 10].

5.2.1 Microbubble as Contrast Agent

Microbubbles are commonly used US contrast agents in the size range of 1–4 microns and are colloids of gas-liquid emulsion enclosed by a shell [11]. The gaseous core of the MBs activates a large echogenic response when sonicated with US, resulting in elevated contrast to tissue background ratio on US images. The MB shell blocks the accumulation and leakage of gas from MBs. The MBs are produced by blending of shell components and gas components as core. Commonly used shell components include galactose, albumin, lipids, or different polymers, and gaseous components as cores include air or high-molecular-weight gases (perfluorocarbon [PFC], sulfur hexafluoride, nitrogen, etc.). The high-molecular weight gases are sparingly soluble in aqueous media and dissolute into the adjacent environment at slow rate, thus extending the lifetime of MBs [12, 13]. Upon intravenous (i.v.) administration of MBs, they are quickly cleared by the reticuloendothelial system (RES), owing to their short half-life time in serum [14–20]. The quick dissolution of gas from MBs in biological fluids, the large size of MBs, and the extent of time they stay before dissolution restricts their application as an US contrast agent. Gas bubbles of MBs needed to be adequately small, so that the bubble suspension risk of embolism can be averted upon administration. Due to their size of a few microns, MBs do not leak out into the extra-vascular space, thus remain within the vascular compartment. This property restricts their use for imaging of several diseases such as angiogenesis, inflammation, and thrombus formation [10]. To upsurge the circulation time and diminish the risk associated with merged gas bubbles, MBs are often stabilized by shell materials [21–27]. In addition to the shell, the gaseous component of MBs affects the stability and circulation time. Sparingly soluble gases (such as perfluorochemicals) have demonstrated superior circulation and stability of MBs under in vivo conditions [28]. The shell of MBs controls the degree of oscillation while insonication [29]. Thus, it is possible to differentiate soft- and hard-shelled MBs, where soft-shelled MBs are supple and appropriate for harmonic non-linear imaging and hard-shelled MBs are appropriate for destructive US imaging, such as Power Doppler US. In addition to the intravascular nature of the MB agents, the size of these agents renders them vulnerable to inertial forces generated by blood flow. Therefore, the use of targeting ligands that display very high affinity is useful. Targeted imaging with US depends on the conjugation of MB contrast agents with targeting ligands that bind specifically to the overexpressed molecular signatures of various diseases. Several ligands, such as antibodies, peptides, and aptamers, have been attached to MBs. The targeted image contrast form MBs enhances the site-specific accumulation in the targeted sites and offers the non-invasive visualization of the diseased condition [30–32]. A MB US contrast agent was developed by the covalent coupling of recombinant single-chain vascular endothelial growth factor construct (scVEGF) that demonstrated preferential binding of scVEGF-MB to recombinant VEGFR-2 and VEGFR-2-expressing endothelial cells and enhanced US imaging of tumor angiogenesis in vivo [33].

The US is being used to track atherosclerosis and heart-related diseases. The major events in the progression of atherosclerosis are the binding and recruitment of circulating leukocytes to the vascular endothelium and subsequent migration to sub-endothelial spaces. These events are facilitated by a distinct family of cellular adhesion molecules, such as intracellular adhesion molecule-1 (ICAM-1), vascular cell adhesion protein-1 (VCAM-1), and endothelial-leukocyte adhesion molecule-1 (E-selectin, P-selectin), which are expressed on vascular endothelial cell surfaces [34]. These molecules have been observed in atherosclerotic plaque and they perform as efficient markers for subclinical coronary heart disease. US imaging of marker, such as E-selectin, P-selectin, ICAM-1, or VCAM-1, has been efficaciously used for endothelial imaging during inflammation. Ischemic myocardium was identified and imaged by MB labeled with selectins, owing to its accumulation in affected tissue [35, 36]. VCAM-1-targeted MB was attached to endothelium upon inflammation in aortic plaque. The binding of targeted MBs in endothelium was associated with disease staging as demonstrated in mouse model with atherosclerosis [35]. Signal augmentation for VCAM-1-targeted MBs improved from wild-type mice on hypercholesterolemic diet (HCD) to $ApoE^{-/-}$ mice on chow to $ApoE^{-/-}$ mice fed on HCD.

Contrast that enhanced US imaging of VCAM-1 is efficient for the quantification of changes in vascular inflammation in development of atherosclerosis. MBs targeted to glycoprotein IIb/IIIa receptors were used for the detection of thrombus [37]. Gas-filled MBs with lipid-shell were linked to the antibody that was specific for activated glycoprotein IIb/IIIa, and MBs that linked to a non-specific antibody functioned as control. Targeted MBs were used to track the in vitro and allowed real-time molecular imaging of thrombus size reduction in carotid arteries of mice. In another study, MBs specific for VCAM-1- and P-selectin-displayed enhanced sensitivity for detecting atherogenic phenotype before the appearance of disruptive atherosclerotic lesions [38]. It was demonstrated contrast that enhanced US and VCAM-1-targeted MB could be used to observe vascular inflammation in pigs. MB conjugated to recombinant P-selectin glycoprotein ligand-1 (PSGL-1) analog presented superior binding to the endothelium post intramuscular administration of endotoxin than sialyl Lewis X-contained or antibody-coupled MB [39].

Molecular US imaging is also used for tracking the development of blood vessel (angiogenesis) in tumor tissue and for the validation of effect of drugs on the vascular related diseases. Some of the examples of MBs for the examining of angiogenesis and the progress of anti-angiogenic therapy or chemotherapy include the study with MBs targeted against VEGF/VEGFR-complex, VEGFR2, and endoglin [40]. The molecular US was used for explaining the expression of angiogenic markers in vivo. Cyanoacrylate MBs linked to VEGFR2 and $\alpha v\beta 3$ integrin binding ligands and quantified the accumulation of MBs in squamous cell carcinoma xenografts in mice based on quantitative volumetric US scanning method. Specificity of targeted MBs was demonstrated by the accumulation of targeted MBs in tumors and variation in expression of markers with matrix metalloproteinase inhibitor treatment was also examined [41]. Significant decrease in marker was observed after the treatment of matrix metalloproteinase inhibitor. Targeted US was beneficial for longitudinal molecular profiling of angiogenesis in tumor and for assessing therapy in in vivo. The dual-targeted MBs integrated with both VEGFR2 and αvβ3 integrin have been studied for the improved in vivo imagining of tumor angiogenesis in a human ovarian cancer xenograft tumor model in mice [42].

BR55 is one of the attractive MBs to which a heterodimeric peptide against VEGFR2 has been attached and has been used for specific tumor detection [43, 44]. Enhanced accumulation of BR55 in the tumor vasculature of prostate and breast cancer xenografts was observed. BR55 was used to assess the angiogenesis in breast cancers of small size. Furthermore, US imaging with BR55 was extremely specific and sensitive to distinguish both benign and malignant breast cancer in in vivo mouse model with breast cancer with high accuracy [45]. A clinical trial has been implemented in subjects with BR55 for the detection of VEGFR2 positive in prostate cancer [46]. Another contrast agent, lipid (DSPC, palmitic acid, and DSPE-PEG2000)-capped MB, has been used to monitor inflammatory bowel disease in mice. The MB was targeted for E-/P-selectin and both the degree of inflammation and the response to anti-inflammatory treatments was quantified [47]. Another example of E-selectin-targeted poly(n-butyl cyanoacrylate) (PBCA) MBs with E-selectin-specific peptide showed the binding efficiency of these MBs under in vitro studies on human umbilical vein endothelial cells (HUVEC) stimulated with TNF-α and also in vivo studies in human ovarian carcinoma induced mice [48].

5.2.2 Different Nanobubbles as Contrast Agents

Non-MB or nanobubbles are liquid–liquid emulsions, liposomes or solid NPs in size the range of 10–1000 nm [49–51]. These contrast agents can serve as an efficient tool for US imaging only if they show good biocompatibility, easy route of administration, and should provide contrast to the surrounding tissue with good contrast-to-tissue ratio (CTR). Non-MB-based contrast agents infiltrate extravascular space offering the prospect of acquisition of images apart from vascular compartment. Tiny free gas bubbles used in US imaging dissolve quickly in the media and hence the contrast arises near to the site of infusion. Owing to the insignificant intrinsic acoustic reflectivity, non-MB-based imaging agents are not spotted independently by US imaging. Therefore, alternative strategies for visualization of these contrast agents with US have been recommended. In present scenario, targeted contrast agents are in use and specifically target the molecular markers and are efficiently internalized to site of interest leading to molecular US imaging. Many passive and active targeting mechanisms are plotted and are utilized for this purpose successfully.

Different types of nanosized particles have been developed for US imaging based on their composition such as PFC emulsion nanodroplets, echogenic liposomes (ELIPs), polymer-based nanobubbles, and solid NPs.

5.2.2.1 PFC emulsion nanodroplets

PFC nanodroplets are made up of liquid–liquid emulsions of a liquid PFC as a core material, which, in turn, is encapsulated by phospholipid, with size ranging from 200 nm to 400 nm. PFC nanodroplets accumulate at biological target and turn to be acoustically detectable when exposed to US. The liquid composition offers PFC mechanical stress and makes it resistant to pressure. Fibrin-targeted PFC nanodroplet was developed and used for the successful detection of blood clots and also intravascular US in carotid arteries of dogs [52]. The administered PFC nanodroplets were accumulated at blood clots, thus enhancing the echogenicity in the clots. Injured pig carotid arteries were detected with intravascular US by using a stretch-induced tissue factor targeted PFC nanodroplets in in vivo studies [53]. Enhanced echogenicity was detected at injury site, owing to the build-up of PFC nanodroplets. A new technology to enhance the echogenicity of PFC nanodroplets has been introduced [54]. Acoustic energy was used to vaporize PFC nanodroplets in the core to gas and the resultant contrast agent had superior acoustic reflectivity. However, it was challenging to vaporize high-boiling point liquid PFC, such as perflurohexyl bromide, since high energy was required to vaporize into a gaseous phase [54]. In depth research on biodistribution and possibility for in vivo molecular US imaging using PFC nanodroplets and vaporization method is required for enhancing the echogenicity.

5.2.2.2 Echogenic Liposomes

Echogenic liposomes (ELIPs) are versatile ultrasound contrast agents (UCAs) composed of air and an aqueous core encapsulated by a lipid shell. The ELIP shell is mechanically stabilized with biocompatible phospholipids for improved persistence in the biological fluids. The air pockets present in the lipid bilayer create acoustic reflectivity [55]. Targeted ELIPs for US detection and enhancement of vasoactive atheroma and pathologic constituents of endothelium have been developed (Figure 5.2). Atheroma is the degeneration of the walls of the arteries affected by the accumulation of fat and scar tissue that ultimately leads to the restriction of the circulation of blood and resulting in the risk of thrombosis. ELIP developed against several inflammatory markers, such as anti-intercellular adhesion molecule 1 (anti-ICAM1), anti-vascular cell adhesion molecule 1 (anti-VCAM1), anti-fibrin, anti-fibrinogen, and anti-tissue factor (anti-TF) has been reported. Liposomes that were targeted accrued at atheroma

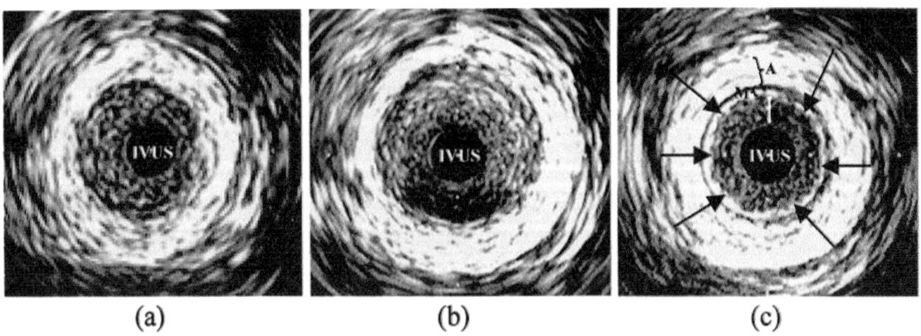

Figure 5.2 The intravascular US images of left carotid artery presenting enhancement of atheroma by VCAM conjugated ELIP after administration of (a) saline (b) 5 min after unconjugated ELIP, and (c) 5 min after VCAM-conjugated ELIP. The arrows denote enhanced atheroma. Reproduced with permission from Ref. [56].

and signal augmentation was observed 5 min after i.v. injection in the swine model [56]. Efficient recognition of molecular constituents permits staging of atheroma, which can improve the therapeutic strategies.

5.2.2.3 Polymer-based nanobubbles

Polymer-based nanobubbles, generally called nanobubbles, are emulsions of gas–liquid encapsulated by polymer such as polylactic acid (PLA). The diameter of such nanobubbles generally ranges between 40 nm and 200 nm. Uniform dispersed nanobubble comprised of gaseous aqueous phases with sodium dodecyl sulfate (SDS) as surfactant enclosed in Shirasu-porous-glass (SPG) membrane was developed [57]. Nanobubbles, with an average diameter of 360–720 nm, were synthesized by pressurizing air in SDS. When aqueous solution was sonicated with a palladium electrode, nanobubbles of different sizes were developed within a few minutes. The size of nanobubbles was tuned by varying the concentration of salt and surfactants during reaction [58]. Nanobubbles mediated US imaging and delivery of the anticancer drug (doxorubicin) in xenograft tumors have been checked in mice [59]. Nanobubbles were accumulated in the tumor, owing to the enhanced permeability and retention (EPR) effect. The nanobubbles extravasate, accumulate, and merge into bigger gas bubbles in the tumor areas offering intense signal for US imaging [60]. When exposed to therapeutic US, the gas bubbles oscillated and collapsed discharging the entrapped drug at tumor site thus enabling the therapy too.

5.2.2.4 Solid NPs

Solid NPs/amorphous solid NPs have demonstrated the generation of backscattering for US imaging owing to the gas pockets in their pores and crevices that they possess. Owing to the acoustic impedance mismatches that exist between solid and soft tissues, solid NPs can offer better US contrast than nanobubbles. Silica NPs and iron oxide (IO) NPs are some of the commonly studied solid NPs for US imaging. Solid silica NPs improved the US images in both tissue phantoms and in vivo (mouse livers) [61]. The brightness (mean gray scale level, increased with particle size and concentration) of mouse liver increased following particle administration indicated the feasibility of solid NPs as contrast augmenting agents for US imaging. A solid nanoparticulate UCA, based on biodegradable polymer PLA, attached to anti-Her2 antibody (Herceptin) that targets overexpressed Her2 receptors has been developed [62]. The targeting efficiency and US enhancement in Her2-positive and negative cells were evaluated. Her2-positive cells demonstrated significant staining upon incubation with NP/antibody conjugates, while in Her2-negative cells, reduced staining was observed, signifying receptor-specific binding of targeted PLA NPs. The high-resolution US images were greyer of the Her2-positive cells after NP treatment, while there was no difference in the Her2-negative cells that clearly shows that targeted NPs enhanced the ultrasonic images in vitro compared to non-targeted NPs.

The clinical impact of US and also the labeling potential of nanomaterials have impelled a great attention in the development of nanosized US contrast agents. Owing to the lower scattering cross sections and low mechanical characteristics, the implementation of nanomaterials in US imaging is not superior to that of MBs mediated US imaging. However, the effective combination of both NPs and MBs for therapy and diagnosis in a single component offers precise therapy and diagnosis together. For example, silica-coated NPs (Au NRs, CdSe/ZnS QDs, Fe_3O_4, and Gd-incorporated silica NPs) were integrated into microfluidic-generated compressible protein lipid, monodisperse, perfluorobutane MBs (less than 3 μm) [63–69]. NPs incorporated MBs were detected in diluted saline under low-pressure US and could be generated at high-throughput, appropriate for in vivo application (10^6 MB/sec) [70].

Microbubbles/nanobubbles can be used as theranostic (therapy + diagnosis) nanocarriers for US contrast agents by the integration of diagnostic probes with therapeutic modalities such as US-induced apoptosis, thermal ablation, drug delivery, sonodynamic therapy, sonoporation/sonotransfection, etc. [71–80]. Mesoporous silica nanocapsules were studied extensively as multifunctional inorganic theranostic platform for contrast-intensified US imaging, US-induced cytolysis and drug release, and high intensity focused ultrasound (HIFU) tumor ablation therapy [81]. Mesoporous silica nanoparticles (MSNs) are versatile category of NPs that were developed as theranostic nanoplatform for stem cell research. Insulin-like growth factor (IGF) was used as a marker for pro-survival agent, owing to its involvement in cardiac stem cell therapy. IGF was delivered to Mesenchymal stem cells (MSCs) in a sustained manor and it stimulated the rate of cell survival even in hypoxic or necrotic parts of the heart. Dual modal US/MRI and drug delivery with MSNs were attained. The US functioned as image-guided delivery of NPs to cells and for quantitation, whereas during follow-up studies,

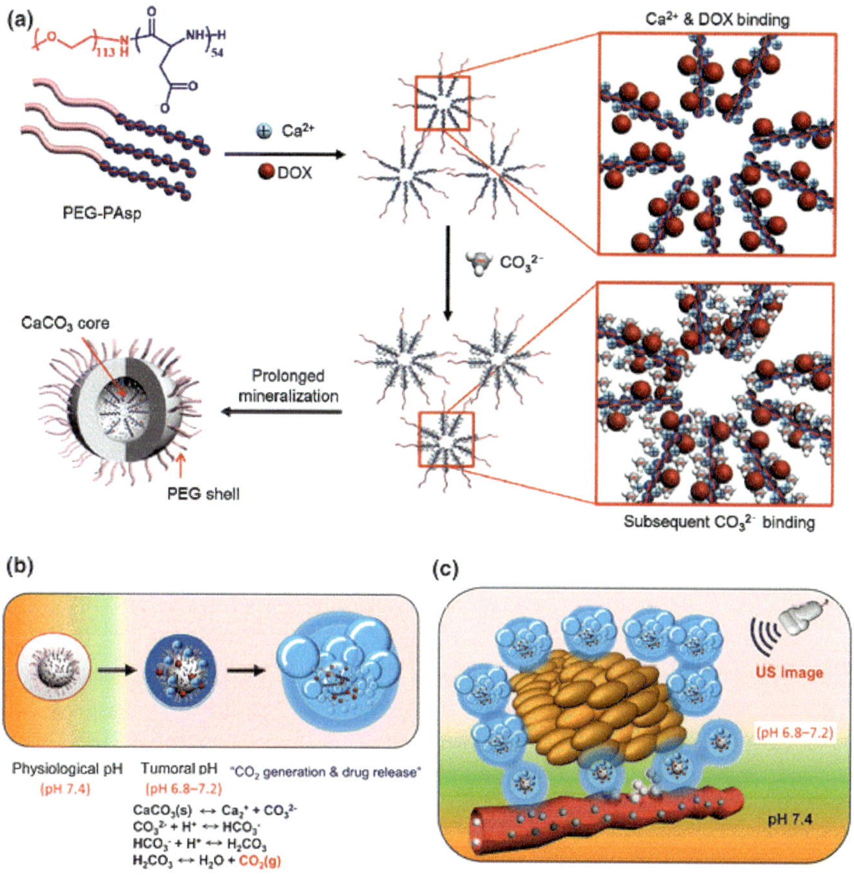

Figure 5.3 (a) Schematic of synthesis of DOX-$CaCO_3$-MNPs, (b) schematic of generation of CO_2 and controlled release of DOX, and (c) performance after accumulation of NPs in tumor. Reproduced with permission from Ref. [83].

MRI offered higher resolution images. The incorporated drug prevented the death of stem cells [82]. Another example for theranostic NPs for US imaging and therapy was the development of doxorubicin-loaded inorganic nanosystem [Doxorubicin (DOX)-loaded mineralized NP (DOX-$CaCO_3$-MNPs)] using block copolymer-templated mineralization of $CaCO_3$ [83]. The nanoconjugate produced CO_2 MBs in acidic tumoral microenvironment and exhibited enhanced US backscattering. Increased release rate of DOX inside endo/lysosomes by the structural breakdown of $CaCO_3$ core was also observed. The DOX-$CaCO_3$-MNPs exhibited antitumor activity in tumor-bearing mice (Figure 5.3). A thermally responsive CO_2-generating system was developed based on ammonium bicarbonate (NH_4HCO_3)-loaded liposomes [84]. When the NPs in water were heated to 42 °C, enhanced US imaging was observed. The cavitation effect triggered by the annihilation of CO_2 bubbles might have instigated cellular necrosis by discharging proteolytic enzymes from lysosomes. Temperature dependent release of DOX was also studied in the nanosystem. CO_2-generation facilitates simultaneous US imaging and therapeutic efficiency by stimuli responsive drug release resulting in superior therapeutic effects with reduced side effects [84].

5.3 Multimodal Imaging: US Imaging with MRI, PET, SPECT, and PA

UCAs, as multimodal imaging agents, are predominantly advantageous for whole body imaging, biodistribution studies, and histological analyses. Multimodal UCAs encompass attachment, encapsulation, or physical adsorption of additional imaging moieties, such as NPs, radiotracers, and fluorescent dyes, in or on the shell of MBs. The polymeric shells of 50–500 nm thickness can effectively entrap or attach additional imaging agents. The production of multimodal UCAs is based on the integration of diagnostic moieties for various imaging modalities during the synthetic process of MB by mechanical agitation, microfluidic devices, chemical bonding, or entrapment [26, 70, 85–89].

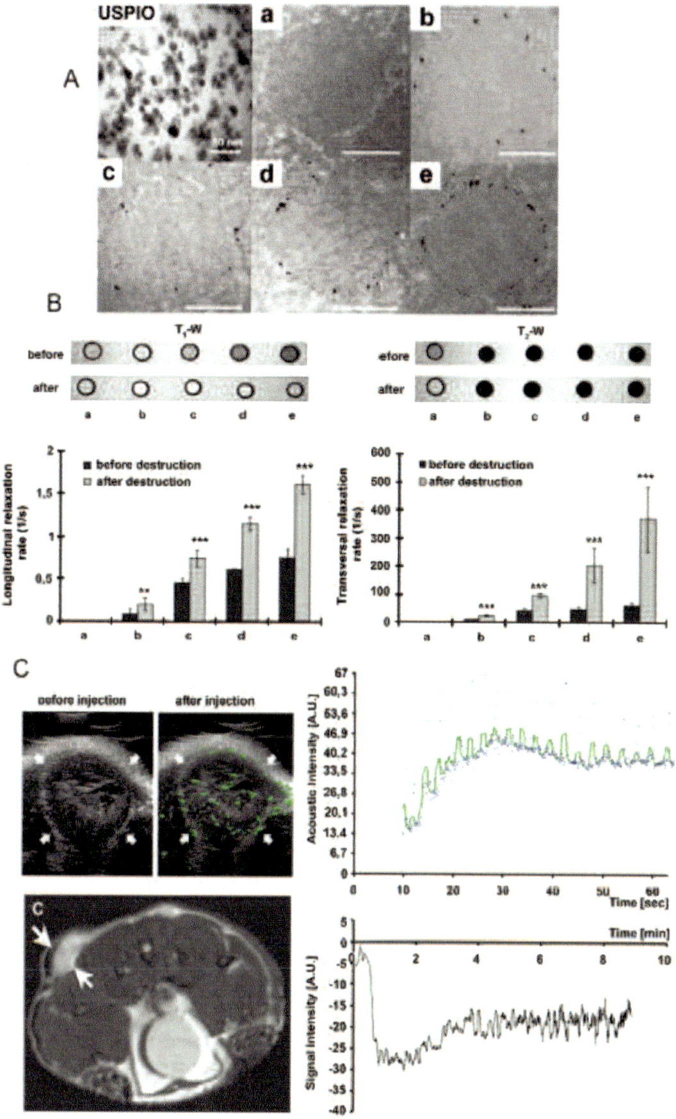

Figure 5.4 USPIO-loaded PBCA MB contrast agents for MRI/US. (A) TEM pictures of PBCA MB with increasing iron content as analyzed by ICP-MS (a: 0, b: 29.67 mg/L, c: 58.88 mg/L, d: 92.55 mg/L, and e: 113.43 mg/L). (B) Phantom MR imaging displaying enhancement in MRI signal that intensifies with the destruction of MB. (C) In vivo signal enhancement, both in US and MRI, upon the administration of USPIO-loaded MB in tumor bearing mice. Reproduced with permission from Ref. [86].

Contrast agents for dual modal US/MRI imaging incorporate magnetic NPs on MBs. By incorporating ultra-small superparamagnetic iron oxide (USPIO) NPs in poly (butyl cyanoacrylate) (PBCA)-based polymeric MBs, appropriate contrast agents for dual US/MRI have been developed for in vitro and in vivo applications (Figure 5.4) [86]. Alternatively, layer-by-layer deposition method and double emulsion polymerization process were used for the development of USPIO-containing MBs. UPSIO-loaded MBs were detected with a strong non-linear response upon US insonification than normal MBs that may be accredited to the augmented resistance to compress the NP-loaded MBs, leading to non-linear oscillations of UPSIO-loaded MBs [89, 90]. Gd^{3+}-DOTA loaded MBs were developed by modification after the synthesis of lipid-shelled MBs. Loading of 7.5×10^5 Gd^{3+} ions per μm^2 of MBs surface was attained and signal enhancement in MRI was observed after destruction of MBs, which could be correlated to the restricted entry of water to lipid head-groups holding Gd^{3+} in MBs than lipid fragments [91, 92].

Radiolabeled MBs for single photon emission computed tomography/positron emission tomography-US (PET/SPECT–US) imaging are commonly used for examining the biodistribution profile of MB upon i.v. injection [93]. Researchers labeled PBCA-MBs with ^{111}In by integrating with DTPA that was supported to the shell of MBs. The activity of radiolabeled MBs in various organs of mice up to 48 h. post-injection was measured after i.v. injection. The activity was relatively high in liver and spleen, while the activity in blood and lungs started reducing. Signal intensity was less in tumor tissue, gastrointestinal tract, and kidney in comparison to liver and spleen. Although less activity was observed in kidneys, urine did not exhibit any activity, suggesting clearance of PBCA MB via kidney. US-sensitive particle acoustic quantification (US-SPAQ) imaging in kidney 48 h post-injection confirmed the absence of MB signals in the fragments. In another study, ^{18}F-labeled lipid ([^{18}F] fluorodipalmitin), combined in MB shells, was employed as dynamic microPET probe for quantitative 90 min biodistribution measurements in male Fischer-344 rats [94]. Lipids accumulated in the liver and spleen were compared to other vital organs. The local deposition of

^{18}F labeled lipids in kidney was detected on US arbitrated destruction of MBs. (^{18}F) fluorodipalmitin was studied for the characterization of biodistribution of contrast MB shells and the deposition of lipid was shown to be locally increased after insonication.

Photoacoustic (PA) imaging can be considered either as an US facilitated electromagnetic imaging modality or an US imaging modality incorporated with electromagnetic enhanced contrast agent. PA imaging is based on short-pulsed laser beam on the excitation of chromophores present in body tissue (Figure 5.5). It results in the thermoelastic expansion of the body tissue, which, in turn, results in the wide band ultrasonic emission. The acoustic waves generated from the initial acoustic source reach tissue surface with various time delays. The resultant ultrasonic waves that are emitted are taken up by ultrasonic transducer, which are used in image reconstruction. The commonly used PA contrast agents are indocyanine green (ICG), hemoglobin, India ink, and gold nanorods. Research on integration of US with PA imaging is in developing phase, but has been demonstrated its ability in sentinel lymph node mapping and staging deep vein thrombosis [95–97].

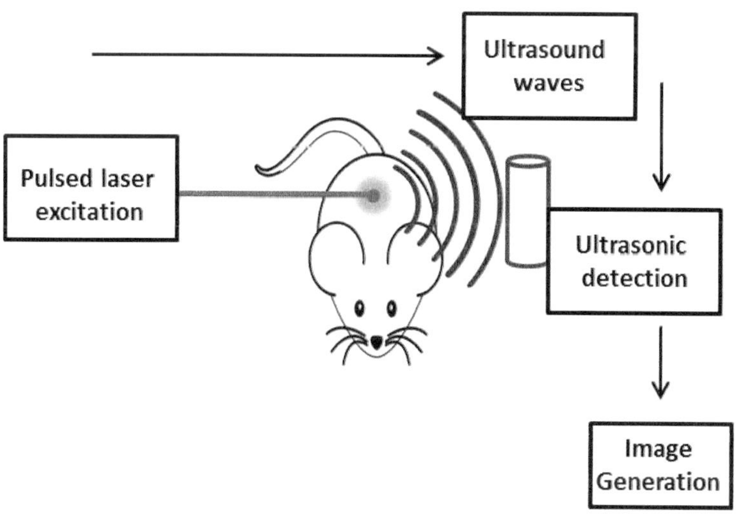

Figure 5.5 Principle of PA imaging.

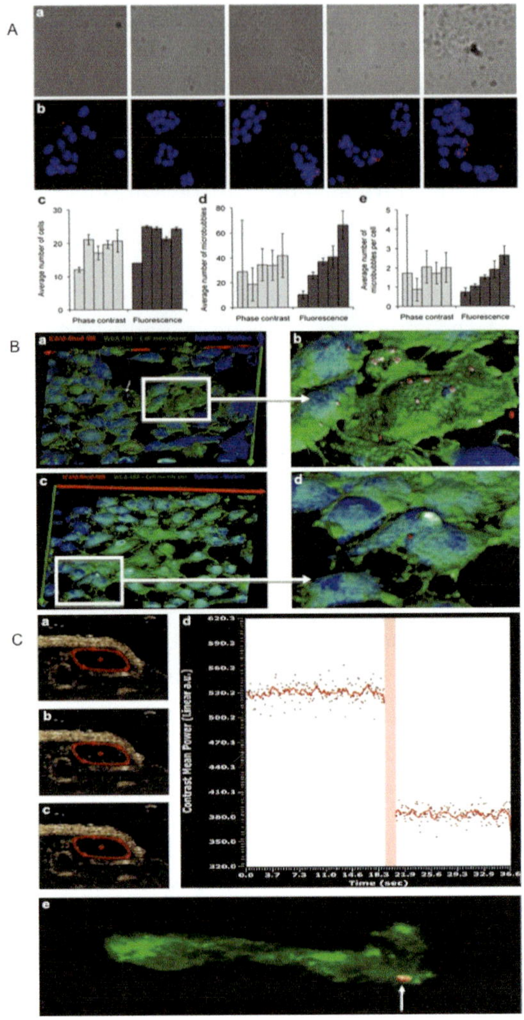

Figure 5.6 Rhodamine-b loaded PBCA MBs as contrast agents for OI/US. (A) In vitro comparative study of fluorescent and non-fluorescent MBs for studying binding of targeted MBs in cells. (B) Two-photon microscopic images of ICAM-1 targeted MBs binding to activated (upper panel) vs non-activated (bottom panel) HUVEC. (C) Application of fluorescent MBs in vivo and ex vivo molecular US studies. (a–c) Display tumor sections during molecular imaging; (d) displays the evaluation of MBs that were bound by destruction replenishment analysis; (e) ex vivo fluorescence microscopy-based confirmation of MB binding (in red; see arrow) to angiogenic tumor endothelium (counterstained using fluorescein-labeled lectin; in green). Reproduced with permission from Ref. [99].

The incorporation of optical imaging contrast agents during the synthesis of MBs or into pre-synthesized MB resulting in the encapsulation within the shell of MB has been demonstrated and used for in vitro studies [98, 99]. The presence of targeting ligands in US/optical imaging probes are being used for the study of binding kinetics of MBs to the targeted cells and for the examination of the fate of bubbles upon i.v. administration. The fluorescence analysis for quantifying the bound target-specific MBs to cells in culture (Figure 5.6A) and the binding of rhodamine encapsulated, ICAM-1-labeled MB to activated HUVEC with two-photon laser scanning microscopy was reported (Figure 5.6B). Targeted rhodamine-loaded MBs in histological sections of tumor endothelium were confirmed by fluorescence microscopy (Figure 5.6C) [99]. Fluorophore quencher-labeled MB for the quantification of external pressure has been developed [100] in which, pressure variations as small as 1 mm Hg could be quantified. Such nanosystems could be a remarkable tool in cancer research for the study of influence of pressure on tumor progression.

5.4 Conclusion and Future Prospects

In summary, US imaging is a well-established molecular imaging technology that has swiftly got importance among different molecular imaging modalities within several years [101–104]. The advantageous features, including non-invasiveness, high spatial and temporal resolution, lack of ionizing irradiation, real-time imaging, low costs, and wide availability, have made this imaging modality convenient in pre-clinical and clinical researches. The improvement in the design of contrast agents for US imaging with nanotechnology has immense application in biomedical arena, especially for theranostics.

References

1. Becher H, Burns PN. Handbook of contrast echocardiography. 1st En. Berlin, Germany, Springer. **2000**.
2. Cosgrove D, Lassau N. Imaging of perfusion using ultrasound. *Eur. J. Nucl. Med. Mol. Imaging*. 37, **2010**, S65–S85.

3. Schellpfeffer MA. Ultrasound imaging in research and clinical medicine. *Birth Defects Res. C Embryo Today*. 99, **2013**, 83–92.
4. Kossoff G. Basic physics and imaging characteristics of ultrasound. *World J Surg*. 24, **2000**, 134–142.
5. Maulik D. (ed.). Doppler ultrasound in obstetrics and gynecology. 2nd Ed. Springer Verlag, New York, Berlin, Heidelberg. **2005.**
6. Hedrick WR, Hykes DL, Starchman DE. Ultrasound physics and instrumentation. Chapter 16 Doppler imaging/Chapter 22 Biologic effects/ Chapter 23 Clinical safety. 4th Ed. Elsevier, Mosby, St. Louis, MO. **2005**.
7. Otto CM. Principles of echocardiographic image acquisition and doppler analysis. In: Textbook of Clinical Ecocardiography. 2nd Ed. Philadelphia, PA, WB Saunders. **2000**, 1–29.
8. Yuan B, Rychak J. Tumor functional and molecular imaging utilizing ultrasound and ultrasound-mediated optical techniques. *Am. J. Pathol.* 182, **2013**, 305–311.
9. Willmann JK, van Bruggen N, Dinkelborg LM, Gambhir SS. Molecular imaging in drug development. *Nat. Rev. Drug Discov.* 7, **2008**, 591–607.
10. Deshpande N, Needles A, Willmann JK. Molecular ultrasound imaging: Current status and future directions. *Clin. Radiol.* 65, **2010**, 567–581.
11. Jokerst JV, Gambhir SS. Molecular imaging with theranostic nanoparticles. *Acc. Chem. Res.* 44, **2011**, 1050–1060.
12. Klibanov AL. Molecular imaging with targeted ultrasound contrast microbubbles. *Ernst Schering Res. Found. Workshop*. 49, **2005**, 171–191.
13. McCulloch M, Gresser C, Moos S, Odabashian J, Jasper S, Bednarz J, Burgess P, Carney D, Moore V, Sisk E, Waggoner A, Witt S, Adams D. Ultrasound contrast physics: A series on contrast echocardiography, article 3. *J Am Soc Echocardiogr*. 13, **2000**, 959–967.
14. Willmann JK, Cheng Z, Davis C, Lutz AM, Schipper ML, Nielsen CH, Gambhir SS. Targeted MBs for imaging tumour angiogenesis: Assessment of whole-body biodistribution with dynamic micro-PET in mice. *Radiology*. 249, **2008**, 212–219.
15. Wang LV, Hu S. Photoacoustic tomography: In vivo imaging from organelles to organs. *Science*. 335, **2012**, 1458–1462.
16. Gessner R, Dayton PA. Advances in molecular imaging with ultrasound. *Mol. Imaging*. 9, **2010**, 117–127.

17. Corlu A, Choe R, Durduran T, Rosen MA, Schweiger M, Arridge SR, Schnall MD, Yodh AG. Three-dimensional in vivo fluorescence diffuse optical tomography of breast cancer in humans. *Opt. Express.* 15, **2007**, 6696–6716.

18. Zhu Q, Hegde PU, Ricci A Jr, Kane M, Cronin EB, Ardeshirpour Y, Xu C, Aguirre A, Kurtzman SH, Deckers PJ, Tannenbaum SH. Early-stage invasive breast cancers: Potential role of optical tomography with US localization in assisting diagnosis. *Radiology.* 256, **2010**, 367–378.

19. Ferrara K, Pollard R, Borden M. Ultrasound microbubble contrast agents: Fundamentals and application to gene and drug delivery. *Annu. Rev. Biomed. Eng.* 9, **2007**, 415–447.

20. Culver J, Akers W, Achilefu S. Multimodality molecular imaging with combined optical and SPECT/PET modalities. *J. Nucl. Med.* 49, **2008**, 169–172.

21. Kiessling F, Fokong S, Bzyl J, Lederle W, Palmowski M, Lammers T. Recent advances in molecular, multimodal and theranostic ultrasound imaging. *Adv. Drug Deliv. Rev.* 72, **2014**, 15–27.

22. Bokor D, Chambers JB, Rees PJ, Mant TG, Luzzani F, Spinazzi A. Clinical safety of SonoVue, a new contrast agent for ultrasound imaging, in healthy volunteers and in patients with chronic obstructive pulmonary disease. *Invest Radiol.* 36, **2001**,104–109.

23. Korpanty G, Grayburn PA, Shohet RV, Brekken RA. Targeting vascular endothelium with avidin microbubbles. *Ultrasound Med Biol.* 31, **2005**, 1279–1283.

24. Cui W, Bei J, Wang S, Zhi G, Zhao Y, Zhou X, Zhang H, Xu Y. Preparation and evaluation of poly (L-lactide-co-glycolide) (PLGA) microbubbles as a contrast agent for myocardial contrast echocardiography. *J. Biomed. Mater. Res. Part B. Appl. Biomater.* 73, **2005**, 171–178.

25. Fokong S, Siepmann M, Liu Z, Schmitz G, Kiessling F, Gatjens J. Advanced characterization and refinement of poly N-butyl cyanoacrylate Microbubbles for ultrasound imaging. *Ultrasound Med Biol.* 37, **2011**, 1622–1634.

26. Sirsi S, Borden M. Microbubble compositions, properties and biomedical applications. *Bubble Sci. Eng. Technol.* 1, **2009**, 3–17.

27. Borden MA, Caskey CF, Little E, Gillies RJ, Ferrara KW. DNA and polylysine adsorption and multilayer construction onto cationic lipid-coated microbubbles. *Langmuir.* 23, **2007**, 9401–9408.

28. Schutt EG, Klein DH, Mattrey RM, Riess JG. Injectable microbubbles as contrast agents for diagnostic ultrasound imaging: The key role of perfluorochemicals. *Angew. Chem. Int. Ed. Engl.* 42, **2003**, 3218–3235.
29. Postema M, Schmitz G. Ultrasonic bubbles in medicine: Influence of the shell. *Ultrason. Sonochem.* 14, **2007**, 438–444.
30. Kiessling F, Bzyl J, Fokong S, Siepmann M, Schmitz G, Palmowski M. Targeted ultrasound imaging of cancer: an emerging technology on its way to clinics. *Curr. Pharm. Des.* 18, **2012**, 2184–2199.
31. Unnikrishnan S, Klibanov AL. Microbubbles as ultrasound contrast agents for molecular imaging: Preparation and application. *AJR. Am. J. Roentgenol.* 199, **2012**, 292–299.
32. Klibanov AL. Microbubble contrast agents: Targeted ultrasound imaging and ultrasound-assisted drug-delivery applications. *Invest Radiol.* 41, **2006**, 354–362.
33. Anderson CR, Rychak JJ, Backer M, Backer J, Ley K, Klibanov AL. scVEGF microbubble ultrasound contrast agents: A novel probe for ultrasound molecular imaging of tumor angiogenesis. *Invest Radiol.* 45, **2010**, 579–585.
34. Hamilton AJ, Huang SL, Warnick D, Rabbat M, Kane B, Nagaraj A, Klegerman M, McPherson DD. Intravascular ultrasound molecular imaging of atheroma components in vivo. *J. Am. Coll. Cardiol.* 43, **2004**, 453–460.
35. Kaufmann BA, Sanders JM, Davis C, Xie A, Aldred P, Sarembock IJ, Lindner JR. Molecular imaging of inflammation in atherosclerosis with targeted ultrasound detection of vascular cell adhesion molecule-1. *Circulation.* 116, **2007**, 276–284.
36. Villanueva FS, Lu E, Bowry S, Kilic S, Tom E, Wang J, Gretton J, Pacella JJ, Wagner WR. Myocardial ischemic memory imaging with molecular echocardiography. *Circulation.* 115, **2007**, 345–352.
37. Wang X, Hagemeyer CE, Hohmann JD, Leitner E, Armstrong PC, Jia F, Olschewski M, Needles A, Peter K, Ahrens I. Novel single-chain antibody-targeted MBs for molecular ultrasound imaging of thrombosis: Validation of a unique noninvasive method for rapid and sensitive detection of thrombi and monitoring of success or failure of thrombolysis in mice. *Circulation.* 125, **2012**, 3117–3126.
38. Masseau I, Davis MJ, Bowles DK. Carotid inflammation is unaltered by exercise in hypercholesterolemic swine. *Medicine and science in sports and exercise.* 44, **2012**, 2277–2289.

39. Bettinger T, Bussat P, Tardy I, Pochon S, Hyvelin JM, Emmel P, Henrioud S, Biolluz N, Willmann JK, Schneider M, Tranquart F. Ultrasound molecular imaging contrast agent binding to both E- and P-selectin in different species. *Invest. Radiol.* 47, **2012**, 516–523.

40. Deshpande N, Ren Y, Foygel K, Rosenberg J, Willmann JK. Tumor angiogenic marker expression levels during tumor growth: longitudinal assessment with molecularly targeted MBs and US imaging. *Radiology.* 258, **2011**, 804–811.

41. Palmowski M, Huppert J, Ladewig G, Hauff P, Reinhardt M, Mueller MM, Woenne EC, Jenne JW, Maurer M, Kauffmann GW, Semmler W, Kiessling F. Molecular profiling of angiogenesis with targeted ultrasound imaging: Early assessment of antiangiogenic therapy effects. *Mol. Cancer Ther.* 7, **2008**, 101–109.

42. Willmann JK, Lutz AM, Paulmurugan R, Patel MR, Chu P, Rosenberg J, Gambhir SS. Dual-targeted contrast agent for US assessment of tumor angiogenesis in vivo. *Radiology.* 248, **2008**, 936–944.

43. Tardy I, Pochon S, Theraulaz M, Emmel P, Passantino L, Tranquart F, Schneider M. Ultrasound molecular imaging of VEGFR2 in a rat prostate tumor model using BR55. *Invest. Radiol.* 45, **2010**, 573–578.

44. Pochon S, Tardy I, Bussat P, Bettinger T, Brochot J, von Wronski M, Passantino L, Schneider M. BR55: A lipopeptide-based VEGFR2-targeted ultrasound contrast agent for molecular imaging of angiogenesis. *Invest. Radiol.* 45, **2010**, 89–95.

45. Bachawal SV, Jensen KC, Lutz AM, Gambhir SS, Tranquart F, Tian L, Willmann JK. Earlier detection of breast cancer with ultrasound molecular imaging in a transgenic mouse model. *Cancer Res.* 73, **2013**, 1689–1698.

46. Wijkstra II, Smeenge M, Rosette JJ, de la Pochon S, Tranquart F. Targeted MB prostate cancer imaging with BR55. *Proceedings of the 17th European Symposium on Ultrasound imaging.* **2012**.

47. Deshpande N, Lutz AM, Ren Y, Foygel K, Tian L, Schneider M, Pai R, Pasricha PJ, Willmann JK. Quantification and monitoring of inflammation in murine inflammatory bowel disease with targeted contrast-enhanced US. *Radiology.* 262, **2012**, 172–180.

48. Fokong S, Fragoso A, Rix A, Curaj A, Wu Z, Lederle W, Iranzo O, Gatjens J, Kiessling F, Palmowski M. Ultrasound molecular imaging of e-selectin in tumor vessels using poly n-butyl cyanoacrylate microbubbles covalently coupled to a short targeting peptide. *Invest. Radiol.* 48, **2013**, 843–850.

49. Lanza GM, Wickline SA. Targeted ultrasonic contrast agents for molecular imaging and therapy. *Curr. Probl. Cardiol.* 28, **2003**, 625–653.

50. Gref R, Dombb A, Quelled P, Blunk T, Miillerd RH, Verbavatz JM, Langerf R. The controlled intravenous delivery of drugs using PEG-coated sterically stabilized nanospheres. *Adv. Drug Deliv. Rev.* 16, **1995**, 215–233.

51. Brigger I, Dubernet C, Couvreur P. Nanoparticles in cancer therapy and diagnosis. *Adv. Drug Deliv. Rev.* 64, **2012**, 24–36.

52. Lanza GM, Wallace KD, Scott MJ, Cacheris WP, Abendschein DR, Christy DH, Sharkey AM, Miller JG, Gaffney PJ, Wickline SA. A novel site-targeted ultrasonic contrast agent with broad biomedical application. *Circulation.* 94, **1996**, 3334–3340.

53. Lanza GM, Abendschein DR, Hall CS, Scott MJ, Scherrer DE, Houseman A, Miller JG, Wickline SA. In vivo molecular imaging of stretch-induced tissue factor in carotid arteries with ligand-targeted nanoparticles. *J. Am. Soc. Echocardiogr.* 13, **2000**, 608–614.

54. Kripfgans OD, Fabiilli ML, Carson PL, Fowlkes JB. On the acoustic vaporization of micrometer-sized droplets. *J. Acoust. Soc. Am.* 116, **2004**, 272.

55. Buchanan KD, Huang S, Kim H, Macdonald RC, McPherson DD. Echogenic liposome compositions for increased retention of ultrasound reflectivity at physiologic temperature. *J. Pharm. Sci.* 97, **2007**, 2242–2249.

56. Hamilton AJ, Huang SL, Warnick D, Rabbat M, Kane B, Nagaraj A, Klegerman M, McPherson DD. Intravascular ultrasound molecular imaging of atheroma components in vivo. *J. Am. Coll. Cardiol.* 43, **2004**, 453–460.

57. Kukizaki M, Goto M. Size control of nanobubbles generated from shirasu-porous-glass (SPG) membranes. *J. Member. Sci.* 281, **2006**, 386–396.

58. Cho SH, Kim JY, Chun JH, Kim JD. Ultrasonic formation of nanobubbles and their zeta-potentials in aqueous electrolyte and surfactant solutions. *Colloids and Surfaces A: Physicochem. Eng. Aspects.* 269, **2005**, 28–34.

59. Rapoport N, Gao Z, Kennedy A. Multifunctional nanoparticles for combining ultrasonic tumour imaging and targeted chemotherapy. *J. Natl. Cancer Inst.* 99, **2007**, 1095–1106.

60. Maeda H, Wu J, Sawa T, Matsumura Y, Hori K. Tumour vascular permeability and the EPR effect in macromolecular therapeutics: A review. *J. Control. Release.* 65, **2000**, 271–284.

61. Liu J, Levine AL, Mattoon JS, Yamaguchi M, Lee RJ, Pan X, Rosol TJ. Nanoparticles as image enhancing agents for ultrasonography. *Phys. Med. Biol.* 51, **2006**, 2179–2189.

62. Liu J, Li J, Rosol TJ, Pan X, Voorhees JL. Biodegradable nanoparticles for targeted ultrasound imaging of breast cancer cells in vitro. *Phys. Med. Biol.* 52, **2007**, 4739–4747.

63. Unger EC, Porter T, Culp W, Labell R, Matsunaga T, Zutshi R. Therapeutic applications of lipid-coated microbubbles. *Adv. Drug Delivery Rev.* 56, **2004**, 1291–1314.

64. Price RJ, Skyba DM, Kaul S, Skalak TC. Delivery of colloidal particles and red blood cells to tissue through microvessel ruptures created by targeted microbubble destruction with ultrasound. *Circulation.* 98, **1998**, 1264–1267.

65. Yang F, Li L, Li Y, Chen Z, Wu J, Gu N. Superparamagnetic nanoparticle-inclusion microbubbles for ultrasound contrast agents. *Phys. Med. Biol.* 53, **2008**, 6129–6141.

66. Lum AF, Borden MA, Dayton PA, Kruse DE, Simon SI, Ferrara KW. Ultrasound radiation force enables targeted deposition of model drug carriers loaded on microbubbles. *J. Control. Release.* 111, **2006**, 128–134.

67. Yang F, Gu AY, Chen ZP, Gu N, Ji M. Multiple emulsion microbubbles for ultrasound imaging. *Mater. Lett.* 62, **2008**, 121–124.

68. Stride E, Pancholi K, Edirisinghe MJ, Samarasinghe S. Increasing the nonlinear character of microbubble oscillations at low acoustic pressures. *J. R. Soc. Interface.* 5, **2008**, 807–811.

69. Yang F, Li Y, Chen Z, Zhang Y, Wu J, Gu N. Superparamagnetic iron oxide nanoparticle-embedded encapsulated microbubbles as dual contrast agents of magnetic resonance and ultrasound imaging. *Biomaterials.* 30, **2009**, 3882–3890.

70. Seo M, Gorelikov I, Williams R, Matsuura N. Microfluidic assembly of monodisperse, nanoparticle-incorporated perfluorocarbon microbubbles for medical imaging and therapy. *Langmuir.* 26, **2010**, 13855–13860.

71. Caissie A, Karshafian R, Hynynen K, Czarnota GJ. Ultrasound contrast microbubbles: In vivo imaging and potential therapeutic

applications. *Pan Stanford Ser. Biomed. Nanotechnol.* 2, **2011**, 267- 291.

72. Sheeran PS, Luois SH, Mullin LB, Matsunaga TO, Dayton PA. Design of ultrasonically activatable nanoparticles using low boiling point perfluorocarbons. *Biomaterials.* 33, **2012**, 3262–3269.

73. Tachibana K, Feril LB, Ikeda-Dantsuji Y. Sonodynamic therapy. *Ultrasonics.* 48, **2008**, 253–259.

74. Owen J, Pankhurst Q, Stride E. Magnetic targeting and ultrasound mediated drug delivery: Benefits, limitations and combination. *Int. J. Hyperthermia.* 28, **2012**, 362–373.

75. Xu B, Lu R, Dou H, Tao K, Sun K, Qiu Y, Ding J, Zhang D, Li J, Shi W, Sun K. Exploring the structure-property relationships of ultrasonic/MRI dual imaging magnetite/PLA microbubbles: Magnetite@Cavity versus magnetite@Shell systems. *Colloid Polym. Sci.* 290, **2012**, 1617–1626.

76. Mannell H, Pircher J, Raethel T, Schilberg K, Zimmermann K, Pfeifer A, Mykhaylyk O, Gleich B, Pohl U, Krötz F. Targeted endothelial gene delivery by ultrasonic destruction of magnetic microbubbles carrying lentiviral vectors. *Pharm. Res.* 29, **2012**, 1282–1294.

77. Lin CY, Li JR, Tseng HC, Wu MF, Lin WL. Enhancement of focused ultrasound with microbubbles on the treatments of anticancer nanodrug in mouse tumors. *Nanomedicine.* 8, **2012**, 900–907.

78. He W, Yang F, Wu Y, Wen S, Chen P, Zhang Y, Gu N. Microbubbles with surface coated by superparamagnetic iron oxide nanoparticles. *Mater. Lett.* 68, **2012**, 64–67.

79. Cai X, Yang F, Gu N. Applications of magnetic microbubbles for theranostics. *Theranostics.* 2, **2012**, 103–112.

80. Mannell H, Pircher J, Fochler F, Stampnik Y, Raethel T, Gleich B, Plank C, Mykhaylyk O, Dahmani C, Wörnle M, Ribeiro A, Pohl U, Krötz F. Site directed vascular gene delivery in vivo by ultrasonic destruction of magnetic nanoparticle coated microbubbles. *Nanomedicine.* 8, **2012**, 1309–1318.

81. Wang X, Chen H, Zheng Y, Ma M, Chen Y, Zhang K, Zeng D, Shi J. Au-nanoparticle coated mesoporous silica nanocapsule-based multifunctional platform for ultrasound mediated imaging, cytoclasis and tumor ablation. *Biomaterials.* 34, **2013**, 2057–2068.

82. Kempen PJ, Greasley S, Parker KA, Campbell JL, Chang HY, Jones JR, Sinclair R, Gambhir SS, Jokerst JV. Theranostic mesoporous silica nanoparticles biodegrade after pro-survival drug delivery

and ultrasound/magnetic resonance imaging of stem cells. *Theranostics*. 5, **2015**, 631–642.

83. Min KH, Min HS, Lee HJ, Park DJ, Yhee JY, Kim K, Kwon IC, Jeong SY, Silvestre OF, Chen X, Hwang YS, Kim EC, Lee SC. pH-controlled gas-generating mineralized nanoparticles: A theranostic agent for ultrasound imaging and therapy of cancers. *ACS Nano*. 9, **2015**, 134–145.
84. Chung MF, Chen KJ, Liang HF, Liao ZX, Chia WT, Xia Y, Sung HW. A liposomal system capable of generating CO_2 bubbles to induce transient cavitation, lysosomal rupturing, and cell necrosis. *Angew. Chem. Int. Ed. Engl.* 51, **2012**, 10089–10093.
85. Cavalieri F, Zhou M, Tortora M, Lucilla B, Ashokkumar M. Methods of preparation of multifunctional microbubbles and their in vitro/in vivo assessment of stability, functional and structural properties. *Curr. Pharm. Des.* 18, **2012**, 2135–2151.
86. Liu Z, Lammers T, Ehling J, Fokong S, Bornemann J, Kiessling F, Gatjens J. Iron oxide nanoparticle-containing microbubble composites as contrast agents for MR and ultrasound dual-modality imaging. *Biomaterials*. 32, **2011**, 6155–6163.
87. Huynh E, Lovell JF, Helfield BL, Jeon M, Kim C, Goertz DE, Wilson BC, Zheng G. Porphyrin shell microbubbles with intrinsic ultrasound and photoacoustic properties. *J. Am. Chem. Soc.* 134, **2012**, 16464–16467.
88. Park JI, Jagadeesan D, Williams R, Oakden W, Chung S, Stanisz GJ, Kumacheva E. Microbubbles loaded with nanoparticles: A route to multiple imaging modalities. *ACS Nano*. 4, **2010**, 6579–6586.
89. Barrefelt AA, Brismar TB, Egri G, Aspelin P, Olsson A, Oddo L, Margheritelli S, Caidahl K, Paradossi G, Dahne L, Axelsson R, Hassan M. Multimodality imaging using SPECT/CT and MRI and ligand functionalized 99mTc-labeled magnetic microbubbles. *EJNMMI Res*. 3, **2013**, 12.
90. Yang F, Li Y, Chen Z, Zhang Y, Wu J, Gu N. Superparamagnetic iron oxide nanoparticle-embedded encapsulated microbubbles as dual contrast agents of magnetic resonance and ultrasound imaging. *Biomaterials*. 30, **2009**, 3882–3890.
91. Feshitan JA, Vlachos F, Sirsi SR, Konofagou EE, Borden MA. Theranostic Gd(III)-lipid microbubbles for MRI-guided focused ultrasound surgery. *Biomaterials*. 33, **2012**, 247–255.

92. Feshitan JA, Boss MA, Borden MA. Magnetic resonance properties of Gd(III)-bound lipid-coated microbubbles and their cavitation fragments. *Langmuir*. 28, **2012**, 15336–15343.

93. Palmowski M, Morgenstern B, Hauff P, Reinhardt M, Huppert J, Maurer M, Woenne EC, Doerk S, Ladewig G, Jenne JW, Delorme S, Grenacher L, Hallscheidt P, Kauffmann GW, Semmler W, Kiessling F. Pharmacodynamics of streptavidin-coated cyanoacrylate microbubbles designed for molecular ultrasound imaging. *Invest Radiol*. 43, **2008**, 162–169.

94. Tartis MS, Kruse DE, Zheng H, Zhang H, Kheirolomoom A, Marik J, Ferrara KW. Dynamic microPET imaging of ultrasound contrast agents and lipid delivery. *J. Control Release*. 131, **2008**, 160–166.

95. Xu RX. Multifunctional microbubbles and nanobubbles for photoacoustic imaging. *Contrast Media Mol. Imaging*. 6, **2011**, 401– 411.

96. Karpiouk AB, Aglyamov SR, Mallidi S, Shah J, Scott WG, Rubin JM, Emelianov SY. Combined ultrasound and photoacoustic imaging to detect and stage deep vein thrombosis: Phantom and ex vivo studies. *J. Biomed. Opt*. 13, **2008**, 054061.

97. Erpelding TN, Kim C, Pramanik M, Jankovic L, Maslov K, Guo Z, Margenthaler JA, Pashley MD, Wang LV. Sentinel lymph nodes in the rat: Noninvasive photoacoustic and US imaging with a clinical US system. *Radiology*. 256, **2010**, 102–110.

98. Fokong S, Theek B, Wu Z, Koczera P, Appold L, Jorge S, Resch-Genger U, van Zandvoort M, Storm G, Kiessling F, Lammers T. Image-guided, targeted and triggered drug delivery to tumors using polymer-based microbubbles. *J. Control Release*. 163, **2012**, 75–81.

99. Koczera P, Wu Z, Fokong S, Theek B, Appold L, Jorge S, Möckel D, Liu Z, Curaj A, Storm G, van Zandvoort M, Kiessling F, Lammers T. Fluorescently labeled microbubbles for facilitating translational molecular ultrasound studies. *Drug Deliv. Transl. Res*. 2, **2012**, 56–64.

100. Yuan B. Sensitivity of fluorophore-quencher labeled micro-bubbles to externally applied static pressure. *Med. Phys*. 36, **2009**, 3455–3469.

101. Yildirim A, Blum NT, Goodwin AP. Colloids, nanoparticles, and materials for imaging, delivery, ablation, and theranostics by focused ultrasound (FUS). *Theranostics*. 9, **2019**, 2572–2594.

102. Jamburidze A, Huerre A, Baresch D, Poulichet V, De Corato M, Garbin V. Nanoparticle-coated microbubbles for combined ultrasound imaging and drug delivery. *Langmuir*. 35, **2019**, 10087–10096.

103. Li X, Sui Z, Li X, Xu W, Guo Q, Sun J, Jing F. Perfluorooctylbromide nanoparticles for ultrasound imaging and drug delivery. *Int J Nanomedicine*. 13, **2018**, 3053–3067.

104. Sun IC, Emelianov S. Gas-generating nanoparticles for contrast-enhanced ultrasound imaging. *Nanoscale*.11, **2019**, 16235–16240.

Chapter 6

X-Ray Computed Tomography and Nanomaterials as Contrast Agents for Tumor Diagnosis

R. G. Aswathy[a] and D. Sakthi Kumar[b]

[a]*Future Industries Institute, Mawson Lakes Campus, University of South Australia, Adelaide, South Australia 5095, Australia*
[b]*Bio Nano Electronics Research Center, Graduate School of Interdisciplinary New Science, Toyo University, Kawagoe, Saitama 350-8585, Japan*

aswathy.ravindrangirija@unisa.edu.au, aswathyrg@gmail.com, sakthi@toyo.jp

6.1 Introduction

Over the last decades, there have been rapid progresses in X-ray imaging, especially with computed tomography (CT) for the diagnosis and treatment of various diseases, such as tumors, cardiovascular diseases, etc. X ray-computed tomography (X-ray CT) is the second most efficient and cost-effective imaging modality after optical imaging. X-ray CT is one of the proficient medical imaging modality using tomography generated by computer processing. Digital geometry processing creates a three-dimensional (3D) image of the interior of the body

Bionanotechnology in Cancer: Diagnosis and Therapy
Edited by D. Sakthi Kumar and Aswathy Ravindran Girija

ISBN 978-981-4800-30-3 (Hardcover), 978-0-429-42291-1 (eBook)
www.jennystanford.com

from a two-dimensional (2D) X-ray image. CT generates data, which could be operated, through 'windowing' process, to determine structures based on blocking X-ray beam. Apart from medicine, CT is also used for non-destructive materials testing. For example, in archeological field, X-ray CT is used for imaging the contents of sarcophagi, or the DigiMorph project that uses a CT scanner to study biological and paleontological specimens.

X-ray imaging is based on photon attenuation among materials. The major factors for the estimation of penetration attenuation and energy deposition of materials are mass-attenuation and mass energy-absorption coefficients. Mass-attenuation coefficient quantifies X-ray opaqueness of the material and mass energy-absorption coefficient signifies the fractional ionization by incident X-rays in the material. Hence, the effectiveness of X-ray imaging contrast agents is based on these coefficients [1].

CT is based on the principle of X-ray [2]. For the generation of images based on X-rays, the X-rays penetrated through the body are either absorbed or attenuated at various intensities generating a profile or matrix of X-ray beams of different strengths. The X-ray profile is recorded on film, resulting in the development of an image. In CT imaging, a detector that quantifies the X-ray profile substitutes the film. CT is based on the density of tissue penetration by X-ray beam and its quantification from attenuation coefficient. Attenuation coefficient is determined by measuring how easily a material can be penetrated by the beam. It measures how much the beam is weakened by the material while it is penetrating through (attenuated). In CT, cross-sectional images (slices) of the body are generated and they are recreated from the attenuation coefficients of X-ray beams of the material under study. In CT, the X-ray emitter revolves around the subject and detector is positioned in diametrically opposite side of the emitter. Each time, the X-ray tube and detector make a 360° rotation around the object, the detector takes several shots (profiles) of the attenuated X-ray beam. To generate tomographic images from the data set in "raw" scan, the computer depends on complex mathematical algorithms for the reconstruction of image. Although images generated were in different planes, such as axial, transverse, or orthogonal, scanners reformat the data in many planes or as volumetric (3D) illustrations of the structures (Figure 6.1).

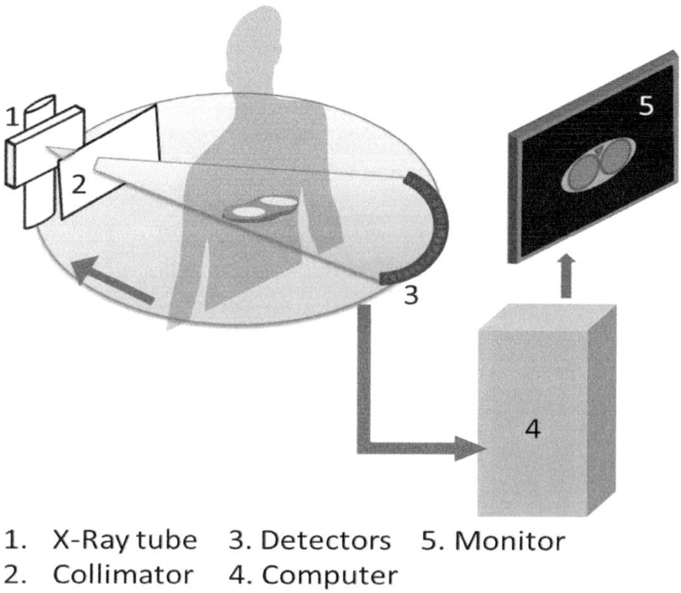

Figure 6.1 Schematic of relationship of X-ray tube, subject, detector, image reconstruction computer, and display monitor.

6.2 Iodinated and Gold-Based Nano Contrast Agents

Although CT or micro-CT is being used to acquire high-resolution volumetric images in short scan time, the imaging of soft tissues or blood vessels remains a challenge. In the application of conventional contrast agents, low-molecular-weight hydrophilic iodinated molecules possess several limitations, such as fast clearance, kidney toxicity, propensity of extravasation, and possibility of allergy. The development of novel contrast agents, exhibiting stealth properties in blood and long retention time with high biocompatibility, is highly necessary for the successful application of this technology for flawless imaging. Blood pool contrast agents (BPCAs) incorporate nanoparticles (NPs) with heavy atoms, which are covalently attached, solubilized, or encapsulated [3]. A few of them include nano-emulsions, chylomicron, liposomes, micelles, polymeric NPs, or dendrimers. Nano-emulsification of lipophilic iodinated complexes via spontaneous emulsification method is the promising and simple method for the preparation of BPCA, thus allowing high iodine loading, up to 85 mgI/mL. However, iodinated nano-emulsions also have a few limitations in some aspects of pharmacokinetics, tolerance, toxicity, in vivo behavior, lower X-ray absorption coefficient, etc. [4–6].

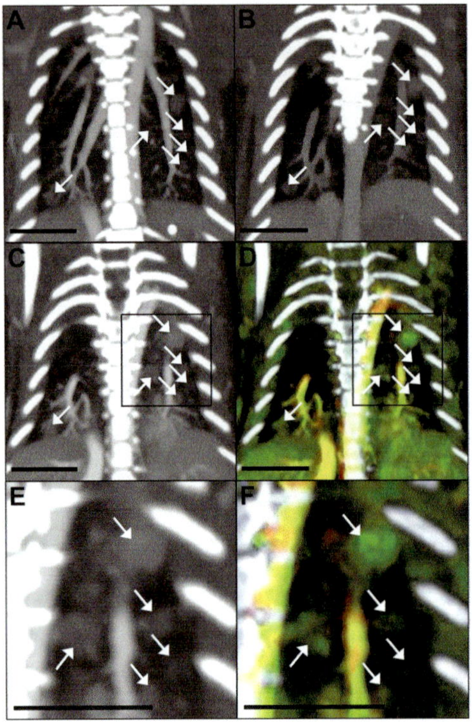

Figure 6.2 Coronal CT images displaying filtered 80 kVp dataset from CT scan 1 (A), CT scan 2 (B), CT scan 3 (C), and overlay of the DE decomposition of CT scan 3 (D) with bones segmented out (bones in white). (E) is a zoomed-in image of the right side of CT scan 3, and (F) is a zoomed in image of the right side of the DE overlay of CT scan 3. The boxes in (C) and (D) show the regions that are enlarged in (E) and (F), respectively. The image depicts CT images on the lungs at 3 time points and overlay of the gold and iodine DE decomposition maps where multiple tumors (size of 0.2–1.5 mm^3, see arrows) in both lungs can be observed. Tumors can be visualized in green owing to the accumulation of gold. Iodine present in the blood stream can be observed as red and the overlay as yellow observed in the liver and soft tissues because of the extravasation of gold and iodine. Reprinted with permission from Ref. [9].

An aggressive and indolent lung tumor was distinguished with iodine-based liposome (Lip-I) as contrast agent [7]. Liposomal-iodinated CT contrast agent differentiated slow and rapidly proliferating pulmonary nodules in mice bearing primary non-small cell lung cancer. Early phase imaging with contrast agent permitted visualization of blood supply to nodule, while delayed-phase imaging displayed the differentiation of slow and rapidly growing nodules based on signal enhancement. Superior accumulation and uptake of contrast agent was observed in rapidly growing pulmonary nodules, thus facilitating early detection of pulmonary lesions. The gap among initial and late phase imaging was overcome by using two discrete NPs and dual-energy (DE) CT. DE-CT offered substantial advancement in clinical imaging of tumors [8]. X-ray absorption of the contrast agents is based on their atomic weight and incident X-ray energy. Therefore, elements of dissimilar atomic weight are distinguished upon their distinctive attenuation coefficients at various X-ray energies. DE-CT permits the discerning imaging and quantification of several contrast agents in one scan. DE micro-CT imaging in lung cancer in mice with gold and iodine NPs is depicted in Figure 6.2 [9]. DE micro-CT permitted added functional data for the characterization of tumor in murine lung tumors. Gold and iodine NPs offered quantification of blood volume in tumor and vascular permeability. Distinguishing of both gold and iodine contrast agents and uptake of gold NPs in vivo in the liver, spleen, and tumors was observed in DE-CT imaging. The tumor blood volume measurements exhibited high correlation between fractional blood volume and histology-based micro vascular density. DE-CT could play a significant role in the characterization of lung tumor.

Heparin-coated gold nanoparticle (AuNP) demonstrated its capability as non-invasive CT imaging agent as well as a therapeutic agent [10]. The simple surface modification strategy via mussel-inspired heparin–3,4-dihydroxy-L-phenylalanine (DOPA) conjugates was biocompatible and displayed liver-specific delivery of the AuNPs resulted in improved liver-specific CT images in in vivo studies. However, higher dose of AuNPs for X-ray absorption and toxicity associated with high concentration of NPs was the major limitation of the study. A nano-liposomal, blood pool agent exhibiting long circulating time, encapsulated with iodinated contrast agent was developed [11]. It was studied for micro-CT imaging of rats bearing breast cancer. The liposomal NPs were used to visualize 3D vascular structure of tumors with 100 micron isotropic resolution. The study establishes the efficacy of the NP-based contrast agent for imaging the development of blood vessels (angiogenesis) in mammary carcinoma. Enhancement of contrast agents in blood vessels was observed 4 h after the injection of contrast agents, and a residence time of at least 4 h in the blood pool was observed compared to non-encapsulated contrast agents. The nanoprobe endorsed a difference in attenuation between the tumor blood vessels and muscular tissue of 326 ± 151 Hounsfield unit (HU), enabling imagining of the 3D vascular network. The difference in attenuation facilitates the evaluations of vascular percentages and volume of tumor. The liposomal wrapping encapsulates the contrast agent that results in the clearance via liver and spleen than renal filtration. The approach holds huge potential for the study of angiogenesis in tumor and evaluation of anti-angiogenic therapy associated with tumor.

The perfect macromolecular CT contrast agent should be stable, monodispersed, hydrophilic, showed reduced viscosity, highly biocompatible, higher concentrations of iodine or any radiodense atoms, and exhibited intravascular distribution and clearance from the body. The synthesis and characterization of water-soluble iodinated dendritic NP with generation 4 (G4) starburst polyamidoamine (PAMAM) dendrimer as core attached to 3-*N*-[(*N*′,*N*′-dimethylaminoacetyl)amino]-α-ethyl-2,4,6-triiodobenzenepropanoic acid (DMMA-IPA) to form G-4-(DMAA-IPA)$_{37}$ were reported [12]. The iodine atoms per dendrimers unit were 111. Care was maintained while designing DMAA-IPA for not disturbing the properties such as total charge and solubility of dendrimers after the conjugation of iodine and G-4-(DMAA-IPA)$_{37}$ was the initially used dendritic NPs for CT imaging.

Poly (ethylene glycol) (PEG) holds numerous advantageous attributes for such applications, and dendritic iodinated contrast agents with PEG cores were developed [13]. The PEGylation of biomacromolecules upsurged blood half-life reduced the immunogenicity. In aqueous environment, PEG repels adjacent biomacromolecules and enhances solubility. A method to amplify the termini of PEG for conjugation to be used as a dendrimer core via carbamate linkages was developed [13]. Triiodophthalamide groups were linked to multiple free amino terminals of paired symmetrical dendritic polylysines originated from either ends of PEG core, to develop iodinated contrast agent for CT imaging. The developed material was projected for intravascular application and marked innovative improvement in dendrimer-centered contrast agents with prospective applications in the diagnosis of diseases such as ischemic injury and cancer and inflammation.

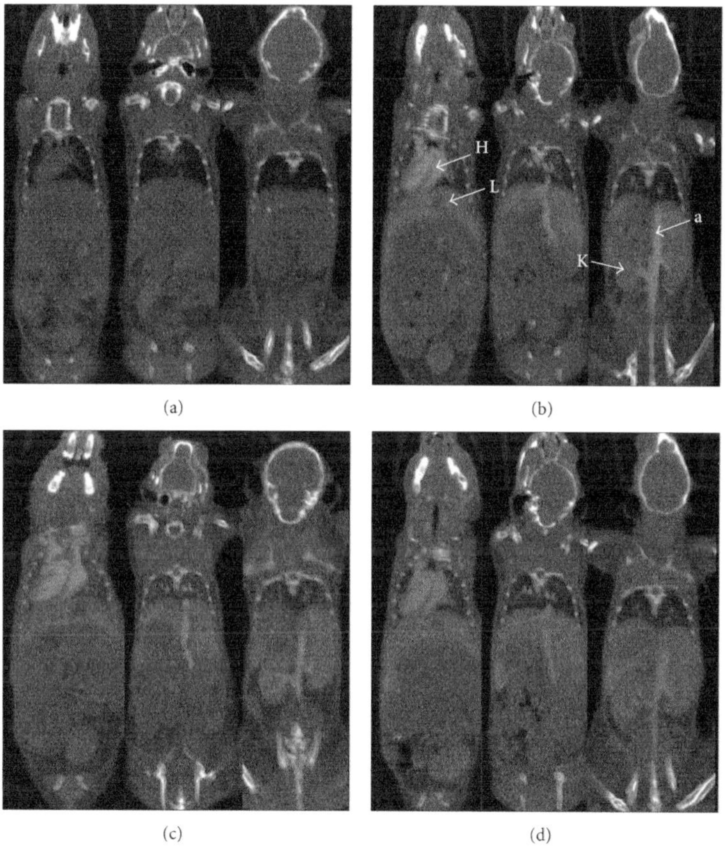

Figure 6.3 X-ray CT images in mice model: (a) before injection, (b) immediately, (c) 30 m, and (d) 24 h after post-tail vein administration of methoxy PEG-iodine-capped AuNPs. The arrows indicate H: heart, L: liver, K: kidney, and a: aorta, respectively. Reprinted with permission from Ref. [14].

The concept of gold NPs covered with iodine and PEG using chemisorption of iodine and replacement of methoxy PEG-SH to the surface of gold NPs was developed [14]. Under in vitro conditions, methoxy PEG-iodine-capped AuNPs exhibited superior enhancement than PEG-coated gold NPs with same quantity of gold. When methoxy PEG-iodine-capped AuNP was administrated in mice, an intense enhancement of contrast in organs, such as heart, aorta, liver, and kidney, was detected (Figure 6.3). The particles were biocompatible and contrast sustained up to 5 days. The methoxy PEG-iodine-capped AuNP could be an excellent CT contrast agent for imaging blood pool. The need of large quantity of the polyiodinated contrast agents and very short circulation time in blood are the drawbacks of conventional contrast agents in X-ray imaging. The iodine capped gold NPs are promising NP candidate for resolving these difficulties.

6.3 Radiopaque Polymeric NPs

There have been great interests in the development of radiopaque polymers, as X-ray contrast agent for imaging for long-term image tracking has been suggested as an alternative contrast agent [15–19]. Emulsion polymerization method with monomer triiodophenyl methacrylate, 2-methacryloyloxyethyl (2,3,5-triiodobenzoate) (MAOETIB) and potassium persulfate (PPS) as an initiator and sodium dodecyl sulfate (SDS) as a surfactant used for the synthesis of iodinated radiopaque polymeric NPs [20]. The effect of parameters influencing polymerization, such as concentration of MAOETIB, SDS, and PPS on the molecular weight, distribution of size, and polymerization yield, was elucidated. The polymeric NPs comprise 58% by weight of iodine, hence it retained a substantial radiopaque nature. The radio opacity of the iodinated NPs dispersed in water and dry state under in vitro condition was confirmed with CT scanner. Intravenous injection of radiopaque NPs in canine model demonstrated enhanced visibility of liver, lymph node, kidney, and spleen in CT-imaging. Radiopaque NPs prepared from poly (allylamine) derivatives were developed and to make the polymer X-ray detectable. Poly (allylamine) was tailored by attaching

4-iodobenzoyl chloride or 2,3,5-triiodobenzoyl chloride [21]. Nanoprecipitation of poly (allylamine) resulted in the stable suspensions of spherical NPs. NPs suspensions of 4.5 wt% presented X-ray visibility ranging from 185 ± 235 HU. The non-agglomeration property of novel poly (allylamine) radiopaque NPs indicates their administration through a catheter without any blockage.

6.4 Inorganic Nanomaterials

Currently used X-ray contrasting agents for clinical applications are based on small hydrophilic iodinated molecules, which are considered as safe at low dosage and are effective. Apart from these advantages, these contrast agents (iodinated molecules) possess a few drawbacks, such as complex synthetic and processing procedure, vascular permeation, rapid clearance from kidney, kidney associated toxicity, and low specificity [22–24]. As an alternative to traditional X-ray iodinated contrast agents, nano-engineered inorganic nanomaterials-based contrast agents have been developed. The inorganic NPs comprise heavy atoms, exhibit strong X-ray radiation absorption, and increase imaging contrast in CT by several folds even at low X-ray doses. The exposure time of X-ray can be drastically reduced, thus minimizing the risks associated with highly ionizing radiations. The simple synthetic techniques and the possibility of the modification of reactions and efficient surface modification, thus tailoring the properties and morphology of NPs, allow them to be used as potential candidates for biological applications. NPs of higher atomic number elements are prospective CT contrast agents to bypass the limitations of iodinated compounds. For example, nanoparticulate Bi_2S_3, TaO_x ($x \approx 1$), Au, Pt, $NaYbF_4$, $NaGdF_4$, $NaLuF_4$, etc. were investigated as novel CT contrast agents and exhibited apparent contrast effect in vivo [25–36].

Polymer-encapsulated bismuth sulfide (Bi_2S_3) NPs have revealed excellent functioning than iodine as CT contrast agent and established a safe in vivo profile [25, 28]. Bismuth sulfide NPs offered high absorption of X-ray (Bi exhibits fivefold X-ray absorption than iodine), high stability of Bi at large concentrations (0.25 M Bi^{3+}), prolonged circulation time (>2 h), and high biocompatibility than iodinated contrast agents. The study demonstrated the efficacy of polymer encapsulated Bi_2S_3 NPs for improved vascular, lymph nodes, and liver imaging in rodents. However, surface modification of Bi_2S_3 NPs, effort in size control of NPs, and risks associated with toxicity of Bi still remain.

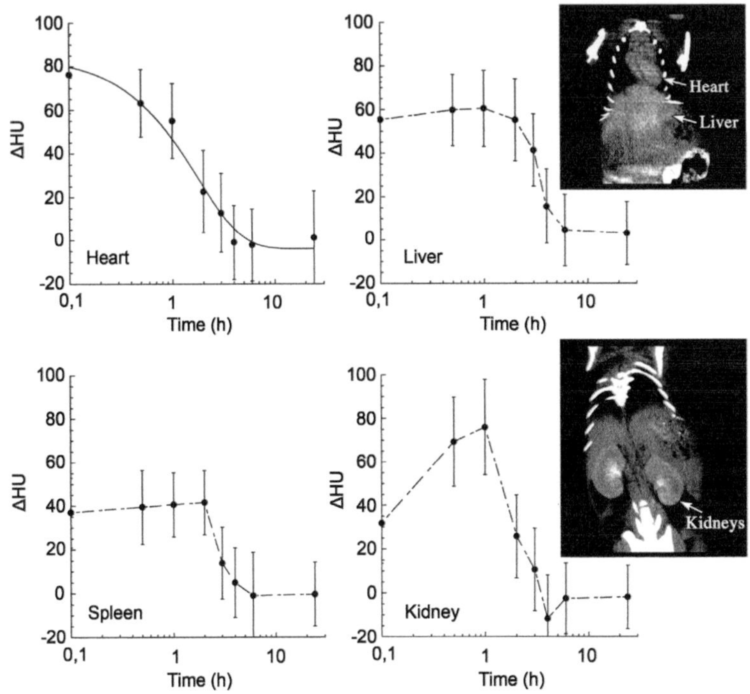

Figure 6.4 In vivo imaging performed in mice by Micro-CT. Contrast enhancement after intravenous administration of PCL-coated WO_3 NPs for region of interest (ROI) in the left ventricle (heart), the hepatic parenchyma (liver), the spleen, and the left kidney. The development of X-ray attenuation was followed-up during 24 h after administration. Pictures show representative average intensity projection of vascular contrast enhancement immediately after injection (top) and of kidneys 1 h after administration (bottom). Reprinted with permission from Ref. [37].

A formulation of proficient contrast agent for X-ray CT imaging based on tungsten that overcame the limitations of the classical iodinated contrast agents was developed [37]. The NPs, of size around 100 nm, were prepared using tungsten oxide nanocrystals (WO_3) and coated with biodegradable polymer (poly-ε-caprolactone, (PCL)) and PEGylated surfactants, to stabilize them in aqueous medium. The in vivo imaging experiments in a micro-CT scanner in mice allowed the evaluation of these NPs to display elevated X-ray attenuation and circulation time (up to 3 h), allowing blood pool imaging along with their pharmacokinetics in blood, liver, spleen, and kidney (Figure 6.4).

Tantalum oxide has garnered greater attention owing to its biocompatibility and relatively high X-ray attenuation coefficient that has been exploited for clinical applications [26]. Bioinert tantalum oxide NP that has the potential for high-performance X-ray CT imaging has been synthesized using microemulsion technique. The surfaces of NPs were modified using silane derivatives via in situ sol-gel reaction [26]. PEGylation and incorporation of fluorescent dye were facilitated by the silane-modified surface in NPs that enabled the NPs for fluorescence imaging (FI) and X-ray CT imaging. The NPs presented significant enhancement in in vivo X-ray CT angiography and dual modal image-guided lymph node imaging. The in vivo studies with tantalum oxide NPs demonstrated the biocompatibility of NPs and did not adversely affect the normal function of vital organs.

6.5 Multi-Modal Imaging of X-Ray CT with MRI, PET, and FI

An approach for the design of Fe_3O_4@Au nanocomposite particles (NCPs) as dual modal contrast agents for magnetic resonance imaging (MRI) and CT imaging has been performed via a combination of a layer-by-layer self-assembly process and dendrimer chemistry (Figure 6.5) [38]. The NPs were stable and biocompatible. The developed Fe_3O_4@Au nanocomposites were used for dual modal imaging of cancer cells in in vitro studies and liver imaging in MRI and imaging of subcutaneous tissue in CT in in vivo models. The results indicated that the fabricated Fe_3O_4@Au nanocomposites could be used as a T_2 negative-contrast agent for in vivo MRI of mouse liver. In CT images, the soft tissues parenchyma, blood vessels, and muscles were observed in dark, whereas bones were visible as white, due to high density and stronger X-ray absorption.

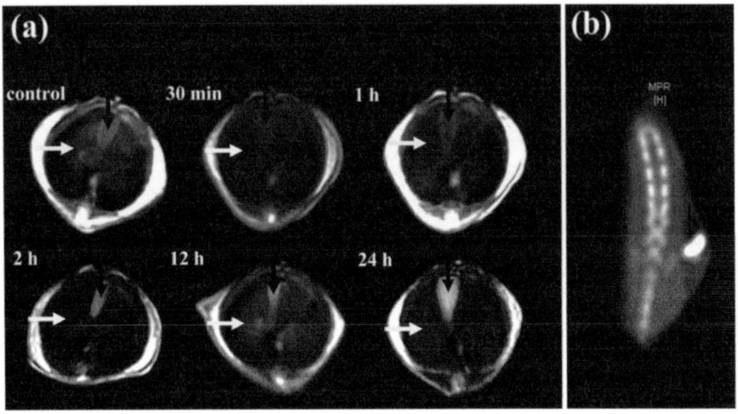

Figure 6.5 (a) T_2-weighted images of mouse liver before and after the administration of Fe_3O_4@Au NCPs. The black arrow denotes gallbladder and white arrow signifies liver. (b) CT image of a mouse administrated with Fe_3O_4@Au NCPs. The black arrow points the site of injection. Reprinted with permission from Ref. [38].

A dual modal molecular imaging was studied in mice with FePt NPs integrated to anti-Her2 antibody targeting the Her2/neu oncogene expression in cancer [39]. FePt NPs of three different sizes were used for biodistribution studies and the particles cleared from the body one week after administration. High serum concentration and half-life during circulation were observed for 12 nm-FePt NPs that were followed by 6 nm- and 3 nm-FePt NPs. The 3 nm-FePt exhibited higher concentration in brain that was perfect for multimodal brain imaging. Anti-Her2 antibody attached to FePt NPs presented molecular expression dependent CT/MRI dual imaging in in vitro studies. Enhancement of contrast of Her2/neu overexpression in cancer in CT and MRI was observed in in vivo models after the administration of NPs. The study signifies the prospect of FePt NPs as dual modal molecular imaging contrast agents in clinical studies.

Iron oxide (IO) NPs labeled with ^{18}F, a radioisotope was developed to reduce the amount of NPs for tumor imaging [40]. The NPs were effectively imaged with different modalities, including positron emission tomography (PET), CT, and FI. Click chemistry was used for the addition of ^{18}F to NPs under mild reaction conditions, maintaining the biocompatibility. The radiolabeled NPs were detected with PET/CT after administration in mice. The incorporation of tumor targeting agents to the NPs improved their application of tumor imaging.

Polysaccharide and diethylenetriaminepentaacetic acid (DTPA)-incorporated magnetofluorescent NPs labeled with the ^{64}Cu to generate PET, MR, and optically active contrast agent were developed [41]. Mice lacking apolipoprotein E (ApoE) was intravenously administrated with NPs and the PET activity after 24 h recorded the high plaque areas and the image correlated with CT, MRI, and optical imaging. X-ray imaging of aortas confirmed the uptake of NPs in atheromata, the NPs accumulated mainly in macrophages were confirmed by FI, and flow cytometry of cells were detached from aortas. The study determines the ability of a trimodal imaging of NP for macrophage detection in atherosclerotic plaques.

The development of all-in-one contrast agents that enable the detection with multiple imaging techniques, including radiolabelled imaging modalities (PET and single photon emission tomography (SPECT)), magnetic-based MRI, CT, or the fluorescence-based optical imaging, is an extreme challenge in biomedical diagnostic research. PEGylated gold/silica based NPs with fluorescent and paramagnetic lipid coatings were developed and studied the trimodal imaging capability of integrated contrast agents for imaging macrophages in vitro via MRI, CT, and FI, and in vivo imaging of mice livers via MRI and CT [42]. The confocal image displayed the red fluorescence signal originating from the Cy5.5-incorporated NPs and that image overlaid with blue fluorescence from cell nuclei stained with 4′,6-diamidino-2-phenylindole (DAPI) and bright field images of macrophages. MRI and CT imaging displayed enhancement in contrast in cell treated with NPs. Under in vivo studies, MRI of the mice livers exhibited enhancement in the mean intensity (24%) post NPs injection, and in CT images mean X-ray attenuation was increased to 50% post NPs injection.

Most of the NPs designed for imaging are multi probe (either dual or trimodal): CT/MRI or CT/NIR-FI [39, 43–47]. The integration of lanthanide ions, including Lu^{3+}, Yb^{3+}, Tm^{3+}, and Gd^{3+}, into a single nanoconjugate, $NaLuF_4{:}Yb^{3+}$, $Tm^{3+}@SiO_2$-

GdDTPA, offers a facile design approach to construct multimodal imaging agents that offers a high signal-to-noise ratio (SNR) and deep tissue imaging for small animal imaging [48–53]. A dual modal imaging system based on PEGylated $NaYbF_4$:Tm^{3+} NPs ranging in the size of 20 nm for CT and near-infrared (NIR)-FI was developed [54]. The Yb^{3+} ions in $NaYbF_4$:Tm^{3+} NPs function both as efficient CT contrast and as a sensitizer offering strong NIR-fluorescent emissions. The nanoprobes exhibited exceptional in vitro and in vivo contrast in the bimodal imaging, high biocompatibility with very less tissue damage up to one month. These NPs were cleared mostly via feces without demonstrable traces in animal bodies. The presence of a large number of unpaired electrons and high atomic weight makes Gadolinium as an excellent contrast agent for MRI and CT imaging. As an example a multifunctional lanthanide ($NaLuF_4$) NP for NIR-to-NIR upconversion luminescence (UCL), X-ray CT and T_1-enhanced MR exhibiting trimodal in vivo imaging were developed [53]. The property of NIR-to-NIR UCL augments the high SNR of in vivo imaging of small animal. The Gd^{3+} grafting on the surface of NPs produces high r_1 relaxivity for the core–shell NPs that become appropriate for T_1 MR imaging. The NP with the core of Yb^{3+}- and Tm^{3+}-doped $NaLuF_4$ displayed a high X-ray attenuation that was useful for CT imaging. Additionally, SiO_2 shell diminished the toxicity of lanthanide-based NPs for small animals that was established by 3-(4,5-dimethylthiazol-2-yl)-2,5-diphenyltetrazolium bromide (MTT) assay and histological analyses.

Yb-based nanoparticulate CT contrast agents have been developed [55] and owing to the attenuation characteristics of Yb-based NPs, they offer higher contrast than clinical iodinated contrast agents. They tend to be efficient than presently employing Au-, Bi-, Pt-, and Ta-based NP-based CT contrast agents operating at 120 kilo voltage peak (kVp). The perfection is chiefly credited to Yb's K-edge energy that is within the higher energy section of X-ray spectrum. The K-edge energy of an element is the rapid enhancement in the attenuation coefficient of photons at energy beyond the binding energy of K-shell electron of atoms within the X-ray beam. When encapsulated in a PEG shell, the NPs showed enhanced biocompatibility and prolonged circulation time in vivo [56]. These NPs exhibited high performance in in vivo angiography and image guided lymph node mapping. Furthermore, NIR-to-visible (NIR-to-Vis) or NIR- to-NIR UCL property of the Yb-NPs was used for deep tissue imaging and preventing the tissue damage related with UV excitation [57]. Doping of Er^{3+} and Gd^{3+} in these NPs imparted them with MRI capability [58, 59].

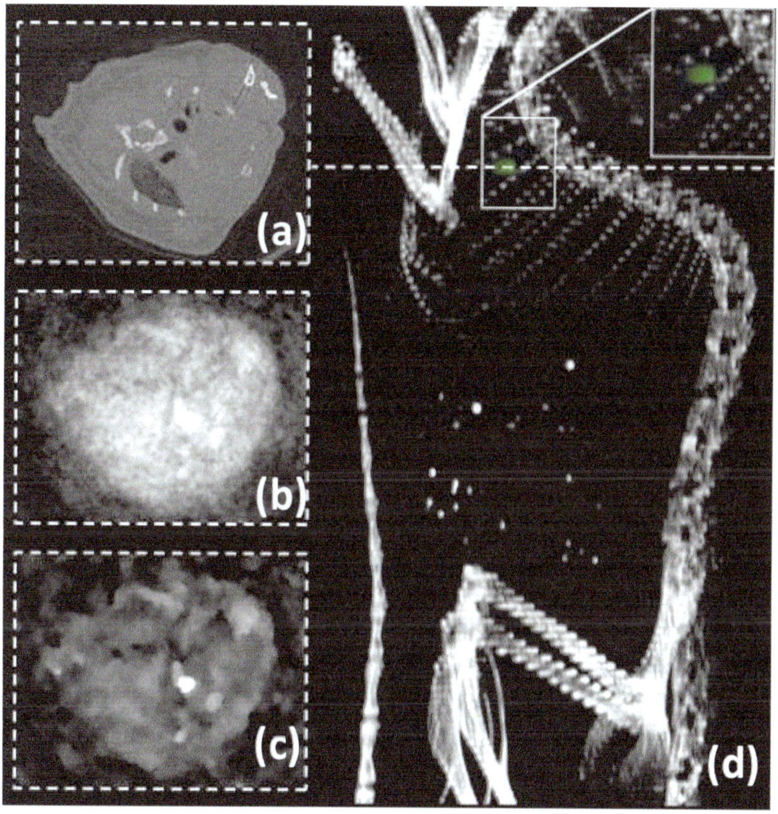

Figure 6.6 Blood pool imaging in mouse after bolus administration of nontargeted Yb nanocolloids (a) pseudo-conventional CT image constituted from spectral measurements, the dashed line represents slice through heart. Reconstruction of image with Yb signal after 1 (b) and 20 iterations. (c) The volume rendered conventional CT image with super positioned Yb signal (green) displaying Yb accumulation in heart (d) and separation from bone (d). Reprinted with permission from Ref. [62].

Ytterbium-based molecular imaging agent was developed to be used with spectral "multicolor" CT. In addition to the rapid tomographic X-ray imaging offered by the traditional CT, spectral CT (multicolored CT) simultaneously differentiates metallic contrast agents depending on metal's distinctive X-ray K-edge energy [60, 61]. The method involved an organically soluble organometallic Yb (III) complex to yield noncrystalline nanocolloids of Yb integrating high density of Yb (>500 K/NP) into a metal particle [62]. High Yb content in hydrophobic metal complexes was stably focused into lipid-encapsulated nanocolloids and offered novel molecular imaging for spectral "multicolor" CT. Difference in contrast owing to Yb was hardly evident in CT image (Figure 6.6a), whereas the K-edge image with spectral CT and iterative reconstruction presented Yb NC high SNR (Figure 6.6b,c). A high concentration of contrast in heart was separated from background noise. It was suggested that the reduced concentrations of NPs in the vascular system are likely to develop in future with further improvements in spectral CT prototype. Bio-distribution and bio-elimination studies projected the safety and clearance of NPs in rodent model.

6.6 Conclusion and Future Prospects

There has been substantial improvement in the development of NPs as contrast agents for CT imaging. Targeted cellular tracking contrast agents for specific detection and blood pool imaging are possible with X-ray CT imaging modality. NPs based on gold, tantalum, and bismuth with enhanced contrast with the potential of renal excretion are being developed, which display longer retention time thus enhancing the imaging capability [63–71]. Considerable advancement in DE-CT cancer imaging in clinical studies permitted the discerning imagining and quantification of several contrast agents. Apart from the advancements, progress in the synthetic strategy involving large-scale development of high contrast density NPs needs to be addressed. Interdisciplinary research involving material scientists and medical professionals including radiologist is crucial for the successful development and application of contrast agents. The toxicity profile due to the use of heavy metals has to be addressed effectively to potentially reduce side effects. Owing to the extensive research in this field, huge progress with the multimodal imaging capabilities will be attained soon for clinical application.

References

1. Seltzer SM, Berger MJ. Bremsstrahlung spectra from electron interactions with screened atomic nuclei and orbital electrons. *Nucl. Instrum. Methods Phys. Res. B.* 12, **1985,** 95–134.
2. Goldman LW. Principles of CT and CT technology. *J. Nucl. Med. Technol.* 35, **2007,** 115–128.
3. Clough RE, Hussain T, Uribe S, Greil GF, Razavi R, Taylor PR, Schaeffter T, Waltham M. A new method for quantification of false lumen thrombosis in aortic dissection using magnetic resonance imaging and a blood pool contrast agent. *J. Vasc. Surg.* 54, **2011,** 1251–1258.
4. Jaspers K, Versluis B, Leiner T, Dijkstra P, Oostendorp M, van Golde JM, Post MJ, Backes WH. MR angiography of collateral arteries in a hind limb ischemia model: Comparison between blood pool agent Gadomer and small contrast agent Gd-DTPA. *PLoS One.* 6, **2011,** e16159.
5. Liu Y, Ai K, Lu L. Nanoparticulate X-ray computed tomography contrast agents: From design validation to in vivo applications. *Acc. Chem. Res.* 45, **2012,** 1817–1827.
6. He W, Ai K, Lu L. Nanoparticulate X-ray CT contrast agents. *Sci. China Chem.* 58, **2015,** 753–760.
7. Badea CT, Athreya KK, Espinosa G, Clark D, Ghafoori AP, Li Y, Kirsch DG, Johnson GA, Annapragada A, Ghaghada KB. Computed tomography imaging of primary lung cancer in mice using a liposomal-iodinated contrast agent. *PloS one.* 7, **2012,** e34496.
8. Fornaro J, Leschka S, Hibbeln D, Butler A, Anderson N, Pache G, Scheffel H, Wildermuth S, Alkadhi H, Stolzmann P. Dual- and multi-energy CT: Approach to functional imaging. *Insights Imaging.* 2, **2011,** 149–159.
9. Ashton JR, Clark DP, Moding EJ, Ghaghada K, Kirsch DG, West JL, Badea CT. Dual-energy micro-CT functional imaging of primary lung cancer in mice using gold and iodine nanoparticle contrast agents: A validation study. *PLoS One.* 9, **2014**, e88129.

10. Sun IC, Eun DK, Na JH, Lee S, Kim IJ, Youn IC, Ko CY, Kim HS, Lim D, Choi K, Messersmith PB, Park TG, Kim SY, Kwon IC, Kim K, Ahn CH. Heparin-coated gold nanoparticles for liver-specific CT imaging. *Chemistry*. 15, **2009,** 13341–13347.

11. Samei E, Saunders RS, Badea CT, Ghaghada KB, Hedlund LW, Qi Y, Yuan H, Bentley RC, Mukundan S Jr. Micro-CT imaging of breast tumors in rodents using a liposomal, nanoparticle contrast agent. *Int. J. Nanomedicine*. 4, 2**009**, 277–282.

12. Yordanov AT, Lodder AL, Woller EK, Cloninger MJ, Patronas N, Milenic D, Brechbiel MW. Novel iodinated dendritic nanoparticles for computed tomography (CT) imaging. *Nano Lett*. 2, **2002,** 6595–6599.

13. Fu Y, Nitecki DE, Maltby D, Simon GH, Berejnoi K, Raatschen HJ, Yeh BM, Shames DM, Brasch RC. Dendritic iodinated contrast agents with PEG-cores for CT imaging: Synthesis and preliminary characterization. *Bioconjug. Chem*. 17, **2006**, 1043–1056.

14. Kim SH, Kim EM, Lee CM, Kim DW, Lim ST, Sohn MH, Jeong HJ. Synthesis of PEG-iodine-capped gold nanoparticles and their contrast enhancement in in vitro and in vivo for X-Ray/CT. *Journal of Nanomaterials*. 2012, **2012,** 9.

15. Mottu F, Rufenacht DA, Doelker E. Radiopaque polymeric materials for medical applications: Current aspects of biomaterial research. *Invest. Radiol*. 34, **1999,** 323–335.

16. Torchilin VP. PEG-based micelles as carriers of contrast agents for different imaging modalities. *Adv. Drug Deliv. Rev*. 54, **2002,** 235–252.

17. Saralidze K, Aldenhoff YBJ, Knetsch MLW, Koole LH. Injectable polymeric microspheres with X-ray visibility: Preparation, properties, and potential utility as new traceable bulking agents. *Biomacromolecules*. 4, **2003,** 793–798.

18. Emans PJ, Saralidze K, Knetsch MLW, Gijbels JJ, Kuijer R, Koole LH. Development of new injectable bulking agents: Biocompatibity of radiopaque polymeric microspheres studies in a mouse model. *J. Biomed. Mater. Res. A*. 73, **2005,** 430–436.

19. Nirmala RJ, Juby P, Jayakrishnan A. Polyurethanes with radiopaque properties. *Biomaterials*. 27, **2006,** 160–166.

20. Galperin A, Margel D, Baniel J, Dank G, Biton H, Margel S. Radiopaque iodinated polymeric nanoparticles for X-ray imaging applications. *Biomaterials*. 28, **2007,** 4461–4468.

21. Mawad D, Lauto A, Penciu A, Mehier H, Fenet B, Fessi H, Chevalier Y. Synthesis and characterization of novel radiopaque poly(allyl amine) nanoparticles. *Nanotechnology*. 21, **2010**, 335603.

22. Lusic H, Grinstaff MW. X-Ray computed tomography contrast agents. *Chem. Rev.* 113, **2013,** 1641–1666.

23. Haller C, Hizoh I. The cytotoxicity of iodinated radiocontrast agents on renal cells in vitro. *Invest Radiol.* 39, **2004,** 149–154.

24. Hizoh I, Haller C. Radiocontrast-induced renal tubular cell apoptosis: Hypertonic versus oxidative stress. *Invest. Radiol.* 37, **2002,** 428–434.

25. Rabin O, Perez JM, Grimm J, Wojtkiewicz G, Weissleder R. An X-ray computed tomography imaging agent based on long circulating bismuth sulphide nanoparticles. *Nat. Mater.* 5, **2006,** 118–122.

26. Oh MH, Lee N, Kim H, Park SP, Piao Y, Lee J, Jun SW, Moon WK, Choi SH, Hyeon T. Large-scale synthesis of bioinert tantalum oxide nanoparticles for X-ray computed tomography imaging and bimodal image-guided sentinel lymph node mapping. *J. Am. Chem. Soc.* 133, **2011,** 5508–5515.

27. Bonitatibus JPJ, Torres AS, Goddard GD, FitzGerald PF, Kulkarni AM. Synthesis, characterization, and computed tomography imaging of a tantalum oxide nanoparticle imaging agent. *Chem. Commun.* 46, **2010,** 8956–8958.

28. Kim D, Park S, Lee JH, Jeong YY, Jon S. Antibiofouling polymer-coated gold nanoparticles as a contrast agent for in vivo X-ray computed tomography imaging. *J. Am. Chem. Soc.* 129, **2007,** 7661–7665.

29. Alric C, Taleb J, Duc GL, Mandon C, Billotey C, Meur-Herland AL, Brochard T, Vocanson F, Janier M, Perriat P, Roux S, Tillement O. Gado- linium chelate coated gold nanoparticles as contrast agents for both X-ray computed tomography and magnetic resonance imaging. *J. Am. Chem. Soc.* 130, **2008,** 5908–5915.

30. Hainfeld JF, Slatkin DN, Focella TM, Smilowitz HM. Gold nanoparticles: A new X-ray contrast agent. *Br. J. Radiol.* 79, **2006,** 248–253.

31. Popovtzer R, Agrawal A, Kotov NA, Popovtzer A, Balter J, Carey TE, Kopelman R. Targeted gold nanoparticles enable molecular CT imaging of cancer. *Nano Lett.* 8, **2008,** 4593–4596.

32. Kim Y-H, Jeon J, Hong SH, Rhim W-K, Lee Y-S, Youn H, Chung JK, Lee MC, Lee DS, Kang KW, Nam JM. Tumor targeting and imaging using cyclic RGD-pegylated gold nanoparticle probes with directly conjugated iodine-125. *Small.* 7, **2011,** 2052–2060.

33. He M, Huang P, Zhang C, Hu H, Bao C, Gao G, He R, Cui D. Dual phase-controlled synthesis of uniform lanthanide-doped $NaGdF_4$ upconversion nanocrystals via an OA/ionic liquid two-phase system for in vivo dual-modality imaging. *Adv. Funct. Mater.* 21, **2011,** 4470–4477.

34. Liu Y, Ai K, Liu J, Yuan Q, He Y, Lu L. A high-performance ytterbium-based nanoparticulate contrast agent for in vivo X-ray computed tomography imaging. *Angew. Chem. Int. Ed.* 51, **2012,** 1437–1442.

35. Xing H, Bu W, Ren Q, Zheng X, Li M, Zhang S, Qu H, Wang Z, Hua Y, Zhao K, Zhou L, Peng W, Shi J. A $NaYbF_4:Tm^{3+}$ nanoprobe for CT and NIR-to-NIR fluorescent bimodal imaging. *Biomaterials.* 33, **2012**, 5384–5393.

36. Zhu X, Zhou J, Chen M, Shi M, Feng W, Li F. Coreeshell Fe_3O_4@$NaLuF_4$: Yb, Er/Tm nanostructure for MRI, CT and upconversion luminescence tri-modality imaging. *Biomaterials.* 33, **2012,** 4618–4627.

37. Jakhmola A, Anton N, Anton H, Messaddeq N, Hallouard F, Klymchenko A, Mely Y, Vandamme TF. Poly-ε-caprolactone tungsten oxide nanoparticles as a contrast agent for X-ray computed tomography. *Biomaterials.* 35, 2014, 2981–2986.

38. Cai H, Li K, Shen M, Wen S, Luo Y, Peng C, Zhang G, Shi X. Facile assembly of Fe_3O_4@Au nanocomposite particles for dual mode magnetic resonance and computed tomography imaging applications. *J. Mater. Chem.* 22, **2012,** 15110–15120.

39. Chou SW, Shau YH, Wu PC, Yang YS, Shieh DB, Chen CC. In vitro and in vivo studies of FePt nanoparticles for dual modal CT/MRI molecular imaging. *J. Am. Chem. Soc.* 132, **2010,** 13270–13278.

40. Devaraj NK, Keliher EJ, Thurber GM, Nahrendorf M, Weissleder R. ^{18}F labeled nanoparticles for in vivo PET-CT imaging. *Bioconju. Chem.* 20, **2009,** 397–401.

41. Nahrendorf M, Zhang H, Hembrador S, Panizzi P, Sosnovil DE, Aikawa E, Libby P, Swirski FK, Weissleder W. Nanoparticle PET-CT imaging of macrophages in inflammatory atherosclerosis. *Circulation.* 117, **2008**, 379–387.

42. Van Schooneveld MM, Cormode DP, Koole R, Van Wijngaarden JT, Calcagno C, Skajaa T, Hilhorst J, 't Hart DC, Fayad ZA, Mulder WJM, Meijerink A. A fluorescent, paramagnetic and PEGylated gold/silica nanoparticle for MRI, CT and fluorescence imaging. *Contrast Media Mol. Imaging.* 5, **2010,** 231–236.

43. Mulder WJM, Strijkers GJ, van Tilborg GAF, Cormode DP, Fayad ZA, Nicolay K. Nanoparticulate assemblies of amphiphiles and

diagnostically active materials for multimodality imaging. *Acc. Chem. Res.* 42, **2009,** 904–914.

44. Zhou M, Zhang R, Huang MA, Lu W, Song SL, Melancon MP, Tian M, Liang D, Li C. A chelator free multifunctional [^{64}Cu]CuS nanoparticle platform for simultaneous micro-PET/CT imaging and photothermal ablation therapy. *J. Am. Chem. Soc.* 132, **2010,** 15351–15358.

45. Lee S, Xie J, Chen XY. Peptides and peptide hormones for molecular imaging and disease diagnosis. *Chem. Rev.* 110, **2010,** 3087–3111.

46. Xia A, Gao Y, Zhou J, Li CY, Yang TS, Wu DM, Wu L, Li F. Core-shell $NaYF_4$:Yb^{3+}, Tm^{3+}, @Fe_xO_y nanocrystals for dual-modality T_2 enhanced magnetic resonance and NIR-to-NIR upconversion luminescent imaging of small-animal lymphatic node. *Biomaterials.* 32, **2011,** 7200–7208.

47. Yang XQ, Hong H, Grailer JJ, Rowland IJ, Javadi A, Hurley SA, Xiao Y, Yang Y, Zhang Y, Nickles RJ, Cai W, Steeber DA, Gong S. cRGD-functionalized, DOX-conjugated, and ^{64}Cu-labeled superparamagnetic iron oxide nanoparticles for targeted anticancer drug delivery and PET/MR imaging. *Biomaterials.* 32, **2011,** 4151–4160.

48. Cheng L, Yang K, Li YG, Chen JH, Wang C, Shao MW, Lee ST, Liu Z. Facile preparation of multifunctional upconversion nanoprobes for multimodal imaging and dual-targeted photothermal therapy. *Angew. Chem. Int. Ed. Eng.* 50, **2011,** 7385–7390.

49. Nam SH, Bae YM, Park YI, Kim JH, Kim HM, Choi JS, Lee KT, Hyeon T, Suh YD. Long-term real-time tracking of lanthanide ion doped upconverting nanoparticles in living cells. *Angew. Chem. Int. Ed. Eng.* 50, **2011,** 6093–6097.

50. Zhou JC, Yang ZL, Dong W, Tang RJ, Sun LD, Yan CH. Bioimaging and toxicity assessments of near-infrared upconversion luminescent $NaYF_4$:Yb, Tm nanocrystals. *Biomaterials.* 32, **2011,** 9059–9067.

51. Hu H, Yu MX, Li FY, Chen ZG, Gao X, Xiong LQ, Huang CH. Facile epoxidation strategy for producing amphiphilic up-converting rare-earth nanophosphors as biological labels. *Chem. Mater.* 20, **2008,** 7003–7009.

52. Liu Q, Sun Y, Yang TS, Feng W, Li CG, Li FY. Sub-10 nm hexagonal lanthanide-doped $NaLuF_4$ upconversion nanocrystals for sensitive bioimaging in vivo. *J. Am. Chem. Soc.* 133, **2011,** 17122–17125.

53. Xia A, Chen M, Gao Y, Wu D, Feng W, Li F. Gd^{3+} complex-modified $NaLuF_4$-based upconversion nanophosphors for trimodality imaging of NIR-to-NIR upconversion luminescence, X-ray computed

tomography and magnetic resonance. *Biomaterials*. 33, **2012,** 5394–5405.

54. Xing H, Bu W, Ren Q, Zheng X, Li M, Zhang S, Qu H, Wang Z, Hua Y, Zhao K, Zhou L, Peng W, Shi J. A $NaYbF_4$:Tm^{3+} nanoprobe for CT and NIR-to-NIR fluorescent bimodal imaging. *Biomaterials*. 33, **2012,** 5384–5393.
55. Liu Y, Liu J, Ai K, Yuan Q, Lu L. Recent advances in ytterbium-based contrast agents for in vivo X-ray computed tomography imaging: Promises and prospects. *Contrast Media Mol. Imaging*. 9, **2014,** 26–36.
56. Bardhan R, Lal S, Joshi A, Halas NJ. Surface-engineered magnetic nanoparticle platforms for cancer imaging and therapy. *Acc. Chem. Res*. 44, **2011,** 893–902.
57. Li ZQ, Zhang Y, Jiang S. Multicolor core/shell-structured upconversion fluorescent nanoparticles. *Adv. Mater*. 20, **2009,** 4765–4769.
58. Liu YL, Ai KL, Liu JH, Yuan QH, He YY, Lu LH. A high-performance ytterbium-based nanoparticulate contrast agent for in vivo X-ray computed tomography imaging. *Angew. Chem. Int. Ed*. 51, **2012,** 1437–1442.
59. Liu Z, Li ZH, Liu JH, Gu S, Yuan QH, Ren JS, Qu XG. Long-circulating Er^{3+} doped Yb_2O_3 upconversion nanoparticle as an in vivo X-ray CT imaging contrast agent. *Biomaterials*. 33, **2012,** 6748–6757.
60. Pan D, Schirra CO, Wickline SA, Lanza GM. Multicolor computed tomographic molecular imaging with non-crystalline high metal density nanobeacons. *Contrast Media Mol. Imaging*. 9, **2014,** 13–25.
61. Roessl E, Brendel B, Engel KJ, Schlomka JP, Thran A, Proksa R. Sensitivity of photon-counting based K-edge imaging in X-ray computed tomography. *IEEE Trans. Med. Imaging*. 30, **2011,** 1678–1690.
62. Pan D, Schirra CO, Senpan A, Schmieder AH, Stacy AJ, Roessl E, Thran A, Wickline SA, Proska R, Lanza GM. An early investigation of ytterbium nanocolloids for selective and quantitative "multi-color" spectral CT imaging. *ACS Nano*. 6, **2012,** 3364–3370.
63. Chakravarty S, Hix JML, Wiewiora KA, Volk MC, Kenyon E, Shuboni-Mulligan DD, Blanco-Fernandez B, Kiupel M, Thomas J, Sempere LF, Shapiro EM. Tantalum oxide nanoparticles as versatile contrast agents for X-ray computed tomography. *Nanoscale*. 12, **2020**, 7720–7734.

64. FitzGerald PF, Butts MD, Roberts JC, Colborn RE, Torres AS, Lee BD, Yeh BM, Bonitatibus PJ Jr. A proposed computed tomography contrast agent using carboxybetaine zwitterionic tantalum oxide nanoparticles: Imaging, biological, and physicochemical performance. *Invest Radiol.* 51, **2016**, 786–796.

65. Dong YC, Hajfathalian M, Maidment PSN, Hsu JC, Naha PC, Si-Mohamed S, Breuilly M, Kim J, Chhour P, Douek P, Litt HI, Cormode DP. Effect of gold nanoparticle size on their properties as contrast agents for computed tomography. *Sci. Rep.* 9, **2019**, 14912.

66. Ashton JR, Castle KD, Qi Y, Kirsch DG, West JL, Badea CT. Dual-energy CT imaging of tumor liposome delivery after gold nanoparticle-augmented radiation therapy. *Theranostics.* 8, **2018**, 1782–1797.

67. Sun N, Zhao L, Zhu J, Li Y, Song N, Xing Y, Qiao W, Huang H, Zhao J. ^{131}I-labeled polyethylenimine-entrapped gold nanoparticles for targeted tumor SPECT/CT imaging and radionuclide therapy. *Int. J. Nanomedicine.* 14, **2019**, 4367–4381.

68. Hainfeld JF, Ridwan SM, Stanishevskiy Y, Smilowitz NR, Davis J & Smilowitz HM. Small, long blood half-life iodine nanoparticle for vascular and tumor imaging. *Sci. Rep.* 8, **2018**, 13803.

69. Fu JJ, Guo JJ, Qin AP, Yu XY, Zhang Q, Lei XP, Humag YG, Chen MY, Li JX, Liu JP, Dang YY, Wu D, Zhao XY, Lin ZX, Lin YL, Li SP, Zhang LY. Bismuth chelate as a contrast agent for X-ray computed tomography. *J. Nanobiotechnol.* 18, **2020**, 110.

70. Bi H, He F, Dong Y, Yang D, Dai Y, Xu L, Lv R, Gai S, Yang P, Lin J. Bismuth nanoparticles with "light" property served as a multifunctional probe for X-ray computed tomography and fluorescence imaging. *Chem. Mater.* 30, **2018**, 3301–3307.

71. Bulmahn JC, Tikhonowski G, Popov AA, Kuzmin A, Klimentov SM, Kabashin AV, Prasad PN. Laser-ablative synthesis of stable aqueous solutions of elemental bismuth nanoparticles for multimodal theranostic applications. *Nanomaterials.* 10, **2020**, 1463.

Chapter 7

Optical Nanoprobes for Diagnosis

R. G. Aswathy[a] and D. Sakthi Kumar[b]

[a]*Future Industries Institute, Mawson Lakes Campus,*
University of South Australia,
Adelaide, South Australia 5095, Australia
[b]*Bio Nano Electronics Research Center,*
Graduate School of Interdisciplinary New Science,
Toyo University, Kawagoe, Saitama 350-8585, Japan

aswathy.ravindrangirija@unisa.edu.au, aswathyrg@gmail.com, sakthi@toyo.jp

7.1 Introduction

The advances in bioimaging technology have led to successful clinical diagnosis, resulting in significant progress in healthcare. Commonly employed clinical imaging modalities for diagnosis include magnetic resonance (MR), X-ray computer tomography (X-ray CT), ultrasound (US) imaging, etc. All these imaging modalities offer precise and high quality structural and anatomical data of the diseased tissue or organ. However, for detailed information at molecular and cellular level, optical imaging with superior contrast plays a critical role that is essential for the early detection and progression of several diseases (e.g., cancer), precise diagnosis, and for image-guided treatment of diseases (theranostic applications). The advancement in nanotechnology

Bionanotechnology in Cancer: Diagnosis and Therapy
Edited by D. Sakthi Kumar and Aswathy Ravindran Girija

ISBN 978-981-4800-30-3 (Hardcover), 978-0-429-42291-1 (eBook)
www.jennystanford.com

has led to several varieties of nanomaterials with unique optical properties leading to the development of optical nanoprobes. The versatility of these optical nanoprobes is improved by the integration of specific ligands, which, in turn, transforms the chemical data to optical signals that further enable the detection through various modes. Nanomaterial-based optical nanoprobes have several advantages, including superior sensitivity and specificity, ease of quantitative and qualitative detection of species in analysis, and early detection of several diseases. This chapter discusses different types of optical nanoprobes for diagnosis.

7.2 Optical Imaging with Nanoparticles

The research on biomedical optics for imaging has developed rapidly within a decade and is projected to continue its development and contribution in coming years. Application of photons as a data source to diagnose diseases has been practiced since many years. Optical imaging has been transformed as an indispensable tool for preclinical experimental diagnosis and in medical diagnosis. The optical properties (absorption and emission) of nanoparticles (NPs) in the ultraviolet-visible-infrared (UV-VIS-IR) region of the electromagnetic spectrum are exploited in optical imaging (Figure 7.1).

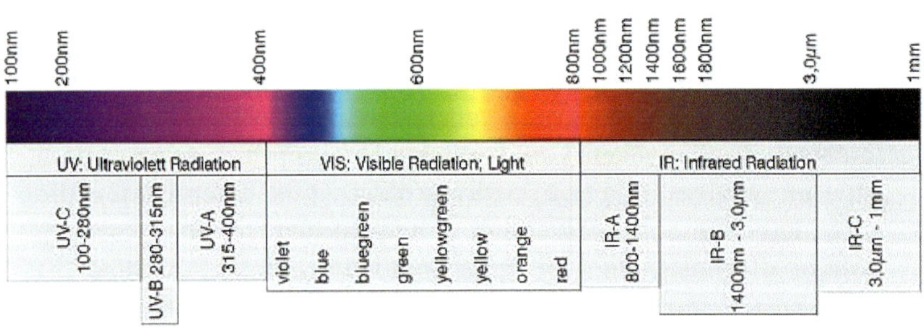

Figure 7.1 Electromagnetic spectrum in UV-VIS-IR region.

Optical imaging is the prospective method for non-invasive detection of molecular targets in live animals. It is used for studying the effects of drugs on target; and tracking the advancement of diseases like cancer, pharmacokinetics of drug molecules, and efficacy of biomarkers in disease and therapy. The fundamental principle of optical imaging trusts on the interaction of photons with an exogenous fluorophore to generate pictorial information through light emission. Photons are mainly produced by the fluorescent material that emits light of one wavelength in response to excitation by light with another wavelength (Figure 7.2). They are also generated from sources (bioluminescent) that produce light, owing to chemical reaction.

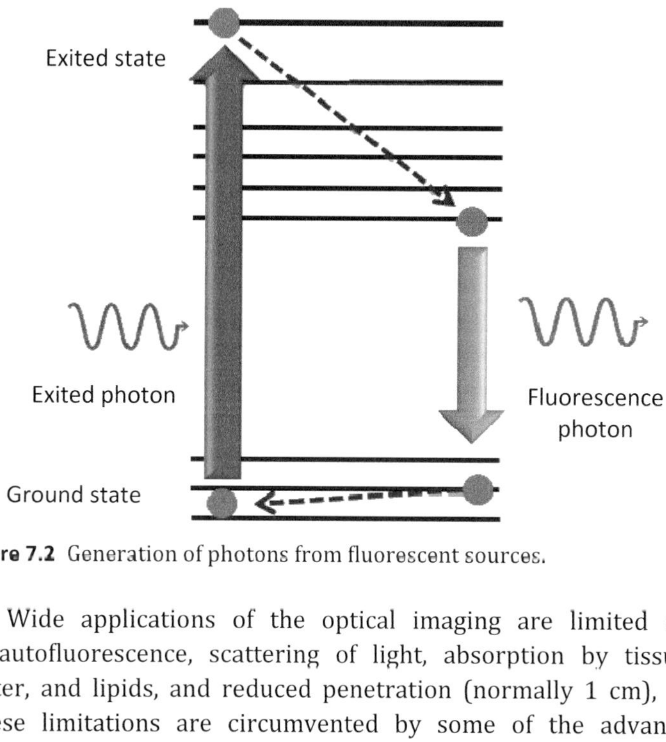

Figure 7.2 Generation of photons from fluorescent sources.

Wide applications of the optical imaging are limited due to autofluorescence, scattering of light, absorption by tissues, water, and lipids, and reduced penetration (normally 1 cm), etc. These limitations are circumvented by some of the advanced techniques in fluorescence, bioluminescence, diffused optical, and optical coherence tomography (OCT). Specifically, near-infrared fluorescent (NIRF) imaging is currently extensively used for in vivo imaging.

In fluorescence imaging, fluorophores are excited with an external light source and their emission is detected by charge-coupled device (CCD) camera. Fluorophores may be either endogenous molecules (such as hemoglobin), or exogenous agents (fluorescein isothiocyanate, rhodamine, etc.). The resolution of the image acquired can be disturbed by autofluorescence caused from the tissue, owing to the presence of endogenous chromophores. This effect can be curtailed by employing near-infrared (NIR) light source defining the range of wavelengths from 650 nm to 1350 nm, known as the diagnostic window or biological window (absorbance is very low at this region), including the first NIR optical window with wavelengths from 650 nm to 950 nm and the second NIR spectral window ranging from 1100 nm to 1350 nm. Currently, the research is focused on materials that can emit fluorescence of wavelength NIR, particularly around 800 nm to find out their advantages for bioimaging applications. NIRF with nanomaterials facilitate the in vivo imaging and provide more information on parameters such as physiological, metabolic, and molecular functions. The light scattering and the absorption properties of tissue have significant impact on the penetration of photons in the tissues. The NIR section of the electromagnetic spectrum has numerous benefits for the penetration of photons originated from organic and inorganic fluorescence contrast materials. The tissue penetration of visible light is limited, which is normally below wavelength around 700 nm, owing to the scattering properties and presence of endogenous absorbers (such as oxyhemoglobin, deoxyhemoglobin, lipids, and water). Figure 7.3 demonstrates the absorption profile of different biologically significant components, effectively limiting the excitation/fluorescence window to 650–1350 nm [1–3].

During the past decade, fluorescence tagging of biological moieties for cellular labeling, tissue imaging, and developing immunoassays has been mainly relied on conventional organic fluorophores and fluorescent proteins. Conventional molecular (organic) fluorophores have numerous limitations, such as lower brightness, broad absorption/emission spectra, color-matched lasers requirement, fluorescence bleaching, and lack of discriminatory capacity of multiple dyes. Fluorescent NPs can significantly overcome these limitations of organic dyes and can provide a key progress toward clinical applicability to image tumors and other fatal diseases in vivo. NPs, such as quantum dots (QDs), Au, and Ag, have high molar extinction coefficients, size/shape dependent tunable absorbance/scattering/

fluorescence, and resistance to photo bleaching. Optical imaging with nanomaterial offers numerous advantages, such as low cost, high sensitivity, minimal cell damage, and nonionizing radiation, compared with other imaging modalities. However, the NPs used for imaging also have some limitations, such as large size of NPs compared to molecular probes, which may inhibit cellular interaction, and also the mechanism of cellular uptake. NPs used for optical imaging should be designed appropriately to attain the desired goals for precise imaging. An ideal NP for cellular imaging must possess the following properties: (a) smaller size (typically 1–100 nm), (b) high biocompatibity, (c) solubility in aqueous medium, buffer solution, cell culture media, and also in body fluid, (d) bright and demonstrable with established optical methods (e.g., fluorescence/dark field microscope), (e) biofunctionalization with functional groups on their surface for site-specific labeling (active targeting), and (f) low non-specific binding. In the following context, we describe the state-of-the-art of optical imaging with different NPs.

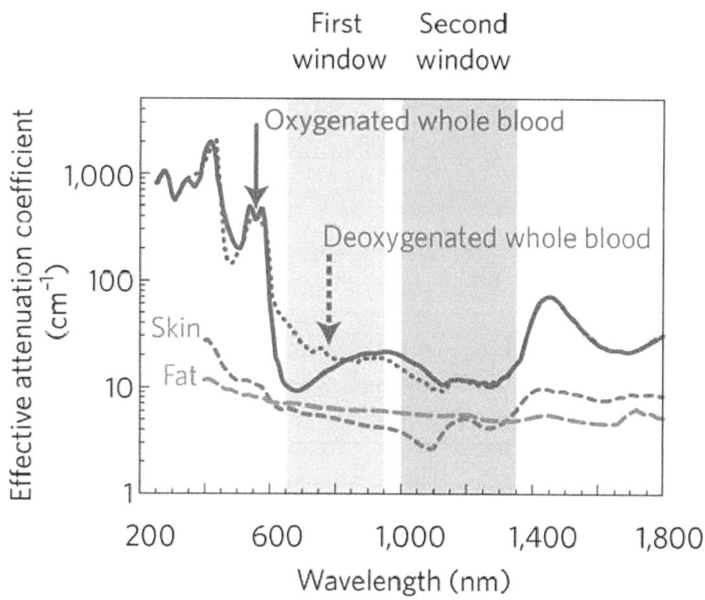

Figure 7.3 The spectral response of biological tissue as a function of wavelength. Reprinted with permission from Ref. [1].

7.2.1 Semiconductor QDs

Semiconductor QDs are nanometer-sized inorganic fluorophores that serve as multipurpose nanomaterials for imaging biological materials [4–7]. QDs belong to groups II–VI (e.g., Cd, Zn, Se, and Te) or III–V (e.g., In, P, and As) elements in the periodic table and are defined as NPs with physical dimensions less than the Exciton Bohr Radius. The chemical composition and ultra-small size of QDs impart excellent optical properties that can be tailored by tuning the size or composition of the dots. QDs absorb light, and rapidly re-emit it in another color. While several organic and inorganic nanomaterials exhibit similar phenomenon of fluorescence, an ideal fluorophore would offer non-photo bleaching and bright with narrow and symmetric emission spectra. They also exhibit multiple colors that can be excited simultaneously using a single excitation wavelength. The most interesting property of QD is that the color in absorption and emission processes can be "finely tuned" to any wavelength by altering the particle size. The basic principle behind this exceptional property is the quantum confinement effect. As the size becomes small, it approaches the size of the material's Exciton Bohr Radius. The electron energy levels are no longer being considered as continuous, instead they are treated as discrete, and small and finite separation exist between the energy levels. The development of discrete energy levels in QDs due to their small size is called quantum confinement effect. The dependence of emission of light on the size of particle permits the advancement of novel fluorescence emitters with accurately modified emission wavelengths (Figure 7.4). QDs enable a combination of distinctive physico-chemical and optical properties, which support the detailed study of the interaction of NP with biological systems via real-time monitoring of intracellular uptake, drug release, biodistribution, and also the fate of the NPs.

These unique properties lead to QDs of different-size emitting lights of different wavelengths from ultraviolet to infrared spectrum. For example, the bulk bandgap of semiconductor cadmium selenide (CdSe) is 1.7 eV corresponding to 730 nm light emission. By changing the nanocrystal diameter of CdSe from 2 nm to 7 nm, it can be tuned to emit light with wavelength between 450 nm and 650 nm. The material composition can also be used as another factor for the modification of band gap of the semiconductor. By changing the composition of the alloy $CdSe_xTe_{1-x}$ of 5 nm size can be tuned to emit light in 610–800 nm. The size and shape of semiconductor QDs can be specifically controlled by factors such as experimental duration, temperature, and concentration of precursor molecules in the synthesis.

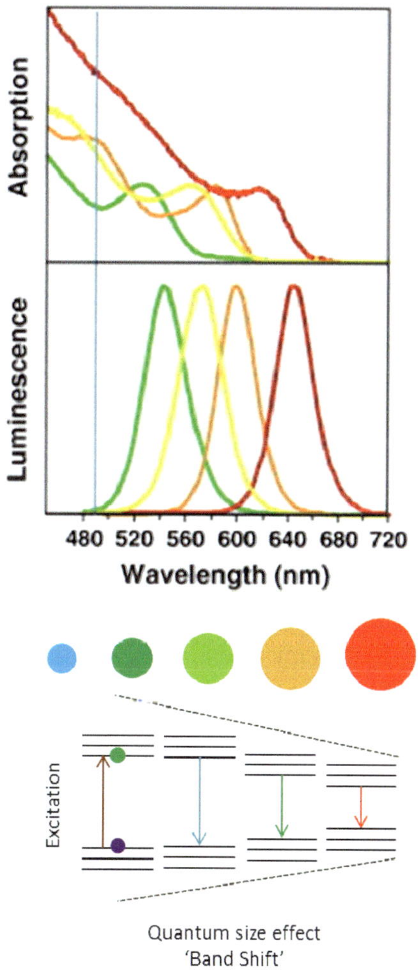

Figure 7.4 Emission of light based on size of QDs. Reprinted with permission from Ref. [4].

The structure of QDs includes a core material that is often coated with an outer shell material. For instance, the most commonly studied QDs consist of a CdSe core and a shell layer made of ZnS or CdS. The core material governs the fluorescence properties and the shell layer diminishes surface defects. The surface defects in the core crystal structure act as temporary "traps" for the electron or hole, thus inhibit the radiative recombination and the intermittent fluorescence (blinking), and result in the reduction of quantum yield (QY). QY is generally represented as the ratio of emitted photons to absorbed photons. The outer shell shelters surface atoms from oxidation and also prevents nonradiative decay. The shell is often designed with the material of large band gap above the core to obtain fluorescence QY of nearly 90%. It leads to significant improvement in QDs photo stability than conventional organic fluorophores dyes.

For biological applications, an extra hydrophilic coating has been provided to QD core/shell to render its solubility in aqueous environments. Since most of the QDs are synthesized in nonpolar organic solvents, the hydrophobic surfaces are to be replaced by hydrophilic nature to solubilize them in aqueous buffers. There are several approaches for QD solubilization. The most commonly employed strategies are ligand exchange, encapsulation by a layer of polymer, layer by layer method of the arrangement of different molecules providing colloidal stability to QDs, etc. [8–12]. Ligand exchange is often implemented with thiol-containing molecules or oligomeric phosphines, peptides, or with organic dendrons. Amphiphilic diblock or triblock copolymers, phospholipid micelles, polymer beads, polymer shells, silica shells, and amphiphilic polysaccharides are commonly used for encapsulation of QDs. Aqueous synthesis of QDs is a substitute method in which they can be employed for biological applications directly after bioconjugation.

In order to make water-soluble QDs functional for molecular imaging, QDs are conjugated with functional ligands without disrupting the bio-function of biomolecules. Biomolecules such as peptides, proteins, antibodies, aptamers, and olgionucleotides have been linked to QDs via strategies such as electrostatic interaction, covalent bonding, ligand [mercapto (–SH)] exchange, and non-specific adsorption. Most of the colloidal semiconductor QDs bare a negative charge and hence positively charged biomolecules can be attached to the surface. The targeting protein with a positively charged domain, for example poly histidine, can be interacted electrostatically with negatively charged dihydrolipoic acid-capped QD [13]. The protein-QD complex was stable and QY was higher than nonconjugated QDs. Thiols (–SH) bind to the surface of QDs and can be conjugated to biological molecules. Colloidal semiconductor QDs treated with thiolated ligands having surfactant molecules present on the surface of nanomaterial will be replaced by ligands. In this method, a part of the hydrophilic surfactant stabilizes QD in aqueous media and substituted by protein. The fundamental disadvantage of this method is the reduced stability of the NP biomolecule complexes and it is often solved by the use of multiple mercapto groups as anchor molecules [14–17].

Biomolecules are covalently attached to QDs by using cross-linkers. Hydrophilic surfactants with reactive groups of –COOH, $-NH_2$, and –SH are required for covalent attachment with functional groups on biomolecules. The procedure includes activation of carboxylic acid groups on one unit, which is followed by formation of amide bond to amino groups present on other entity or by the bonding between thiol and amino groups with cross-linkers, such as 4-(*N*-maleimidomethy) cyclohexane1-carboxylic acid *N*-hydroxysuccinimide ester (SMCC). Streptavidin–biotin system is also used as an alternative to direct conjugation of biomolecules to the hydrophilic shell of QDs. Streptavidin is covalently bound to QD surface, and QDs are reactive toward biomolecules that are bound to biotin moiety. Streptavidin-coated QDs can be easily linked to biotinylated proteins or antibodies. QDs surface can also be grafted with different functionalities to tailor QD as multifunctional probe for biological investigations. For example, in addition to a targeting ligand, QDs can also be integrated with a membrane crossing or cell-penetration functionality, or with an enzyme.

Recently, majority of studies have focused on the application of the distinctive optical properties of QDs in the bioimaging sector. Currently, most of optical bioimaging is based on traditional dyes, although they possess several drawbacks associated with their application. The autofluorescence in the visible spectrum leads to several problems, such as masking signals from organic dye molecules and results in an unstable dye upon photo-irradiation. In addition to these drawbacks, conventional dye molecules possess narrow excitation window that makes simultaneous excitation of multiple dyes difficult and most of the dyes have a broad emission spectrum with a long tail at red wavelengths that generate spectral cross talk between different detection channels, thus accessing and quantitating the amounts of different probes difficult. On the other side, optical and electrical properties such as higher extinction coefficients, tailoring of emissions and absorbance profile based on size, broad excitation and narrow emission peaks, superior QYs, reduced photo bleaching, application of multiple QDs in a single assay with negligible interference, less toxic that is less than conventional organic dyes, and functionalization of QDs with different bio-active agents make QDs role significant in bioimaging. Apart from the normal QDs that emitting light in the visible spectrum, NIR QDs are employed for deep tissue imaging that avoid interference from the autofluorescence [18–21]. Thus, exploiting the magnificent properties of QDs, such as tunable fluorescence in UV-VIS range, charge transfer through fluorescence resonance energy transfer (FRET), surface enhanced Raman spectroscopy (SERS), radio opacity and paramagnetic properties, and MRI contrast agent, it has been demonstrated that QDs can be used efficiently for bioimaging purpose.

A few of the early and effective applications of QDs have been in immunofluorescence labeling of fixed cells and tissues, immuno-staining of membrane proteins, cytoskeleton, and nuclear antigens; and fluorescence in situ hybridization on chromosomes or DNA. The specificity and efficacy for labeling molecular targets at the subcellular level have been demonstrated by QDs of two colors (630 nm and 535 nm) [22]. QDs were encapsulated within a polymer-shell that was biofunctionalized with streptavidin and immunoglobulin (HER2 to label breast cancer cells). The NPs were used to target the cancer cell surface receptor, actin and microtubules, and to identify nuclear antigen in fixed and living cells. NPs were compared to Alexa dye and the QDs were photostable than the Alexa dye. Two cellular targets were detected with one excitation wavelength using QDs and confirmed the advantages over organic dyes for multiplex target detection.

The specificity of peptide conjugated CdSe/ZnS QDs to endothelial cells in lung blood vessel and lymphatic vessel in breast cancer has been demonstrated [23]. Histology studies of various organs revealed that the targeted CdSe/ZnS QDs enhanced the accumulation of NP at vascular regions in in vivo models upon intravascular injection. In fact, the study established the significance of QDs for the detection in molecular scale in addition to imaging. In another study, microinjection into *Xenopus* embryo with CdSe/ZnS coated with a micelle forming with block copolymer of polyethylene glycol (PEG)-phospatidyl ethanol amine and PEG-phosphatidyl choline was used for fluorescence imaging [24]. When QD micelles were injected into *Xenopus* embryos, the conjugate found to be stable, biocompatible (5×10^9 per cell), and was slow to photo bleach. Nanocrystal fluorescence could be followed to the tadpole stage, thereby allowing lineage-tracing experiments in embryogenesis.

QD was integrated with A10 Ribonucleic acid (RNA) aptamer that recognized the extracellular domain of the prostate specific membrane antigen (PSMA) was accumulated in PSMA expressing prostate cancer cells than PSMA negative cells demonstrating the targeting specifity of biofuncionalized QDs [25]. Employing in vivo imaging techniques, researchers observed QDs that were detectable at one hour-, one day-, three days-, and seven days-post intracortical injection [26]. QDs were administrated to mice and peak fluorescence from brain was observed three days-post injection and continued for a week. This work had various significant features. The subcellular resolution was attained and permitted the location of the QD conjugates. The cellular uptake of QD conjugates in different cell-types was demonstrated. In vivo imaging in neural cells and neuronal internalization of QD were challenging that had great relevance to toxicity of QD, and permitted the exploration of real-time neuronal immune response [26]. In addition to the use of QD conjugates, transgenic animals with astrocyte-responsive luciferase reporter were also used for imaging studies. Astrocyte response was detected as an enhancement in bioluminescence in response to NPs in mice.

Very small NIR QDs were developed for the mapping of sentinel lymph node mapping [27]. InAs/ZnSe QDs of sub-10 nm in size were synthesized and dihydrolipoic acid-PEG was used for rendering hydrophilicity and the QDs were employed for multiple lymph node imaging. When these QDs were injected in paw of mouse, QDs migrated to lymphatic system. Lymph nodes and lymphatic channels between the lymph nodes were imaged with QDs. Intravenous injection of QDs revealed the circulation of QD in blood vessel for several minutes and extravasated to the interstitial fluid. Multiplexed targeted in vivo imaging was also achieved with NIR QDs [28]. An iron transferring protein (transferrin) was attached to QDs, and after injection of two QDs of different wavelengths emitting 660 nm and 750 nm in mouse, fluorescence images were obtained. Bright fluorescence signal was distinguished with bandpass filters and multiplex studies with QDs enabled real-time monitoring in live animals. Multiplex imaging with QDs can be a beneficial strategy for the in vivo trafficking of immune cells for immunology studies in the future.

Novel drug nanocarriers "quantoplexes" were developed for imaging studies [29]. It was synthesized with NIR CdTe QDs, polyethyleneimine (PEI), plasmid DNA, and a model drug. Intravenous administration of quantoplexes revealed the accumulation in liver, spleen, and lungs, and fluorescence signal was identified up to one week. Based on the topology of PEI and formulation of quantoplex, reallocation of quantoplex from the lung to liver was traced within 5 min. PEG modified quantoplex was used for passive tumor accumulation of the quantoplex in tumor by passive mode and for blood circulation studies. Quantoplex with NIR QDs offered an innovative labeling strategy for drugs for their screening and biodistribution studies.

Dual-model see-through display was developed for intraoperative fluorescence imaging [30]. The NIR fluorescence goggle system was used as a wearable imaging and display system that can be used to guide surgical operation in real time. It was developed based on complementary metal–oxide–semiconductor (CMOS) active pixel sensor and transparent display technology.

The wearable goggle could detect indocyanine green (ICG) fluorescence in picomolar range. This work demonstrated the possibility to guide sentinel lymph node mapping with NIR QDs (Figure 7.5), the detection and guiding surgical removal of breast cancer metastases in liver with tumor-specific NIR fluorescence probes. The novel goggle system aids to improve clinical surgery in the future.

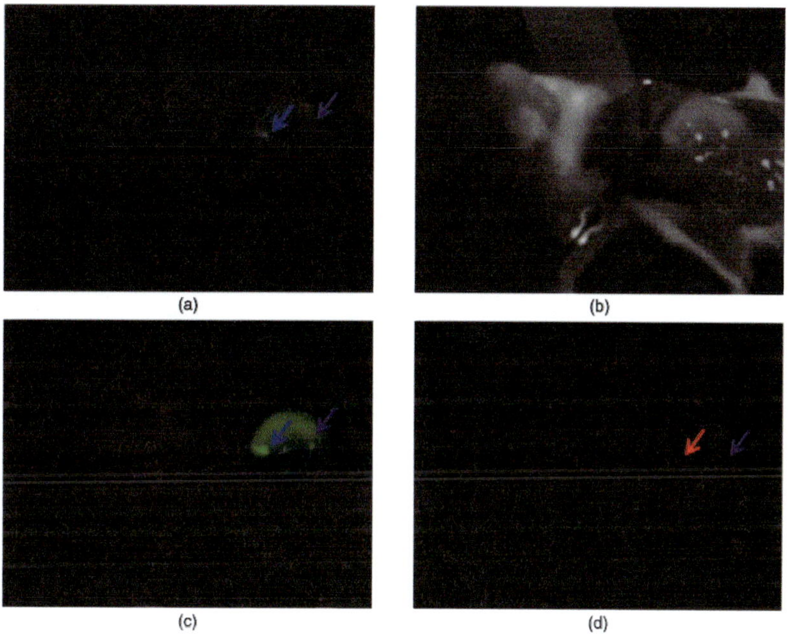

Figure 7.5 Images of intraoperative breast cancer metastases in the liver with goggle system and LS301. (a) NIR fluorescence image, (b) reflectance image, and (c) merged image of (a) and (b) of a mouse at 24 h after administration. (d) Fluorescence image of metastases after surgical resection. Blue, red, and purple arrows denote metastasis, the surgical bed after resection, and the liver respectively. Reprinted with permission from Ref. [30].

In addition to all the above-mentioned reports, there are several reports regarding the in vivo application of QD bioimaging. Some of the significant reports are tabulated under non-targeted and targeted animal imaging studies in Table 7.1.

Table 7.1 Non-targeted and targeted animal imaging

QD	Animal model	Application	Remark	Refs
Non-targeted animal imaging				
CdSe/ZnS	Mouse	In vivo vascular imaging	Visualized green-light emitting QDs in capillaries of adipose tissue and skin of a live mouse followed by IV injection (in capillaries hundreds of micrometers deep) by using two-photon microscopy	[31]
Aqueous soluble 752 nm CdTe (CdSe)	Male sprague – Dawley rats		Imaged the coronary vasculature of heart of rat	[32]
NIR and visible QDs	Chicken embryos		Imaged the blood vessels of chicken embryos	[33]
CdSe/ZnS-carboxyl and methoxy	Mouse	Imaging of the lymphatic system	Carboxyl- and methoxyl-QDs were injected to tumors and sentinel lymph nodes were mapped. The approach permitted monitoring of in vivo fluorescence as the QD conjugates moved away from the tumor	[34]
Type II NIR QDs	Mice and pigs		Non-invasive, real-time sentinel lymph node mapping, and rapid uptake (5 min) of QDs into nearby lymph nodes were imaged virtually background free	[35]
InAs/ZnSe QDs	Rat and pig		Four to five lymph nodes and extravasation of QDs into interstitial fluid were observed by injecting QDs	[27]

QD	Animal model	Application	Remark	Refs
CdTe/CdSe	Pigs		Identified sentinel lymph node within 5 min in 100% of pigs	[36]
CdTe/CdSe	Pigs		Demonstrated that station 1 lymph nodes are the sentinel lymph nodes of the pleural space in rats and pigs	[37]
CdSe/ZnS coated with micelle of *n*-poly(ethyleneglycol) phosphatidylethanolamine (PEG-PE) and phosphatidylcholine (PC)	Frog embryos	In vivo tracking of QD-loaded cells	As the embryos matured during cell division, each cell daughter retained a portion of fluorescent cytoplasm that was imaged in real time	[24]
CdSe/ZnS	Immunecompromized mouse		The cancer cells division in solid tumor was observed through the mouse skin	[38]
QD 655 ITK carboxyl quantum dots	Canine hearts		Human mesenchymal stem cells loaded with QDs was implanted into extracellular matrix patch as a regenerative implant for canine hearts with a surgically induced defect. The cellular tracking was possible with fluorescence from QDs eight weeks following implantation	[39]
Targeted animal imaging				
QD-peptide conjugates	Mice	In vivo tumor imaging	Histological sections of different organs revealed accumulation of QDs in tumor vessels guided by the peptides and not in adjacent tissues	[23]

(*Continued*)

Table 7.1 (*Continued*)

QD	Animal model	Application	Remark	Refs
Antibodies specific to PSMA conjugated red fluorescent QDs	Mouse		The QD tagged PSMA antibodies specifically bind to tumor cells and were imaged in vivo	[38]
QDs conjugated to an antibody against alpha-fetoprotein	Mice		Used to target and image mouse with human liver cancer	[40]
NIR QDs amine-modified QD705 (emission maximum at 705 nm) conjugated to $\alpha v \beta 3$ integrin targeting cyclic RGD peptide epidermal growth factor (EGFR) conjugated NIR (800 nm emission peak) QDs	Mice bearing colorectal cancer cell xenografts		In vivo imaging and quantitative determination of EGFR receptor with NIR QDs to differentiate EGFR receptor overexpressing cancer cells and normal cells	[41]
QDs conjugated human EGFR receptor 2 (HER2) antibodies	Mice		Visualization of individual QDs during circulation extravasation into tumor, receptor bound tumor cells, and translocation into perinuclear region of cells	[42]
CdSe QDs conjugated with folate receptors	Mice		In vivo targeting of mouse lymphoma cells where folate-QD conjugates were spotted in cancer cells than normal cells	[43]
QDs conjugated to three different cell adhesion molecules (CAMs)	Diabetic rat		Fluorescence angiography of retinal blood vessel demonstrated CAM-specific QDs enhanced the fluorescence and permitted imaging of inflammatory leukocytes	[44]

QD	Animal model	Application	Remark	Refs
NIR emitting QDs conjugated with cyclic RGD peptide	Mice	In vivo vascular imaging	RGD-conjugated QDs were used for imaging blood vessels, liver, bone marrow, spleen, and lymph nodes	[45]
Carboxylated QDs, with emission wavelengths of 565, 605, 655, 705, and 800 nm	Mice	Multiplexed imaging	Cancer metastasis into lymph nodes was predicted by the simultaneous visualization of QDs with five-emission wavelength as five different colors	[46]
Transferrin labeled NIR QDs with QDs emitting wavelength of 660 nm and 750 nm	Mice		Bright fluorescence signal of QDs was differentiated using bandpass filters, and real-time tracking of multiple biomolecules in live animals was attained	[28]
NIR CdTe QDs and PEI DNA complexes (quantoplexes)	Mice		Quantoplexes were accumulated in lungs, liver, and spleen upon administration	[29]
Anti-vascular endothelial growth factor (VEGF) siRNA coupled QD-hylauronic acid conjugates with PEI	Mice	Gene nano carriers	Suppression of tumor growth and reduced production of VEGF were observed	[47]

7.2.2 Metallic NPs (Gold NPs)

Owing to the potential of multiple applications, such as labeling, delivery, heating, and sensing, gold NPs (GNPs) have garnered considerable attention among metal NPs, as shown in Figure 7.6 [48]. The unique properties of GNP, such as inertness and excellent biocompatibility, make it an attractive material for biomedical applications as well as in therapy and imaging [49–55]. There are many subclasses of GNPs based on shape, size, and physical properties, as provided in Figure 7.7 [56]. GNPs with different shape and size, such as nanorods, nanospheres, nanoshells, and nanocages, are being widely investigated to attain desired optical properties appropriate for biological applications [57–60]. Owing to the extensive research in synthetic methods over past two decades, currently GNPs are fabricated with ordered size distribution, with spectacular accuracy; for example, the development of gold nanocages [61, 62].

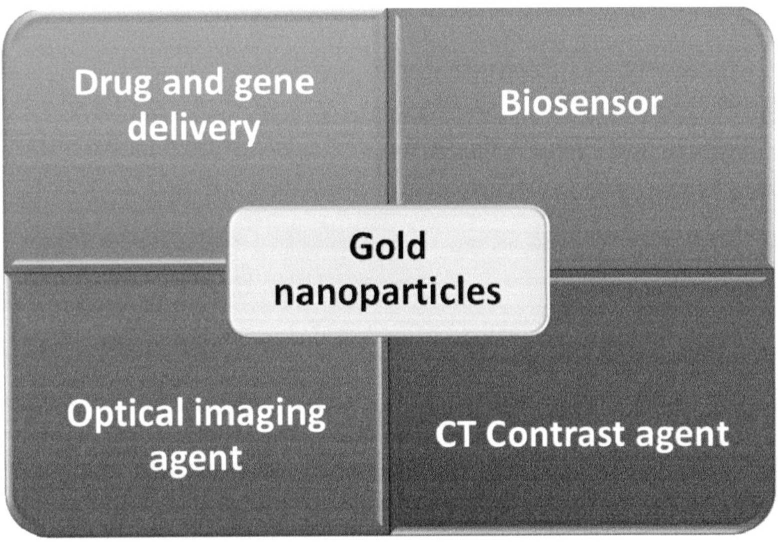

Figure 7.6 Significance properties and biomedical applications of GNPs.

Gold nanospheres are generally known as gold colloids ranging in the size of 2–100 nm, and particles that are synthesized normally by the reduction reaction of aqueous $HAuCl_4$ solution with reducing agents. The most commonly employed reducing agent is citrate, which results in nearly monodisperse gold nanospheres [63–65]. The ratio of citrate and gold is varied to obtain NPs with different size. The reduced yield and limitation of using water as solvent are the major limitations of this method. A two-phase method has been adopted for the synthesis of thermally stable gold nanospheres with well-controlled size distribution and reduced dispersity [66]. Reports are available for the application of dendrimers as stabilizers or as templates for the synthesis of gold nanospheres [67–69]. Several block copolymers have been used for the development of gold nanospheres that could readily control the size and shape by the modification synthetic parameters, including the composition of block copolymers, gold precursor $HAuCl_4$, and reaction conditions [70–74].

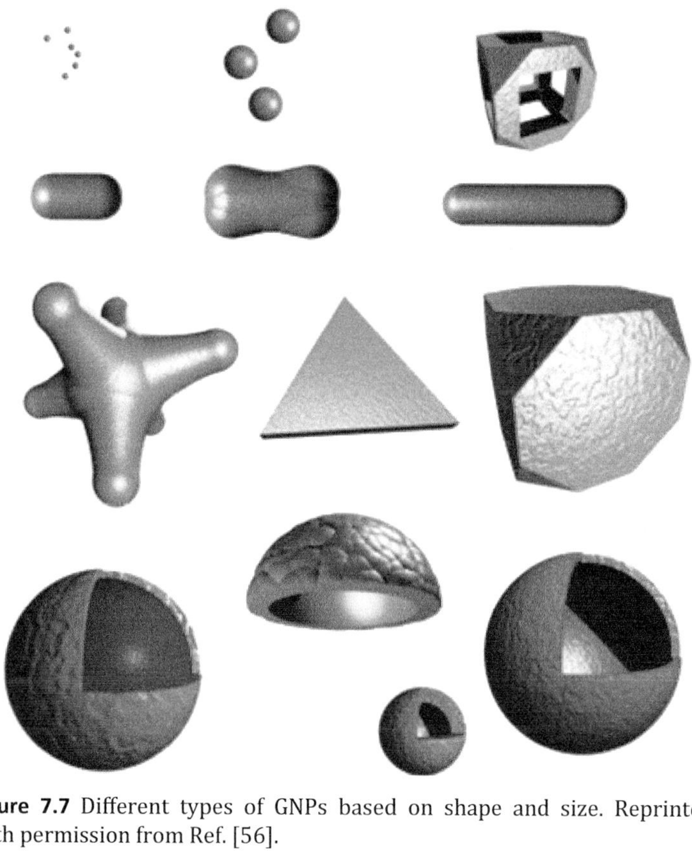

Figure 7.7 Different types of GNPs based on shape and size. Reprinted with permission from Ref. [56].

Gold nanorods are generally prepared by utilizing template method and seed-mediated synthesis where gold is electrochemically deposited inside the pores of nanoporous template membranes (polycarbonate or alumina) [75–78]. The seed-mediated synthesis delivers nanorods with high aspect ratios than other methods. Synthesis of gold nanorods through electrochemical routes also tried. The advantages of electrochemical synthesis of gold nanorods include high yield, ease of isolation, and simple control of particle size of nanorods [79, 80]. Apart from these methods, there are numerous strategies for the synthesis of gold nanorods, such as biological reduction and photochemical synthesis [81–83].

By changing the chemical contents and proportions of layers, gold nanoshells can be developed with surface plasmon resonance (SPR) peaks, extending from visible to NIR region of the electromagnetic spectrum [84, 85]. Gold nanoshells exhibiting SPR in NIR region can also be synthesized by layering polymer or silica as core with gold shells of variable thickness. The size of silica core influences the size of the gold nanoshell and the thickness of gold nanoshell is in turn controlled by the extent of gold deposited on the core surface [86]. Thermosensitive core-shell particles are also used as template for in situ gold nanoshell formation on GNPs [87]. The employment of microgel as core offers reduced particle aggregation. A virus platform was used to construct gold nanoshells with cores of narrow size distribution and small diameters (80 nm) than silica [88]. Galvanic replacement reactions were employed for the development of gold nanocages with controllable pores [89, 90]. Among that, one reaction was between the truncated silver nanocubes and aqueous $HAuCl_4$, where the silver nanocubes were synthesized by the reduction of Ag(I) salts in presence of polyol (e.g., ethylene glycol). The silver nanostructures formed are used as a suicidal template that was converted to hollow gold nanocages by galvanic replacement. GNPs offer several advantages for optical imaging compared to other agents, such as very strong light scattering, resistance to photobleaching, and easily detectable at lower concentrations as low as 10^{-16} M. An anti-EGFR antibody-conjugated GNP was developed for the detection of cervical cancer. Laser scanning confocal reflectance was used for cervical cancer cells detection, owing to SPR scattering property of GNPs [91–94].

Raman spectroscopy is one of the assured imaging techniques for GNP-based contrast agents. SERS provides several benefits than conventional imaging technologies, such as fluorescence and chemiluminescence. The advantages also include superior sensitivity, with the possibility of multiplex imaging and advanced functioning in biological conditions. It has been reported that GNPs have the property of intensifying the Raman scattering efficiencies to 10^{15}-fold of adsorbed molecules, thereby permitting spectroscopic detection of single molecules under ambient conditions. Gold nanospheres were programmed with Raman probe and stabilized with thiolated PEG [95]. In a study, gold nanospheres of 13 nm were modified with Cy3-labeled, and alkylthiol capped DNA strands were used as probes to observe the existence of target DNA strands [96]. Owing to its large Raman cross-section, Cy3 group was selected as Raman label. As SERS NP composed of Au core, Raman-functioning layer and silica layer were developed. The silica layer ensured inertness to various environmental settings. Biocompatibility was rendered by maleimide activated PEG chains [97].

Noble metal NPs have the ability of scattering light intensely at their SPR frequency. GNPs offer a suitable method for tumor imaging and its detection using an optical microscope fitted with dark field condenser. For other NPs, it requires high-tech instrumentation such as lasers, optical components, and detectors, or multifunctional image processing methods or dark-field imaging. SPR in GNPs is based on the interaction of light and conduction electrons. When light interacts with the metal, conduction electrons oscillate at a resonance frequency relative to the lattice of anions. As a result, scattering photons are released at the same resonance frequency in all directions. The absorption and scattering of GNPs are higher than organic molecular dyes that mark them as exceptional contrast agents for biological imaging applications. Since the SPR effect is very sensitive to changes with respect to morphology of particles and dielectric constant of the adjacent medium, tuning the optical properties of GNPs is attainable. By utilizing gold shell NPs (gold nanoshells) and controlling the shell thickness-to-particle diameter ratio, the SPR peak can also be finely tuned. Among GNPs, gold nanocubes have the potential for tumor imaging due to their maximum photoluminescence QY. Upon exposure to NIR light (650–900 nm), GNPs generate heat due to interactions of electron-phonon and phonon-phonon. The penetration of NIR light in human tissue enhances from half a millimeter to a few centimeters, owing to the minimal absorption of water and bodily fluids, including blood, in the NIR region. GNPs absorb NIR light and end up in the generation of heat to the neighboring tissue, which has been studied for tumor ablation and imaging.

In vivo imaging with GNPs can be attained with several emerging techniques that deliver data with high precision and accuracy. Versatility of GNPs has facilitated the researchers to apply them for optical imaging of cells and phantoms as described in Table 7.2. In vivo imaging with GNPs can be accomplished via OCT imaging, two-photon luminescence (TPL), photoacoustic tomography (PAT), SERS, X-ray CT, etc. [98]. OCT provides resolution in the range of micrometer and with penetration of 2–3 mm. OCT-based imaging modality offers cross-sectional subsurface imaging that is based on NIR interferometry. Irradiation with NIR light permits superior tissue penetration, and generation of image is based on difference in the absorption and scattering properties. TPL has been explored for the diagnosis of precancerous and cancerous tissues. In TPL imaging, excitation of electrons depends on same lower energy of two photons than high-energy single photon. Owing to localized SPR, TPL signal from noble metal NPs, especially GNPs, is highly enhanced. PAT is one of the hybrid imaging techniques, which is based on light and US-based imaging. PAT generates ultrasonic waves that are stimulated by laser exposed to body tissue. In PAT, deep tissue imaging of about 5 cm depth and spatial resolution of about 0.2 mm can be attained. GNPs generate strong photoacoustic signals in the NIR region, and are studied as contrast-enhanced agents for PAT imaging. SERS and Raman spectroscopy have been utilized for tumor tissue imaging.

Table 7.2 GNPs for optical imaging

Technique employed	Type of gold NP	In vivo imaging	Remark	Refs.
OCT	PEG Au nanoshells	Mouse xenograft colon tumor model	Improved brightness in the OCT image due to the accumulation of Au nanoshells in tumor	[99]
	Ab conjugated PEG Au NPs	Mouse xenograft epidermoid tumor model	Improved concentration of targeted Au nanoshells than untargeted nanoshells in tumor	[100]
	Au NPs	Oral tumor model in hamster cheek pouch	Nanospheres delivered by microneedles and US and observed an 150% enhancement of OCT contrast compared to healthy hamsters	[101]
TPL	Au nanorods	Mouse earlobe vessels	Real-time TPL imaging, signifying single nanorods	[102]
PAT	PEG-coated gold nanoshells	Rat brain vasculature	63% enhancement of optical absorption of nanoshells by brain blood vessels	[103]
	PEG-coated gold nanocages	Cerebral cortex of rats	80% augmentation of contrast in cerebral cortex of rats	[104]
	Au nanocages	Lymph node tracer	Mapping and imaging sentinel lymph node below skin surface (3.3 cm).	[105]
	Hollow Au nanospheres	Mouse brain vasculature	The PAT displayed brain vasculature as small as ~100 μm in diameter	[106]
	Au nano beacons	Sentinel lymph node	Larger nanobeacons with high payloads of gold were not as efficient as smaller size nanobeacons with lower payloads.	[107]
	Antibody conjugated Au nanorods	Squamous tumor	GNPs (20 nm and 50 nm) have high photoacoustic contrast and are visualized in mice upon subcutaneous administration using PAT.	[108]
	PEGylated Au nanospheres	Breast cancer cells implanted mice	PA imaging of human breast cancer cells established in abdominal wall of mice	[109]

Table 7.2 (*Continued*)

Technique employed	Type of gold NP	In vivo imaging	Remark	Refs.
	Au nanocages, conjugated with melanocyte stimulating hormone	Human breast cancer cells	300% contrast enhancement 6 h upon tail vein injection in mice and targeted NPs showed 3.5-fold accumulation than untargeted nanocages	[110]
SERS	PEGylated Au NPs	Mice	Targeted NPs were 10-fold efficient than non-targeted NPs	[95]
	GNPs enclosed with Raman-active material and silica coating	Mice	Multiplexed imaging with four different SERS NPs exhibiting Raman spectrum in deep-tissue of live mice	[97]
	SERS-coded PEG-Au nanorods	Mice	In vivo multiplexed Raman detection and photothermal heating	[111]
X-ray CT imaging	1.9 nm Au nanospheres	Mammary carcinoma xenograft tumor model and blood vessels	X-ray images demonstrated enhanced contrast for nanospheres than iodine for imaging mammary carcinoma xenograft tumor and blood vessels	[112]
	PEGylated Au NPs	Imaging of heart/great vessels	PEGylated GNPs exhibited much longer blood circulation time (>4 h) than ultravist (<10 min), and CT images of rats using PEG-coated GNPs showed a clear delineation of cardiac ventricles and great vessels	[113]
	PEGylated Au nanorods	MDA-MB-435 bilateral human tumor bearing mice	3D distribution of PEGylated gold nanorods was acquired via X-ray CT	[114]

7.2.3 Carbon Nanotubes

The discovery of carbon nanotubes (CNTs), including single-walled carbon nanotubes (SWCNTs) and multi-walled carbon nanotubes (MWCNTs), in early 1990s began an innovative area of research in nanoscale materials [115–119]. CNTs are cylindrical graphene sheets that hold novel properties, which make them possibly useful in a wide diversity of applications in nanotechnology, energy/data storage devices, sensors, optics, electronics, field effect emitters, and other arenas of material science [120–127]. The distinctive physio-chemical properties of CNTs, such as electrical conductance, transistor behavior, thermal conductivity, high mechanical stiffness, piezo resistance, light weight, expansion of electrochemical bond, luminescence, and flexible chemistry, brand them as excellent nanomaterials for a wide range of applications extending from energy storage to drug/gene delivery vehicles to nanosensors (NSs) and for bioimaging applications. CNTs exhibit impending medical imaging capabilities, and integrated diagnostic and therapeutic efficacies as potential theragnostic agents. In recent years, biomedical applications of CNTs have been extensively explored [128–136]. CNTs are commonly generated by three main techniques: laser ablation method, electric arc discharge method, and thermal or plasma-enhanced chemical vapor deposition (PECVD) method [137–140]. Several purification procedures have been practiced for CNTs including chemical treatment, physical separation, and gas/liquid phase oxidation [141]. Both SWCNTs (1–3 nm in diameter) and MWCNTs (10–150 nm in diameter and 200 nm to a few microns in length) have been explored extensively for various biological applications (Figure 7.8). SWCNTs are studied to be the "darkest" materials with optical absorption ranging from UV to NIR region of the spectrum.

Direct fluorescent labeling of SWCNTs for optical imaging applications has not been very efficacious, owing to the superior light quenching properties. However, enveloping SWCNTs with fluorescently labeled polymer could make SWCNTs visible by a fluorescent microscope. The conjugation of SWCNTs with QDs was able to develop the new imaging probes, which were steadily dispersed and imaged by fluorescence microscopy. The restricted penetration depth of light poses the major difficulty for the deep tissue imaging with fluorescence imaging. The development and implementation of fluorescent probes functioning in the biological transparent NIR diagnostic window are considered as the greatest advancements in biological imaging [2, 142, 143].

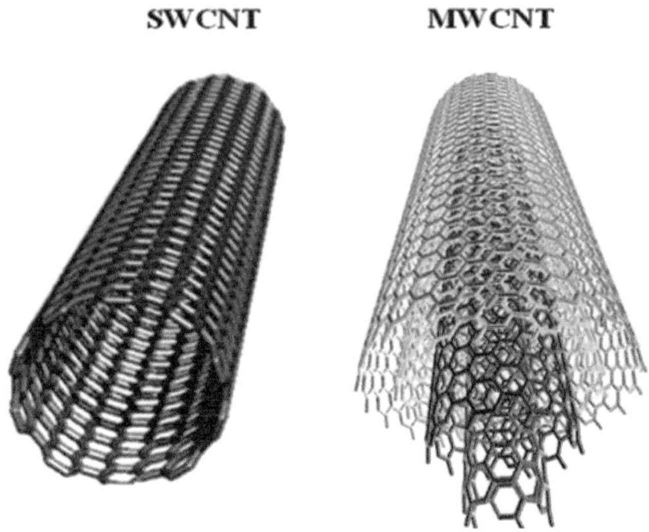

Figure 7.8 SWCNT and MWCNT.

The classical NIR window, or NIR I window ranges from ~650 nm to ~900 nm, in which body fluids including hemoglobin and water displays do not have any significant absorbance of light. Light with longer wavelength than NIR I, from ~1100 to ~1350 nm (NIR II), displays even further reduced tissue absorbance and scattering by body tissues, permitting in vivo fluorescence imaging with enhanced tissue penetration and superior spatial resolutions. SWCNTs exhibit strong light absorption properties in the NIR window. Wavelength range 700–1400 nm has made SWCNTs as favorable NIR fluorescent contrast agent in biological systems, owing to the low autofluorescence in the NIR range and large Stokes shift between the excitation and emission bands [144–146]. This, in turn, permits excitation in the near 800 nm by diminishing background effects of autofluorescence and scattering, thereby allowing penetration of deep tissue in the NIR region beyond 1 mm. CNTs offer an exclusive approach to explore this advantageous emission region. The hyperthermia stemmed from NIR light absorption by SWCNTs has been employed for therapy (photothermal therapy) and diagnosis (photoacoustic imaging). Resonance-SERS of SWCNTs allows detection, tracking, and imaging in both in vitro and in vivo studies [146]. SWCNT serves as an exceptional nanoplatform for multimodal biomedical imaging [147]. For example, MRI imaging with SWCNTs has been reported, by utilizing the metal impurities present in nanotube samples or exploiting gadolinium (Gd^{3+}) attached to nanotubes. SWCNTs are also integrated with radioactive isotopes, such as ^{64}Cu, for positron emission tomography (PET) imaging [148–150].

The cellular uptake of SWCNTs in macrophage-like cells based on the intrinsic NIRF signal was studied [151]. In one study, semiconducting SWCNTs were conjugated to antibodies, specific for various cell surface markers and employed as NIRF probes for selectively targeting cell surface receptors and imaging [152] (Figure 7.9). Passive or nonspecific phagocytic cellular uptake of SWCNTs has also been demonstrated. In another work, real-time NIR photoluminescence was used to demonstrate endocytosis and exocytosis of SWCNTs in NIH-3T3 cells [153]. Although the reports on SWCNTs as suitable NIRF tags for biological imaging, the aqueous SWCNTs suspensions were sensitive to physiological and environmental settings, including pH, concentration, and salinity. It was overcome by developing SWCNTs suspended in sodium dodecylbenzene sulfonate (SDBS) with poly (vinyl pyrrolidone) (PVP) to entrap the SWCNT SDBS micelles developed. The nanoconstruct was stable in various pH ranges and exhibited steady luminescent [154]. The stable nano-suspensions were detected in human embryonic kidney cells. Biocompatible PEGylated SWCNTs conjugated with antibodies as NIR fluorescent probes for targeted cell imaging were developed [155]. Cells were treated with SWCNT-antibody conjugate and NIR fluorescence images of HER2/neu positive BT-474 exhibited SWCNTs photoluminescence signals, compared to HER2 negative MCF-7 cells suggested the selective intake of SWCNTs in cancer cells.

In vivo imaging of live animal with NIR photoluminescence was demonstrated successfully [156]. Drosophila larvae was fed with food containing SWCNTs and imaged by NIR fluorescence microscope. The biodistribution of SWCNTs in live larvae was observed by the fluorescence signals from the nanotube. Biocompatible SWCNTs with high QY are a vital goal to attain. Biocompatible SWCNTs were developed by sonication of

SWCNTs with sodium cholate and surfactant exchange to generate phospholipid–PEG for in vivo imaging [157]. The exchange procedure was superior to direct sonication with phospholipid–PEG, because the former resulted in less damage to the nanotubes and improved the QY. SWCNTs obtained after exchange procedure known as exchange SWCNT conjugates were studied as NIR photoluminescence contrast agents for in vivo imaging in mice. They demonstrated whole-animal in vivo imaging at a relatively low dose of contrast agents and the imaging small vessels inside the tumor.

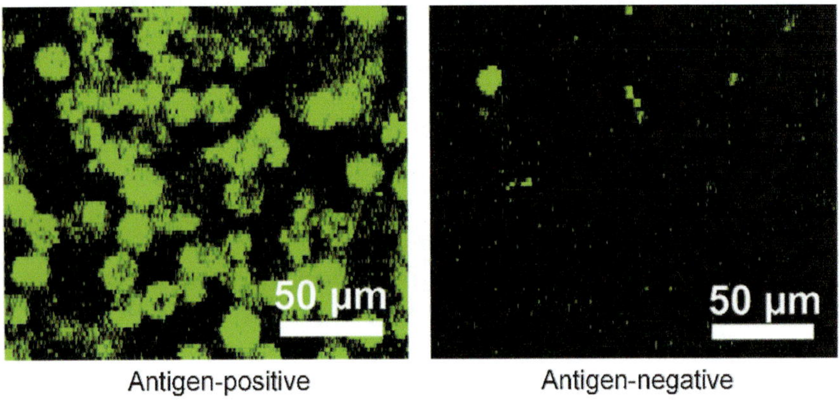

Figure 7.9 NIRF imaging with SWCNTs. (a) NIRF images of antigen-positive and (b) control cells treated with SWCNT-antibody nanoconjugates. Reprinted with permission from Ref. [152].

To study the MRI-NIR based imaging with SWCNT and magnetic NPs, researchers engineered novel FeCO/single-graphitic-shell nanocrystals for both MRI-NIR imaging and therapy [158]. Chemical vapor deposition (CVD) technique was used to produce graphitic shell nanocrystals, and multi-functionalize the core shell materials, which were made for magnetism and NIR optical absorbance. The engineered NPs exhibited high saturation magnetization and r_1 and r_2 relaxivities in both T_1-weighted and T_2-weighted MR images than commercial MRI contrast agents. FeCO/single-graphitic-shell nanocrystals were administrated in rabbit for in vivo animal imaging studies. An enhanced positive contrast and longer circulation at lower dose was observed. The optical properties of engineered FeCO/single-graphitic-shell nanocrystals under λ = 808 nm NIR laser radiation exhibited considerable temperature increase over time, which indicated the efficiency of the transformation of NIR photon energy into thermal energy. This study revealed the potential applications of FeCO/single-graphitic-shell nanocrystals for in vivo multimodal diagnostic imaging, and NIR responsive tumor destruction. During the synthesis of CNTs, metal catalysts such as Fe and CO are often employed. The presence of impurities of iron can offer the magnetic potential to the CNTs. Although the impurities are removed, the trace amount of iron in sample can offer enhanced T_2 contrast for MR imaging. The biodistribution of SWCNTs by quantifying the MR signals was reported. SWCNTs with associated metal impurities were distinguished in vivo by non-invasive MR techniques. Combined helium-3 and standard proton MR was used to quantify pulmonary and systemic biodistribution of raw SWCNTs demonstrated multimodal imaging potential, respectively, while no signal modifications were observed in super purified SWCNTs treated group [159, 160].

SWCNTs exhibit multiple Raman peaks. The radial breathing mode (RBM) and tangential G-band are the most significant characteristic peaks in a SWCNT Raman spectrum. Both the G-band and RBM SWCNT Raman peaks have been utilized for biological Raman imaging applications. Raman imaging of SWCNTs in live cells was reported in 2005 [161]. Hipco SWCNTs was functionalized with DNA and incubated with 3T3 fibroblast and myoblast stem cells. For Raman spectroscopic mapping, a 785 nm laser was used as the excitation source. A high SWCNT Raman signal inside cells was the proof for cellular entry of nanotube than using extra fluorescent probes. Superior photo-stability of SWCNTs in several orders of magnitudes than organic dyes and NIR QDs was observed. Owing to the indolent chemical structure of SWCNTs, the long-term tracking of SWCNTs was highly possible. In vivo Raman imaging of SWCNTs in animals was demonstrated as the proof-of-principle study using Raman spectroscopy [162]. In that study, Arg-Gly-Asp (RGD) peptide was conjugated to Hipco SWCNTs functionalized by DSPE-PEG to bind integrin αvβ3 receptors expressed in tumor vasculature and U87MG human glioblastoma cancer cells. SWCNT-RGD conjugate was intravenously injected to U87MG tumor bearing mice. Whole mouse Raman imaging was performed and temporal information of SWCNT levels in tumors was attained. Strong SWCNT Raman signals were observed on tumors of mice injected with peptide conjugated SWCNTs, owing to the specific binding between the RGD peptide and the integrin αvβ3 protein expressed on cancer cells and the vasculature, but not from those injected with plain SWCNTs.

Multi-color Raman imaging has been achieved by employing isotopically modified SWCNTs [163]. When carbon isotope from normal ^{12}C is changed to ^{13}C, the Raman G-band peak was lowered by nearly 62 cm^{-1}. Also, the G-band frequency shifts with ^{13}C isotope ratios in nanotube samples. Pure ^{12}C, pure ^{13}C, and mixed $^{12}C/^{13}C$ SWCNTs displaying Raman G-band peaks at 1590 cm^{-1}, 1528 cm^{-1}, and 1544 cm^{-1}, respectively, were synthesized and three different Raman 'colors' for multiplexed Raman imaging was studied [164]. The nanotubes were functionalized and labeled with three distinctive targeting moieties. The Raman spectra was deconvoluted into the spectra of the three SWCNT samples to analyze the relative amounts of each color in any mixture and on live cells. The labeled SWCNT Raman 'colors' in multi-color Raman images differentiated three different cancer cell lines displaying different receptors, demonstrating the ability of multiplexed Raman imaging of SWCNTs with different isotope compositions.

Five different SWCNTs $^{13}C/^{12}C$ isotopic formulations with distinctive Raman peaks were synthesized and conjugated with five different targeting ligands for multiplexed Raman imaging (165). An ex vivo multiplexed Raman imaging of tumor samples was performed to demonstrate the pre-clinical application (Figure 7.10). LS174T tumor cells displayed enhanced EGFR/Her1 (yellow) and CEA (blue) receptors. Integrin αvβ3 expression on tumor blood vessels was imaged by Color5 Raman staining (red). Zero background interference was attained, owing to the sharp Raman peak exclusive to nanotubes than autofluorescence background signal from biological tissue. SWCNTs with resonance-enhanced Raman signals were integrated with SERS to enhance Raman signals for improving sensitivity and reduction in imaging time. In short, SWCNTs are promising candidates for high-performance NIR Raman probes for multiplexed optical imaging and detection.

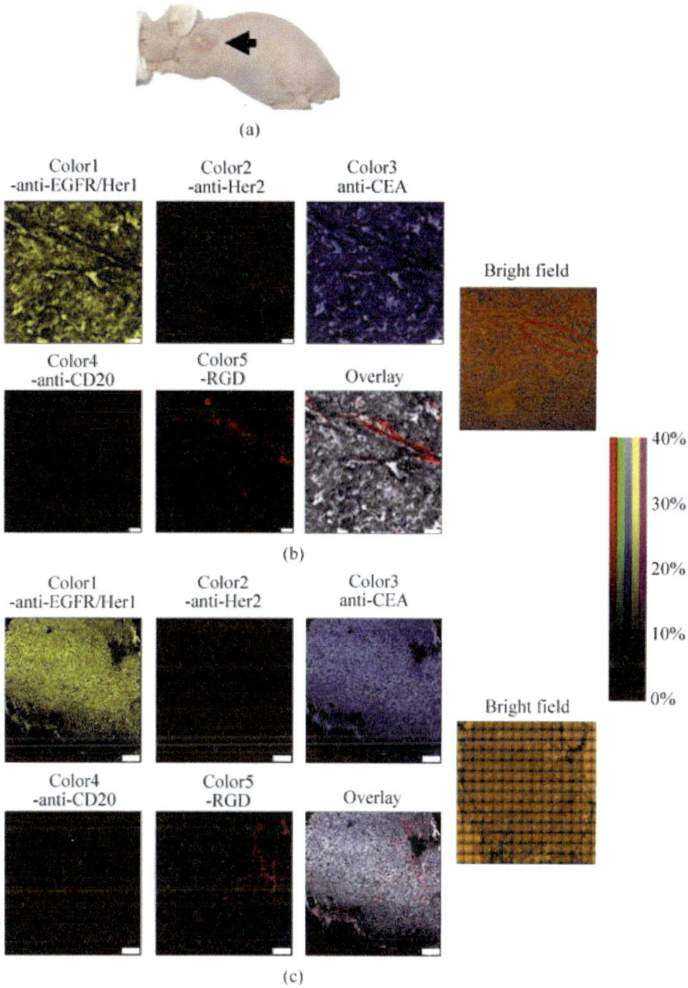

Figure 7.10 Multiplexed ex vivo Raman imaging. Hipco SWNTs (C12) with a high Raman scattering intensity factor was used as Color5. (a) Picture of LS174T tumor-bearing mouse with tumor. (b) Five color Raman images of LS174T human colon tumor section from mouse model. The overlapped image denotes high correlation between CEA expression and EGFR expression. LS174T cells in tumor exhibited high levels of EGFR/Her1 (yellow) and CEA (blue) receptors. Integrin expression on tumor vessels was imaged by Color5-RGD Raman staining (red). Bright field optical image of the tumor is shown in the right and tumor vessel is highlighted. Scale bar = 50 μm. (c) Five-color Raman images of large area whole tumor section. Scale bar = 1 mm. Reprinted with permission from Ref. [165].

NIRF imaging with SWCNTs has presented a huge potential in bioimaging, signifying distinctive benefits than other imaging modalities. The development of SWCNTs with high QY and purified chiralities offering bright fluorescence will be highly beneficial in the bioimaging field. Also, the potential of multi-color NIRF imaging with SWCNTs of purified single chiralities exhibiting distinctive excitation and emission wavelengths has to be addressed for bioimaging studies.

7.2.4 Polymeric Nanoparticles

Polymeric nanoparticles (PNPs) have gained greater attention, owing to the immense flexibility presented by macromolecular synthetic approaches, adaptability in the modification of structures, a range of polymers based on the composition, properties and easiness for functionalization, efficient delivery of cargoes to the required site of action, and effective tracking of the NPs and their response to specific physiological stimuli [166–173]. The chemistry of PNPs affects their stability, biocompatibility, biodegradability, biodistribution, and cellular and subcellular fate [172–174]. The design and development of PNPs generally depend on the applications, target site, and route of administration. These NPs are being integrated with components to permit the delivery to different biological obstructions, including skin, mucus, extracellular matrix, and cellular and subcellular barriers [172]. Polymer-based engineered NPs offer the most promising candidates for diagnosis and therapy of numerous disorders including cancer. The key building blocks of PNPs are (a) core-forming polymers, which may be either charged or hydrophobic, (b) shell-forming polymer, which may be either neutral or hydrophilic, (c) targeting moiety for cell specific uptake and accumulation, and (d) stimuli-responsiveness cargo release (e.g., pH, temperature, light, enzymatic, reductive, or oxidative).

Most commonly studied NPs based on PNPs include the following:

(a) Polyesters: Poly(D,L-lactic-co-glycolic acid) (PLGA) NPs, polystyrene, PEG covered or PEGylated NPs, poly(ethylene glycol)-block-poly(aspartic acid), (PEG-PAA)-coated calcium phosphate, and poly-vinyl chloride (PVC)

(b) Polysaccharides: Chitosan and Cyclodextrin
(c) Poly (amino acid), polypeptides, and proteins
(d) Lipids and lipoproteins
(e) NPs containing DNA in addition to inorganic molecules or non-nucleic acid polymers: PEG/DNA NPs, poly(methyl methacrylate)/poly(ethyleneimine)-NP/pDNA complexes, and poly-L-Lysine-DNA complexes
(f) Several fluorescent polymer NPs

Polyesters exemplify the most extensively employed class of polymers, owing to their biodegradable and biocompatible qualities. These include polylactide (PLA), polyglycolide (PGA), poly(e-caprolactone) (PCL), and poly(g-valerolactone) (PVL), and most of them are generally synthesized by ring-opening polymerization of lactide, lactide/glycolide, e-caprolactone, and gvalero-lactone, respectively. PLGA copolymer of lactide and glycolide is the Food and Drug Administration (FDA) approved polymer for drug delivery applications. Amphiphilic PLGA-b-PEG or PLA-b-PEG copolymers, comprising hydrophobic PLGA or PLA block and a hydrophilic PEG, have been used to prepare NPs, polymersomes, or micelles, where the hydrophobic part (PLGA or PLA) and hydrophilic PEG results in the formation of core and shell of nanocarriers, respectively. PEGylated NPs demonstrate the potential to be functionalized with various ligands, such as small molecules, peptides, antibodies, and aptamers conjugated to PEG chain. There have been several reports on the application of PNPs as diagnostic as well as therapeutic (theragnostic) multifunctional agents. PLGA NPs carrying various drugs (curcumin, paclitaxel, gemcitabine, 5-Fluorouracil (5FU), etc.) with different targeting moieties (folate, aptamer, peptide, transferrin, etc.) against pancreatic as well as breast cancer cells have been reported [175–178].

Functionalized nanoconjugates have been developed by an aptamer conjugated dual drug (curcumin and gemcitabine-loaded) and magnetic nanoparticle (MNP) encapsulated in PLGA NPs (Apt-Cur-Gem-PLGA-MNPs) for pancreatic cancer therapy [176]. The encapsulated curcumin served both as an imaging agent and efficient therapeutic agent. The presence of MNP in the nanoconjugate was exploited for magnetic hyperthermia and magnetic ablation of cancer cells by a rotating magnetic field. Thus, Apt-Cur-Gem-PLGA-MNPs could deliver an effective and targeted delivery of therapeutics for pancreatic cancer treatment. The efficiency in targeting and therapeutic potential of the combination of drugs (curcumin and 5FU) and magnetic NPs encapsulated PLGA NPs, functionalized with two cancer-specific ligands (folate and transferrin) was demonstrated [177]. Upon magnetic hyperthermia, the advantage of incorporating MNPs was attained by the synergistic action of drugs for the destruction of cancer cells within short time. The significant multimodal efficacy of polymeric magnetic therapeutic nanoformulation delivers the prospective for targeting, imaging, and treatment of cancer within 120 min by instigating early and late apoptosis. PLGA-lecithin-PEG NPs functionalized with AS1411 anti-nucleolin aptamers for site-specific targeting and encapsulated with anticancer drug against tumor cells were developed [179]. The NP was labeled with Nile red (NR) and the confocal laser scanning microscopy (CLSM) images confirmed the internalization of targeted NPs in tumor cells. PLGA-lecithin-PEG NPs displayed enhanced drug encapsulation efficiency and sustained release of anticancer drug than the drug encapsulated in PLGA NPs. Aptamer-targeted NPs demonstrated superior cellular toxicity than the non-targeted NPs. The study demonstrated that AS1411 aptamer-PLGA-lecithin-PEG NPs were prospective nanocarrier for targeted drug delivery.

Polysaccharides are the most prevalent polymers for the synthesis of nanocarriers for imaging and drug/gene delivery applications, owing of their exceptional physical and biological properties. The presence of several functional groups on the polysaccharide polymers makes them vulnerable for easy biochemical modification. Polysaccharides are stable, hydrophilic, and biocompatible, and many of them are easily biodegradable. There are several reports on the application of various polysaccharides, including chitosan, carboxymethyl chitosan, carboxymethyl cellulose, and cyclodextrin. In addition to the various plant-based polysaccharides, microbial polysaccharides are also being exploited. Carboxymethyl cellulose-based QDs for imaging cancer cells were developed [180]. In another study, fluorescein isothiocyanate was integrated into carboxymethyl cellulose NPs with MNPs and the uptake of targeted NPs in cancer cells was observed [181]. Microbial polysaccharide especially Mauran from *Halomonas maura* was studied and was employed for biomedical research. The NP based on Mauran was labeled with Sypro ruby and conjugated with QDs and then the uptake of the same by cancer cells was studied. Other polysaccharides, such as heparin and dextran, have also been used for the development of targeted NPs, drug delivery, and imaging studies [182].

Proteins also signify as an essential category of natural polymers, owing to their inherent biocompatibility, and have been extensively studied for diagnostic and drug delivery applications. Ligand functionalization can be attained through lysine existing at protein surface. Proteins, which are commonly used as nanocarriers, include albumin, gelatin, gliadin, etc. Protein NPs are often prepared by different methods, such as desolvation, solvent displacement, direct solubilization, self-assembly, etc. The conjugation of QDs with gelatin and Zein and their biocompatibility and the uptake of NPs by cancer cells have been reported [183, 184].

PNPs for biomedical imaging studies are being synthesized by covalent conjugation (Type I) or encapsulations (Type II) of dye or contrast agents to polymer matrix [185]. Type I method is based on the covalent linking of contrast agents to polymeric unit. The contrast agent-grafted polymers were then transformed into NPs by suitable methods. The main benefit of Type I nanoconstructs is the absence of burst release due to the covalent bonding between the contrast agent and the polymer. However, the reduced loading property and random dispersal of imaging agent in polymer matrix are considered to be the major disadvantages. On the other hand, Type II nanoconstructs are generated by encapsulation of contrast agent in PNPs and, hence, they demonstrate superior loading efficiency and uniform dispersal in polymeric matrix. However, the rapid burst discharge of imaging agent is the major disadvantage of Type II nanoconstructs.

Non-invasive optical imaging together with polymeric nanotechnology is offering an exceptional prospect in cancer diagnostics and therapy. Molecular level recognition of tumor is currently being possible with polymer-based fluorescent NPs margins. Fluorescent-based molecular imaging mainly relates to NIR dyes (NIRF) capable for deep tissue penetration. Therefore,

NIRF is the most appropriate for clinical imaging [186–188]. NIRFs are categorized into several subclasses, such as cyanine, squaraine, phthalocyanine, and porphyrin derivatives. Apart from organic dyes, inorganic nanomaterials such as QDs and upconversion nanoparticles (UCNPs) also display exceptional fluorescence. However, their cellular toxicity, biodistribution, and cellular kinetics are to be thoroughly addressed before being applied for human.

In Type I nanoconstructs, the polymeric matrix is covalently integrated to 'Cy' analogs of cyanine NIRF dye group (Cy5, Cy5.5, and Cy7). Among the cyanine NIRF dye family, Cy5.5 has been widely studied for optical imaging. Some of the Type I nanoconstructs include Cy5-PLA NPs, Cy5.5-PEI, Cy5.5-glycol-chitosan, Cy5.5-PEG, and Cy7-polymethylmethacrylate copolymer [186, 189–195]. The influence of molecular weight of PNPs was studied with glycol-chitosan NPs conjugated with Cy5.5 of various sizes 20 kDa–231 nm, 100 kDa–271 nm, and 250 kDa–310 nm on in vivo imaging. Upon intravenous administration of NPs to nude mice with subcutaneous SCC7 tumors, researchers observed enhanced blood circulation time for 250 kDa NPs and it sustained for about 3 days than with 20KDa and 100KDa NPs [192]. The study signifies that 250 kDa NPs were 4.1- and 2.4-times superior to 20 kDa and 100 kDa, respectively. Thus, NIRF-polymeric NPs can be used for imaging applications and for studying the basics of performance of NPs in physiological environments.

A self-assembled polymeric NP-based probe (NS) functioning as activatable optical contrast agent for imaging protease-associated diseases was developed [195]. The NS comprises intensely quenched matrix metalloproteinase (MMP)-specific NIR fluorogenic peptides [Cy5.5-Gly-Pro-Leu-Gly-Val-Arg-Gly-Lys(BHQ3)-Gly-Gly] on the surface of tumor-homing polymeric NPs as a carrier. The nanoconstruct results in the cleavage of the NIR-dye substrate, owing to the specific substrate recognition by the MMPs. The released Cy5.5 dye results in fluorescence dequenching (Figure 7.11). The nanoconstruct was administrated to MMP-2/9-positive SCC7 tumor-bearing mice directly without pretreatment and with the pretreatment of inhibitor for MMP-

2/9. The tumors treated with NS without pretreatment with inhibitor was observed due to fluorescence dequenching. The results signify enhanced fluorescence in the direct mode (without pretreatment with inhibito), whereas pretreated mode displayed substantially reduced fluorescence as demonstrated in Figure 7.11.

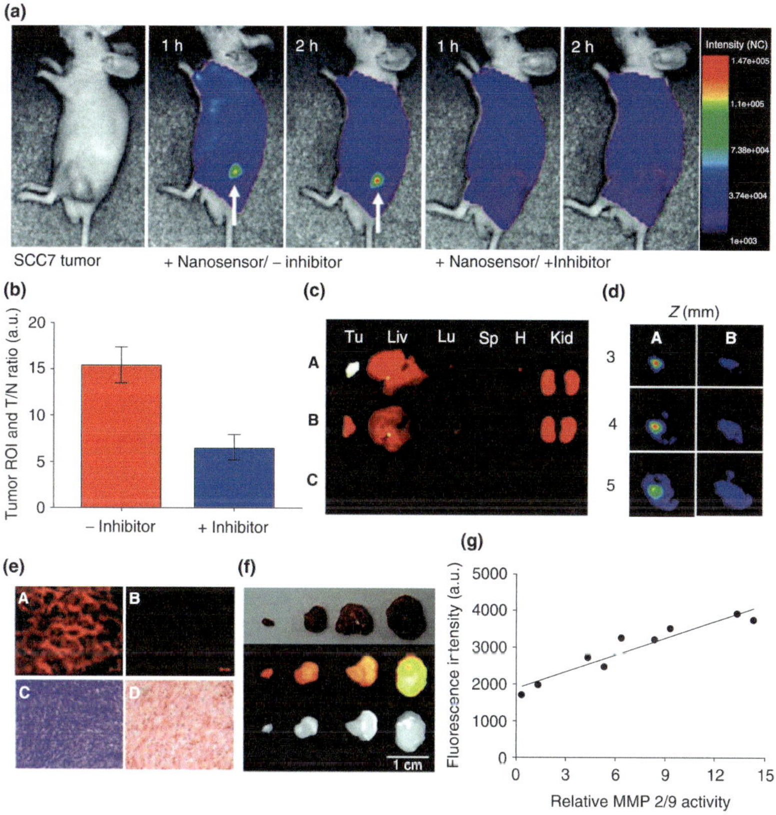

Figure 7.11 (a) In vivo NIRF tomographic images of mice bearing SCC7 tumor after intravenous administration of NS in the presence and absence of inhibitor. (b) Tumor region of interest (ROI) and T/N ratio. (c) Ex vivo NIRF images of organs and tumor. (d) 2D images of tumor. (e) NIRF microscopic image of tumors treated with nanoconstruct. (f) NIRF images of excised NS-treated tumors in mice with different tumor size. (g) Correlation between total fluorescence intensity and relative MMP-2/9 activity of excised tumors. Reprinted with permission from Ref. [195]. Copyright © 2009, American Chemical Society.

A new optical molecular imaging technique with small PNPs incorporating synthetic dyes, known as CLF dyes owing to the chemiluminescence and fluorescence, was developed [196]. Squaraine catenane endoperoxide (SCEP) conjugated carboxyl functional polystyrene NP was developed via swelling technique and the NP was investigated for whole-body fluorescence imaging and chemiluminescence imaging in mice. The study proposes a novel paradigm for optical molecular imaging using PNPs incorporating CLF dyes. Whole-body chemiluminescence imaging was effective than fluorescence for identifying lung accumulation by the NPs. The fluorescent intensity was greatly 'surface weighted', signifying the organs nearer to probe zone exhibited enhanced fluorescent intensity, whereas deep organs and tissues generated moderately less in intensity and fluorescence imaging exhibited superior uptake in liver and reduced accumulation of NPs in lungs. The results displayed that chemiluminescence imaging could be potentially employed to accurately identify accumulation of NPs in the lungs of live mice. Squaraine dyes, such as N-propanesulfonate-benzothiazolium (NPBT) squaraine and N-propanesulfonate-benzoindolium (NPBI) squaraine, labeled to bovine serum albumin (BSA) NPs produced by desolvation technique, have also been investigated for optical imaging [197–198].

Type II nanoconstructs are encapsulated with ICG (a tricarbo cyanine, US-FDA approved NIRF dye) in different polymers, including poly (allylamine hydrochloride), PLCL-Poloxamer, poly (lactide co-glycolide) Eudragit-based NPs, etc. [199–203]. ICG encapsulated and stabilized polymeric micelles were developed based on thermo-sensitive block copolymer pluronic F-127 and poly(ethylene oxide)-poly(propylene oxide)-poly(ethylene oxide), and were synthesized via solvent evaporation method. The biodistribution and plasma clearance rate and ICG-loaded micelles were determined in colon carcinoma tumor-bearing mice. Whole-body NIR imaging was implemented for the detection of tumor. The studies presented the accumulation of both ICG and ICG-nanoconstruct in liver and kidney than tumor. However, the fluorescent images documented 36 h post injection displayed superior accumulation of nanoconstruct in tumor than liver

[204], thus validating higher retention efficiency of nanoconstruct within tumor sites.

Studies on Type II nanoconstructs with the derivatives of porphyrin and phthalocyanine (meso-tetra(4-hydroxyphenyl) porphyrin, hexadecafluoro zinc phthalo-cyanine, zinc(II) phthalocyanine, etc. incorporated into PLGA and PLA NPs via different methods, including spontaneous emulsification solvent diffusion method and salting-out technique, have been reported [205–208]. Although porphyrin and phthalocyanine are studied under photodynamic therapy (PDT), the inherent NIRF properties of these compounds are to be exploited and a comprehensive analysis of nanoformulation with porphyrin or phthalocyanine with superior blood circulation can be employed as ideal candidate for optical imaging [205–208].

Owing to the rapid development of novel NIRF imaging instruments, there is a significant growth in non-invasive optical imaging. Fluorescence-assisted resection and exploration (FLARETM) intraoperative NIRF imaging system was used for mapping sentinel lymph nodes in breast cancer [209]. ICG diluted in human serum albumin (HSA) was used as an NIR fluorescent lymphatic tracer and was administrated via subcutaneously and intratumorally in six patients. In addition to ICG-HSA, the subjects also received $^{(99m)}$Tc-sulfur colloid lymphoscintigraphy. The FLARETM imaging system was used for the optically imaging of sentinel lymph node. In the pilot clinical trial, nine sentinel lymph nodes were identified by $^{(99m)}$Tc-lymphoscintigraphy and nine sentinel lymph nodes were identified by NIR fluorescence too, though results fluctuated in two patients. The promising results are anticipated to create a great revolution in optical imaging. In another study, a first human clinical trial in perforator flap breast reconstruction was done with the aid of an intraoperative, NIR fluorescence imaging system. Based on the study, NIR assessment of perforator flap breast reconstruction was possible with a light-emitting diode-based system. Design and development of novel imaging probes would certainly complement existing technology and for advanced NIRF imaging systems and biomedical imaging applications.

7.2.5 UCNPs

Although fluorescent dyes have significant applications in biomedical imaging, the present status of them for visualizing biological tissue is seriously restricted, owing to autoflourescence stemming from fluorescent dyes. The design and development of next-generation fluorescence probes based on nanoscience and technology is crucial in current technology. With the advancement of nanotechnology, semiconductor QDs have developed as a substitute and favorable fluorescent nano labels than organic dyes [210]. QDs have been studied in FRET analysis, imaging cells and tissues, and gene/drug delivery [211]. The rising concerns on chemical instability and intrinsic cytotoxicity of QDs have restricted the application for biological usage [212]. The major drawback of QDs includes the difficulty in deep tissue imaging, low signal-to-noise ratio (SNR), and possible impairment of biomolecules due to long-term irradiation. Hence, there is an increasing need for the development of more proficient biological labels to overcome the above-explained limitations.

The conversion of NIR radiations into visible light through nonlinear optical routes and designated upconversion (UC) has engendered substantial attention in biomedical imaging. Rare earth (RE) element-doped NIR-to-Vis UCNPs perform as an exceptional alternative fluorescent probe. Currently RE-doped UCNPs, converting longer wavelength into short-wavelength fluorescence through two-photon or multiple photon process, are used for developing novel fluorescent biolabels (213–218). UCNP does not undergo photobleaching as QDs and it does not exhibit blinking during imaging. Multiple color emission can be attained by changing the dopant and host crystal. Also, the emission wavelengths of the UCNPs are not dependent on size as in QDs. NIR radiations are often used to excite RE-doped UCNPs and hold several advantages such as (a) better SNR and detection sensitivity due to the lack of autofluorescence, (b) deep tissue imaging due to the deeper NIR light penetration, thus instigating reduced light-induced damage to biological samples, (c) low-power NIR laser excitation, (d) large Stokes shifts and narrow emission peaks, (e) superior physicochemical stability, and (f) reduced toxicity. Thus, RE-doped UCNPs are capable substitutes to conventional fluorescent dyes for biomedical applications.

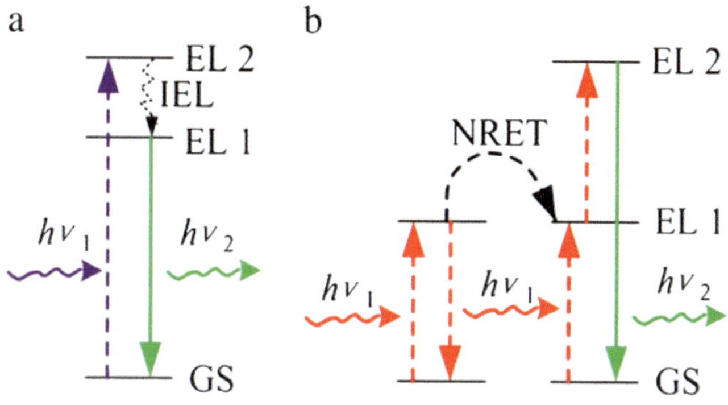

Figure 7.12 Schematic of (a) downconversion and (b) energy transfer UP mechanism. IEL: internal energy loss; GS: ground state; EL: energy level; NRET: non-radiative energy transfer; $h\nu_1$: incident light; $h\nu_2$: emission light. Reprinted with permission from Ref. [219]. Copyright © 2012 Elsevier Inc.

Upconversion luminescence (UCL) is a distinctive process in which longer wavelength (low-energy photons) is transformed to shorter wavelength of high-energy photons by consecutive absorption process and by energy transfer [213–220]. Conventional fluorophores display downconversion where high-energy photons are absorbed, whereas low-energy photons are discharged due to internal energy loss (IEL) (Figure 7.12a). UC is a process that results in the emission of high-energy photons through successive absorption of low-energy photons. The mechanism of UC process is mostly divided into three classes: (a) excited state absorption (ESA), (b) energy transfer upconversion (ETU), and (3) photon avalanche. ETU has been extensively used to acquire high-UC efficiency, comprising the absorption of a pump phonon of similar energy by two adjacent ions (Figure 7.12b). The non-radiative energy transfer stimulates one ion to higher energy level (EL2), whereas another ion relaxes to the ground state (GS). Consequently, relaxation from higher energy level (EL2) marks in high energy photon emission.

Generally, UCNP is made up of an inorganic host material to which lanthanide dopant ions are inserted in the lattice of host component [216]. To augment UC efficacy, Yb^{3+} with a larger absorption cross-section in the NIR spectral region is commonly doped as a sensitizer in combination with the activators. The dopant range is maintained less than 2 mol % to reduce excitation energy loss by cross-relaxation process [221]. UC processes are based on the hierarchical organization of energy levels of dopant lanthanide ions. In addition, optical property and crystal structure of host component has significant role for efficient UC processes. The host material absorbs energy from the dopant ion's excited energy by lattice vibrations. Difference in the crystal structure of the host component modifies crystal field of dopant ions, resulting in the distinctive optical properties of UCNPs. Host materials should possess sufficient transparency within the wavelength range of interest, high optical damage threshold, low-phonon energy, and lattice matching to dopant ions for attaining superior level of doping. Inorganic RE compounds, alkaline earth ions, and several transition metal ions, such as Mn^{2+}, Zr^{4+}, and Ti^{4+}, are appropriate host materials for lanthanide doping [222–226]. The intra-configurational $4f^n$ electron transitions of lanthanide ions influence their absorption and emission profile. Since 4f electrons are covered by completely filled $5s^2$ and $5p^6$ subshells, $4f^n$ electron rarely undergo interactions with host lattice. Owing to this reason, the absorption and emission profiles of lanthanide-doped NPs display sharp lines. The emission wavelength of UCNPs does not depend on the chemical or physical properties of the host materials and the emission colors can be engineered by controlling emission wavelength or emission intensities by controlling the ratio of host and dopant mixtures and concentration of dopant [225–234]. As in QDs, these UCs also require surface treatments prior to conjugation with biomolecules for targeting applications. Surface modification can be of anyone among ligand engineering, ligand attraction, surface polymerization, and layer by layer assembly.

The UCNP imaging is superior to commonly used NIR multiphoton microscopy that necessitates complicated experimental setup and pulsed lasers. On the other hand, a common optical microscope equipped with xenon lamp modified to diode laser is sufficient for UCNP imaging that makes the NPs specifically suitable contrast agents for optical imaging. Submicron-sized Y_2O_2S:Yb/Tm particles were employed to evaluate the bodistribution of prostate-specific antigen in human prostate tissue using immune histological techniques [235]. Unlike certain dyes used for bioimaging, UCNP does not photo bleach upon continuous exposure of high excitation energy levels, avoids the use of short wavelength for excitation that may results in the non-specific autofluorescence, and permits single-molecule imaging. Currently, UC imaging has been extensively studied for the high-resolution imaging of cells. UCNPs are also functionalized with ligands for targeted imaging. $NaYF_4$:Yb/Er NPs were attached with antibody and used for imaging of HeLa cells with antigen overexpressed on cell membrane. The application of Y_2O_3:Er NPs that conjugated to cyclic RGD for cellular imaging studies was demonstrated [236]. Small interference RNA (siRNA) was conjugated to silica-coated $NaYF_4$:Yb/Er NP, which was attached with anti-Her2 antibody and uptake of the NPs was studied [237]. Effect of gene silencing by siRNA was evaluated in luciferase assay and real-time monitoring of the intracellular release of siRNA from UCNPs was reported [238]. siRNA was labeled with BOBO-3 intercalating dye and FRET between UCNPs and BOBO-3 was measured that indicated siRNA release from UCNPs.

Recently, in vivo imaging with UCNPs has been receiving great attention. The study of UCNPs for imaging live animal has been demonstrated [239]. Y_2O_3:Yb/Er NPs of 50–150 nm in size were administrated in live *Caenorhabditis elegans* (C. elegans) worms and the digestive system was imaged upon excitation at 980 nm. In vivo wholebody imaging of a Balb-c mouse with NIR-to-NIR UCNPs has been demonstrated [240]. The outstanding advantage of the study was the excitation and emission in NIR range, permitting deep tissue imaging. In another study, in vivo vascular imaging of nude mice was reported [241] using Y_2O_3:Yb/Er NPs coated with PEG polymer. The multi-channel label was injected to mice for the examination of their effectiveness in vascular imaging under clinically significant low power density laser excitation. The colocalization of the UC and cyanine fluorescence suggested the polymer coating was intact in vivo (Figure 7.13). Emission instigating from UC was bright to support real-time imaging stipulating a nanoplatform for in vivo imaging. The feasibility of UCNPs in transillumination imaging has been reported [242]. Images acquired by the transillumination technique were compared with NIR cyanine dyes and the results propose the better-quality and contrast of UCNP-based imaging moieties.

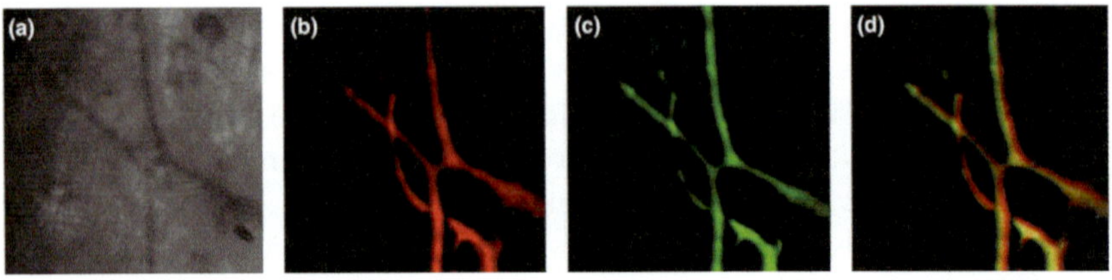

Figure 7.13 Optical imaging of mouse blood vessels after intravenous administration of NPs (10 mg): (a) image under blue light filter, (b) UC image with excitation at 980 nm, (c) fluorescence image of the carbocyanine dye, and (d) merged image of UC and fluorescence signals. Reprinted with permission from Ref. [241]. Copyright © 2009, Royal Society of Chemistry.

Real-time monitoring application based on UCNPs has immense potential in biomedical imaging [243], in which dual-color in situ imaging of lymph was attained by using NIR- and green-emitting UCNPs. In vivo targeted imaging was achieved by conjugating folic acid (FA) and RGD peptide to UCNPs [244, 245] for the detection of HeLa and U87MG tumors in nude mice. UCNP-loaded live myoblast cells were injected in living mouse with cryo damaged hind limb, and in vivo imaging was performed to validate the dispersal and bioactivity of the cells delivered along with NPs. Greater penetration depth mediated UCL NPs for bioimaging in vivo models has been demonstrated. UCNP of 20 nm $NaLuF_4$:Yb,Tm could penetrate >1.5 cm depth of pork pieces [246]. Yb,Tm-codoped UCNPs have been effectively studied for in vivo imaging of different types of mice (Kunming mouse with fur and black mouse) (Figure 7.14a–c), however, the fluorescence signal was not promising for whole-body imaging of the both the mice with fur and black mouse [247–249]. The same research group has studied that hexagonal $NaLuF_4$:Yb,Tm NPs of sub-10 nm size have emissions in 800 nm, penetrated well through a black mouse with a depth of 2 cm, upon excitation with 980 nm laser (Figure 7.14b,c). This was further proved in the in vivo imaging of a rabbit with $NaLuF_4$:Yb,Tm NPs (sub-20 nm in size) as a fluorescent reporter (Figure 7.14e–g). UCL NP of LaF_3:20% Yb, 1% Tm was developed for in vivo imaging of lymphatic system with enhanced SNR of the UCL signal (λ_{em} = 800 nm). A strong signal of UCL was detected from the excised lymph node, while any observable signal of UCL was noticed outside draining lymphatic area [250].

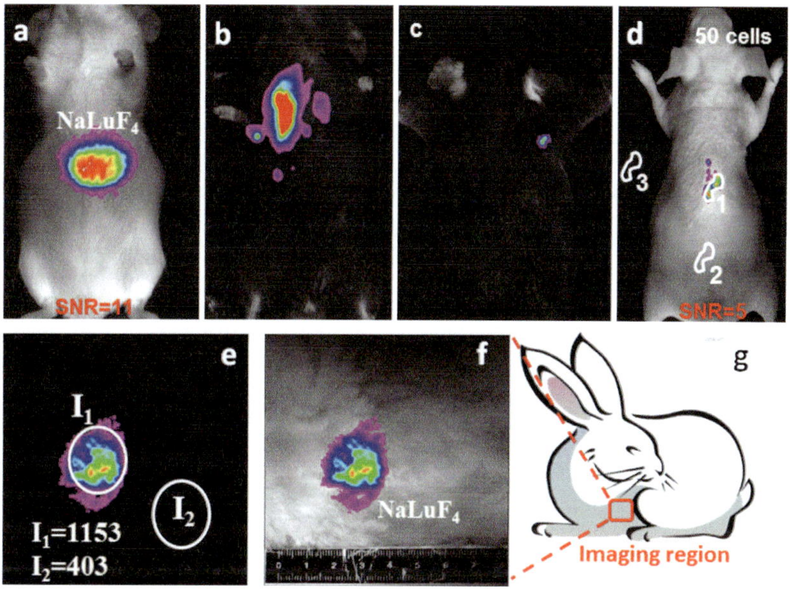

Figure 7.14 UCNPs for in vivo imaging of naked mouse, Kunming mouse, black mouse, and fur-rich rabbit. Reprinted with permission from Ref. [248, 249]. Copyright © 2011, American Chemical Society, Copyright © 2012, Elsevier Ltd.

7.2.6 Ceramic NPs

The design and development of ceramic materials for biomedical application is gaining considerable attention [251]. Ceramics have been studied extensively in bone tissue engineering, owing to their osteo inductive and biocompatible properties [252]. Ceramics are non-metallic inorganic crystalline compounds, often in the form of oxides, made by heating and subsequent cooling of the materials [253]. The significant qualities of inorganic ceramics include high abundance in nature and simple synthetic strategies with tunable properties. Nanoscale ceramics, such as hydroxyapatite (HA), zirconia (ZrO_2), silica (SiO_2), titanium oxide (TiO_2), and layered double hydroxide (LDH), were prepared to advance their physical-chemical properties with high biocompatibility in biological systems [254]. Various ceramic NPs have been studied in vitro and in vivo as theranostics (therapy + diagnosis) as briefly discussed below.

Carbon-, sulfide-, and oxide-based ceramics are inorganic nanomaterials that have created substantial attention in nanomedicine as targeted theranostics, especially as image able gene/drug carriers. These nanostructures may vary in their shape, size, porosity, and core-shell configurations, integrated with functional organic groups such as peptides, nucleic acids, antibodies, and small ligands, and serve as efficient bioimaging agents.

Built-in multifunctional LDH structure within a single-phase NP before incorporating, conjugating with other phases is gaining greater interest [255]. LDHs or anionic clays or hydrotalcite-like compounds are clay-like materials that exhibit exciting properties for many applications. LDHs are a broad class of inorganic lamellar compounds with high capacity for intercalation of anion. The LDH structure is represented by $[M^{2+}{}_{1-x}M^{3+}x(OH)_2]^{x+}(A^{n-})_{x/n}\cdot mH_2O$ that results from the stacking of brucite-like layers ($[Mg(OH)_2]$) with extra positive charge, owing to the partial isomorphous substitution of M^{2+} by M^{3+} [256–259]. The positive excess charge is well balanced by anions that are present in interlamellar spaces. The frequently used divalent cations are Mg^{2+}, Cu^{2+}, Ni^{2+}, Co^{2+}, Zn^{2+}, Fe^{2+}, and Mn^{2+}, and trivalent cations include Al^{3+}, Ni^{3+}, Fe^{3+}, Mn^{3+}, Gd^{3+}, and Cr^{3+}. The natural mineral, hydrotalcite ($Mg_6Al_2(OH)_{16}CO_3\cdot 4H_2O$), with space group R-3 m, has a typical LDH assembly, shown in Figure 7.15. The partial substitution of octahedral Mg^{2+}, in the layered brucite ($Mg(OH)_2$ stacks), with Al^{3+} generates positive charge on the piled layers. Specifically required anions are intercalated during synthesis and substitution of interlayer anions occurs after the synthesis through ion exchange. Charge neutrality is supported by the intercalation of bound anions through electrostatic binding in the interlayer space, with water through hydrogen bonding to hydroxyls. The commonly bound anions include nitrates, carbonates, organic acids, glycerol, porphyrins, nucleotides, DNA, ATP, vitamins, and drugs.

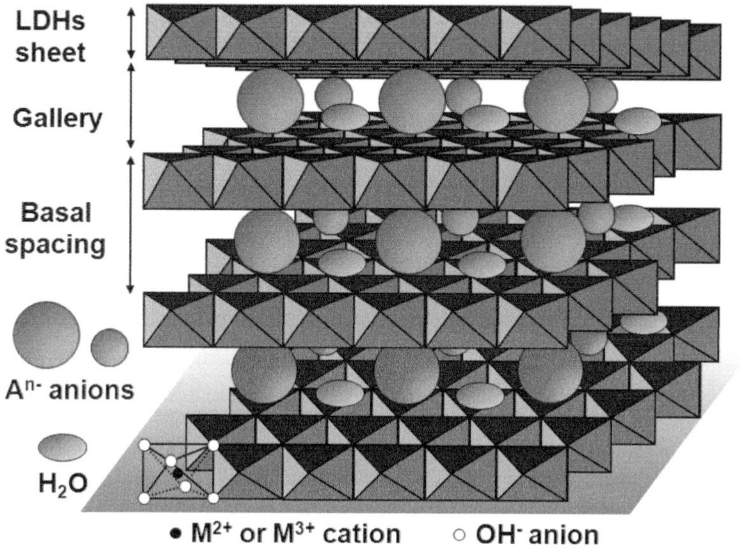

Figure 7.15 Structure of LDH. Reprinted with permission from Ref. [259].

Some of the exciting properties of these materials include high specific surface area, swelling property, memory effect, high anion exchange capacity, and dispersive property. These interesting properties along with low cytotoxicity and good biocompatibility mark LDHs appropriate candidates for biomedical applications. LDH structures have been used as ion exchangers, catalysts, stabilizers, adsorbents, and polymer composites, and currently studies have focused on the drug delivery applications of these NPs. Application of LDHs in medical diagnostic systems, such as biosensing and medical imaging systems, is also reported [260, 261]. These applications trust on acid buffering effect and anion exchange property. Therefore, LDH is being intercalated with several anionic biomolecules, such as siRNA, DNA and radio labeled ATP (γ-^{32}P) radioisotopes ($^{57}Co^{2+}$, $^{67}Ga^{3+}$), fluorescein isothiocyanate, small drugs molecule, and small proteins at physiological pH [262–275]. In vitro studies have indicated the cellular uptake of these NPs by endocytosis followed by the disintegration of LDH host in the lysosomes with acidic environment resulting in cellular delivery and distribution of drugs. In vivo studies on toxicity, pharmacokinetics, transfection, and therapeutic effectiveness of LDH nanoplatforms have been reported [276].

Paramagnetic gadolinium complexes are often used as clinical contrast agents. Gadolinium (Gd) is a metal with large magnetic moment due to seven unpaired electrons in its 4f orbital. The magnetic dipoles of these electrons interact with neighboring nuclear dipoles. The interaction instigates the nuclear dipoles to align more quickly in the presence of an external magnetic field. The proton relaxation enhancement, in addition to the slow electron relaxation time, marks Gd as a promising agent for MR enhancement. Gd has the largest T1-relxation enhancement effect of all the known elements [277]. Gd chelates have played a significant role in MR imaging in the past two decades. However, the efficiency of Gd chelates is low in terms of the detection of molecular events. Therefore, new MR contrast agents with Gd and LDH are being developed for molecular imaging purposes.

In a study, Gd-DTPA was intercalated in an Mg-Al LDH using an anion exchange method [278]. It was demonstrated that the LDH-NP with 1.6 mM $[Gd(DTPA)]^{2-}$ displayed longitudinal proton relaxivity r_1 of 6 $mM^{-1}s^{-1}$ and a transverse proton relaxivity r_2 of 50 $mM^{-1}s^{-1}$ and magnetic field of 190 MHz. It signified an enhancement of r_1 to four times and r_2 to 12 times than free $[Gd(DTPA)]^{2-}$ under similar reaction settings. The modified LDH-Gd(DTPA) NPs were bar-like with a width of 30–60 nm and a length of 50–150 nm. Different amounts of Gd-DTPA loaded in LDH resulted in various interlayer spaces ranging from 3 nm to 4 nm. Gd-DTPA is a large ion with restricted negative charge that can compensate the part of the positive charges of LDH layers. Hence, about 40% equivalent small anions, like Cl^- and CO_3^{2-}, intercalate in LDH. Thus an Mg_2Al–[Gd (DTPA)]$^{2-}$/Cl^-/CO_3^{2-}–LDH hybrid phase was synthesized and characterized by X-ray diffraction (XRD). The researchers designed LDH NPs into excellent contrast agents with strong relaxivity for their applications in MRI.

The toxicity and biodistribution of hybrid Mg, Al LDH, and Mg-Gd-LDH have been demonstrated in FVB/N inbred mice [279]. Apart from their role as carriers for controlled drug release, layered nanohybrids (LNH) served as ideal candidates for biomedical imaging, including multiphoton microscopy and MRI. The LDH nanohybrids were administrated to ovarian bursa, and observed them in the ovarian epithelial cells four days post administration. The ovary and the ovarian bursa were also observed in MRI, three days after the injection. According to these evaluations, there was an only minor pathologic response when mice were subjected to intravenous, intraperitoneal, and subcutaneous injections of LDH NPs, which explain LDHs safety in biomedical applications. Mg-Gd LDH was then detected with MRI even three days after injection in mice. Negatively charged fluorescence dye, Lucifer Yellow, was intercalated in Mg-Al LDH and demonstrated its potential for in vitro and in vivo imaging by multiphoton microscopy. Remarkably, the fluorescence of LDH was detectable even 35 days after injection. This signifies a possible long-time screening of drug that can lead to an improved understanding of the delivery systems. Overall LDH NPs revealed negligible toxicity and were promising for dual modal imaging studies.

7.3 Conclusion and Future Prospects

A diverse range of NPs have been designed and developed as nanoprobes for the improvement of optical bioimaging application. Nanoprobes offer excusive information on the molecular and cellular events that can be eventually correlated to the early diagnosis and also for the progression of several disorders. The nanoprobes remain versatile for the integration of other conventional imaging modalities, such as X-ray, MRI, CT, and US, to offer multimodal imaging. The ability to potentially target the nanoprobes finds immense application in the precise delivery of drugs and for accurate imaging and image-guided surgery [280–288]. These optical nanoprobes can also be finely tuned as theranostic nanoprobes engaging diagnosis and treatment in the same go. However, proper address on the biocompatibility criteria of optical nanoprobes may enhance their routine application in clinical imaging in the near future.

References

1. Smith AM, Mancini MC, Nie S. Second window for in vivo imaging. *Nature nanotechnology.* 4, **2009**, 710–711.
2. Weissleder R. A clearer vision for in vivo imaging. *Nat. Biotechnol.* 19, **2001**, 316–317.
3. Frangioni JV. In vivo near-infrared fluorescence imaging. *Curr. Opin. Chem. Biol.* 7, **2003**, 626–634.
4. Mattoussi H, Palui G, Bin Na H. Luminescent quantum dots as platforms for probing in vitro and in vivo biological processes. *Adv. Drug Deliv. Rev.* 64, **2012**, 138–166.
5. Probst CE, Zrazhevskiy P, Bagalkot V, Gao X. Quantum dots as a platform for nanoparticle drug delivery vehicle design. *Adv. Drug Deliv. Rev.* 65, **2013**, 703–718.
6. Jin Z, Hildebrandt N. Semiconductor quantum dots for in vitro diagnostics and cellular imaging. *Trends. Biotechnol.* 30, **2012**, 394–403.
7. Jin S, Hu Y, Gu Z, Liu L, Wu HC. Application of quantum dots in biological imaging. *J. Nanomater.* Article ID 834139, **2011**, 13.
8. Pisanic II TR, Zhang Y, Wang TH. Quantum dots in diagnostics and detection: Principles and paradigms. *Analyst.* 139, **2014**, 2968–2981.
9. Mirzaei J, Reznikov M, Hegmann T. Quantum dots as liquid crystal dopants. *J. Mater. Chem.* 22, **2012**, 22350–22365.
10. Mazumder S, Dey R, Mitra MK, Mukherjee S, Das GC. Biofunctionalized quantum dots in biology and medicine. *J. Nanomater.* Article ID 815734, **2009**, 17.
11. Smith AM, Duan H, Mohs AM, Nie S. Bioconjugated quantum dots for *in vivo* molecular and cellular imaging. *Adv Drug Deliv Rev.* 60, **2008**, 1226–1240.
12. Medintz IL, Uyeda HT, Goldman ER, Mattoussi H. Quantum dot bioconjugates for imaging, labelling and sensing. *Nat. Mater.* 4, **2005**, 435–446.
13. Clapp AR, Goldman ER, Mattoussi H. Capping of CdSe–ZnS quantum dots with DHLA and subsequent conjugation with proteins. *Nat. Protoc.* 1, **2006**, 1258–1266.
14. Dameron CT, Winge DR. Peptide-mediated formation of quantum semiconductors. *Trends Biotechnol*, 8, **1990**, 3–6.

15. Sukhanova A, Venteo L, Devy J, Artemyev M, Oleinikov V, Pluot M, Nabiev I. Highly stable fluorescent nanocrystals as a novel class of labels for immunohistochemical analysis of paraffin-embedded tissue sections. *Lab. Invest,* 82, **2002**, 1259–1261.
16. Bae W, Mehra RK. Cysteine-capped ZnS nanocrystallites: Preparation and characterization. *J. Inorg. Biochem.* 70, **1998**, 125–135.
17. Sukhanova A, Devy J, Venteo L, Kaplan H, Artemyev M, Oleinikov V, Klinov D, Pluot M, Cohen JH, Nabiev I. Biocompatible fluorescent nanocrystals for immunolabeling of membrane proteins and cells. *Anal. Biochem.* 324, **2004**, 60–67.
18. Cassette E, Helle M, Bezdetnaya L, Marchal F, Dubertret B, Pons T. Design of new quantum dot materials for deep tissue infrared imaging. *Adv. Drug Deliv. Rev.* 65, **2013**, 719–731.
19. Cai W, Shin DW, Chen K, Gheysens O, Cao Q, Wang SX, Gambhir SS, Chen X. Peptide-labeled near-infrared quantum dots for imaging tumor vasculature in living subjects. *Nano Lett.* 6, **2006**, 669–676.
20. Aswathy RG, Yoshida Y, Maekawa T, Sakthi Kumar D. Near-infrared quantum dots for deep tissue imaging. *Anal. Bioanal. Chem.* 397, **2010**, 1417–1435.
21. Rogach AL, Orgis M. Near-infrared-emitting semiconductor quantum dots for tumor imaging and targeting. *Curr. Opin. Mol. Ther.* 12, **2010**, 331–339.
22. Wu XY, Liu HJ, Liu JQ, Haley KN, Treadway JA, Larson JP, Ge NF, Peale F, Bruchez MP. Immunofluorescent labeling of cancer marker Her2 and other cellular targets with semiconductor quantum dots. *Nat. Biotechnol.* 21, **2003**, 41–46.
23. Akerman ME, Chan WCW, Laakkonen P, Bhatia SN, Ruoslahti E. Nanocrystal targeting in vivo. *Proc. Natl. Acad. Sci. USA.* 99, **2002**, 12617–12621.
24. Dubertret B, Skourides P, Norris DJ, Noireaux V, Brivanlou AH, Libchaber A. In vivo imaging of quantum dots encapsulated in phospholipid micelles. *Science.* 298, **2002**, 1759–1762.
25. Bagalkot V, Zhang L, Levy-Nissenbaum E, Jons, Kantoff PW, Langer R, Farokhzad OC. Quantum dotaptamer conjugates for synchronous cancer imaging, therapy, and sensing of drug delivery based on bi-fluorescence resonance energy transfer. *Nano Lett.* 7, **2007**, 3065–3070.
26. Maysinger D, Behrendt M, Hébert ML, Kriz J. Real-time imaging of astrocyte response to quantum dots: In vivo screening model

system for biocompatibility of nanoparticles. *Nano Lett.* 7, **2007**, 2513–2520.

27. Zimmer JP, Kim SW, Ohnishi S, Tanaka E, Frangioni JV, Bawendi MG. Size series of small indium arsenide-zinc selenide core-shell nanocrystals and their application to in vivo imaging. *J. Am. Chem. Soc.* 128, **2006**, 2526–2527.
28. Jiang W, Singhal A, Kim BYS, Zheng J, Rutka JT, Wang C, Chan WCW. Assessing near-infrared quantum dots for deep tissue, organ, and animal imaging applications. *J. Assoc. Lab. Autom.* 13, **2008**, 6–12.
29. Zintchenko A, Susha AS, Concia M, Feldmann J, Wagner E, Rogach AL, Ogris M. Drug nanocarriers labeled with near-infrared-emitting quantum dots, (quantoplexes): Imaging fast dynamics of distribution in living animals. *Mol. Ther.* 17, **2009**, 1849–1856.
30. Liu Y, Njuguna R, Matthews T, Akers WJ, Sudlow GP, Mondal S, Tang R, Gruev V. Near-infrared fluorescence goggle system with complementary metal–oxide–semiconductor imaging sensor and see-through display. *J. Biomed. Opt.* 18, **2013**, 101303.
31. Larson DR, Zipfel WR, Williams RM, Clark SW, Bruchez MP, Wise FW, Webb WW. Water-soluble quantum dots for multiphoton fluorescence imaging in vivo. *Science.* 300, **2003**, 1434–1436.
32. Lim YT, Kim S, Nakayama A, Stott NE, Bawendi MG, Frangioni JV. Selection of quantum dot wavelengths for biomedical assays and imaging. *Mol. Imaging.* 2, **2003**, 50–64.
33. Smith JD, Fisher GW, Waggoner AS, Campbell PG. The use of quantum dots for analysis of chick CAM vasculature. *Microvasc. Res.* 73, **2007**, 75–83.
34. Ballou B, Lagerholm BC, Ernst LA, Bruchez MP, Waggoner AS. Noninvasive imaging of quantum dots in mice. *Bioconjug Chem.* 15, **2004**, 79–86.
35. Kim S, Lim YT, Soltesz EG, De Grand AM, Lee J, Nakayama A, Parker JA, Mihaljevic T, Laurence RG, Dor DM, Cohn LH, Bawendi MG, Frangioni JV. Near-infrared fluorescent type II quantum dots for sentinel lymph node mapping. *Nat. Biotechnol.* 22, **2004**, 93–97.
36. Soltesz EG, Kim S, Laurence RG, DeGrand AM, Parungo CP, Dor DM, Cohn LH, Bawendi MG, Frangioni JV, Mihaljevic T. Intraoperative sentinel lymph node mapping of the lung using near-infrared fluorescent quantum dots. *Ann. Thorac. Surg.* 79, **2005**, 269–277.

37. Parungo CP, Colson YL, Kim SW, Kim S, Cohn LH, Bawendi MG, Frangioni JV. Sentinel lymph node mapping of the pleural space. *Chest.* 127, **2005**, 1799–1804.

38. Gao X, Cui Y, Levenson RM, Chung LWK, Nie S. In vivo cancer targeting and imaging with semiconductor quantum dots. *Nature Biotechnol.* 22, **2004**, 969–976

39. Rosen AB, Kelly DJ, Schuldt AJ, Lu J, Potapova IA, Doronin SV, Robichaud KJ, Robinson RB, Rosen MR, Brink PR, Gaudette GR, Cohen IS. Finding fluorescent needles in the cardiac haystack: tracking human mesenchymal stem cells labeled with quantum dots for quantitative in vivo three-dimensional fluorescence analysis. *Stem Cells.* 25, **2007**, 2128–2138.

40. Yu X, Chen L, Li K, Li Y, Xiao S, Luo X, Liu J, Zhou L, Deng Y, Pang D, Wang Q. Immunofluorescence detection with quantum dot bioconjugates for hepatoma in vivo. *J. Biomed. Opt.* 12, **2007**, 014008.

41. Diagaradjane P, Orenstein-Cardona JM, Colón Casasnovas NE, Deorukhkar A, Shentu S, Kuno N, Schwartz DL, Gelovani JG, Krishnan S. Imaging epidermal growth factor receptor expression *in vivo*: Pharmacokinetic and biodistribution characterization of a bioconjugated quantum dot nanoprobe. *Clin. Cancer. Res.* 14, **2008**, 731–741.

42. Tada H, Higuchi H, Wanatabe TM, Ohuchi N. In vivo real-time tracking of single quantum dots conjugated with monoclonal anti-HER2 antibody in tumors of mice. *Cancer Res.* 67, **2007**, 1138–1144.

43. Schroeder JE, Shweky I, Shmeeda H, Banin U, Gabizon A. Folate-mediated tumor cell uptake of quantum dots entrapped in lipid nanoparticles. *J. Control Rclcasc.* 124, **2007**, 28–34.

44. Jayagopal A, Russ PK, Haselton FR. surface engineering of quantum dots for *in vivo* vascular imaging. *Bioconjug. Chem.* 18, **2007**, 1424–1433.

45. Hu R, Yong KT, Roy I, Ding H, Law WC, Cai H, Zhang X, Vathy LA, Bergey EJ, Prasad PN. Functionalized near-infrared quantum dots for *in vivo* tumor vasculature imaging. *Nanotechnology.* 21, **2010**, 145105.

46. Kobayashi H, Hama Y, Koyama Y, Barrett T, Regino CAS, Urano Y, Choyke PL. Simultaneous multicolor imaging of five different lymphatic basins using quantum dots. *Nano Lett.* 7, **2007**, 1711–1716.

47. Jiang G, Park K, Kim J, Kim KS, Hahn SK. Target specific intracellular delivery of siRNA/PEI-HA complex by receptor mediated endocytosis. *Mol. Pharm.* 6, **2009**, 727–737.

48. Kim D, Jon S. Gold nanoparticles in image-guided cancer therapy. *Inorg. Chim. Acta.* 393, **2012**, 154–164.

49. Sperling RA, Rivera P, Zhang F, Zanella M, Parak WJ. Biological applications of gold nanoparticles. *Chem. Soc. Rev.* 37, **2008**, 1896–1908.

50. Faraji AH, Wipf P. Nanoparticles in cellular drug delivery. *Bioorg. Med. Chem.* 17, **2009**, 2950–2962.

51. Stewart ME, Anderton CR, Thompson LB, Maria J, Gray SK, Rogers JA, Nuzzo RG. Nanostructured plasmonic sensors. *Chem. Rev.* 108, **2008**, 494–521.

52. Jain PK, Huang XH, El-Sayed IH, El-Sayed MA. Noble metals on the nanoscale: Optical and photothermal properties and some applications in imaging, sensing, biology, and medicine. *Acc. Chem. Res.* 41, **2008**, 1578–1586.

53. Anker JN, Hall WP, Lyandres O, Shah NC, Zhao J, Van Duyne RP. Biosensing with plasmonic nanosensors. *Nat. Mater.* 7, **2008**, 442–453.

54. Chourpa I, Lei FH, Dubois P, Manfait M, Sockalingum GD. Intracellular applications of analytical SERS spectroscopy and multispectral imaging. *Chem. Soc. Rev.* 37, **2008**, 993–1000.

55. Ghosh P, Han G, De M, Kim CK, Rotello VM. Gold nanoparticles in delivery applications. *Adv. Drug Deliv. Rev.* 60, **2008**, 1307–1315.

56. Pissuwan D, Cortie CH, Valenzuela SM, Cortie MB. Functionalised gold nanoparticles for controlling pathogenic bacteria. *Trends Biotechnol.* 28, **2010**, 207–213.

57. Gao Y, Li Y. Gold nanostructures for cancer imaging and therapy, Chapter 2, Springer-Verlag Berlin Heidelberg, 2016, Z. Dai (ed.), *Advances in Nanotheranostics I*, Springer Series in Biomaterials Science and Engineering 6.

58. Dykman L, Khlebtstov N. Gold nanoparticles in biomedical applications: Recent advances and perspectives. *Chem. Soc. Rev.* 41, **2012**, 2256–2282.

59. Zhang X. Gold nanoparticles: Recent advances in the biomedical applications. *Cell Biochem. Biophys.* 72, **2015**, 771–775.

60. Cabuzu D, Cirja A, Puiu R, Grumezescu AM. Biomedical applications of gold nanoparticles. *Curr. Top. Med. Chem.* 15, **2015**, 1605–1613.

61. Au L, Chen J, Wang LV, Xia Y. Gold nanocages for cancer imaging and therapy. *Methods Mol. Biol.* 624, **2010**, 83–99.

62. Skrabalak SE, Au L, Lu X, Li X, Xia Y. Gold nanocages for cancer detection and treatment. *Nanomedicine (Lond).* 2, **2007**, 657–668.

63. Turkevich J, Stevenson PC, Hillier J. A study of the nucleation and growth processes in the synthesis of colloidal gold. *Discuss Faraday Soc.* 11, **1951**, 55–75.

64. Neiman B, Grushka E, Lev O. Use of gold nanoparticles to enhance capillary electrophoresis. *Anal. Chem.* 73, **2001**, 5220–5227.

65. Hall SR, Davis SA, Mann S. Cocondensation of organosilica hybrid shells on nanoparticle templates: A direct synthetic route to functionalized core–shell colloids. *Langmuir.* 16, **2000**, 1454–1456.

66. Giersig M, Mulvaney P. Preparation of ordered colloid monolayers by electrophoretic deposition. *Langmuir.* 9, **1993**, 3408–3413.

67. Shi X, Ganser T, Sun K, Balogh LP, Baker Jr JR. Characterization of crystalline dendrimer-stabilized gold nanoparticles. *Nanotechnology.* 17, **2006**, 1072–1078.

68. Knecht MR, Cracia-Martinez JC, Crooks RM. Hydrophobic dendrimers as templates for Au nanoparticles. *Langmuir.* 21, **2005**, 11981–11986.

69. Shi X, Sun K, Baker Jr JR. Spontaneous formation of functionalized dendrimer-stabilized gold nanoparticles. *J. Phys. Chem. C. Nanomater. Interfaces.* 112, **2009**, 8251–8258.

70. Sakai T, Alexandridis P. Mechanism of gold metal ion reduction, nanoparticle growth and size control in aqueous amphiphilic block copolymer solutions at ambient conditions. *J. Phys. Chem. B.* 109, **2005**, 7766–7777.

71. Yuan JJ, Schmid A, Armes SP. Facile synthesis of highly biocompatible poly(2-(methacryloyloxy)ethylphosphorylcholine)-coated gold nanoparticles in aqueous solution. *Langmuir.* 22, **2006**, 11022–11027.

72. Polte J, Emmerling F, Radtke M, Reinholz U, Riesemeier H, Thunemann AF. Real-time monitoring of copolymer stabilized growing gold nanoparticles. *Langmuir.* 26, **2010**, 5889–5894.

73. Goy-Lopez S, Castro E, Taboada P, Mosquera V. Block copolymer-mediated synthesis of size-tunable gold nanospheres and nanoplates. *Langmuir.* 24, **2008**, 13186–13196.

74. Tang C, Sosa CL, Pagels RF, Priestley RD, Prud'homme RK. Efficient preparation of size tunable PEGylated gold nanoparticles. *J. Mater. Chem. B.* 4, **2016**, 4813–4817.

75. Martin CR. Nanomaterials: A membrane-based synthetic approach. *Science.* 266, **1994**, 1961–1966.

76. Van der Zande BMI, Boehmer MR, Fokkink LGJ, Schonenberger. Aqueous gold sols of rod-shaped particles. *J. Phys. Chem. B.* 101, **1997**, 852–854.

77. Xu W, Luo T, Li P, Zhou C, Cui D, Pang B, Ren Q, Fu S. RGD-conjugated gold nanorods induce radiosensitization in melanoma cancer cells by downregulating $\alpha(v)\beta_3$ expression. *Int. J. Nanomedicine.* 7, **2012**, 915–924.

78. Pan B, Ao L, Gao F, Tian H, He R, Cui D. End-to-end self-assembly and colorimetric characterization of gold nanorods and nanospheres via oligonucleotide hybridization. *Nanotechnology.* 16, **2005**, 1776–1780.

79. Chang SS, Shih CW, Chen CD, Wei-Cheng L, Chris Wang CR. The shape transition of gold nanoparticles. *Langmuir.* 15, **1999**, 701–709.

80. Reetz MT, HElbig W. Size-selective synthesis of nanostructured transition metal clusters. *J. Am. Chem. Soc.* 116, **1994**, 7401–7402.

81. Kim F, Song JH, Yang P. Photochemical synthesis of gold nanorods. *J. Am. Chem. Soc.* 124, **2002**, 14316–14317.

82. Placido T, Comparelli R, Giannici F, Cozzoli PD, Capitani G, Striccoli M, Agostiano A, Curri ML. Photochemical synthesis of water-soluble gold nanorods: The role of silver in assisting anisotropic growth. *Chem. Mater.* 21, **2009**, 4192–4202.

83. Canizal G, Ascencio JA, Torresday JG, Yacamán MJ. Multiple twinned gold nanorods grown by bio-reduction techniques. *J. Nanopart. Res.* 3, **2001**, 475–481.

84. Caruso F, Spasova M, Salgueirino-Maceira V, Liz-Marzan LM. Multilayer assemblies of silica-encapsulated gold nanoparticles on decomposable colloid templates. *Adv. Mater.* 13, **2001**, 1090–1094.

85. Oldenburg SJ, Averitt RD, Westcott SL, Halas NJ. Nanoengineering of optical resonances. *Chem. Phys. Lett.* 288, **1998**, 243–247.

86. Oldenburg SJ, Jackson JB, Westcott SL, Halas NJ. Infrared extinction properties of gold nanoshells. *Appl. Phys. Lett.* 75, **1999**, 2897–2899.

87. Suzuki D, Kawaguchi H. Gold nanoparticle localization at the core surface by using thermosensitive core-shell particles as a template. *Langmuir.* 21, **2005**, 12016–12024.

88. Radloff C, Vaia RA, Brunton J, Bouwer GT, Ward VK. Metal nanoshell assembly on a virus bioscaffold. *Nano Lett.* 5, **2005**, 1187–1191.

89. Sreejith R, Anindito S, Maekawa T, Sakthikumar DS. Ultra-fast microwave aided synthesis of gold nanocages and structural maneuver studies. *Nano Research.* 10, **2017**, 1078–1091.

90. Chen J, Saeki F, Wiley BJ, Cang H, Cobb MJ, Li ZY, Au L, Zhang H, Kimmey MB, Li X, Xia Y. Gold nanocages: Bioconjugation and their potential use as optical imaging contrast agents. *Nano Lett.* 5, **2005**, 473–477.

91. Sokolov K, Follen M, Aaron J, Pavlova I, Malpica A, Lotan R, Richards-Kortum R. Real-time vital optical imaging of precancer using anti-epidermal growth factor receptor antibodies conjugated to gold nanoparticles. *Cancer Res.* 63, **2003**, 1999–2004.

92. El-Sayed IH, Huang XH, El-Sayed MA. Surface plasmon resonance scattering and absorption of anti-EGFR antibody conjugated gold nanoparticles in cancer diagnostics: Applications in oral cancer. *Nano Lett.* 5, **2005**, 829–834.

93. Tearney GJ, Brezinski ME, Bouma BE, Boppart SA, Pitris C, Southern JF, Fujimoto JG. In vivo endoscopic optical biopsy with optical coherence tomography. *Science.* 276, **1997**, 2037–2039.

94. Nel A, Xia T, Madler L, Li N. Toxic potential of materials at the nanolevel. *Science.* 311, **2006**, 622–627.

95. Qian X, Peng XP, Ansari DO, Goen QY, Chen GZ, Shin DM, Yang L, Young AN, Wang MD, Nie S. In vivo tumor targeting and spectroscopic detection with surface-enhanced Raman nanoparticle tags. *Nature Biotechnol.* 26, **2008**, 83–90.

96. Cao YC, Jin R, Mirkin CA. Nanoparticles with Raman spectroscopic fingerprints for DNA and RNA detection. *Science.* 297, **2002**, 1536–1540.

97. Keren S, Zavaleta C, Cheng Z, et al. Noninvasive molecular imaging of small living subjects using Raman spectroscopy. *Proc. Natl. Acad. Sci. USA.* 105, **2008**, 5844–5849.

98. Cai W, Gao T, Hong H, Sun J. Applications of gold nanoparticles in cancer nanotechnology. *Nanotechnol. Sci. Appl.* **2008**, 2008, 17–32.

99. Gobin AM, Lee MH, Halas NJ, James WD, Drezek RA, West JL. Near-infrared resonant nanoshells for combined optical imaging and photothermal cancer therapy. *Nano Lett.* 7, **2007**, 1929–1934.

100. Kah CY, Olivo M, Chow TH, Song KS, Koh KZY, Mhaisalkar S, Sheppard CJR. Control of optical contrast using gold nanoshells for optical coherence tomography imaging of mouse xenograft tumor model in vivo. *J. Biomed. Opt.* 14, **2009**, 054015.

101. Kim CS, Wilder-Smith P, Ahn YC, Liaw LHL, Chen ZP, Kwon YJ. Enhanced detection of early-stage oral cancer in vivo by optical coherence tomography using multimodal delivery of gold nanoparticles. *J. Biomed. Opt.* 14, **2009**, 034008.

102. Wang HF, Huff TB, Zweifel DA, He W, Low PS, Wei A, Cheng JX. In vitro and in vivo two-photon luminescence imaging of single gold nanorods. *Proc. Natl. Acad. Sci. USA.* 102, **2005**, 15752–15756.

103. Wang YW, Xie XY, Wang XD, Ku G, Gill KL, O'Neal DP, Stoica G, Wang LV. Photoacoustic tomography of a nanoshell contrast agent in the in vivo rat brain. *Nano Lett.* 4, **2004**, 1689–1692.

104. Yang X, Skrabalak SE, Li ZY, Xia Y, Wang LV. Photoacoustic tomography of a rat cerebral cortex in vivo with au nanocages as an optical contrast agent. *Nano Lett.* 7, **2007**, 3798–3802.

105. Song KH, Kim CH, Cobley CM, Xia YN, Wang LV. Near-infrared gold nanocages as a new class of tracers for photoacoustic sentinel lymph node mapping on a rat model. *Nano Lett.* 9, **2009**, 183–188.

106. Lu W, Huang Q, Geng KB, Wen XX, Zhou M, Guzatov D, Brecht P, Su R, Oraevsky A, Wang LV, Li C. Photoacoustic imaging of living mouse brain vasculature using hollow gold nanospheres. *Biomaterials.* 31, **2010**, 2617–2626.

107. Pan DPJ, Pramanik M, Senpa A, Ghosh S, Wickline SA, Wang LV, Lanza GM. Near infrared photoacoustic detection of sentinel lymph nodes with gold nanobeacons. *Biomaterials.* 31, **2010**, 4088–4093.

108. Li PC, Wang CRC, Shieh DB, Wei CW, Liao CK, Poe C, Jhan S, Ding AA, Wu YN. In vivo photoacoustic molecular imaging with simultaneous multiple selective targeting using antibody-conjugated gold nanorods. *Opt. Express.* 16, **2008**, 18605–18615.

109. Zhang Q, Iwakuma N, Sharma P, Moudgil BM, Wu C, McNeill J, Jiang H, Grobmyer SR. Gold nanoparticles as a contrast agent for in vivo tumor imaging with photoacoustic tomography. *Nanotechnology.* 20, **2009**, 1.

110. Kim C, Cho EC, Chen JY, Song KH, Au L, Favazza C, Zhang QA, Cobley CM, Gao F, Xia YN, Wang LHV. In vivo molecular photoacoustic tomography of melanomas targeted by bioconjugated gold nanocages. *ACS Nano.* 4, **2010**, 4559–4564.

111. Von Maltzahn G, Centrone A, Park JH, Ramanathan R, Sailor MJ, Hatton TA, Bhatia SN. SERS-coded gold nanorods as a multifunctional platform for densely multiplexed near-infrared imaging and photothermal heating. *Adv. Mater.* 21, **2009**, 3175–3180.

112. Hainfeld JF, Slatkin DN, Focella TM, Smilowitz HM. Gold nanoparticles: A new X-ray contrast agent. *Br. J. Radiol.* 79, **2006**, 248–253.

113. Kim D, Park S, Lee JH, Jeong YY, Jon S. Antibiofouling polymer-coated gold nanoparticles as a contrast agent for in vivo X-ray computed tomography imaging. *J. Am. Chem. Soc.* 129, **2007**, 7661–7665.

114. Von Maltzahn G, Park JH, Agrawal A, Bandaru NK, Das SK, Sailor MJ, Bhatia SN. Computationally guided photothermal tumor therapy using long-circulating gold nanorod antennas. *Cancer Res.* 69, **2009**, 3892–3900.

115. Iijima S. Helical microtubules of graphitic carbon. *Nature.* 354, **1991**, 56–58.

116. Iijima S, Ichihashi T. Single-shell carbon nanotubes of 1 nm diameter. *Nature.* 363, **1993**, 603–605.

117. Bethune DS, Kiang CH, de Vries MS, Gorman G, Savoy R, Vazquez J, Beyers R. Cobalt-catalysed growth of carbon nanotubes with single-atomic-layer walls. *Nature.* 363, **1993**, 605–607.

118. Dresselhaus MS, Dai H. Carbon nanotubes: Continued innovations and challenges. *MRS Bull.* 29, **2004**, 237–243.

119. Dai HJ. Carbon nanotubes: Synthesis, integration, and properties. *Acc. Chem. Res.* 35, **2002**, 1035–1044.

120. Liu XM, Huang ZD, Oh SW, Zhang B, Ma PC, Yuen MMF, Kim JK. Carbon nanotube (CNT)-based composites as electrode material for rechargeable Li-ion batteries: A review. *Compos. Sci. Technol.* 72, **2012**, 121–44.

121. Rowell MW, Topinka MA, McGehee MD, Prall HJ, Dennler G, Sariciftci NS, Hu L, Gruner G. Organic solar cells with carbon nanotube network electrodes. *Appl. Phys. Lett.* 88, **2006**, 233506.

122. Gao C, Guo Z, Liu JH, Huang XJ. The new age of carbon nanotubes: An updated review of functionalized carbon nanotubes in electrochemical sensors. *Nanoscale.* 4, **2012**, 1948–1963.

123. Ago H, Petritsch K, Shaffer MS, Windle AH, Friend RH. Composites of carbon nanotubes and conjugated polymers for photovoltaic devices. *Adv. Mater.* 11, **1999**, 1281–1285.

124. Park JG, Cheng Q, Lu J, Bao JW, Li S, Tian Y, Liang Z, Zhang C, Wang B. Thermal conductivity of MWCNT/epoxy composites: The effects of length, alignment and functionalization. *Carbon.* 50, **2012**, 2083–2090.

125. Ouyang M, Huang JL, Lieber CM. Fundamental electronic properties and applications of single-walled carbon nanotubes. *Acc. Chem. Res.* 35, **2002**, 1018–1025.

126. Guo W, Liu C, Sun X, Yang ZB, Kia HG, Peng HS. Aligned carbon nanotube/polymer composite fibers with improved mechanical strength and electrical conductivity. *J. Mater. Chem.* 22, **2012**, 903–908.

127. Lee NS, Chung DS, Han IT, Kang JH, Choi YS, Kim HY, Park SH, Jin YW, Yi WK, Yun MJ, Jung JE, Lee CJ, You JH, Jo SH, Lee CG, Kim JM. Application of carbon nanotubes to field emission displays. *Diam. Relat. Mater.* 10, **2001**, 265–270.

128. Zhang Y, Bai Y, Yan B. Functionalized carbonnanotubes for potential medicinal applications. *Drug Discov. Today.* 15, **2010**, 428–435.

129. Sailor MJ, Park JH. Hybrid nanoparticles for detection and treatment of cancer. *Adv. Mater.* 24, **2012**, 3779–3802.

130. Lu F, Gu L, Meziani MJ, Wang X, Luo PG, Veca LM, Cao L, Sun YP. Advances in bioapplications of carbon nanotubes. *Adv. Mater.* 21, **2009**, 139–152.

131. Kostarelos K, Bianco A, Prato M. Promises, facts and challenges for carbon nanotubes in imaging and therapeutics. *Nat. Nanotechnol.* 4, **2009**, 627–633.

132. Gulati N, Gupte H. Two faces of carbon nanotube: Toxicities and pharmaceutical applications. *Crit. Rev. Ther. Drug Carrier Syst.* 29, **2012**, 65–88.

133. Lacerda L, Bianco A, Prato M, Kostarelos K. Carbon nanotubes as nanomedicines: From toxicology to pharmacology. *Adv. Drug Deliv. Rev.* 58, **2006**, 1460–1470.

134. Fisher C, Rider AE, Han ZJ, Kumar S, Levchenko I, Ostrikov K. Applications and nanotoxicity of carbon nanotubes and graphene in biomedicine. *J. Nanomater.* Article ID 315185, **2012**, 19.

135. Harris DL, Bawa R. The carbon nanotube patent landscape in nanomedicine: An expert opinion. *Expert. Opin. Ther. Pat.* 17, **2007**, 1165–1174.

136. Saito N, Usui Y, Aoki K, Narita N, Shimizu M, Hara K, Ogiwara N, Nakamura K, Ishigaki N, Kato H, Taruta S, Endo M. Carbon nanotubes: Biomaterial applications. *Chem. Soc. Rev.* 38, **2009**, 1897–1903.

137. Yan Y, Chan-Park MB, Zhang Q. Advances in carbon-nanotube assembly. *Small.* 3, **2007**, 24–42.

138. Journet C, Maser WK, Bernier P, Loiseau A, de la Chapelle ML, Lefrant S, Deniard P, Lee R, Fischer JE. Large-scale production of single-walled carbon nanotubes by the electric-arc technique. *Nature.* 388, **1997**, 756–758.

139. Scott CD, Arepalli S, Nikolaev P, Smalley RE. Growth mechanisms for single-wall carbon nanotubes in a laser-ablation process. *Appl. Phys. A. Mater. Sci. Process.* 72, **2001**, 573–580.

140. Huang ZP, Wu JW, Ren ZF, Wang JH, Siegal MP, Provencio PN. Growth of highly oriented carbon nanotubes by plasma-enhanced hot filament chemical vapor deposition. *Appl. Phys. Lett.* 73, **1998**, 3845–3847.

141. Pillai SK, Ray SS, Moodley M. Purification of single-walled carbon nanotubes. *J. Nanosci. Nanotechnol.* 7, **2007**, 3011–3047.

142. Konig K. Multiphoton microscopy in life sciences. *J. Microsc.* 200, **2000**, 83–104.

143. Robinson JT, Welsher K, Tabakman SM, Sherlock SP, Wang H, Luong R, Dai H. High performance in vivo near-IR (>1 μm) imaging and photothermal cancer therapy with carbon nanotubes. *Nano Res.* 3, **2010**, 779–793.

144. Welsher K, Sherlock SP, Dai H. Deep-tissue anatomical imaging of mice using carbon nanotube fluorophores in the second near-infrared window. *Proc. Natl. Acad. Sci.* 108, **2011**, 8943–8948.

145. Heller DA, Baik S, Eurell TE, Strano MS. Single-walled carbon nanotube spectroscopy in live cells: Towards long-term labels and optical sensors. *Adv. Mater.* 17, **2005**, 2793–2799.

146. Hong G, Lee JC, Robinson JT, Cooke JP, Dai H, Raaz U, Xie L, Huang NF, Cooke JP, Dai H. Multifunctional in vivo vascular imaging using near-infrared II fluorescence. *Nat. Med.* 18, **2012**, 1841–1846.

147. Hong H, Gao T, Cai W. Molecular imaging with single-walled carbon nanotubes. *Nano Today.* 4, **2009**, 252–261.

148. de Roos A, Doornbos J, Baleriaux D, Bloem HL, Falke TH. Clinical applications of gadolinium-DTPA in MRI. *Magn. Reson. Annu.* 4, **1988**, 113–145.

149. Sitharaman B, Kissell KR, Hartman KB, Tran LA, Baikalov A, Rusakova I, Sun Y, Khant HA, Ludtke SJ, Chiu W, Laus S, Tóth E, Helm L, Merbach AE, Wilson LJ. Superparamagnetic gadonanotubes

are high-performance MRI contrast agents. *Chem. Commun.* 31, **2005**, 3915–3917.

150. Liu Z, Cai W, He L, Nakayama N, Chen K, Sun X, Chen X, Dai H. In vivo biodistribution and highly efficient tumour targeting of carbon nanotubes in mice. *Nat. Nanotechnol.* 2, **2007**, 47–52.
151. Cherukuri P, Bachilo SM, Litovsky SH, Weisman RB. Near-infrared fluorescence microscopy of single-walled carbon nanotubes in phagocytic cells. *J. Am. Chem. Soc.* 126, **2004**, 15638–15639.
152. Welsher K, Liu Z, Daranciang D, Dai H. Selective probing and imaging of cells with single walled carbon nanotubes as near-infrared fluorescent molecules. *Nano Lett.* 8, **2008**, 586–590.
153. Jin H, Heller DA, Strano MS. Single-particle tracking of endocytosis and exocytosis of single-walled carbon nanotubes in NIH-3T3 cells. *Nano Lett.* 8, **2008**, 1577–1585.
154. Duque JG, Cognet L, Parra-Vasquez AN, Nicholas N, Schmidt HK, Pasquali M. Stable luminescence from individual carbon nanotubes in acidic, basic, and biological environments. *J. Am. Chem. Soc.* 130, **2008**, 2626–2633.
155. Liu Z, Tabakman SM, Chen Z, Dai H. Preparation of carbon nanotube bioconjugates for biomedical applications. *Nat. Protoc.* 4, **2009**, 1372–1382.
156. Leeuw TK, Reith RM, Simonette RA, Harden ME, Cherukuri P, Tsyboulski DA, Beckingham KM, Weisman RB. Single-walled carbon nanotubes in the intact organism: Near-IR imaging and biocompatibility studies in Drosophila. *Nano Lett.* 7, **2007**, 2650–2654.
157. Welsher K, Liu Z, Sherlock SP, Robinson JT, Chen Z, Daranciang D, Dai H. A route to brightly fluorescent carbon nanotubes for near-infrared imaging in mice. *Nat. Nanotechnol.* 4, **2009**, 773–780.
158. Seo WS, Lee JH, Sun X, Suzuki Y, Mann D, Liu Z, Terashima M, Yang PC, McConnell MV, Nishimura DG, Dai H. FeCo/graphitic-shell nanocrystals as advanced magnetic-resonance-imaging and near-infrared agents. *Nat. Mater.* 5, **2006**, 971–976.
159. Al Faraj A, Cieslar K, Lacroix G, Gaillard S, Canet-Soulas E, Crémillieux Y. In vivo imaging of carbon nanotube biodistribution using magnetic resonance imaging. *Nano Lett.* 9, **2009**, 1023–1027.
160. Ananta JS, Matson ML, Tang AM, Mandal T, Lin S, Wong K, Wong ST, Wilson LJ. Single-walled carbon nanotube materials as T_2-weighted MRI contrast agents. *J. Phys. Chem. C.* 113, **2009**, 19369–19372.

161. Heller DA, Baik S, Eurell TE, Strano MS. Single-walled carbon nanotube spectroscopy in live cells: Towards long-term labels and optical sensors. *Adv. Mater.* 17, **2005**, 2793–2799.

162. De La Zerda A, Zavaleta C, Keren S, Vaithilingam S, Bodapati S, Liu Z, Levi J, Smith BR, Ma TJ, Oralkan O, Cheng Z, Chen X, Dai H, Khuri-Yakub BT, Gambhir SS. Carbon nanotubes as photoacoustic molecular imaging agents in living mice. *Nat. Nanotech.* 3, **2008**, 557–562.

163. Liu Z, Li X, Tabakman SM, Jiang K, Fan S, Dai H. Multiplexed multicolor Raman imaging of live cells with isotopically modified single walled carbon nanotubes. *J. Am. Chem. Soc.* 130, **2008**, 13540–13541.

164. Rao AM, Richter E, Bandow S, Chase B, Eklund PC, Williams KA, Fang S, Subbaswamy KR, Menon M, Thess A, Smalley RE, Dresselhaus G, Dresselhaus MS. Diameter-selective Raman scattering from vibrational modes in carbon nanotubes. *Science.* 275, **1997**, 187.

165. Liu Z, Tabakman S, Sherlock S, Li X, Chen Z, Jiang K, Fan S, Dai H. Multiplexed five-color molecular imaging of cancer cells and tumor tissues with carbon nanotube raman tags in the near-infrared. *Nano Res.* 3, **2010**, 222–233.

166. Sukhorukov GB, Rogach AL, Zebli B, Liedl T, Skirtach AG, Köhler K, Antipov AA, Gaponik N, Susha AS, Winterhalter M, Parak WJ. Nanoengineered polymer capsules: Tools for detection, controlled delivery, and site-specific manipulation. *Small.* 1, **2005**, 194–200.

167. Sinha R, Kim GJ, Nie S, Shin DM. Nanotechnology in cancer therapeutics: Bioconjugated nanoparticles for drug delivery. *Mol. Cancer Ther.* 5, **2006**, 1909–1917.

168. Bajpai AK, Shukla SK, Bhanu S, Kankane S. Responsive polymers in controlled drug delivery. *Prog. Polym. Sci.* 33, **2008**, 1088–1118.

169. de Sousa A, Maria DA, de Sousa RG, Barros de Sousa EM. Synthesis and characterization of mesoporous silica/poly(*N*-isopropylacrylamide) functional hybrid useful for drug delivery. *J. Mater. Sci.* 45, **2009**, 1478–1486.

170. Gao W, Chan JM, Farokhzad OC. pH-Responsive nanoparticles for drug delivery. *Mol. Pharm.* 7, **2010**, 1913–1920.

171. Zhu LJ, Tu CL, Zhu BS, Su Y, Pang Y, Yan DY, Wu JL, Zhu XY. Construction and application of pH-triggered cleavable hyperbranched polyacylhydrazone for drug delivery. *Polym. Chem.* 2, **2011**, 1761–1768.

172. Elsabahy M, Wooley KL. Design of polymeric nanoparticles for biomedical delivery applications. *Chem. Soc. Rev.* 41, **2012**, 2545–2561.

173. Nicolas J, Mura S, Brambilla D, Mackiewicz N, Couvreur P. Design, functionalization strategies and biomedical applications of targeted biodegradable/biocompatible polymer-based nanocarriers for drug delivery. *Chem. Soc. Rev.* 42, **2013**, 1147–1235.

174. Kamaly N, Xiao Z, Valencia PM, Radovic-Moreno AF, Farokhzad OC. Targeted polymeric therapeutic nanoparticles: design, development and clinical translation. *Chem. Soc. Rev.* 41, **2012**, 2971–3010.

175. Sivakumar B, Aswathy RG, Nagaoka Y, Iwai S, Hasumura T, Venugopal K, Kato K, Yoshida Y, Maekawa T, Sakthikumar DN. Augmented cellular uptake and antiproliferation against pancreatic cancer cells induced by targeted curcumin and SPION encapsulated PLGA nanoformulation. *Mater. Express.* 4, **2014**, 183–195.

176. Sivakumar B, Aswathy RG, Nagaoka Y, Iwai S, Suzuki M, Venugopal K, Kato K, Yoshida Y, Maekawa T, Sakthi kumar DN. Aptamer targeted theragnostic multifunctional magnetic nanoparticles as nanoplatform for pancreatic cancer therapy. *RSC Adv.* 3, **2013**, 20579–20598.

177. Balasubramanian S, Girija AR, Nagaoka Y, Iwai S, Suzuki M, Kizhikkilot V, Yoshida Y, Maekawa T, Nair SD. Curcumin and 5-FU loaded, folate and transferrin decorated polymeric magnetic nanoformulation: A synergistic cancer therapeutic approach, accelerated by magnetic hyperthermia. *Int. J. Nanomedicine.* 9, **2014**, 437–459.

178. Aravind A, Nair R, Raveendran S, Veeranarayanan S, Nagaoka Y, Fukuda T, Hasumura T, Morimoto H, Yoshida Y, Maekawa T, Sakthi Kumar D. Aptamer conjugated paclitaxel and magnetic fluid loaded fluorescently tagged PLGA nanoparticles for targeted cancer therapy. *J. Magn. Magn. Mater.* 334, **2013**, 116–123.

179. Aravind A, Jeyamohan P, Nair R, Veeranarayanan S, Nagaoka Y, Yoshida Y, Maekawa T, Kumar DS. AS1411 aptamer tagged PLGA-lecithin-PEG nanoparticles for tumor cell targeting and drug delivery. *Biotechnol. Bioeng.* 109, **2012**, 2920–2931.

180. Aswathy RG, Sivakumar B, Brahatheeswaran D, Sreejith R, Ukai T, Fukuda T, Yoshida Y, Maekawa T, Sakthi Kumar D. Multifunctional biocompatible fluorescent carboxymethyl cellulose nanoparticles. *J. Biomater. Nanobiotechnol.* 3, **2012**, 254–261.

181. Sivakumar B, Aswathy RG, Nagaoka Y, Suzuki M, Fukuda T, Yoshida Y, Maekawa T, Sakthikumar DN. Multifunctional carboxymethyl cellulose-based magnetic nanovector as a theragnostic system for folate receptor targeted chemotherapy, imaging, and hyperthermia against cancer. *Langmuir.* 29, **2013**, 3453–3466.

182. Sivakumar B, Aswathy RG, Sreejith R, Nagaoka Y, Iwai S, Suzuki M, Fukuda T, Hasumura T, Yoshida Y, Maekawa T, Sakthikumar DN. Bacterial exopolysaccharide based magnetic nanoparticles: A versatile nanotool for cancer cell imaging, targeted drug delivery and synergistic effect of drug and hyperthermia mediated cancer therapy. *J. Biomed. Nanotechnol.* 10, **2014**, 885–899.

183. Aswathy RG, Sivakumar B, Brahatheeswaran D, Fukuda T, Yoshida Y, Maekawa T, Sakthi Kumar D. Biocompatible fluorescent zein nanoparticles for simultaneous bioimaging and drug delivery application. *Adv. Nat. Sci. Nanosci. Nanotechnol.* 3, **2012**, 025006.

184. Aswathy RG, Sivakumar B, Ukai T, Brahatheeshwaran D, Yoshida Y, Maekawa T, Kumar SD. Biocompatible luminescent jelly quantumdots for bioimaging. *Mater. Express.* 1, **2011**, 291–298.

185. Srikar R, Upendran A, Kannan R. Polymeric nanoparticles for molecular imaging. *Wiley Interdiscip. Rev. Nanomed. Nanobiotechnol.* 6, **2014**, 245–267.

186. Chaney EJ, Tang L, Tong R, Cheng J, Boppart SA. Lymphatic biodistribution of polylactide nanoparticles. *Mol. Imaging.* 9, **2010**, 153–162.

187. Yaseen MA, Yu J, Wong MS, Anvari B. In-vivo fluorescence imaging of mammalian organs using charge assembled mesocapsule constructs containing indocyanine green. *Opt. Express.* 16, **2008**, 20577–20587.

188. Sevick-Muraca EM, Houston JP, Gurfinkel M. Fluorescence enhanced, near infrared diagnostic imaging with contrast agents. *Curr. Opin. Chem. Biol.* 6, **2002**, 642–650.

189. Kim K, Lee M, Park H, Kim JH, Kim S, Chung H, Choi K, Kim IS, Seong BL, Kwon IC. Cell-permeable and biocompatible polymeric nanoparticles for apoptosis imaging. *J. Am. Chem. Soc.* 128, **2006**, 3490–3491.

190. Lee CM, Jeong HJ, Cheong SJ, Kim EM, Kim DW, Lim ST, Sohn MH. Prostate cancer-targeted imaging using magnetofluorescent polymeric nanoparticles functionalized with bombesin. *Pharm. Res.* 27, **2010**, 712–721.

191. Yang Z, Zheng S, Harrison WJ, Harder J, Wen X, Gelovani JG, Qiao A, Li C. Long-circulating near-infrared fluorescence core-cross-linked polymeric micelles: Synthesis, characterization, and dual nuclear/optical imaging. *Biomacromolecules.* 8, **2007**, 3422–3428.

192. Park K, Kim JH, Nam YS, Lee S, Nam HY, Kim K, Park JH, Kim IS, Choi K, Kim SY, Kwon IC. Effect of polymer molecular weight on the tumor targeting characteristics of self-assembled glycol chitosan nanoparticles. *J. Control. Release.* 122, **2007**, 305–314.

193. Veiseh O, Sun C, Gunn J, Kohler N, Gabikian P, Lee D, Bhattarai N, Ellenbogen R, Sze R, Hallahan A, Olson J, Zhang M. Optical and MRI multifunctional nanoprobe for targeting gliomas. *Nano Lett.* 5, **2005**, 1003–1008.

194. Cheng Z, Levi J, Xiong Z, Gheysens O, Keren S, Chen X, Gambhir SS. Near-infrared fluorescent deoxyglucose analogue for tumor optical imaging in cell culture and living mice. *Bioconjug. Chem.* 17, **2006**, 662–669.

195. Lee S, Ryu JH, Park K, Lee A, Lee SY, Youn IC, Ahn CH, Yoon SM, Myung SJ, Moon DH, Chen X, Choi K, Kwon IC, Kim K. Polymeric nanoparticle-based activatable near-infrared nanosensor for protease determination in vivo. *Nano Lett.* 9, **2009**, 4412–4416.

196. Lee JJ, White AG, Rice DR, Smith BD. In vivo imaging using polymeric nanoparticles stained with near-infrared chemiluminescent and fluorescent squaraine catenane endoperoxide. *Chem. Commun.* 49, **2013**, 3016–3018.

197. Zhang Y, Yue X, Kim B, Yao S, Bondar MV, Belfield KD. Bovine serum albumin nanoparticles with fluorogenic near-IR-emitting squaraine dyes. *ACS Appl. Mater. Interfaces.* 5, **2013**, 8710–8717.

198. Li FQ, Su H, Wang J, Liu JY, Zhu QG, Fei YB, Pan YH, Hu JH. Preparation and characterization of sodium ferulate entrapped bovine serum albumin nanoparticles for liver targeting. *Int. J. Pharm.* 349, **2008**, 274–282.

199. Liu P, Yue C, Shi B, Gao G, Li M, Wang B, Ma Y, Cai L. Dextran based sensitive theranostic nanoparticles for near-infrared imaging and pho-tothermal therapy in vitro. *Chem. Commun.* 49, **2013**, 6143–6145.

200. Lim YT, Noh YW, Han JH, Cai QY, Yoon KH, Chung BH. Biocompatible polymer-nanoparticle-based bimodal imaging contrast agents for the labeling and tracking of dendritic cells. *Small.* 4, **2008**, 1640–1645.

201. Larush L, Magdassi S. Formation of near-infrared fluorescent nanoparticles for medical imaging. *Nanomedicine.* 6, **2011**, 233–240.

202. El-Daly SM, Gamal-Eldeen AM, Abo-Zeid MAM, Borai IH, Wafay HA, Abdel-Ghaffar A-RB. Photo-dynamic therapeutic activity of indocyanine green entrapped in polymeric nanoparticles. *Photodiagn. Photodyn. Ther.* 10, **2013**, 173–185.

203. Tang Y, Lei T, Manchanda R, Nagesetti A, Fernandez-Fernandez A, Srinivasan S, McGoron AJ. Simultaneous delivery of chemotherapeutic and thermal-optical agents to cancer cells by a polymeric (PLGA) nanocarrier: An in vitro study. *Pharm. Res.* 27, **2010**, 2242–2253.

204. Kim TH, Chen Y, Mount CW, Gombotz WR, Li X, Pun SH. Evaluation of temperature-sensitive, indocyanine green-encapsulating micelles for non-invasive near-infrared tumor imaging. *Pharm. Res.* 27, **2010**, 1900–1913.

205. Allemann E, Rousseau J, Brasseur N, Kudrevich SV, Lewis K, Van Lier JE. Photodynamic therapy of tumors with hexadecafluoro zinc phthalocyanine formulated in PEG-coated poly(lactic acid) nanoparticles. *Int. J. Cancer.* 66, **1996**, 821–824.

206. Konan YN, Cerny R, Favet J, Berton M, Gurny R, Allemann E. Preparation and characterization of sterile sub-200 nm meso-tetra(4-hydroxylphenyl)porphyrin-loaded nanoparticles for photodynamic therapy. *Eur. J. Pharm. Biopharm.* 55, **2003**, 115–124.

207. Ricci-Junior E, Marchetti JM. Zinc(II)phthalocyanine loaded PLGA nanoparticles for photodynamic therapy use. *Int. J. Pharm.* 310, **2006**, 187–195.

208. Konan YN, Berton M, Gurny R, Allemann E. Enhanced photodynamic activity of meso-tetra(4-hydroxyphenyl)porphyrin by incorporation into sub-200nm nanoparticles. *Eur. J. Pharm. Sci.* 18, **2003**, 241–249.

209. Troyan SL, Kianzad V, Gibbs-Strauss SL, Gioux S, Matsui A, Oketokoun R, Ngo L, Khamene A, Azar F, Frangioni JV. The FLARE intraoperative near-infrared fluorescence imaging system: A first-in-human clinical trial in breast cancer sentinel lymph node mapping. *Ann. Surg. Oncol.* 16, **2009**, 2943–2952.

210. Chan WCW, Nie SM. Quantum dot bioconjugates for ultrasensitive nonisotopic detection. *Science.* 28, **1998**, 2016–2018.

211. Smith AM, Duan HW, Mohs AM, Nie SM. Bioconjugated quantum dots for in vivo molecular and cellular imaging. *Adv. Drug Deliv. Rev.* 60, **2008**, 1226–1240.

212. Chang E, Thekkek N, Yu WW, Colvin VL, Drezek R. Evaluation of quantum dot cytotoxicity based on intracellular uptake. *Small.* 2, **2006**, 1412–1417.

213. Auzel F. Upconversion and anti-Stokes processes with f and d ions in solids. *Chem. Rev.* 104, **2004**, 139–173.

214. Wang F, Liu XG. Recent advances in the chemistry of lanthanide-doped upconversion nanocrystals. *Chem. Soc. Rev.* 38, **2009**, 976–989.

215. Vetrone F, Capobianco JA. Lanthanide-doped fluoride nanoparticles: Luminescence, upconversion, and biological applications. *Int. J. Nanotechnol.* 5, 2008, 1306–1339.

216. Wang F, Banerjee D, Liu YS, Chen XY, Liu XG. Upconversion nanoparticles in biological labeling, imaging, and therapy. *Analyst.* 135, **2010**, 1839–1854.

217. Li CX, Lin J. Rare earth fluoride nano-/microcrystals: Synthesis, surface modification and application. *J. Mater. Chem.* 20, **2010**, 6831–6847.

218. Bunzli JCG. Lanthanide luminescence for biomedical analyses and imaging. *Chem. Rev.* 110, **2010**, 2729–2755.

219. Lin M, Zhao Y, Wang SQ, Liu M, Duan ZF, Chen YM, Li F, Xu F, Lu TJ. Recent advances in synthesis and surface modification of lanthanide-doped upconversion nanoparticles for biomedical applications, *Biotechnol. Adv.* 30, **2012**, 1551–1561.

220. Wang C, Tao HQ, Cheng L, Liu Z. Near-infrared light induced in vivo photodynamic therapy of cancer based on upconversion nanoparticles. *Biomaterials.* 32, **2011**, 6145–6154.

221. Jiang C, Xu WB. Theoretical model of Yb^{3+}-Er^{3+}-Tm^{3+}-codoped system for white light generation. *J. Disp. Technol.* 5, **2009**, 312–318.

222. Heer S, Lehmann O, Haase M, Güdel HU. Blue, green, and red upconversion emission from lanthanide-doped $LuPO_4$ and $YbPO_4$ nanocrystals in a transparent colloidal solution. *Angew. Chem. Int. Ed.* 42, **2003**, 3179–3182.

223. Heer S, Kömpe K, Güdel HU, Haase M. Highly efficient multicolour upconversion emission in transparent colloids of lanthanide-doped $NaYF_4$ nanocrystals. *Adv. Mater.* 16, **2004**, 2102–2105.

224. Patra A, Friend CS, Kapoor R, Prasad PN. Effect of crystal nature on upconversion luminescence in Er^{3+}:ZrO_2 nanocrystals. *Appl. Phys. Lett.* 83, **2003**, 284–286.

225. Yi GS, Lu HC, Zhao SY, Yue G, Yang WJ, Chen DP, Guo LH. Synthesis, characterization, and biological application of size-

controlled nanocrystalline $NaYF_4$:Yb, Er infrared-to-visible up-conversion phosphors. *Nano Lett.* 4, **2004**, 2191–2196.

226. Krämer KW, Biner D, Frei G, Güdel HU, Hehlen MP, Lüthi SR. Hexagonal sodium yttrium fluoride based green and blue emitting upconversion phosphors. *Chem. Mater.* 16, **2004**, 1244–1251.

227. Mahalingam V, Mangiarini F, Vetrone F, Venkatramu V, Bettinelli M, Speghini A, Capobianco JA. Bright white upconversion emission from $Tm^{3+}/Yb^{3+}/Er^{3+}$-Doped $Lu_3Ga_5O_{12}$ nanocrystals. *J. Phys. Chem. C.* 112, **2008**, 17745–1774.

228. Wang F, Liu XG. Upconversion multicolor fine-tuning: Visible to near-infrared emission from lanthanide-doped $NaYF_4$ nanoparticles. *J. Am. Chem. Soc.* 130, **2008**, 5642–5643.

229. Chen GY, Liu Y, Zhang YG, Somesfalean G, Zhang ZG, Sun Q, Wang FP. Bright white upconversion luminescence in rare-earth-ion-doped Y_2O_3 Nanocrystals. *Appl. Phys. Lett.* 91, **2007**, 133103.

230. Yang LW, Han HL, Zhang YY, Zhong JX. White emission by frequency up-conversion in Yb^{3+}-Ho^{3+}-Tm^{3+} triply doped hexagonal $NaYF_4$ nanorods. *J. Phys. Chem. C.* 113, **2009**, 18995–18999.

231. Yang J, Zhang CM, Peng C, Li CX, Wang LL, Chai RT, Lin J. Controllable red, green, blue (RGB) and bright white upconversion luminescence of Lu_2O_3:$Yb^{3+}/Er^{3+}/Tm^{3+}$ nanocrystals through single laser excitation at 980 nm. *Chem. Eur. J.* 15, **2009**, 4649–4655.

232. Dong B, Song HW, Qin RF, Bai X, Lu SZ, Ren XG, Pan GH, Zhang H, Wang F, Fan LB. Upconversion white light devices: Ln^{3+}-Tridoped $NaYF_4$ nanoparticles and PVP modified films. *J. Nanosci. Nanotechnol.* 8, **2008**, 3921–3925.

233. Wang J, Wang F, Xu J, Wang Y, Liu Y, Chen H, Chen X, Liu X. Lanthanide-doped $LiYF_4$ nanoparticles: Synthesis and multicolor upconversion tuning. *C. R. Chim.* 13, **2010**, 731–736.

234. Wang F, Wang JA, Xu J, Xue XJ, Liu XG. Tunable upconversion emissions from lanthanide-doped monodisperse-$NaYF_4$ nano-particles. *Spectrosc. Lett.* 43, **2010**, 400–405.

235. Zijlmans H, Bonnet J, Burton J, Kardos K, Vail T, Niedbala RS, Tanke HJ. Detection of cell and tissue surface antigens using up-converting phosphors: A new reporter technology. *Anal. Biochem.* 267, **1999**, 30–36.

236. Zako T, Nagata H, Terada N, Utsumi A, Sakono M, Yohda M, Ueda H, Soga K, Maeda M. Cyclic RGD peptide-labeled upconversion nanophosphors for tumor cell targeted imaging. *Biochem. Biophys. Res. Commun.* 381, **2009**, 54–58.

237. Jiang S, Zhang Y, Lim KM, Sim EKW, Ye L. NIR-to-visible upconversion nanoparticles for fluorescent labeling and targeted delivery of siRNA. *Nanotechnology.* 20, **2009**, 155101.

238. Jiang S, Zhang Y. Upconversion nanoparticle-based FRET system for study of siRNA in live cells. *Langmuir.* 26, **2010**, 6689–6694.

239. Lim SF, Riehn R, Ryu WS, Khanarian N, Tung CK, Tank D, Austin RH. In vivo and scanning electron microscopy imaging of upconverting nanophosphors. *Caenorhabditis elegans. Nano Lett.* 6, **2006**, 169–174.

240. Nyk M, Kumar R, Ohulchanskyy TY, Bergey EJ, Prasad PN. High contrast in vitro and in vivo photoluminescence bioimaging using near infrared to near infrared up-conversion in Tm^{3+} and Yb^{3+} doped fluoride nanophosphors. *Nano Lett.* 8, **2008**, 3834–3838.

241. Hilderbrand SA, Shao FW, Salthouse C, Mahmood U, Weissleder R. Upconverting luminescent nanomaterials: Application to *in vivo* bioimaging. *Chem. Commun.* **2009**, 4188–4190.

242. Vinegoni C, Razansky D, Hildebrand SA, Shao F, Ntziachristos V, Weissleder R. Transillumination fluorescence imaging in mice using biocompatible upconverting nanoparticles. *Optics Lett.* 34, **2009**, 2566–2568.

243. Kobayashi H, Kosaka N, Ogawa M, Morgan NY, Smith PD, Murray CB, Ye XC, Collins J, Kumar GA, Bell H, Choyke PL. In vivo multiple color lymphatic imaging using upconverting nanocrystals. *J. Mater. Chem.* 19, **2009**, 6481–6484.

244. Xiong LQ, Chen ZG, Yu MX, Li FY, Liu C, Huang CH. Synthesis, characterization, and in vivo targeted imaging of amine-functionalized rare-earth up-converting nanophosphors. *Biomaterials.* 30, **2009**, 5592–5600.

245. Xiong LQ, Chen ZG, Tian QW, Cao TY, Xu CJ, Huang CH. High contrast upconversion luminescence targeted imaging in vivo using peptide-labeled nanophosphors. *Anal. Chem.* 81, **2009**, 8687–8694.

246. Liu Q, Feng W, Li F. Water-soluble lanthanide upconversion nanophosphors: Synthesis and bioimaging applications in vivo. *Coord. Chem. Rev.* 273–274, **2014**, 100–110.

247. Zhou J, Sun Y, Du X, Xiong L, Hu H, Li FY. Dual-modality *in vivo* imaging using rare-earth nanocrystals with near-infrared to near-infrared (NIR-to-NIR) upconversion luminescence and magnetic resonance properties. *Biomaterials.* 31, **2010**, 3287–3295.

248. Liu Q, Sun Y, Yang TS, Feng W, Li CG, Li FY. Sub-10 nm hexagonal lanthanide-doped $NaLuF_4$ upconversion nanocrystals for sensitive bioimaging in vivo. *J. Am. Chem. Soc.* 133, **2011**, 17122–17125.

249. Yang TS, Sun Y, Liu Q, Feng W, Yang PY, Li FY. Cubic sub-20 nm $NaLuF_4$-based upconversion nanophosphors for high-contrast bioimaging in different animal species. *Biomaterials.* 33, **2012**, 3733–3742.

250. Cao TY, Yang Y, Gao Y, Zhou J, Li, ZQ, Li FY. High-quality water-soluble and surface-functionalized upconversion nanocrystals as luminescent probes for bioimaging. *Biomaterials.* 32, **2011**, 2959–2968.

251. Jamal SA. Application of nanoparticles of ceramics, peptides, silicon, carbon, and diamonds in tissue engineering. *Chemical Sciences Journal.* 91, **2013**, 1–10.

252. Wang L, Detamore MS. Tissue engineering the mandibular condyle. *Tissue Eng.* 13, **2007**, 1955–1971.

253. Webster TJ, Ergun C, Doremus RH, Siegel RW, Bizios R. Specific proteins mediate enhanced osteoblast adhesion on nanophase ceramics. *J. Biomed. Mater. Res.* 51, **2000**, 475–483.

254. Moreno-Vega AI, Gómez-Quintero T, Nuñez-Anita RE, Acosta-Torres LS, Castaño V. Polymeric and ceramic nanoparticles in biomedical applications. *J. Nanotechnol.* Article ID 936041, **2012**, 10.

255. Cavani F, Trifiro F, Vaccari, A. Hydrotalcite-type anionic clays: Preparation, properties and applications. *Catal. Today.* 11, **1991**, 173–301.

256. Pesic L, Salipurovic S, Markovic V, Vucelic D, Kagunya W, Jones W. Thermal characteristics of a synthetic hydrotalcite-like material. *J. Mater. Chem.* 2, **1992**, 1069–1073.

257. Rives V, Labajos FM, Ulibarri MA, Malet P. A new hydrotalcite-like compound containing vanadium(3+) ions in the layers. *Inorg. Chem.* 32, **1993**, 5000–5001.

258. Khan AI, O'Hare D. Intercalation chemistry of layered double hydroxides: Recent developments and applications. *J. Mater. Chem.* 12, **2002**, 3191–3198.

259. Xue B, Hui Z, Liguang D. Layered double hydroxide-based nanocarriers for drug delivery. *Pharmaceutics.* 6, **2014**, 298–332.

260. Park DH, Hwang SJ, Oh JM, Yang JH, Choy JH. Polymer-inorganic supramolecular nanohybrids for red, white, green, and blue applications. *Prog. Polym. Sci.* 38, **2013**, 1442–1486.

261. Ladewig K, Xu ZP, Lu GQ. Layered double hydroxide nanoparticles in gene and drug delivery. *Expert Opin. Drug Deliv.* 6, **2009**, 907–922.

262. Wong Y, Cooper HM, Zhang K, Chen M, Bartlett P, Xu ZP. Efficiency of layered double hydroxide nanoparticle-mediated delivery of

siRNA is determined by nucleotide sequence. *J. Colloid Interface Sci.* 369, **2012**, 453–459.

263. Choy JH, Kwak SY, Jeong YJ, Park JS. Inorganic layered double hydroxides as nonviral vectors. *Angew. Chem. Int. Ed.* 39, **2000**, 4042–4045.

264. Wong YY, Markham K, Xu ZP, Chen M, Lu GQ, Bartlett PF, Cooper HM. Efficient delivery of siRNA to cortical neurons using layered double hydroxide nanoparticles. *Biomaterials.* 31, **2010**, 8770–8779.

265. Kwak SY, Jeong YJ, Park JS, Choy JH. Bio-LDH nanohybrid for gene therapy. *Solid State Ion.* 151, **2002**, 229–234.

266. Ladewig K, Niebert M, Xu ZP, Gray PP, Lu GQM. Efficient siRNA delivery to mammalian cells using layered double hydroxide nanoparticles. *Biomaterials.* 31, **2010**, 1821–1829.

267. Park DH, Cho J, Kwon OJ, Yun CO, Choy JH. Biodegradable inorganic nanovector: Passive versus active tumor targeting in siRNA transportation. *Angew. Chem. Int. Ed.* 55, **2016**, 4582–4586.

268. Shi S, Fliss BC, Gu Z, Zhu Y, Hong H, Valdovinos HF, Hernandez R, Goel S, Luo H, Chen F, Barnhart TE, Nickles RJ, Xu ZP, Cai W. Chelator-free labeling of layered double hydroxide nanoparticles for *in vivo* PET imaging. *Sci. Rep.* 5, **2015**, 16930.

269. Musumeci AW, Schiller TL, Xu ZP, Minchin RF, Martin DJ, Smith SV. Synthesis and characterization of dual radiolabeled layered double hydroxide nanoparticles for use in in vitro and in vivo nanotoxicology studies. *J. Phys. Chem. C.* 114, **2010**, 734–740.

270. Nagaraj VJ, Sun X, Mehta J, Martin M, Ngo T, Dey SK. Synthesis, characterization, and in vitro drug delivery capabilities of (Zn, Al)-based layered double hydroxide nanoparticles. *J. Nanotechnol.* 2015, **2015**, 350370.

271. Oh JM, Park M, Kim ST, Jung JY, Kang YG, Choy JH. Efficient delivery of anticancer drug MTX through MTX-LDH nanohybrid system. *J. Phys. Chem. Solids.* 67, **2006**, 1024–1027.

272. Oh JM, Choi SJ, Kim ST, Choy JH. Cellular uptake mechanism of an inorganic nanovehicle and its drug conjugates: Enhanced efficacy due to clathrin-mediated endocytosis. *Bioconjug. Chem.* 17, **2006**, 1411–1417.

273. Choi SJ, Oh JM, Choy JH. Anticancer drug-layered hydroxide nanohybrids as potent cancer chemotherapy agents. *J. Phys. Chem. Solids.* 69, **2008**, 1528–1532.

274. Weingart JD, Thompson RC, Tyler B, Colvin OM, Brem H. Local-delivery of the topoisomerase-I inhibitor camptothecin sodium

prolongs survival in the rat intracranial 9l gliosarcoma model. *Int. J. Cancer.* 62, **1995**, 605–609.

275. Li A, Qin L, Wang W, Zhu R, Yu Y, Liu H, Wang S. The use of layered double hydroxides as DNA vaccine delivery vector for enhancement of anti-melanoma immune response. *Biomaterials.* 32, **2011**, 469–477.

276. Zhang K, Xu ZP, Lu J, Tang ZY, Zhao HJ, Good DA, Wei MQ. Potential for layered double hydroxides-based, innovative drug delivery systems. *Int. J. Mol. Sci.* 15, **2014**, 7409–7428.

277. Lee BI, Lee KS, Lee JH, Lee IS, Byeon SH. Synthesis of colloidal aqueous suspensions of a layered gadolinium hydroxide: A potential MRI contrast agent. *Dalton Trans.* 14, **2009**, 2490–2495.

278. Xu ZP, Kurniawan ND, Bartlett PF, Lu GQ. Enhancement of relaxivity rates of Gd-DTPA complexes by intercalation into layered double hydroxide nanoparticles. *Chemistry.* 13, **2007**, 2824–2830.

279. Flesken-Nikitin A, Toshkov I, Naskar J, Tyner KM, Williams RM, Zipfel WR, Giannelis EP, Nikitin AY. Toxicity and biomedical imaging of layered nanohybrids in the mouse. *Toxicol. Pathol.* 35, **2007**, 804–881.

280. May JN, Golombek SK, Baues M, Dasgupta A, Drude N, Rix A, Rommel D, von Stillfried S, Appold L, Pola R, Pechar M, van Bloois L, Storm G, Kuehne AJC, Gremse F, Theek B, Kiessling F, Lammers T. Multimodal and multiscale optical imaging of nanomedicine delivery across the blood-brain barrier upon sonopermeation. *Theranostics.* 10, **2020**, 1948–1959.

281. Shi S, Chen F, Goel S, Graves SA, Luo H, Theuer CP, Engle JW, Cai W. In vivo tumor-targeted dual-modality PET/optical imaging with a yolk/shell-structured silica nanosystem. *Nano-Micro Lett.* 10, **2018**, 65.

282. Qu Z, Shen J, Li Q, Xu F, Wang F, Zhang X, Fan C. Near-IR emissive rare-earth nanoparticles for guided surgery. *Theranostics.* 10, **2020**, 2631–2644.

283. Wojtynek NE, Mohs AM. Image-guided tumor surgery: The emerging role of nanotechnology. *Wiley Interdiscip Rev Nanomed Nanobiotechnol.* 12, **2020**, e1624.

284. Song J, Zhang N, Zhang L, Yi H, Liu Y, Li Y, Li X, Wu M, Hao L, Yang Z, Wang Z. IR780-loaded folate-targeted nanoparticles for near-infrared fluorescence image-guided surgery and photothermal therapy in ovarian cancer. *Int. J. Nanomedicine.* 14, **2019**, 2757–2772.

285. Wojtynek NE, Olson MT, Bielecki TA, An W, Bhat AM, Band H, Lauer SR, Silva-Lopez E, Mohs AM. Nanoparticle formulation of indocyanine green improves image-guided surgery in a murine model of breast cancer. *Mol Imaging Biol.* 22, **2020**, 891–903.

286. On KC, Rho J, Yoon HY, Chang H, Yhee JY, Yoon JS, Jeong SY, Kim HK, Kim K. Tumor-targeting glycol chitosan nanoparticles for image-guided surgery of rabbit orthotopic VX2 lung cancer. *Pharmaceutics.* 12, **2020**, 621.

287. Xu H, Han Y, Zhao G, Zhang L, Zhao Z, Wang Z, Zhao L, Hua L, Naveena K, Lu J, Yu R, Liu H. Hypoxia-responsive lipid–polymer nanoparticle-combined imaging-guided surgery and multitherapy strategies for glioma. *ACS Appl. Mater. Interfaces.* 12, **2020**, 52319–52328.

288. I YL, Jiang M, Xue Z, Zeng S. 808 nm light triggered lanthanide nanoprobes with enhanced down-shifting emission beyond 1500 nm for imaging-guided resection surgery of tumor and vascular visualization. *Theranostics.* 10, **2020**, 6875–6885.

NANOTHERAPY

Chapter 8

Nanomaterials in Chemotherapy

P. K. Hashim[a] and Anjaneyulu Dirisala[b]

[a]*Laboratory of smart molecules,*
Research Institute for Electronic Science,
Hokkaido University, Hokkaido 001-0020 Japan
[b]*Innovation Center of NanoMedicine,*
Kawasaki Institute of Industrial Promotion,
Kawasaki-ku, Kawasaki 210-0821, Japan

hashim@es.hokudai.ac.jp, dirisala-a@kawasaki-net.ne.jp

8.1 Introduction

Several nanotechnology methods were utilized for the development of versatile nano drug delivery systems (DDSs) together with their behavior in the biological fluid, accumulation at the target site, and drug release mechanisms. The DDSs are made of lipids, polymers, and organic and inorganic materials. Although the phrase "chemotherapy" is applicable for both small molecule- and macromolecule-based drugs, here discussions will be mainly restricted to DDS comprising small molecule drugs. In the end, clinically approved DDSs for cancer therapeutics are also briefly discussed.

Bionanotechnology in Cancer: Diagnosis and Therapy
Edited by D. Sakthi Kumar and Aswathy Ravindran Girija

ISBN 978-981-4800-30-3 (Hardcover), 978-0-429-42291-1 (eBook)
www.jennystanford.com

8.1.1 DDS

A drug is defined as a natural or artificial pharmacologically active ingredient used for the detection, diagnosis, and treatment of a disease. A drug can be either hydrophobic (e.g., paclitaxel, cisplatin, and methotrexate) or hydrophilic (e.g., gemcitabine, L-asparaginase, and antibodies) depending on its aqueous solubility [1]. Also, a drug can be negatively charged (e.g., gene therapeutics such as DNA, messenger (m)RNA, and short interfering (si)RNA), positively charged (e.g., doxorubicin (DOX)), and neutral (e.g., cisplatin) based on the charged species present in the drug. A drug also can be a small molecule, which has a low molecular weight compound below 900 daltons (e.g., DOX and cisplatin), or a high molecular weight molecule (e.g., DNA, RNA, and proteins). The structural characteristics of drugs are often related to their functions. For instance, a small molecule drug usually binds to a target gene/protein as a first step in the complicated drug action mechanisms. Small molecule drugs often induce off-target adverse effects because of their lack of specificity. In contrast to small molecule drugs, macromolecule drugs (e.g., genes and proteins) typically bind to their target either expressed on the cell surface or intracellular component with high specificity, and correct or program the disease-causing elements. Advantageously, both small molecule- and macromolecule-drugs can be used for cancer therapy. Collectively, more than 150 anticancer drugs have been approved by the US Food and Drug Administration (FDA) for various types of cancer [2].

Drug delivery is an important pharmaceutical aspect that deal with formulations of a drug for maximizing therapeutic efficacy and safety by improving their adsorption, distribution, metabolism, and excretion (ADME) properties. In a broad sense, a DDS can be defined as a formulation/device/carrier that packages/encapsulates the drug and delivers it to a disease site with minimal off-targeting to unintended tissues/cells [3]. Conceptually, commercial drug formulations such as tablets, capsules, suspensions, and powders are one type of DDS and are generally known as "macro-DDS" due to their macroscopic dimension. For instance, Matulane® is a capsule type macro-DDS containing procarbazine drug molecule, which is commercialized as an oral medication for the treatment of Hodgkin's lymphoma and brain cancers. However, macro-DDSs are generally associated with many delivery challenges including poor solubility and bioavailability, reduced half-life, in vivo instability, lack of selectivity, and severe side effects [4]. In an attempt to address some of these limitations, micro-DDS with an average diameter of 1–300 μm (e.g., microemulsions and microsphere-depots) has been developed [5]. An example of micro-DDS is Decapeptyl® developed for prostate cancer treatment that consists of triptorelin drug molecule encapsulated in a degradable poly(lactic-*co*-glycolic acid) (PLGA) microsphere depot [6]. However, immediate release, repeated dosing, patient's inconvenience, and noncompliance are some of the major issues to be solved. In this regard, nano-DDS with several nanometers size regime, slightly above the size of drug molecules, is ideal to address many of the existing drawbacks associated with conventional drug delivery methods.

8.1.2 DDS Evolution

As early as 1906, As early as 1906, Nobel laureate Nobel laureate Paul Ehrlich proposed the design of a "magic bullet" to alter the drug distribution by conjugating toxic drugs to site-specific targeting antibodies, which can selectively target the cancer cells to improve therapeutic efficacy with minimal toxicity to healthy cells [7]. In 1955, the first polymer–drug conjugate was reported by von Horst Jatzkewitz [8]. In 1964, a silicone rubber (Silastic) prosthesis was demonstrated as an implantable carrier for prolonged drug therapy [9]. In the same year, liposome structure was also reported [10] followed by an exploration of its potential for drug delivery [11], which became one of the breakthroughs in the field of DDS. Since then, a myriad of nanoscale DDSs composed of various kinds of lipids encapsulated with different therapeutic agents as well as diagnostic agents together with methods for ligand functionalization have been developed [12]. In 1973, Gregory has demonstrated that the liposome could be used for anticancer drug loading [13]. In 1976, Kopf et al. reported the first polymeric micelle nanocarrier for the delivery of low molecular weight drugs [14] and in the same year, Langer et al. demonstrated the first sustained release polymer system for delivery of macromolecular drugs [15]. In 1995, Doxil® (PEGylated liposomal DOX) was approved by the FDA, the first nano-DDS for ovarian cancer, AIDS-related Kaposi's sarcoma, and multiple myeloma [16]. A brief historical timeline of major advancements in the field of cancer nanomedicine is shown in Figure 8.1.

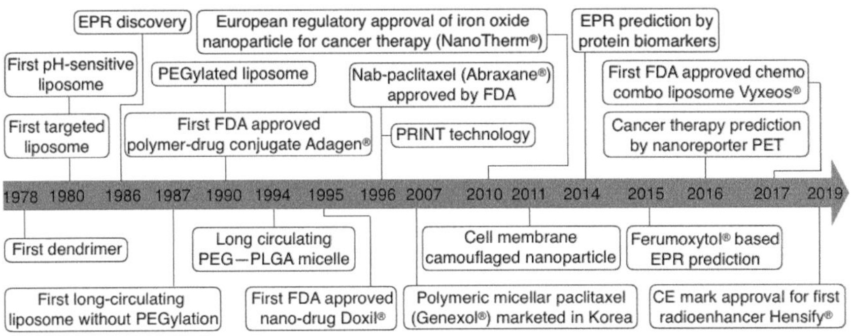

Figure 8.1 Timeline of major advancements in the history of cancer nanomedicine development.

8.1.3 Nanomedicine and DDS

Nanotechnology showed significant progress in the past few decades as a new interdisciplinary field, allowing a clear understanding of several phenomena at the nanoscale as well as the construction of varieties of nanoscale architectures for material applications. One of the important areas of nanotechnology is "nanomedicine," which refers to highly specific medical intervention at the nanometric scale to advance the detection, diagnosis, prediction, prevention, and treatment of diseases. A significant portion of the nanomedicine is DDS involving nano-sized soft nanoparticles (e.g., liposomes, lipid nanoparticles, polymeric micelles, polymer–drug conjugates, dendrimers, hydrogels, etc.) and hard nanoparticles (e.g., carbon nanotubes (CNTs), gold nanoparticle (AuNP), metal-organic framework (MOF), etc.) (Figure 8.2). Advantageously, due to a large surface-to-volume ratio, DDS with nanoscale dimension offers several interesting properties compared to the macro/micro-scale DDS [17–20]. Following are some of the beneficial aspects of DDS:

(1) Improved solubility of poorly water-soluble drugs (a 50,000-fold enhancement in the solubility is observed when a highly hydrophobic aryl imidazole drug, ML220, was encapsulated in liposome) [21]

(2) High-drug loading, owing to the availably of numerous chemically active sites

(3) Prolonged drug half-life by improving the protection of a drug against harsh biological milieu (a 10,000-fold higher protection of mRNA drug against ribonucleases is observed when packaged in a polyplex micelle) [22]

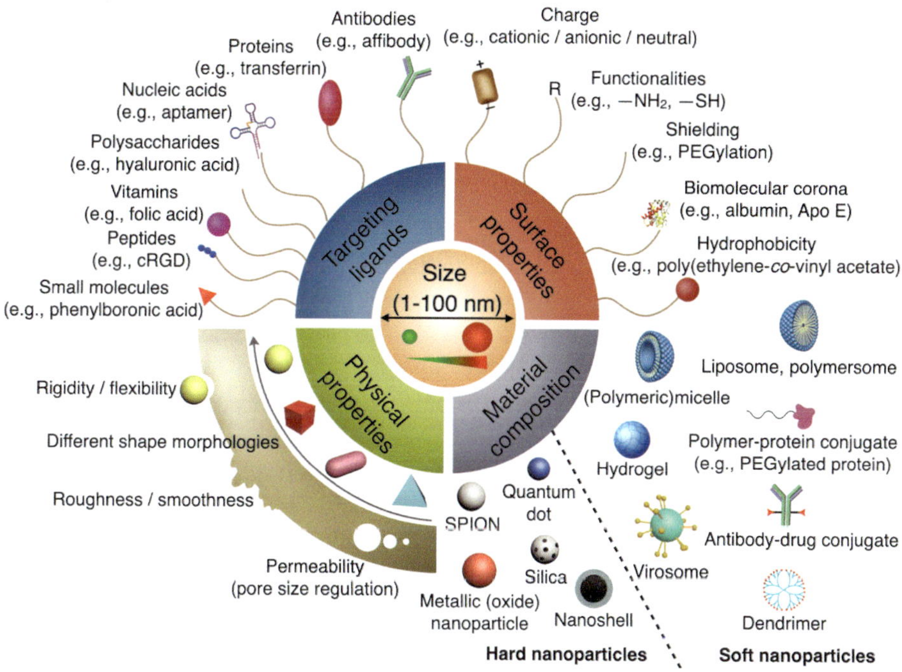

Figure 8.2 Schematic representations of DDS engineering for cancer therapy. The material composition indicates various kinds of chemical compounds that can be used for DDS fabrication. Physical properties, such as rigidity, roughness, and permeability, can be tuned by altering the composition of materials to improve cellular uptake and drug release kinetics. Surface properties of DDS can be modulated by functionalizing a myriad of stealth materials, which can be useful for improving the stability in the bloodstream. Targeting ligands can also be introduced on the surface of DDSs to selectively deliver a drug at the diseased site.

(4) Controlled drug release at a precise dose to achieve effective drug concentration at the diseased site, thereby maximizing the localized drug effect
(5) Crossing the endothelial and epithelial barriers by transcytosis
(6) Targeted drug delivery into tissues, cells, and organelles
(7) Co-delivery of multiple therapeutic agents for combinatorial therapy and/or imaging agents (e.g., fluorescent probes) for real-time tracking of the disease progress [23, 24]
(8) Maximizing therapeutic efficacy, and minimizing toxicity and immunogenicity of drugs

8.2 DDS Behavior in Bloodstream

The therapeutic efficacy of a DDS is often decided by its behavior in body fluids as well as specific tissue level interactions at the diseased site. The route of administration of DDS also significantly affects the bioavailability of a drug, which determines both the onset and the duration of drug efficacy. Systemic administration is often preferred due to its non-invasive nature. Intravenous injection is highly advantageous due to the direct accessibility of a wide variety of organs/tissues through the blood circulatory system along with bypassing the digestive system. Upon intravenous injection, an ideal DDS should stably circulate in the bloodstream without a premature drug release, avoid nonspecific clearance by the reticuloendothelial system (RES) organs/cells [25], and eventually accumulate at the targeted site of action followed either by the internalization into target cells or release the drug in their vicinity. DDSs allow a low volume of distribution (Vd), indicating the limited extravasation to the healthy tissues because of the continuous and tight lining of blood vessels (Figure 8.3A) [26–28]. For instance, Doxil® (V_d = 2.8 l/m^2) reduces the Vd of DOX ~357 times lower compared to the free DOX (Vd ~1000 l/m^2) by restricting the distribution within the plasma. Compared to bare drugs, DDS substantially decreases the renal clearance of the loaded drugs due to the threshold size limit of approximately 6.5 nm [29] (equivalent to 50 kDa nominal M_w threshold cutoff) at the kidney glomerular filtration. Surface modification of DDS by non-ionic stealth polymers such as poly(ethylene glycol) (PEG) is inevitable not only to avoid adsorption of opsonic proteins (e.g., immunoglobulins; IgG, IgM, and complement proteins; C3b, C1q), but also to provide colloidal stability [26–28]. These stealth-engineered DDSs protect the loaded drugs from recognition and elimination by RES organs/cells, resulting in a prolonged blood half-life of drugs for hours to days. Besides surface shielding, size [30] and surface charge [31] modulation, installation of "don't eat me" signal CD47 [32], and protective protein corona (e.g., apolipoprotein E) formation [33] on DDSs also proven to be effective in avoiding the nonspecific clearance by RES and thus results in a prolonged blood half-life of loaded drugs. Unlike conventional surface shielding directly onto DDS, in-situ selective stealth coatings of liver sinusoidal scavenger cells using PEGylated oligopeptides also have recently developed that substantially elongates the blood circulation of DDS [34].

Besides stable prolonged circulation in the bloodstream, DDS must access the diseased site (e.g., tumor) for its therapeutic action, which is usually occurs via passive or active method. Advantageously for tumor chemotherapy, DDS can passively accumulate at the tumor site via a special phenomenon known as enhanced permeability and retention (EPR) effect [35, 36], which is known to be the central dogma of cancer nanomedicine (Figure 8.3B). The enhanced permeability of DDS to tumors is thought to result from the malformed vascular walls with leaky inter-endothelial gaps and trans-endothelial channels [37] and the retention is thought to result from the suppressed lymphatic drainage [35, 36]. In order to exploit the EPR-based targeting, the DDS should have a prolonged blood half-life with small size characteristics to extravasate from the bloodstream to the tumor site [12, 26]. In some cases, the size of DDS becomes particularly important in the extravasation to the tumor site; for instance, a sub-100 nm DDS extravasates into hyper-vascularized tumors as in the case of Doxil® accumulation at Kaposi's sarcoma [38], while a sub-50 nm DDS extravasates into hypo-vascularized tumors [39–41] (typical one is a PEGylated polymeric micelle accumulation at the pancreatic tumor).

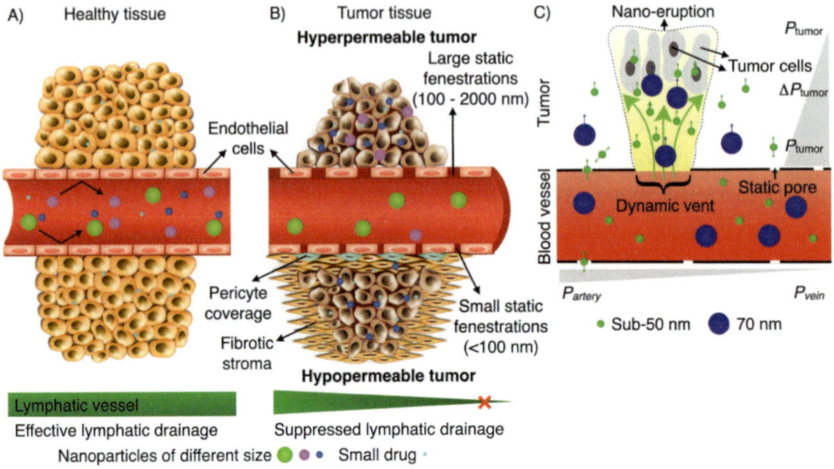

Figure 8.3 (A) Small molecule drugs can leak extensively to healthy tissues (extravasation) due to their low molecular weight, resulting in toxic adverse effects. On the contrary, DDS does not extravasate to healthy tissues because of their large molecular weight and/or size. (B) DDS selectively accumulates in the tumor site by enhanced permeability resulting from the leaky vasculature of tumors with poorly aligned fenestrated endothelium. Enhanced retention of DDS resulted from the slow venous blood return and suppressed lymphatic clearance in the tumors (EPR effect). In tumors with high permeability (hyperpermeable tumors) such as colon cancer, as a consequence of large fenestrae in the tumor blood vessels, allowing DDS to leak substantially into tumor tissues (upper panel). Pericyte coverage and thick stromal fibrosis limit the permeability and extravasation of DDS in hypopermeable tumors such as pancreatic cancer (lower panel). (C) DDS of both large (70 nm) and small (sub-50 nm) size can also transiently extravasate the tumor interstitial space along with abrupted vigorous outward flow of fluid (nano-eruptions) even in hypopermable tumors due to transient irregular openings in tumor blood vessels (dynamic vents). These dynamic vents allow pressure-driven enhanced accumulation and retention of DDS. A dynamic vent allows the extravasation of both large (70 nm) and small (30 nm) nanoparticles, while a static pore allows the extravasation of only sub-50 nm particles Ref. [42].

The extravasation of large-sized DDS into hypo-vascularized tumors is additionally possible by (a) the formation of transient dynamic vents at tumor vasculature to allow the passage of fluid from bloodstream to the tumor site [42] (Figure 8.3C), (b) the enhanced fluid flow to the tumor from the bloodstream due to pressure differences across the tumor vasculature leads to the accumulation of large nanoparticles (solvent drag) [43], and (c) the use of vascular mediators such as bradykinin and nitric oxide derivatives augment the permeability of tumor blood vessels (EPR enhancers) [44]. Together with the EPR-based passive accumulation, active targeting methods also have been widely investigated. In this case, a surface-functionalized DDS with ligand motifs is expected to selectively bind the specific overexpressed receptors on tumoral vasculature and/or stromal cells [45]. For example, cyclic Arginine-Glycine-Aspartic acid (cRGD) peptide ligand can selectively bind to the $\alpha_v\beta_3$ and $\alpha_v\beta_5$ integrin receptors on the tumor neovasculature and is exploited to extravasate the polymeric micelle from vessels to the tumor parenchyma [45].

Following selective accumulation at the targeted site, the next hurdle is the penetration of DDS into the tissues/cells, which occurs by the various mechanisms. A typical example is an endocytosis process involving a cell membrane incorporated with DDS followed by spontaneous translocation into the interior of the cells. Endosomes possess a vesicular architecture, and hence the encapsulated DDS during endocytosis needs to be delivered in cellular cytoplasm, which is often activated by site-specific stimuli [46–49]. For example, a decrease in pH from early endosomes (pH ~6.9) to late endosomes (pH ~5.5) activates the DDS to disrupt the endosomal membrane and release to the cytosolic compartment. Cell-specific recognition and subcellular targeting can also be directed by conjugating the ligands on the surface of DDS. For example, a triphenylphosphonium equipped DDS can selectively recognize the mitochondria [50].

8.3 Drug Release Mechanisms

The efficacy of a drug can be dominantly affected by the mode of drug release [51]. The DDS releases its payload drug in a spatiotemporally regulated manner to improve patient compliance by lowering the dose and administration frequency [51]. To achieve a spatiotemporal release, DDS needs to be accumulated site specifically, which often occurs via the EPR effect. One important aspect of the spatiotemporal control of drug release

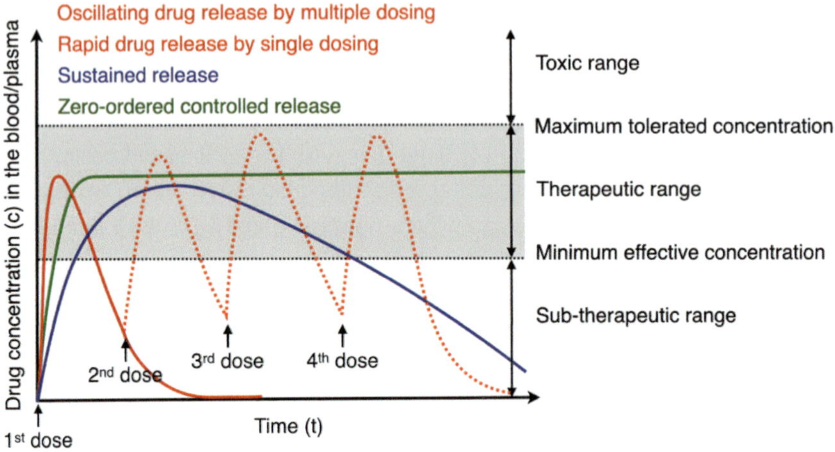

Figure 8.4 Hypothetical drug release profiles in the blood or plasma. Drug concentration should sustain between MEC and MTC. A single conventional dose causes a rapid rise and fall in drug concentration (red solid line). Multiple conventional dosing at regular intervals causes oscillating drug concentration (red solid line followed by a red dotted line). The zero-ordered controlled release aims to maintain a constant drug concentration at a specific controllable rate (drug release is definite per unit time) for a prolonged time, followed by an initial peak (green solid line). Sustained drug release aims to maintain the drug concentration for a prolonged time where release of the drug is not necessarily with a certain rate (blue solid line).

is sustained drug concentration in the bloodstream or targeted tissue between a minimum effective concentration (MEC) and maximum tolerated concentration (MTC) (Figure 8.4) [52]. Drug concentration above MTC and below MEC results in toxic adverse effects and ineffective therapy, respectively. Single dosing of macro-DDSs, such as tablets and capsules, immediately raise the drug concentration and rapidly falls below the MEC, resulting in subtherapeutic concentration. To maintain prolonged drug concentration within the therapeutic range, frequent multiple dosing at regular time intervals is necessary; however, it can lead to patient noncompliance issues. Therefore, it is desirable to engineer a DDS that can supply the drug either sustained release or controlled release with a low dosing frequency. For this purpose, zero-ordered controlled release systems have been developed that can sustain the steady-state drug concentration within the therapeutic window at a definite rate per unit time. Controlled release systems not only offer spatiotemporal drug release for a prolonged period, but also can aid patient compliance by minimizing the repetitive dose regimens. DDSs having controlled release property can be broadly classified into three groups based on their specific release mechanisms such as diffusion-controlled, solvent-controlled, and chemically programmed release [53].

8.3.1 Diffusion-Controlled Release

In this case, drug release is controlled by diffusion through an inert nonbiodegradable polymeric membrane (reservoir system) or a polymeric matrix (monolithic system) [54], and the diffusion process is driven by a concertation gradient across the membrane or matrix. In the reservoir system, drugs are incorporated into an aqueous interior and are segregated from a permeable membrane (Figure 8.5A). Silicone rubber, polyurethane, poly(ethylene-*co*-vinyl acetate) (PEVA), and hydrogel can be used as a drug reservoir. The drug release rate also can be controlled by adjusting the membrane thickness. An example of this category is Progestasert® where a PEVA-based matrix allows the diffusion of drugs for prolonged contraception.

In a polymeric matrix composed of a three-dimensional polymer network, drugs are evenly dispersed throughout the matrix (Figure 8.5B) and do not have a membrane as a diffusion barrier. Thus, this system shows a high initial release followed by a gradual decline in the release. Polymer matrices, hydroxypropyl methylcellulose, alginates, and scleroglucan can be used as hydrophilic segments, while wax, polyethylene, polypropylene, and ethylcellulose as hydrophobic segments. Nitro-Dur I® is an example of DDS that utilizes the diffusion-controlled release of nitroglycerin drug for the treatment of angina pectoris via a transdermal patch.

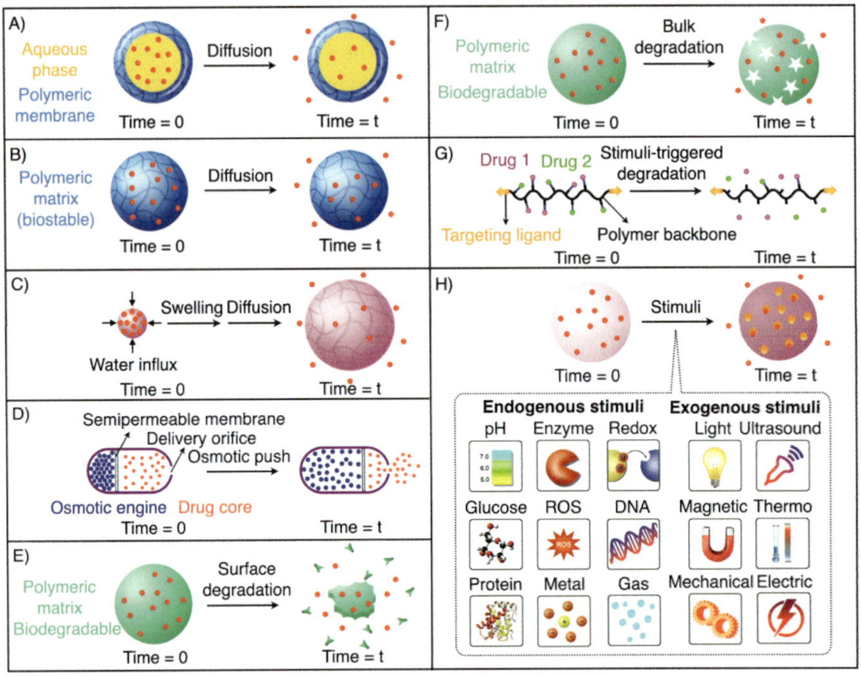

Figure 8.5 Drug release mechanisms of different DDS types. Diffusion-controlled release from (A) reservoir system and (B) monolithic system; solvent-controlled release from (C) osmosis-controlled device and (D) swelling-controlled matrix; and chemically-programmed release from (E) surface biodegradable matrix, (F) bulk biodegradable matrix, (G) polymer–drug conjugate, and (H) stimuli-sensitive system.

8.3.2 Solvent-Controlled Release

In this case, drug release occurs as a result of osmosis or swelling. Osmosis-controlled DDS comprised a drug-loaded core separated from an osmogen reservoir (e.g., high salt concentration) with a semipermeable membrane (Figure 8.5C) [2, 55, 56]. The membrane allows an influx of solvent from outside to a high concentration salt gradient generating an "osmotic pump or engine." The water influx forces the drug suspension out of the orifice, and the drug release rate can be regulated by the diameter of the orifice and thickness of the membrane. Viadur®

is a typical example of DDS with an osmotic pump that triggers the drug release for the treatment of prostate cancers [57].

Swelling-controlled DDS often uses a glassy hydrophilic polymer (dry/non-swollen state) as a three-dimensionally crosslinked matrix, in which the mobility of the polymers is substantially restricted (Figure 8.5D). Solvent entry into the matrix results in relaxation of the matrix (swollen state), eventually pushing the drug molecules to diffuse out of the matrix. The release rate of a drug is controlled by the chain relaxation rate of polymers and the diffusion rate of water [55]. DDSs composed of hydrogels and poly(2-hydroxyethylmethacrylate-*co*-*N*-vinylpyrrolidone) disk are prominent examples that utilized swelling as a trigger for the drug release [56].

8.3.3 Chemically Programmed Release

Drug release can be regulated by tuning the chemical structure of a polymer/macromolecule presents in the DDS [51, 52, 57]. Chemically controlled release DDS can be further classified into three groups: degradation (erosion)-controlled release, scission in polymer–drug conjugates, and stimuli-sensitive release.

In degradation (erosion)-controlled release, a biodegradable polymeric matrix/scaffold is spontaneously degraded in the biological environment releasing the non-covalently incorporated drugs. Biodegradable DDSs are advantageous for practical applications as the carrier component that can be completely cleared from the body leading to negligible chronic toxicity. The degradation process occurs either from the exterior surface (surface degradation) or throughout the whole polymeric matrix/scaffold (bulk degradation), and the rate of drug release is regulated by the rate of degradation. When the rate of water diffusion into the polymer bulk is slower, surface degradation occurs while the faster water diffusion causes bulk degradation (Figure 8.5E). Polyanhydrides and poly(orthoesters)-based DDSs are known to release drugs via surface degradation. DDSs with polymeric matrices based on PLGA, poly(glycolic acid), poly(lactic acid), and polycaprolactone are more prone to undergo bulk degradation (Figure 8.5F).

In polymer–drug conjugates (pendant chain systems), drugs are covalently conjugated to the polymer backbone via a physiologically labile (hydrolytically or enzymatically degradable or exchangeable or stimuli-responsive) spacers (Figure 8.5G), and the drug release is tailored by the rate of scission of the bonds [58, 59].

Drug release from a stimuli-sensitive DDS occurs in response to certain endogenous (or pathophysiological) stimuli available at diseased sites or exogenous stimuli applied from outside the body (Figure 8.5H) [60–62]. Endogenous stimuli at the tissue level are a characteristic of the microenvironmental differences between healthy and pathological sites (ischemia, tumors, inflammatory diseases, and transplant rejection areas). Examples include interstitial pH (healthy tissue ~7.4 versus tumor tissue ~6.5–7.2) [63], reactive oxygen species (ROSs) [64], and elevated expression of enzymes [65] such as matrix metalloproteinases at tumor areas. Some stimuli present at the cellular level also useful for releasing drugs after DDS internalization into the cells. Intracellular pH especially at endosomes/lysosomes is much lower (pH = 6.5–4.5) compared to extracellular pH of 7.4 [46].

Similarly, in redox potential, intracellular glutathione concentration is ~2–10 mM versus extracellular space ~2–20 μM [39, 46], and in adenosine triphosphate (ATP), intracellular space is ~3 mM versus extracellular space ~0.4 μM [66]. Exogenous stimuli include electric and magnetic field, temperature, light pulses (laser beams), ultrasound, and mechanical force [60]. Certain stimuli can also be locally intensified after selective accumulation at the target tumor site [63]. For example, glucose oxidase loaded nanoreactor consumes glucose and significantly increases the tumoral oxidative stress, and finally destroys the cancer cells [67]. Stimuli-sensitive DDSs are much exploited for chemotherapy applications in several cancers.

8.4 Lipid-Based DDSs

8.4.1 Composition and Preparation

Lipids are highly hydrophobic in nature, hence insoluble in water, and typically consist of saturated or unsaturated hydrocarbon motifs often connected to glycerol. For instance, a triglyceride is composed of three fatty acids linked via ester bonds to glycerol. Many of the lipids are isolated from their biological origin, which then purified via chromatographic separations. Progresses in the field of lipid synthesis with high purity diversified the scope of using lipid as a 'solubilizer' in cosmetics and foods, and as a 'building block' for making nanoscale materials. Phospholipids are a class of lipids with amphiphilic characteristics due to the presence of both hydrophobic fatty acid chain and a hydrophilic phosphate group. In an aqueous environment, phospholipid tends to form a bilayer architecture as in the case of the biological cell membrane. Since the phosphate part of the phospholipids can be easily functionalized with organic molecules, such as choline, a wide variety of modified phospholipids have been developed that form interesting nanostructures. Liposome is a representative example formed by the self-assembly of phospholipid derivatives, such as phosphatidylcholine, phosphatidylethanolamine, phosphatidylserine, and phosphatidylglycerol, which have been found applications in various fields including DDSs.

For DDS application, liposomes should be stable at body temperature, 37°C. The stability can be ensured by a careful selection of lipids used for the liposome preparation, where the phase transition temperature (T_M) is above 37°C. The 'flip-flop' motion of phospholipids often destabilizes the liposomes, hence cholesterol, an organic molecule with lipid characteristics, is routinely added to the liposome formulation. Depending on the amount of cholesterol content in a liposome, its dynamic properties, and hence drug encapsulation and releasing ability change [68]. For instance, a high concentration of cholesterol enhances the stability of liposomes in blood and a longer circulation is observed in a mice model, while those of low cholesterol content quickly cleared from the blood circulation [69]. Another important parameter is surface charge of liposomes [70, 71]. Cationic liposomes are commonly used for DDS application because of their electrostatic interactions with the anionic cell membrane, which triggers the endocytic internalization into the cell [72, 73]. However, anionic, zwitterionic, and neutral liposomes also have been developed for specific applications.

8.4.2 Drug Loading into Liposomes

Hydrophilic drugs such as small molecules and biomolecules (e.g., nucleic acid and protein) can be loaded into the aqueous interior of liposomes, while hydrophobic (lipophilic) drugs are placed within the lipid bilayer. This is usually conducted during the preparation steps of liposomes, also known as 'passive loading'. In a typical liposome preparation process by Bangham method or thin lipid film hydration method [74], a mixture of lipids and cholesterol in organic solvents first evaporated followed by a freeze-drying procedure to form a thin film in a round-bottom flask. Upon the addition of aqueous solvents, rehydration takes place producing a multilamellar vesicle (MLV) or a giant unilamellar vesicle (GUV) [75]. To achieve small unilamellar vesicles (SUVs) with homogeneous size distribution, an additional sonication or multiple extrusions through a polycarbonate membrane is necessary [76–78]. Advanced techniques using microfluidics have recently been introduced to prepare monodisperse liposomes. A notable example is a micro hydrodynamic focusing method proposed by Jahn et al. where a flow of phospholipids in an organic phase diffuse to an aqueous laminar flow, perpendicular to each other, causing the phospholipids to assemble into liposomes [79]. In 'passive loading', the degree of drug loading may depend on the type of lipids, type of drugs, liposome composition, as well as preparation methods [80, 81].

Another method is 'active loading', where a transmembrane pH gradient is constructed across the liposome bilayers. In this case, after liposome formation with an aqueous core of known pH, the outside buffer is replaced with a buffer of different pH [82, 83]. A modified version of the 'active loading' method can also generate a salt gradient across the liposome membrane [84, 85], which has been utilized in some commercial drugs. Doxil and Myocet are liposomal formulations of DOX. An ammonium sulfate gradient is utilized across the liposomal membrane to load a high amount of DOX in Doxil [86]. Similarly, using a proton gradient with acidic and basic buffers inside and outside of the liposomes, respectively, can act as the driving force for the DOX loading in Myocet. In the case of nucleic acid drugs (DNA, RNA, and antisense oligonucleotides), loading is usually carried using cationic lipids, which form a liposome/gene conjugate via electrostatic interaction between cationic lipids and anionic genes [87, 88].

Another important aspect is the sterilization of liposomes, which is often conducted to remove the possible pathogens and endotoxins contaminants. Conventional methods, such as sterile filtration and steam sterilization, cannot be used due to the generation of high temperature that potentially damages the liposomes. However, chemical sterilization that utilizes an ethylene oxide purge into a liposome suspension or a lyophilized liposome can effectively remove all contaminants. Because ethylene oxide is a known carcinogen and may form toxic compounds in aqueous buffers, complete removal of ethylene oxide should be ensured after sterilization [89]. As a non-invasive method of sterilization, γ-irradiation to a freeze-dried liposome in the presence of cryoprotectants such as trehalose showed promising results with no damage to the liposome [90]. Alternative methods include aseptic manufacturing where the whole preparation process of liposome is carried out in sterilized conditions using sterile filtered raw materials [91].

8.4.3 Targeting and Drug Release

The therapeutic efficiency of a DDS depends on many factors, such as high loading in a small carrier, drug/carrier stability in the bloodstream, target recognition, cell internalization, and drug release. Liposomes satisfy many of these requirements because the constituent lipids can be modified with chemical motifs that impart specific biological functions, as shown in Figure 8.6. For instance, liposomes with PEG, namely 'stealth liposomes', enhance the circulation time in blood. Liposome accumulates at the diseased site via passive or active mechanisms. Passive permeation occurs mostly due to the EPR effect operating at tumors and inflamed tissue sites [92]. In particular, tumors have a leaky vasculature due to a relatively large gap between the endothelial cell and many drug/carrier conjugates including liposomes can permeate passively through the endothelial gaps. Thus, the EPR effect becomes dominant when the liposomes circulate a longer time in blood. Passive accumulation is a general strategy and cannot recognize a specific diseased site. Liposomes having targeting ligands at their surfaces are shown to recognize specifically at the site of action via a ligand-receptor binding mechanism. Here the ligand, e.g., folic acid appended on a liposome can interact non-covalently with a folate receptor on the cell membrane inducing cell internalization followed by drug release in the cell cytoplasm. Many tumors have an excess amount of folate receptors in their cell membrane, hence folate-targeted liposomes are widely studied in cultured cancer cells as well as in tumor mice models [93, 94]. In other cases, monoclonal antibodies (mAbs), antibody fragments such as antigen-binding fragment (Fab) and crystallizable fragment (Fc), and proteins such as transferrin are functionalized on the liposomes to effectively deliver drug via active targeting method [95, 96].

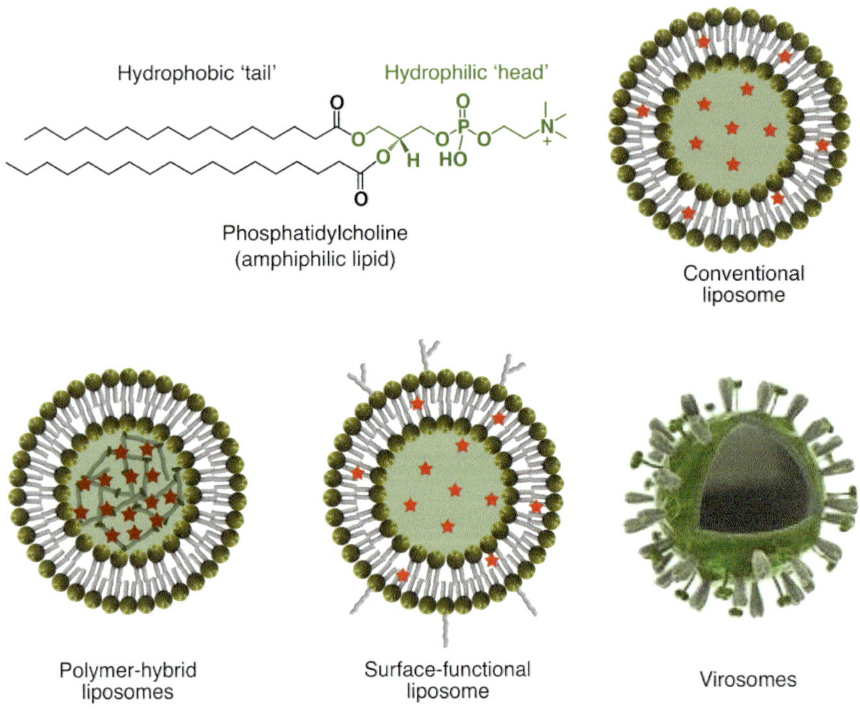

Figure 8.6 Molecular structure of a representative lipid and schematic illustrations of conventional liposomes, polymer-hybrid liposomes, surface-functionalized liposomes, and virosomes. A red-colored star shape represents the drug molecule.

A separate category is a cell-penetrating peptide (CPP) appended liposomes, which recognize specific receptor proteins on the cell membrane and translocate via endocytosis or a direct pathway through the cell membrane [97]. CPPs such as trans-activator of transcription protein, octa-arginine (R8), and amphipathic model peptide are handy and showed their potential for target recognition and delivery when appended on the liposomes [98–100]. In most of the cases, the targeting ligands are directly conjugated on a lipid, while a spacer PEG is utilized in some other cases. For instance, Biswas et al. conjugated an R8 peptide to PEG-phosphatidylethanolamine and then insert it into the DOX-loaded PEGylated liposome by using a special lipid insertion technique [101]. This liposome having both PEG and targeting R8 peptides effectively suppressed the tumor growth in a lung tumor xenograft in mice [102].

Drug release inside the cell involves two steps in a broad sense, endosomal escape and membrane leakage/disassembly of a liposome. Both these characteristics are usually associated with the type of lipids and occur in response to a stimulus that is available in an intracellular environment (e.g., pH) or supplied externally (heat, photon, ultrasound, magnetic field, etc.). Fusogenic (prone to undergo membrane fusion) lipids, such as dioleylphosphoethanolamine and cholesteryl hemisuccinate, are one category of responsive lipids, which undergo conformational changes at low pH causing the destabilization of the endosome membrane. Liposomes containing fusogenic lipids are useful for the delivery of nucleic acid drugs as these drugs specifically function at cytosol, for instance, siRNA knockdowns a target gene via RNA interference.

A notable category is thermoresponsive liposomes that can potentially work for cancers. Hyperthermia is a type of cancer treatment based on the principle of selective cytotoxicity by the heat generated locally at cancer cells [103]. To generate heat (>37 °C) locally at a particular tumor tissue, microwave, radiofrequency, and ultrasound are often used. Hyperthermia can also increase vascular permeability and cellular membrane fluidity, which is useful for DDS extravasation from the bloodstream and diffusion across the cell membrane. The inherent thermoresponsive behavior of liposomes is utilized for designing efficient drug release in a controlled manner. For this purpose, liposomes made of lysolipids, a lipid variant having a single hydrophobic chain, can be used due to a heat-induced conformation change of lysolipids followed by the destabilization of the liposomes [104]. This technology has been tested in clinical phases of DOX loaded liposomes in the name, ThermoDox, where the lipid constituents include a 10% mono-palmitoylphosphatidylcholine lysolipid. However, the clinical results were unsatisfactory probably due to the dissociation of lysolipids from the liposomes after intravenous injection causing a DOX release in the bloodstream [105]. To achieve highly thermoresponsive liposomes, a surfactant Brij78 (4%) or a hydrogenated lipid (18.4%) have been added in lipid composition, which showed an improved in vivo performance of tumor-specific drug release in response to heat energy [106, 107]. In a conceptually different approach, Chen et al. prepared a thermosensitive liposome where an ammonium bicarbonate (NH_4HCO_3) gradient used for drug encapsulation could be tuned to release the drug. In this case, at a temperature around 42 °C, NH_4HCO_3 produces CO_2 bubbles that destabilize the liposome membrane causing a rapid release of encapsulated drug [108]. A slightly different category is a liposome having thermoresponsive polymer such as poly(*N*-isopropylacrylamide) (pNiPAAm), which exhibits a lower critical solution temperature property. At high temperature, pNiPAAM becomes hydrophobic and destabilizes liposome membrane while keeping its hydrophilicity at body temperature [109].

Light responsive liposomes are also considered for efficient drug delivery, which often uses a photosensitizing agent along with a drug. In some cases, light responsive lipids can also be used for drug release from liposomes. For instance, liposomes consisting of dipalmitoylphosphatidylcholine (DPPC) and diacetylene phospholipid with an encapsulated DOX drug showed an enhanced delivery and cytotoxicity in cancer cells upon irradiation at 514 nm laser compared to non-irradiated cells [110]. Similarly, liposome having an embedded benzoporphyrin derivative monoacid ring A (Verteporfin) has been utilized in a commercially available drug, Visudyne, for photodynamic therapy (PDT) of age-related diseases [111].

8.4.4 Lipid-Coated Polymeric Nanoparticles

Liposomes and polymeric nanoparticles are two distinct classes that have been utilized for designing DDS. A hybrid category encompassing the advantages of both liposomes and polymeric nanoparticles, known as lipid–polymer hybrid nanoparticles (LPNs), has the potential to fulfill the requirements for designing an ideal DDS with high biocompatibility and drug loading capacity. LPN consists of a polymeric core with encapsulated drug and a lipid shell that provide biocompatibility. The lipid shell can be a monolayer, bilayer, or with an additional layer of PEG [112]. LPN can be prepared either via a two-step or a single-step approach. The lipid shell and the polymeric core are separately prepared in the two-step approach, which then fused via direct hydration, sonication, or extrusion method. The fusion process in this case mostly triggered via electrostatic interactions between the anionic polymeric core and the cationic lipid vesicle. The single-step approach involves simple mixing of lipid in an aqueous solvent and polymer and drug in an organic solvent, followed by sonication and evaporation of the organic solvent.

LPN has several advantages for cancer drug delivery. Both hydrophobic and hydrophilic drugs can be incorporated into the lipid and polymer layers [113]. LPN shows higher drug loading capacities and greater structural stability in addition to its excellent controlled drug release compared to the conventional liposomes [114]. The slower degradation rate of polymers in the LPN core and also the slower diffusion rate of drugs through the lipid membrane provide a sustained release profile. By using pH-sensitive and redox-sensitive components, stimuli-responsive LPNs also can be designed to deliver drugs specifically to tumors. Similar to liposomes, lipid shells of LPNs can be functionalized with PEG and targeting ligands to increase circulation time in the bloodstream and actively target tumor sites, respectively. Recently, LPN has been tested in all possible therapeutic options for cancers using small molecule cytotoxic drugs, macromolecule-based nucleic acid, and protein drugs. Major applications of LPNs include inhibition of tumor cell proliferation, inhibition of cancer-promoting oncogenes, targeted drug delivery to the tumor vasculature and microenvironment, targeting multi-drug resistant tumors, and transporting imaging agents such as gadolinium and iron oxide nanoparticles for MRI imaging of the tumors.

8.4.5 Hybrid Liposomes (HLs)

When one of the constituents of liposomes is a non-lipid derivative such as micelle forming molecules, anomalous properties can be expected and can be referred to as HLs. The group of Matsumoto and Ueoka developed HLs by using a non-lipid component polyoxyethylene dodecyl ether ($C_{12}(EO)_n$, n = 4 to 25) along with a conventional lipid L-α-dimyristoylphosphatidylcholine (DMPC). In buffer, $C_{12}(EO)_n$ forms micelle, and DMPC forms liposome; however, sonicating a mixture of these two produces an HL. By changing the composition ratio of $C_{12}(EO)_n$ and DMPC, the size, shape, and membrane fluidity of HLs can be controlled. Interestingly, HL itself without any drug molecule showed a therapeutic effect by selectively killing cancer cells in cultured cells as well as in tumor mice model [115, 116]. Multiple factors contributing to the selective cytotoxicity observed in cancers by HLs treatment such as the ability of HLs to fuse cancer cell membrane as well as the greater membrane fluidity of cancer cells compared to normal cells. For instance, an HL consisting of L-α-dimyristoylphosphatidylglycerol and $C_{12}(EO)_{10}$ in a 90:10 with a <100 nm size selectively inhibited tumor growth in lung adenocarcinoma and stomach tumor cells [117, 118].

In a clinical study, HL with DMPC:$C_{12}(EO)_{23}$ ratio 95:5 injected patients (n = 10) showed prolonged survival in one of the patients who was in an advanced stage of B-lymphoma. Another category of HL is consisting of polymers and phospholipids [119]. For example, a mixture of polybutadiene-*b*-PEG and 1-palmitoyl-2-oleoyl-*sn*-glycero-3-phosphatidylcholine (POPC) can form nano-sized HL via a special thin-film hydration technique followed by extrusion [120]. HLs also can be prepared with an encapsulated drug. A melatonin analog drug, 5-methoxycarbonylamino-*N*-acetyltryptamine, encapsulated in HL consisting of mucoadhesive polymers (sodium hyaluronate or carboxymethylcellulose) showed a reduction in the intraocular pressure in rabbit eyes. The therapeutic effect observed in this case was partially due to the behavior of polymer components in HL that provided long-lasting hydration and lubrication at the ocular surface [121]. A related category of HLs includes lipid-coated inorganic nanoparticles [122], cyclodextrin appended liposomes [123], and gel-in-liposomes [124].

Virosomes can be a special category of HLs where the constituents are derived from virus components. Structurally, a virus consists of DNA or RNA gene segments surrounded by capsid proteins. Targeting ligands, such as glycoprotein, present in the virus surface recognize specific receptors on the cell membrane and internalize into the target cells. In some viruses, an additional lipid envelop is also present. Virosomes prepared from the influenza virus, hemagglutinating virus of Japan (HVJ virus), and hepatitis B virus are commonly investigated for drug delivery applications. By adding detergents to the virus-containing solution followed by centrifugation, genetic materials are first removed. Then, conventional phospholipids and cholesterol are added to the solution containing virus protein, where a coassembly process produces virosomes [125]. Depends on the viral protein and phospholipid composition, the size of virosomes can be varied, for instance, virosomes formed from HVJ have a mean diameter of 400–500 nm, while those from influenza are 150–200 nm. Because of ability to trigger membrane fusion, the HVJ-derived virosomes showed potential therapeutic application for the delivery of nucleic acids, proteins, and anti-cancer drugs. Importantly, HVJ-derived virosomes have intrinsic anti-tumor activities, activate multiple anti-tumor immunities, and induce cancer-selective apoptosis [126].

8.5 Polymer-Based DDSs

Polymers are large molecular weight substances consisting of repeating monomeric units that are connected via covalent bonds. Nature derived polymers such as polysaccharides (e.g., cellulose, hyaluronic acid, etc.) and synthetic polymers such as poly(glycolic acid), poly(lactic acid), poly(caprolactone), and polydioxanone have become an inevitable component of DDSs. With the advancements in the field of polymers, many efficient methods of polymer synthesis are now available. Careful selection of monomeric motifs can produce polymers having different structures (linear, branched, and dendritic), configurations (copolymers), and properties (hydrophilic and hydrophobic), as shown in Figure 8.7. Some specific type of polymers such as amphiphilic block copolymers is widely utilized for constructing drug carriers in biomedical applications. Hydrophilic/hydrophobic small molecule drugs, nucleic acids, peptides, and proteins can be loaded into polymeric nanocarriers (e.g., micelles and polymersomes), where the designed carriers sustain in the bloodstream and accumulate in a diseased site followed by cellular internalization and drug release [127, 128].

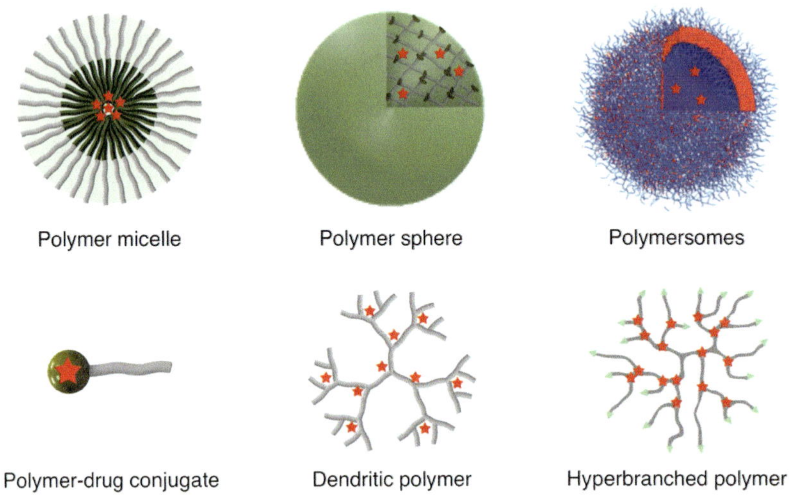

Figure 8.7 Schematic representation of polymer-based DDS such as polymer micelle, polymer sphere, polymersome, polymer–drug conjugate, dendritic polymer, and hyperbranched polymer. A red-colored star shape represents the drug molecule.

8.5.1 Polymeric Micelles

Micelles are non-covalently assembled objects with a hydrophobic core and hydrophilic shell. Block copolymers having distinct hydrophilic and hydrophobic segments can assemble into micelles whose characteristics, such as size, shape, and surface property, are solely dependent on the type of polymer constituents. Critical micelle concentration (CMC) is an important parameter that often indicates the stability of micelles. For applications in DDS, the micelles need to be stable in a highly dilute condition; hence CMC always needs to keep as low as possible. In the case of polymeric micelles, core-forming segments, as well as drugs loaded in the micelles, contribute to a lower value of CMC. Depending on the polymer characteristics and the polymer length of each segment in a block copolymer, the size of the polymeric micelle can be varied between 10 nm and 200 nm [27, 28, 129]. Sub-100-nm-sized micelles are advantageous for tumor penetration via discontinuous endothelium of cancerous tissues, hence high accumulation via either EPR effect [39] or vascular bursts followed by nano-eruption of fluid to the tumor [42] (Figure 8.3C). The selection of each polymer segment is critically important to ensure micelle stability, drug loading efficiency, drug release at the target site, and biosafety profile. As a shell forming segment in micelles, several hydrophilic polymers, either from a natural resource or a synthetic one, have been investigated [130]. Examples include PEG, poly(vinyl alcohol), zwitterionic polymers, polysaccharides, and poly aminoacids. Because PEG is a clinically approved polymer segment in addition to its chain flexibility and high-water solubility characteristics, a large number of micelles have utilized PEG as a shell forming segment. PEG motifs are known to enhance circulation time in the bloodstream, negligible interaction with blood components, poor recognition by immune cells, and minimum side-effects. Biopolymers are highly hydrophilic and thus can functions as shell-forming segments, however, their limited availability in a pure form, as well as plausible immunogenic behavior, restrict their wide usability. Some polysaccharides such as chitosan and hyaluronic acid can function as targeting ligands for improved cellular uptake [131, 132].

Core forming polymer segment provides structural stability as well as a drug loading location in polymer micelles. Several non-covalent interactions, such as hydrophobic, metal complexation, electrostatic interaction, and hydrogen bonding, are responsible for keeping the polymer together in the micellar core as well as for the incorporation of drugs. Common core-forming polymers include polyethers, polyesters, polypeptides, and polycations. For loading anticancer drugs, polyester-based core appeared to be better than those of polyethers. For instance, the loading efficiency of a paclitaxel drug into a micelle containing polyester core was 25% w/w, while those of polyether core loaded a maximum of 1.46% w/w drug [133, 134]. Polypeptides are another core-forming segment, which is biocompatible, and their sidechains can be readily modified. Hence, they are advantageous for designing micelles with appropriate assembly characteristics, efficient drug loading, and drug release [135]. Polycations are mostly used for micelles preparation with negatively charged drugs, especially nucleic acid and protein therapeutic agents, in their core [46, 136]. This category of micelles is known as polyion complex (PIC), in which block copolymers containing neutral and charged polymer segments (block ionomers) assemble in the presence of oppositely charged species via electrostatic interaction [137].

8.5.1.1 Drug loading and release in polymeric micelles

Pioneering works from the group of Kazunori Kataoka in the field of polymer micelles for cancer therapeutics have advanced into clinical phases. They developed DOX and paclitaxel (PTX) drug loaded polymeric micelles by utilizing the hydrophobic interaction of drug and drug-appended polymer segment in the core of the micelles. In their system, a PEG-*block*-poly(aspartic caid) (PEG-*b*-PAsp) copolymer with a covalently conjugated DOX forms a polymeric micelle, where additional DOX can be loaded in the micelle core via π–π stacking [138]. Importantly, the physically entrapped DOX can further stabilize the micelles preventing the micelle disassembly upon dilution in the bloodstream. The optimized version of this DOX-loaded micelles proceeded into clinical trials with a brand name NK911 (Nippon Kayaku, Co.). In a slightly different strategy, polymeric micelles incorporating PTX [139] showed high efficacy in preclinical tumor models with reduced side effects.

PIC micelles utilize electrostatic interaction for drug loading into the micelle core. Although macromolecular drugs are commonly loaded into PIC micelles, small molecule drugs with charged motifs can also be incorporated. For instance, PIC micelles loaded with a photosensitizer molecule can be useful for PDT, where a ROS triggers apoptosis causing cell death [140]. A specially designed dendrimer porphyrin forms PIC micelle with PEG-*block*-polylysine copolymers via electrostatic interactions between positively charged polylysine and negatively charged periphery of the dendrimer porphyrin. This PIC micelle showed longer circulation time in the bloodstream and improved therapeutic efficacy in various tumor models with no damage to healthy tissues [141].

Another strategy for drug loading into polymeric micelle is a direct conjugation of drugs to the core-forming polymer segment via labile bonds. The labile bonds usually cleave at specific conditions such as high pH or a reductive environment. For instance, a DOX-loaded micelle having a hydrazone bond between DOX and PEG-*b*-PAsp polymer constituent can release at acidic pH while negligible release at neutral pH. In a subcutaneous C26 tumor, this system significantly suppressed the tumor growth and a modified version of this pH sensitive DOX-loaded micelle has shown excellent safety, tolerability, and recommended dosage in clinical stages (NC-6300; Nanocarrier Co.) [142–144]. Disulfide bonds are responsive to reductive environments such as cell cytoplasm, thus polymeric micelle with disulfide linkages at its drug loading places [145] or in the main block copolymer chain has shown drug release in cell cytoplasm by micelle disruption. Metal ion-based drugs such as cisplatin and oxaliplatin can also be loaded into polymeric micelles via polymer–metal complexation [146]. By mixing PEG-*block*-poly(L-glutamic acid) copolymer and cisplatin, polypeptide block binds with cisplatin forming α-helical bundles in the micelle core. Because of a strong carboxylate–platinum interaction, the micelle showed stability even in the harsh conditions of the bloodstream, however, the drug can be released in the presence of chloride ions. Polymeric micelles can be functionalized by targeting ligands on their surface. Antibody fragment appended polymeric micelle showed an enhanced delivery of cancer drugs in cultured cells and improved efficacy against different types of tumor-installed mice models. In another study, cRGD peptide appended polymeric micelles efficiently delivered anticancer drugs in mice model of glioblastoma, which is a challenging target due to the poor permeability across the blood-brain barrier [45, 147]. In this case, cRGD present on the micelle periphery selectively binds with the integrin receptors overexpressed in neovasculature and cancer cells.

8.5.2 Polymer Vesicles (Polymersomes)

Amphiphilic block copolymers self-assemble in water resulting in nanoscale vesicles with a polymer bilayer membrane called polymersomes (Figure 8.7). The bilayer thicknesses (5–30 nm) in polymersomes are much higher than those of liposomes (3–5 nm), which provide a solute impermeable property to their membrane [148]. Hydrophilic and hydrophobic drugs can be incorporated in the aqueous interior and polymer bilayer, respectively. Similar to drug loading into liposomes, both active and passive (e.g., salt and pH gradient) methods have been utilized for polymersomes [149]. For instance, PTX loading to a polymersome by active method showed 10-fold higher efficiency than that of a liposome. To achieve improved biological performance, surface stealth such as PEGylation as well as surface-functionalization of polymersomes also has been developed [150]. Drug release from the polymersomes often required a stimuli-responsive moiety in their polymer constituents. A large class of polymersomes introduced pH-sensitive motifs include biodegradable polymers such as poly(lactic acid), poly(ε-caprolactone), acid labile groups (e.g., acetals, orthoesters), and proton buffering functional groups (e.g., ethylenimine) [151]. Reductively cleavable linkages such as disulfide bonds, photo-labile bonds, and thermo-responsive polymers also have been utilized to fabricate polymersomes that can efficiently release the drug in vivo [152]. In all these cases, upon contact with the stimuli, membrane destabilization occurs followed by shrinkage of polymersomes causing the drug release into the environment where the stimuli are present. A related category is hybrid lipo-polymersomes consisting of both polymers and lipids, where the robust structural features of polymersomes are combined with the excellent biocompatibility of liposomes [153].

8.5.3 Polymer–Drug Conjugates

Polymer–drug conjugate (Figure 8.7) may be the smallest DDS because it consists of only two components, and a polymer and a therapeutic agent (e.g., small molecule, peptide, protein, and aptamer), which are connected via a covalent bond. In 1955, von Horst Jatzkewitz reported that the conjugation of an alkaloid drug to a copolymer resulted in extended resident time in mice with a controlled release of the conjugated drug [8]. Together with the growth of polymer science and its synthetic methods [154], the field of polymer–drug conjugate also progressed and a clinically approved polymer–protein conjugate, Adagen, was realized in 1990. The conjugation of chemotherapeutics to polymers is beneficial for the enhancement of drug solubility, which can potentially increase clinical efficacy and improved pharmacokinetics. For instance, a clinically investigated polymer–drug conjugate, PK1 (FCE28068), consists of DOX and *N*-(2-hydroxypropyl)methacrylamide (HPMA) copolymer connected via a bio-cleavable peptidyl linker [155]. PK1 showed high efficacy in preclinical animal models with enhancement in plasma half-life and antitumor efficacy than unconjugated DOX [156]. However, subsequent clinical trials using PK1 in humans did not show high efficacy of antitumor activity, even though an extended circulation time and improved safety was observed [157]. Recently, a four-arm PEG conjugated irinotecan with a cleavable ester bond, Onzeald, has progressed to clinical phase III [158]. At the tumor site, the ester bonds cleave to release irinotecan, which then metabolized into a therapeutic anticancer agent, SN38. Despite the preclinical efficacy of Onzeald in mice models, available clinical data does not support a similar efficacy in human tumors.

A related category is a cyclodextrin (CD) functionalized polymers that can assemble into nanoaggregates. In this case, a polymer segment, such as a PEG, can accommodate multiple CDs via host-guest interaction, and by functionalizing a hydrophobic segment at the termini of PEG can induce a micellar assembly, similar to the assembly process of core-shell micelles. The drugs are either conjugated covalently to the CD segment or encapsulated non-covalently in the assemblage. For instance, a cinnamic-acid-terminated PEG forms a 'polyrotaxane' inclusion complex with α-CDs, which then assembled in the presence of DOX forming a micelle. This DOX-loaded micelle effectively inhibited breast cancer cell growth in vitro as well as antitumor activity in vivo [159]. In an alternate approach, a covalently linked DOX to CD by acid-labile hydrazone bonds forms micelles by mixing with a PEG-*block*-poly(2-methacryloyloxyethyl phosphorylcholine) copolymer, in an aqueous. In this case, each of the CD contains two DOXs and hence a high loading is possible [160]. In a conceptually novel approach, the group of Mark E. Davis developed linear, high molecular weight β-CD-embedded polymers that contain alternating β-CD units with pendant carboxylic functionalities [161]. A promising anti-cancer drug, camptothecin (CPT) can be conjugated to the β-CD-embedded polymers producing a CD–polymer–CPT prodrug, which showed excellent antitumor activity in nude mice bearing human colon carcinoma tumors. An optimized version of this prodrug also advanced to clinical trials.

8.5.4 Dendritic Polymers

Dendrimers are unimolecular, highly branched, and three-dimensional polymers (Figure 8.7), which can be synthesized from monomeric units by adding new branches in a step-by-step manner until a uniform tree-like structure is formed. Depending on the branching architecture, one can design perfect dendrimers, dendrons, dendronized polymers, and hyperbranched polymers. A typical dendrimer possesses unique nanoarchitecture with an average diameter of 1–10 nm, low viscosity, high solubility, high surface functionality, and biocompatibility. Due to the presence of active termini on its surface, easy functionalization is possible in dendrimers. In comparison to linear polymers, dendritic polymer architecture is advantageous for drug delivery applications [162]. For example, due to its defined multivalency and drug conjugation, the addition of targeting ligand and/or solubilizing modalities can be incorporated in a single dendrimer. Dendrimer-based drug carriers showed extended circulation time and altered biodistribution profiles compared to bare drugs in preclinical animal models.

Because of relatively large hydrodynamic radii, dendrimer-based drug carriers exhibit reduced renal clearance and increased plasma exposure than linear polymers of similar molecular mass [163]. Small molecule anticancer agents, e.g., methotrexate (MTX), CPT, PTX, cisplatin, etc., [164] have conjugated to dendrimers for delivery applications. On the other hand, targeting ligands such as folate residues, antibodies, and hormones are also functionalized on a drug-loaded dendrimer surface to evaluate their potential for tumor cell specificity [165]. Cationic amine terminated poly(amidoamine) (PAMAM) dendrimers are widely studied, however, their clinical usage is restricted due to non-specific adhesion originating from the cationic surface as well as the non-biodegradable nature of PAMAM [166]. Biodegradable dendrimers based on peptides, polyurea, and polyester have been developed [167]. Although drugs can be physically entrapped within dendrimers, covalent conjugation is better because of the enhanced drug loading and controlled drug release kinetics. In an extreme case, a PAMAM dendrimer conjugated with a fluocinolone acetonide drug showed a sustained drug release profile of 90 days [168]. In addition, the release of conjugated drugs buried within the dendrimers is possible to modulate further by varying the number of generations as well as the surface chemistry [169]. For enhancing the antitumor activity, drug-loaded in a dendrimer is incorporated into sophisticated nanoarchitcctures, where it effectively accumulates at tumors due to the EPR effect while the small-size dendrimers effectively penetrate the tumors [170].

8.5.5 Hyperbranched Polymers (HBPs)

HBPs possess a three-dimensional globular structure with a high degree of branching as well as a high number of terminal functional groups (Figure 8.7). Compared to linear polymers, HBPs have low viscosity, minimum chain entanglements, smaller hydrodynamic radius, and good solubility. Compared to dendrimers, HBP structures are irregular with randomly distributed dendritic, linear, and terminal units. HBP can be easily synthesized in a one-pot reaction, however with broad molecular weight distribution. In solution, HBPs form cavities around the branching site, which can be used for loading therapeutic drugs such as DOX [171], CPT [172], cisplatin [173], and 5-fluorouracil [174]. Due to terminal functional groups, HBPs are also available for the covalent conjugation of drugs [175]. For instance, up to 70% of ibuprofen drug can be conjugated on a hyperbranched polyglycerol via coupling reaction [176], which can be cleaved by lysosomal enzymes present in the cell. In a specific case, HBP constituents can act as a drug as in an HBP consisting of alternated hydrophobic diselenide and hydrophilic phosphate groups [177]. Phosphate groups acted as branching units with some anticancer activity originated from selenium compounds [178]. Both passive and active targeting methods are investigated for efficient accumulation of HBP-based DDS. The average hydrodynamic size of HBP prepared with a high molecular weight polymer is over 10 nm and surface modification with PEG or other motifs can dramatically increase its hydrodynamic size, which favors a passive EPR accumulation [179]. For active targeting, peptides, aptamers, and folic acid have been conjugated to HBP. Most of the HBPs used for the delivery application are designed with cleavable bonds; hence drug release can be efficiently achieved in an intracellular environment [180].

8.6 Inorganic Nanomaterials

With the advancement in the field of inorganic nanomaterials, a growing interest is to use nanoparticles of unique size and shape for biomedical applications. Gold, silver, platinum, and iron-oxide nanoparticles have been investigated for the purpose of delivering macromolecular drugs, imaging agents, as well as using the nanoparticle itself as a therapeutic drug. Among all the nanomaterials, AuNPs (Figure 8.8) are relatively non-toxic [181] and their surface can be modified via thiol and amine functionalities. AuNPs show a strong surface plasmon resonance (SPR) band that can be varied dependent on the particle size and the SPR property. AuNPs have been investigated for imaging as well as therapeutics in preclinical animal models. AuNP conjugated with an antibody that can bind to epidermal growth factor receptors (EGFR) overexpressed in cervical cancer cells showed an enhanced detection of cancer cells via SPR scattering [182]. In another study, tumor necrosis factor (TNF) and PEG functionalized AuNP showed a 10-fold increase in anticancer activity in animal models. In some specific cases, AuNP can function as an anticancer agent. One example is the inhibition of tumor angiogenesis, which is the formation of new blood vessels and capillary from existing tumor mediated via endothelial growth factors [183]. A significant reduction in cancer cell proliferation was observed in cellular studies as well as in animal models, where AuNP interacts with the heparin-binding domain and inhibits the binding of growth factors.

Photothermal therapy using inorganic nanoparticles is also well studied [184]. In this case, bare nanoparticle or nanoparticle functionalized with a targeting ligand is first supplied to cancers followed by irradiation of suitable wavelength, where the heat energy produced locally kill cancer cells. For AuNP, an SPR absorption can be converted into heat energy within picoseconds, thus an AuNP with surface functionalized EGFR antibodies caused cancer cell death upon light irradiation at 514 nm [185]. For in vivo application of nanoparticle-mediated hyperthermia, light irradiation at a longer wavelength is required. One option is to use a core-shell nanoparticle that can absorb light in the near-infrared (NIR) region. Gold nanorods with silica coating exhibit NIR scattering and hence suitable for photothermal therapy. An exposure of 6 min at NIR caused irreversible tissue damage in human breast cancer cells in vivo.

PDT is an effective strategy for cancer treatment, in which a photosensitizer molecule absorbs light and transfers its energy into the surrounding oxygen producing singlet oxygen that can damage the cells. Zinc (II) phthalocyanine is an FDA-approved photosensitizer, which showed highly specific antitumor activity both in vitro and in vivo [186]. Silica and ceramic nanoparticles encapsulated with the photosensitizer is a promising approach to enhance PDT efficiency and showed encouraging results with cellular experiments [187]. Utilizing a different mechanism, quantum dots with core-shell architecture also have applied for PDT in some types of cancers [188].

8.7 Carbon Nanostructures

Carbon-based nanostructures (Figure 8.8) such as carbon dots, fullerenes, CNTs, and graphene have been investigated for theragnostic applications [189]. These nanostructures are inert in nature with no intrinsic surface charge and hence less toxic to bio-surfaces. Drug loading to the carbon nanostructures is possible both by covalent and non-covalent approaches. Aromatic cancer drugs (DOX) can be easily loaded onto a graphene oxide (GO), a water-soluble oxidized form of graphene, via π-stacking and hydrogen bonding interactions. GO also can be functionalized with several motifs such as carboxyl/hydroxyl/epoxide as well as PEG/polyethylenimine (PEI) molecules, which make it biocompatible for nanomedicine applications.

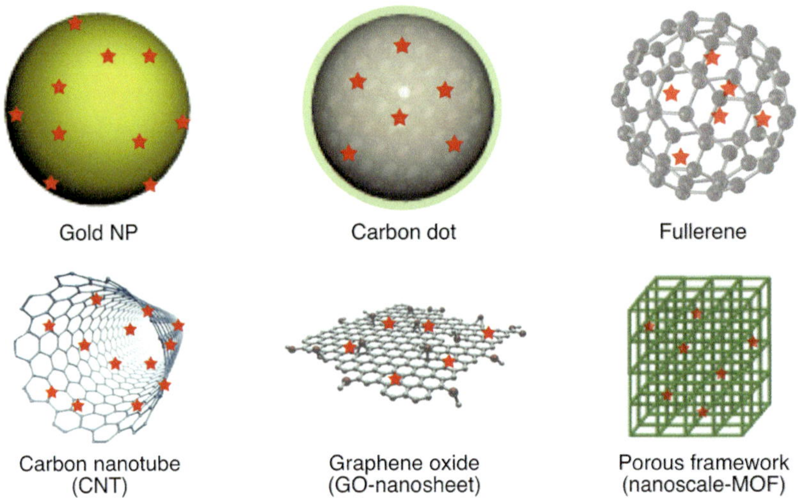

Figure 8.8 Schematic representation of DDS based on AuNP, carbon dot, fullerene, CNTs, GO nanosheet, and nanoscale MOF. A red-colored star shape represents the drug molecule.

The large surface area of GO nanosheet allows the loading of multiple types of drugs. A PEG-functionalized GO loaded with DOX and photosensitizer molecule, chlorin e6, showed an enhanced delivery both in cultured cells and in mice model [190]. By irradiation at 660 nm, a photodynamic effect also was observed. In a different study, GO loaded with DOX and targeting ligand hyaluronic acid (HA) through a hydrazine bond showed a pH-dependent release of DOX [191]. For controlled release, disulfide bonds also have been introduced in separate design of GO-based DDS. For in vivo cancer therapy, a DOX-loaded HA-GO nanosheet upon a subcutaneous injection (6 mg/kg) into a tumor mice model showed efficient anticancer activity with ~50% reduction in tumor volume compared to free DOX (~37%) [192].

In the case of CNTs, the drug loading efficiency often relates to their single-walled or multiwalled structural characteristics. For targeting specific receptors overexpressed on the targets, the outer surface of CNT also can be functionalized by simple chemical reactions [193]. In some cases, the functionalization is done for improving the biocompatibility of CNTs. For instance, a chitooligosaccharide (COS) appended single-walled CNT does not show a reduction in cell viability in breast cancer cells (MCF-7) and cervical cancer cells (HeLa) [194]. However, when a Gliotoxin (GTX) drug, together with targeting ligands such as p53 protein, was loaded on this CNT, cell viability significantly decreased due to the cell-specific penetration of CNTs-COS-GTX-p53 followed by apoptosis. Similarly, a poly(vinyl alcohol) coated multi-walled CNT loaded with CPT anticancer drug exhibited 15 times lower IC_{50} value (50% growth inhibition concentration) compared to the free CPT dose under identical conditions.

Carbon nanohorns, which are a one-side capped version of CNTs, have also been utilized for drug delivery application by conjugating drugs and targeting moieties on both the inner and outer tube walls of carbon nanohorns [195].

Another category is a fullerene-based nanocarrier, for instance, C_{60} fullerenes modified with dendrimer units and alkane chain self-assemble in water forming "buckysomes" [196]. Hydrophobic anticancer drug PTX can be loaded into buckysomes, which showed cytotoxicity in MCF-7 breast carcinoma cells, while empty buckysomes showed negligible cytotoxicity. Both drug and targeting moiety also can be functionalized to C_{60} fullerene. In this case, about 5 DOX molecules were conjugated to the fullerene surface via acid cleavable hydrazone bonds and few folic acid motifs via amide bonds for specific binding to the folate receptors on cancer cells [197]. The cell viability remained the same for cells that do not have folate receptors such as murine fibroblast cells (L929) and adenocarcinomic human alveolar basal epithelial cells (A549), but significantly reduced in folate expressed HeLa cells due to target-specific binding and internalization of multifunctional fullerene nanocarrier followed by the release of DOX. C_{60} fullerenes-based DDSs also have been studied for in vivo anticancer experiments. For this purpose, docetaxel (DTX) drug was loaded onto a fullerene containing both folic acid and PEI, and intravenously injected into a tumor-installed mouse model with a dose of 11.96 mg/kg. They observed a tumor growth inhibition rate of 74.5% without any loss in the mice weight. The high antitumor activity, in this case, was due to the higher uptake of the DTX-FA-PEI-C_{60} via tumor expressing folate receptors [198].

Nanodiamonds with 2–8 average diameters are also studied as drug carriers due to their unique surface electrostatics, functionalization versatility, biocompatibility, and inherent fluorescence [199, 200]. Fluorescent carbon dots that exhibit a high aqueous solubility and biocompatibility along with tunable absorption and emission characteristics also showed promising results for drug delivery applications [201].

8.8 Nanoscale Metal-Organic Frameworks (MOFs)

MOFs, also known as coordination polymers, are a new class of hybrid material formed by the coordinate bonding interaction between metal ions or clusters and organic linkers. Typical MOFs are macro-sized crystals; however, their size regime can be tuned to nanoscales forming a unique family of nanoscale hybrid nanomaterials, referred to as nanoscale MOFs (NMOFs). Due to the porous nature together with intrinsic biodegradability, NMOFs are recently exploited as a nanocarrier in biomedical applications [202]. NMOFs can be synthesized in a controlled manner using appropriate metals and bridging ligands via several methods such as nanoscale precipitation, surfactant-templated, solvothermal, and reverse microemulsion. Drug loading is possible both by direct incorporation of drugs during NMOF synthesis and by a drug encapsulation after NMOF formation. However, loading efficiency is low in the former case due to the potential risk of structural instability at a higher concentration of the drug. In the latter case, the drug is supplied into a NMOF solution where a non-covalent or a designed covalent interaction holds the drug in the pores of NMOF. A sustained drug release is observed in some of the NMOF loaded with therapeutic agents.

For chemotherapy application, many of the conventional small molecule drugs have been incorporated into NMOF, which then functionalized with stabilizing motifs or targeting ligands for enhancing their therapeutic effect. The group of Lin developed a cisplatin drug appended NMOF having an extra layer of silica and silyl derived RGD peptide [203]. This conjugate decomposed in physiological media releasing the cisplatin prodrug and induced cytotoxicity in human colon cancer cells (HT29) and MCF-7 cells. For in vivo application, they also have developed a nanocarrier based on a self-assembled zinc bisphosphonate NMOFs carrying cisplatin (~48% wt/wt) or oxaliplatin (~45% wt/wt) prodrugs with a relatively high drug loading [204]. After coating with lipids and PEG motifs, these NMOFs showed an excellent blood circulation half-life and antitumor efficacy in multiple tumor models, including colon, lung, and pancreatic cancers, at a very low drug dose as compared to their counter free drugs. The drug loading efficiency can be maximized by using the drug motif as one of the constituents of NMOF. MTX has been used as a bridging ligand, which forms NMOF in the presence of Zn^{2+}, Zr^{4+}, or Gd^{3+} with a maximum of 79.1% wt/wt drug loading [205]. When these NMOFs were coated with a lipid bilayer and targeting ligand (anisamide), NMOF efficiently internalized into cancer cells and induced cytotoxicity by the MTX drug. Other chemotherapeutic agents such as DOX and 5-fluorouracil also have been incorporated into NMOF for delivery to cancer cells [206].

NMOF loaded with Hf-porphyrin (77% wt/wt) has been studied for PDT against cancers [207]. The singlet oxygen generated by light irradiation can diffuse through the pores and exhibited an efficient PDT efficacy in vitro and in vivo. A complete tumor eradication is observed in half of the mice treated with a single dose (3.5 mg/kg, local administration) and single light irradiation (630 nm, 100 mW/cm^2, 30 min), showing the potential of this strategy for human translation.

8.9 Clinically Approved Cancer Nano-Chemotherapeutics

Development of any therapeutic drug consists of multiple stages such as laboratory experiments in cells/tissues, pre-clinical studies in small animals, and clinical studies in humans. In 1995, a PEGylated liposome-based DOX, Doxil® (or Caelyx®), received clinical approval in the US by the FDA for the treatment of metastatic ovarian cancer and HIV/AIDS-related Kaposi's sarcoma. Doxil® is the first nano-sized liposome approved for clinical use [86]. Later, DaunoXome®, liposomal daunorubicin, was first marketed in the United Kingdom in 1995 and later approved by the FDA in 1996. Since their inception, more than 45 nanodrug formulations were approved for clinical use by various agencies for different types of cancers (Table 8.1) [208–211] and more than 265 nanoformulations are under clinical investigation for the treatment of various types of cancer [212–214]. Approval of Hensify®/NBTXR3, a selective radio-enhancer for high energy deposition in the tumor only when exposed to ionizing radiations (on/off activity), is a revolutionary approach for the local treatment of solid tumors. Approval of Vyxeos®/CPX-351, a unique dual drug liposomal co-formulation of cytarabine and daunorubicin engineered to ratiometrically deliver the drugs at a synergistic ratio, is a breakthrough concept paves the way for a futuristic combination regimen as standard clinical care for many cancers.

8.10 Conclusion and Future Prospects

In this chapter, we have discussed various nanotechnology approaches for the construction of efficient DDSs, their behavior in the bloodstream, accumulation mechanisms at the diseases site, cellular internalization, and drug release mechanisms. Although numerous types of DDSs for cancer chemotherapy are available, we selected widely investigated DDSs that significantly contributed to the field of "nanotechnology in chemotherapy." These include DDSs based on lipids, polymers [215, 216], dendrimers, organic and inorganic materials, and NMOFs.

Table 8.1 Clinically approved nanoformulations for diagnosis and treatment of cancer.

Formulation (Drug) Approved year (Agency)	Material composition	Advantages of nanoformulation	Indications	Route of administration
(1) Liposomes PEGylated				
Doxil®/Caelyx® (DOX) 1995 (FDA) 1996 (EMA)	PEGylated SUV liposome (80–100 nm) of HSPC, Chol, and DSPE-PEG 2 kDa (55 40:5 M)	RES evasion, passive accumulation via EPR, reduced myelosuppression, and cardiotoxicity	Ovarian cancer, HIV-associated KS, Multiple myeloma	Intravenous
Lipodox® (DOX) 2013 (FDA)	A generic DOX hydrochloride liposome	RES evasion, passive accumulation via EPR, reduced myelosuppression, and cardiotoxicity	Ovarian cancer, HIV-associated KS, Multiple myeloma	Intravenous
Onivyde® (Irinotecan) 2015 (FDA)	PEGylated SUV liposome (110 nm) of DSPC, Chol, and DSPE-PEG 2 kDa (3:2:0.015 M)	RES evasion	Metastatic pancreatic adenocarcinoma	Intravenous
Non-PEGylated				
DepoCyt® (Cytarabine) 1999 (FDA)	Cytarabine loaded in nanochambers of MLV liposome (25 μm) of DOPC, DPPC, Chol, and triolein	Prolonged half-life (40 times longer than cytarabine), sustained-release, reduced systemic toxicity	Lymphomatous malignant meningitis	Intrathecal
Lipusu® (Paclitaxel) 2006 (China FDA)	SUV liposome of lecithin and Chol (8:2 M)	Minimized toxicity and enhanced therapeutic activity	Breast cancer and NSCLC	Intravenous

(*Continued*)

Table 8.1 (*Continued*)

Formulation (Drug) Approved year (Agency)	Material composition	Advantages of nanoformulation	Indications	Route of administration
Marqibo® (Vincristine) 2012 (FDA)	SUV liposome (100 nm) of sphingomyelin and Chol (6:4 M)	Passive accumulation via EPR. Liposome extravasates to the tumor nest from fenestrated bone marrow endothelium	Ph-negative ALL	Intravenous
Mepact® (Mifamurtide) 2009 (EMA)	Mifamurtide is anchored in the bilayer of MLV liposome (<100 nm) of POPC and DOPSE (7:3 M)	Accumulation to RES and stimulates macrophages and monocytes of RES to kill cancer cells	Osteosarcoma	Intravenous
Myocet® (DOX) 2000 (EMA)	OLV liposome (180 nm) of egg PC and Chol (1:1 M)	Accumulation to RES and acts as "RES depot" for persisted release of drug	Metastatic breast cancer	Intravenous
Vyxeos®/CPX-351 (Cytarabine: daunorubicin) 2017 (FDA) 2018 (EMA)	SUV liposome (107 nm) of DSPC, DSPG, and Chol (7:2:1 M) loaded with cytarabine and daunorubicin (5:1 M)	Prolonged blood retention, bypassing P-gp-based efflux, prolonged exposure of synergistic drug ratio	AML	Intravenous
(2) Lipid nanoparticle				
Liporaxel®/DHP107 (Paclitaxel) 2016 (South Korea)	Monoolein, tricaprylin, polysorbate 80 and 1% (w/v) paclitaxel	Formulation adhered to the villi 24 h	Gastric cancer	Oral

Formulation (Drug) Approved year (Agency)	Material composition	Advantages of nanoformulation	Indications	Route of administration
(3) Micellar nanoparticles				
Apealea® (Paclitaxel) 2018 (EMA)	XMeNa and 13XMeNa micelle (20–60 nm)	Increased solubility of the drug	Platinum-sensitive EOC, primary peritoneal cancer, and fallopian tube cancer	Intravenous
Genexol® (Paclitaxel) 2001 (South Korea)	PEG-PDLLA micelle (20–50 nm)	Increased solubility and efficacy, reduced toxicity and hypersensitivity of the drug	r/MBC and NSCLC	Intravenous
Nanoxel® (Paclitaxel) 2006 (DCGI)	Methoxy-PEG-PDLLA micelle (10–50 nm)	Increased solubility and reduced hypersensitivity of the drug	Advanced NSCLC, breast cancer, pancreatic cancer, ovarian cancer, and AIDS-related KS	Intravenous
(4) Protein/Antibody-drug conjugates				
Abraxane® (Paclitaxel) 2005 (FDA) 2008 (EMA)	Nanoparticles of albumin-bound paclitaxel (Nab-paclitaxel) (130 nm)	Improved solubility of paclitaxel. Passive accumulation via EPR. Dissociates into individual paclitaxel-albumin for transcytosis	Metastatic pancreatic and breast cancer, advanced NSCLC	Intravenous
Adcetris® (Brentuximab vedotin) 2018 (FDA)	cAC1 conjugated to MMAE via a protease-cleavable linker	Selective delivery of the drug to tumor cells, while sparing normal cells thus increasing the therapeutic index	Refractory HL and sALCL	Intravenous

(*Continued*)

Table 8.1 (*Continued*)

Formulation (Drug) Approved year (Agency)	Material composition	Advantages of nanoformulation	Indications	Route of administration
Besponsa® (Inotuzumab ozogamicin) 2017 (FDA)	A recombinant humanized IgG4 kappa CD22-mAb conjugated to CalichDMH	Selective delivery of the drug to tumor cells, while sparing normal cells thus increasing the therapeutic index	Refractory CD22-positive B cell precursor ALL and Ph positive refractory B cell precursor ALL	Intravenous
Bexxar® ([^{131}I]-Tositumomab) 2003 (FDA)	Anti-CD20 murine IgG2a λ mAb conjugated to ^{131}I (Radioimmunoconjugate)	Targeted delivery to CD20-positive cells and [^{131}I] emits β radiation and ADCC	CD20-positive, relapsed or refractory, low-grade, follicular, transformed non-HL	Intravenous
Kadcyla® (Ado-trastuzumab emtansine/T-DM1) 2013 (FDA)	Humanized anti-HER2 mAb conjugated with DM1 via a stable thioether linker, SMCC	Anti-HER2 mAb inhibits HER2 receptor signaling and DM1 is a cytotoxin, which inhibits the assembly of microtubules	HER2-positive metastatic breast cancer	Intravenous
Mylotarg® (Gemtuzumab ozogamicin) 2017 (FDA)	Humanized anti-CD33 mAb (hP67.6) conjugated to 2–3 molecules of Calich	Selective delivery of the drug to tumor cells while sparing normal cells thus increasing the therapeutic index	AML	Intravenous
Ontak® (Diphtheria toxin fragment A; Denileukin diftitox) 1999 (FDA)	Recombinant fusion protein with diphtheria toxin fragment A and IL-2 receptor binding domain	Targeted delivery to T-cells which expressed interleukin-2 receptors and lysosomal escape	CD25-positive recurrent cutaneous T-cell lymphoma	Intravenous

Formulation (Drug) Approved year (Agency)	Material composition	Advantages of nanoformulation	Indications	Route of administration
Zevalin® ([^{90}Y]-ibritumomab tiuxetan) 2002 (FDA)	Anti-CD20 murine IgG1 mAb conjugated to ^{90}Y chelating agent MX-DTPA (tiuxetan)	Targeted delivery to CD20-positive cells and [^{90}Y] emits pure high-energy β radiation	Relapsed or refractory, low-grade or follicular B-cell non-HL	Intravenous
(5) Nanocrystals				
Megace ES® (Megestrol) 2005 (FDA)	Nanocrystals of megestrol	Enhanced stability and long-term acting (extended-release)	Palliative treatment of breast and womb cancer	Oral
Hensify®/NBTXR3 (HfO_2) 2019 (CE Mark)	Nanocrystals of HfO_2 (50 nm) coated with negatively charged phosphate	HfO_2 stimulated with external radiation to enhance tumor cell death via electron production	Locally advanced soft tissue sarcoma	Intratumoral
Rapamune® (Rapamycin/ sirolimus) 2015 (FDA)	Nanocrystals of rapamycin (tablets)	Increased solubility and high bioavailability	Lymphangioleiomyomatosis	Oral
Ryanodex® (Dantrolene) 2014 (FDA)	Nanocrystalline suspension of dantrolene	High drug loading, extended-release, and faster administration at higher dose	Malignant hyperthermia	Intravenous
(6) Nanodispersions				
Paclitaxel injection concentrate for nanodispersion (PICN) 2014 (DGCI)	Polymeric-lipidic nanoparticle of paclitaxel stabilized by polyvinylpyrrolidone, cholesteryl sulfate, and caprylic acid (100–150 nm)	High drug solubility and loading, no significant hypersensitivity	Metastatic breast cancer	Intravenous

(*Continued*)

Table 8.1 (*Continued*)

Formulation (Drug) Approved year (Agency)	Material composition	Advantages of nanoformulation	Indications	Route of administration
(7) Polymer-based drugs/polymer nanoparticles/ polymer–drug conjugates				
Copaxone® (Glatiramer acetate/ COP 1) 1996 (FDA)	Glatiramer acetate (M_w 5–9 kDa)	No mechanism attributable to nanosize. Glatiramer competes with myelin antigens for their presentation to T cells	Relapsing-remitting MS	Subcutaneous
Eligard® (Leuprolide/ Leuprorelin) 2002 (FDA)	Leuprolide loaded into PLGA copolymer nanoparticles (Artigel)	Prolonged circulation and sustained release of leuprolide	Advanced prostate cancer	Subcutaneous
Lupron Depot® (Leuprolide/Leuprorelin) 1989 (FDA)	PLGA microsphere	Persisted release	Palliative treatment of advanced prostatic cancer	Intramuscular
Opaxio®/Xyotax® (Paclitaxel) 2012 (FDA)	Paclitaxel is covalently linked via an ester linkage to the γ-carboxylic acid side chains of α-PGA	Passive targeting via EPR, enzymatic hydrolysis of polyglutamate releases the paclitaxel	Glioblastoma	Intravenous
Trelstar® (Triptorelin) 2000 (FDA)	Triptorelin pamoate microgranules incorporated in a biodegradable PLGA	Persisted release	Palliative treatment of advanced prostate cancer	Intramuscular
Zinostatin®/SMANCS® (Neocarzinostatin) 1994 (Japan)	Two strands of PSMA (1.6 kDa) are conjugated to one molecule of NCS	Prolonged half-life leads to passive targeting via EPR	Hepatocellular carcinoma	Intravenous

Formulation (Drug) Approved year (Agency)	Material composition	Advantages of nanoformulation	Indications	Route of administration
Zoladex® (Goserelin) 1989 (FDA)	Goserelin is impregnated in polymer rods of biodegradable PLGA	Extended elimination half-life (4.2 h) compared to natural LHRH (13 min)	Advanced prostate carcinoma, palliative treatment of advanced breast cancer	Subcutaneous
(8) PEGylated proteins, polypeptides				
Asparlas® (Calaspargase pegol-mknl) 2018 (FDA)	Random amine PEGylation with 31–39 strands of linear PEG (5 kDa)	Prolonged blood half-life and reduced immunogenicity	ALL	Intravenous
Neulasta®/Onpro®/ Pegfilgrastim Recombinant methionyl human granulocyte-colony stimulating factor (r-metHu G-CSF/ filgrastim) 2002 (FDA) (EMA)	Selective *N*-terminal PEGylation with a single linear 20 kDa PEG (PEGylated filgrastim)	PEGylated filgrastim presented prolonged blood half-life (15–80 h) than filgrastim (3–4 h)	Chemotherapy-induced febrile neutropenia	Subcutaneous
Oncaspar® (L-asparaginase) 1994 (FDA) 2016 (EMA)	Random amine PEGylation with 69–82 strands of linear PEG (5 kDa)	Prolonged blood half-life and reduced immune recognition	ALL	Intramuscular or intravenous
Plegridy® (Interferon β-1a) 2014 (FDA) (EMA)	Selective *N*-terminal PEGylation with a single linear PEG (20 kDa)	Prolonged blood half-life permits less frequent dosing	Relapsing forms of MS	Subcutaneous

(*Continued*)

Table 8.1 (*Continued*)

Formulation (Drug) Approved year (Agency)	Material composition	Advantages of nanoformulation	Indications	Route of administration
Sylatron® (Interferon α-2b) 2011 (FDA)	Selective C-terminal cysteine PEGylation with a single linear PEG (12 kDa)	Adjuvant therapy	Melanoma	Subcutaneous
(9) Virosomes				
Gendicine® (p53 gene) 2003 (China FDA)	Recombinant adenovirus encoding wildtype human-p53 (rAd-p53)	Protects p53 gene against nucleases and rAd-p53 infects the tumor cells. The expressed p53 induces apoptosis	HNSCC, Nasopharyngeal cancer	Intratumoral, intracavity, intravascular infusion
Rexin-G® (Cyclin G1 gene) 2007 (Philippines)	Leukemia virus vector (100 nm) encoding a gene for cytotoxic cyclin G1 gene	Active targeting to the tumor by the collagen-binding peptide of human vWF	All solid tumors	Intravenous
(10) Metal nanoparticles				
Feridex®/Endorem® (SPION) 1996 (FDA)	Dextran-coated SPION (D_H = 80–150 nm)	Large accumulation to the liver	MRI agent for liver/spleen lesions	Intravenous
Ferumoxtran-10/Combidex®/ Sinerem® (US-SPION) 2005 (FDA) (Holland)	Dextran-coated SPION (D_H = 30 nm)	Lymphotropic MR contrast agent. High surface area allows faster absorption and increases the bioavailability	Imaging agent for lymph node metastases	Intravenous
GastroMARK™/Umirem®/ Ferumoxsil (SPION) 2009 (FDA)	Siloxane coated non-stoichiometric SPION (D_H = >300 nm)	Superparamagnetic character	MRI contrast media for imaging the delineation of the bowel from other organs and tissues	Oral

Formulation (Drug) Approved year (Agency)	Material composition	Advantages of nanoformulation	Indications	Route of administration
NanoTherm® (US-SPION) 2013 (EMA)	Aminosilane-coated SPION (D_H = 15 nm)	Exposure to a magnetic field changes SPION polarity 100,000 times per second resulting in high core temperature	Recurrent glioblastoma multiforme, prostate, and pancreatic cancer	Intratumoral
Resovist®/Cliavist®/ Ferucarbotran (US-SPION) 2011 (EMA)	Carboxy dextran coated SPION (D_H = 20 nm)	Liver specific accumulation	Contrast-enhanced MRI agent for liver lesions	Intravenous

Abbreviations: ADCC: Antibody-dependent cellular cytotoxicity; AML: Acute myeloid leukemia; cAC10-MMAE: Chimeric anti-CD30 mAb conjugated to monomethyl auristatin E; Calich *N*-acetyl-γ-calicheamicin; CalichDMH: *N*-acetyl-γ-calicheamicin dimethyl hydrazide; Chol: Cholesterol; COP1: Copolymer 1; DCGI: Drugs controller general of India; DM1: Maytansine derivative; DOPC: 1,2-dioleoyl-*sn*-glycero-3-phosphocholine; DOPSE: 1,2-dioleoyl-*sn*-glycero-3-phospho-L-serine; DPPG: 1,2-dipalmitoyl-*sn*-glycero-3-phosphatidylglycerol; DSPC: 1,2-distearoyl-*sn*-glycero-phosphatidylcholine; DSPE-PEG: 1,2-distearoyl-*sn*-glycero-3-phosphoethanolamine-conjugated PEG; EOC: Epithelial ovarian cancer; Glatiramer: Random copolymer of L-glutamate, L-alanine, L-lysine, and L-tyrosine with an average molar fraction of 0.141, 0.427, 0.095, and 0.338, respectively; HIV/AIDS: Human immunodeficiency virus/Acquired immunodeficiency syndrome; HfO_2: Hafnium oxide; HL: Hodgkin's lymphoma; HNSCC: Head and neck squamous cell carcinoma; HSPC: Hydrogenated soya phosphatidylcholine; IL-2: Interleukin 2; KS: Kaposi's sarcoma; MBP: Myelin basic protein; MS: Multiple sclerosis; MX-DTPA: 1-*p*-isothiocyanatobenzyl 3-methyl- and 1-*p*-isothiocyanatobenzyl 4-methyldiethylenetriamine pentaacetic acid; MLV: Multilamellar vesicle; *Mw*: Molecular wieght; NCS: Neocarzinostatin; NSCLC: Non-small cell lung carcinoma; OLV: Oligolamellar vesicle; PEG-PDLLA: PEG-poly(D,L-lactide); PGA: Poly-L-glutamic acid; P-gp: P-glycoprotein; Ph-ALL: Philadelphia chromosome-negative (Ph-) acute lymphoblastic leukemia (ALL); PLGA: Poly(D,L-lactic-*co*-glycolic acid); PMSA: Poly(styrene-*co*-maleic acid) (SMA); r/MBC: Recurrent/metastatic breast cancer; sALCL: Systemic anaplastic large cell lymphoma; SMCC: Succinimidyl 4-[*N*-maleimidomethyl] cyclohexane-1-carboxylate; SPIONs: Superparamagnetic iron oxide nanoparticles; SUV: Small unilamellar vesicle; XMeNa: *N*-(all-trans-retinoyl)-L-cysteic acid methyl ester; 13XMeNa: *N*-(13-cis-retinoyl)-L-cysteic acid methyl ester; US-SPIONs: Ultra small SPIONs; vWF: von Willebrand factor; ^{90}Y: Radioisotope yttrium-90.

We have briefly introduced all these types focusing on their fabrication process, stability aspects in the bloodstream, and representative experimental results of in vitro cells as well as in vivo tumor mice models. We have also highlighted DDSs that are tested in clinical phases for cancer chemotherapy and approved for commercialization.

In the past few decades, DDS has been at the forefront of cancer chemotherapy and a few DDSs composed of liposomal formulations are available in the market for cancer chemotherapy such as Doxil (DOX), Onivyde (irinotecan), and Vyxeos (daunorubicin plus cytarabine). Polymeric micelles showed an excellent anticancer effect in laboratory experiments and early clinical tests, however, their efficacy in humans is still under investigation. Now is the time to push forward with the practical aspects of "nanomedicine." A rational and realistic approach with advanced materials and refine treatments should be immediately implemented in cancer nanomedicine. Along this line, patient stratification, drug selection, combination therapies, and immunomodulation are some of the plausible future directions to boost nanomedicine performance [210]. For instance, combination therapy has enormous potential due to the possibility of co-encapsulation of multiple drugs/imaging reagents in the same DDS with stimuli responsive controlled drug release. A combination of detection, diagnosis, and treatment using a single therapeutic method is particularly important for complete tumor eradication due to the easy proliferation of cancers from a minute amount of tumor tissues left at the treated site. In fact, the combination therapies using nanomedicine can be considered as a "seed" for the realization of the visionary goal "In-Body Hospital" which is an imaginary futuristic hospital consisting of smart nanomachines operating inside the body [217].

References

1. Sun T, Zhang YS, Pang B, Hyun DC, Yang M, Xia Y. Engineered Nanoparticles for Drug Delivery in Cancer Therapy. *Angew. Chem. Int. Ed. Engl.* 53, **2014**, 12320–12364.
2. Sun J, Wei Q, Zhou Y, Wang J, Liu Q, Xu H. A Systematic Analysis of FDA-Approved Anticancer Drugs. *BMC Syst. Biol.* 11, **2017**, 87.

3. Allen TM, Cullis PR. Drug Delivery Systems: Entering the Mainstream. *Science*. 303, **2004**, 1818–1822.
4. Park K. The Controlled Drug Delivery Systems: Past Forward and Future Back. *J. Control. Release*. 190, **2014**, 3–8.
5. Anselmo AC, Mitragotri S. An Overview of Clinical and Commercial Impact of Drug Delivery Systems. *J. Control. Release*. 190, **2014**, 15–28.
6. Langer R. Controlled Release of a Therapeutic Protein. *Nat. Med*. 2, **1996**, 742–743.
7. Strebhardt K, Ullrich A. Paul Ehrlich's Magic Bullet Concept: 100 Years of Progress. *Nat. Rev. Cancer*. 8, **2008**, 473–480.
8. Jatzkewitz H. Peptamin (glycyl-l-leucyl-mescaline) Bound to Blood Plasma Expander (polyvinylpyrrolidone) as a New Depot Form of a Biologically Active Primary Amine (mescaline). *Z. Naturforsch*. 10B, **1955**, 27–31.
9. Folkman J, Long DM. The Use of Silicone Rubber as a Carrier for Prolonged Drug Therapy. *J. Surg. Res*. 4, **1964**, 139–142.
10. Bangham AD, Horne RW. Negative Staining of Phospholipids and Their Structural Modification by Surface-Active Agents as Observed in the Electron Microscope. *J. Mol. Biol*. 8, **1964**, 660–668.
11. Bangham AD, Standish MM, Watkins JC. Diffusion of Univalent Ions Across the Lamellae of Swollen Phospholipids. *J. Mol. Biol*. 13, **1965**, 238–252.
12. Shi J, Kantoff PW, Wooster R, Farokhzad OC. Cancer Nanomedicine: Progress, Challenges and Opportunities. *Nat. Rev. Cancer*. 17, **2017**, 20–37.
13. Gregoriadis G. Drug Entrapment in Liposomes. *FEBS Lett*. 36, **1973**, 292–296.
14. Kopf H, Joshi RK, Soliva M, Speiser P. Study on Micelle Polymerization in the Presence of Low Molecular-Weight Drugs. 1. Production and Isolation of Nanoparticles, Residual Monomer Determination, Physical–Chemical Data. *Pharm. Ind*. 38, **1976**, 281–284.
15. Langer R, Folkman J. Polymers for the Sustained Release of Proteins and Other Macromolecules. *Nature*. 263, **1976**, 797–800.
16. Barenholz Y. Doxil®—The First FDA-Approved Nano-drug: Lessons Learned. *J. Control. Release*. 160, **2012**, 117–134.
17. Couvreur P. Nanoparticles in Drug Delivery: Past, Present and Future. *Adv. Drug Delivery Rev*. 65, **2013**, 21–23.

18. Ferrari M. Cancer Nanotechnology: Opportunities and Challenges. *Nat. Rev. Cancer.* 5, **2005**, 161–171.
19. Davis ME, Chen ZG, Shin DM. Nanoparticle Therapeutics: An Emerging Treatment Modality for Cancer. *Nat. Rev. Drug Discov.* 7, **2008**, 771–782.
20. Lammers T, Aime S, Hennink WE, Storm G, Kiessling F. Theranostic Nanomedicine. *Acc. Chem. Res.* 44, **2011**, 1029–1038.
21. Liu J, Lee H, Huesca M, Young A, Allen C. Liposome Formulation of a Novel Hydrophobic Aryl-Imidazole Compound for Anti-Cancer Therapy. *Cancer Chemother. Pharmacol.* 58, **2006**, 306–318.
22. Dirisala A, Uchida S, Tockary TA, Yoshinaga N, Li J, Osawa S, Gorantla L, Fukushima S, Osada K, Kataoka K. Precise Tuning of Disulphide Crosslinking in mRNA Polyplex Micelles for Optimising Extracellular and Intracellular Nuclease Tolerability. *J. Drug Target.* 27, **2019**, 670–680.
23. Liong M, Lu J, Kovochich M, Xia T, Ruehm SG, Nel AE, Tamanol F, Zink JI. Multifunctional Inorganic Nanoparticles for Imaging, Targeting, and Drug Delivery. *ACS Nano.* 2, **2008**, 889–896.
24. Bao G, Mitragotri S, Tong S. Multifunctional Nanoparticles for Drug Delivery and Molecular Imaging. *Annu. Rev. Biomed. Eng.* 15, **2013**, 253–282.
25. Owens III DE, Peppas NA. Opsonization, Biodistribution, and Pharmacokinetics of Polymeric Nanoparticles. *Int. J. Pharm.* 307, **2006**, 93–102.
26. Petros RA, DeSimone JM. Strategies in the Design of Nanoparticles for Therapeutic Applications. *Nat. Rev. Drug Discov.* 9, **2010**, 615–627.
27. Kataoka K, Harada A, Nagasaki Y. Block Copolymer Micelles for Drug Delivery: Design, Characterization and Biological Significance. *Adv. Drug Deliv. Rev.* 47, **2001**, 113–131.
28. Kabanov AV, Batrakova EV, Alakhov VY. Pluronic® Block Copolymers as Novel Polymer Therapeutics for Drug and Gene Delivery. *J. Control Release.* 82, **2002**, 189–212.
29. Choi HS, Liu W, Misra P, Tanaka E, Zimmer JP, Ipe BI, Bawendi MG, Frangioni JV. Renal Clearance of Quantum Dots. *Nat. Biotechnol.* 25, **2007**, 1165–1170.
30. Anraku Y, Kishimura A, Kobayashi A, Oba M, Kataoka K. Size-Controlled Long-Circulating PIC some as a Ruler to Measure Critical Cut-Off Disposition Size into Normal and Tumor Tissues. *Chem. Commun.* 47, **2011**, 6054–6056.

31. Nel AE, Madler L, Velegol D, Xia T, Hoek EM, Somasundaran P, Klaessig F, Castranova V, Thompson M. Understanding Biophysicochemical Interactions at the Nano–Bio Interface. *Nat. Mater.* 8, **2009,** 543–557.

32. Rodriguez PL, Harada T, Christian DA, Pantano DA, Tsai RK, Discher DE. Minimal "Self" Peptides That Inhibit Phagocytic Clearance and Enhance Delivery of Nanoparticles. *Science.* 339, **2013**, 971–975.

33. Lynch I, Dawson KA. Protein-Nanoparticle Interactions. *Nano Today.* 3, **2008,** 40–47.

34. Dirisala A, Uchida S, Toh K, Li J, Osawa S, Tockary TA, Liu X, Abbasi S, Hayash K, Mochida Y, Fukushima S, Kinoh H, Osada K, Kataoka K. Transient Stealth Coating of Liver Sinusoidal Wall by Anchoring Two-Armed PEG for Retargeting Nanomedicines. *Sci. Adv.* 6, **2020**, eabb8133.

35. Matsumura Y, Maeda H. A New Concept for Macromolecular Therapeutics in Cancer Chemotherapy: Mechanism of Tumoritropic Accumulation of Proteins and the Antitumor Agent Smancs. *Cancer Res.* 46, **1986**, 6387–6392.

36. Gerlowski LE, Jain RK. Microvascular Permeability of Normal and Neoplastic Tissues. *Microvasc. Res.* 31, **1986**, 288–305.

37. Hashizume H, Baluk P, Morikawa S, McLean JW, Thurston G, Roberge S, Jain RK, McDonald DM. Openings Between Defective Endothelial Cells Explain Tumor Vessel Leakiness. *Am. J. Pathol.* 156, **2000**, 1363–1380.

38. Gabizon A, Catane R, Uziely B, Kaufman B, Safra T, Cohen R, Martin F, Huang A, Barenholz Y. Prolonged Circulation Time and Enhanced Accumulation in Malignant Exudates of Doxorubicin Encapsulated in Polyethylene-glycol Coated Liposomes. *Cancer Res.* 54, **1994**, 987–992.

39. Cabral H, Matsumoto Y, Mizuno K, Chen Q, Murakami M, Kimura M, Terada Y, Kano MR, Miyazono K, Uesaka M, Nishiyama N, Kataoka K. Accumulation of Sub-100 nm Polymeric Micelles in Poorly Permeable Tumours Depends on Size. *Nat. Nanotechnol.* 6, **2011**, 815–823.

40. Tang L, Yang X, Yin Q, Cai K, Wang H, Chaudhury I, Yao C, Zhou Q, Kwon M, Hartman JA, Dobrucki IT. Investigating the Optimal Size of Anticancer Nanomedicine. *Proc. Natl. Acad. Sci. U.S.A.* 111, **2014**, 15344–15349.

41. Wang J, Mao W, Lock LL, Tang J, Sui M, Sun W, Cui H, Xu D, Shen Y. The Role of Micelle Size in Tumor Accumulation, Penetration, and Treatment. *ACS Nano.* 9, **2015**, 7195–7206.

42. Matsumoto Y, Nichols JW, Toh K, Nomoto T, Cabral H, Miura Y, Christie RJ, Yamada N, Ogura T, Kano MR, Matsumura Y, Nishiyama N, Yamasoba T, Bae YH, Kataoka K. Vascular Bursts Enhance Permeability of Tumour Blood Vessels and Improve Nanoparticle Delivery. *Nat. Nanotechnol.* 11, **2016**, 533–538.

43. Netti PA, Berk DA, Swartz MA, Grodzinsky AJ, Jain RK. Enhancement of Fluid Filtration Across Tumor Vessels: Implication for Delivery of Macromolecules. *Proc. Natl. Acad. Sci. U.S.A.* 96, **1999**, 3137–3142.

44. Fang J, Islam W, Maeda H. Exploiting the Dynamics of the EPR Effect and Strategies to Improve the Therapeutic Effects of Nanomedicines by Using EPR Effect Enhancers. *Adv. Drug Deliv. Rev.* **2020** (DOI: 10.1016/j.addr.2020.06.005) In press.

45. Miura Y, Takenaka T, Toh K, Wu S, Nishihara H, Kano MR, Ino Y, Nomoto T, MatsumotoY, Koyama H, Cabral H, Nishiyama N, Kataoka K. Cyclic RGD-Linked Polymeric Micelles for Targeted Delivery of Platinum Anticancer Drugs to Glioblastoma through the Blood–Brain Tumor Barrier. *ACS Nano.* 7, **2013**, 8583–8592.

46. Miyata K, Nishiyama N, Kataoka K. Rational Design of Smart Supramolecular Assemblies for Gene Delivery: Chemical Challenges in the Creation of Artificial Viruses. *Chem. Soc. Rev.* 41, **2012**, 2562–2574.

47. Chou LYT, Ming K, Chan WCW. Strategies for the Intracellular Delivery of Nanoparticles. *Chem. Soc. Rev.* 40, **2011**, 233–245.

48. Murthy N, Campbell J, Fausto N, Hoffman AS, Stayton PS. Bioinspired pH-Responsive Polymers for the Intracellular Delivery of Biomolecular Drugs. *Bioconjugate Chem.* 14, **2003**, 412– 419.

49. Whitehead KA, Langer R, Anderson DG. Knocking Down Barriers: Advances in siRNA Delivery. *Nat. Rev. Drug Discov.* 8, **2009**, 129–138.

50. Zielonka J, Joseph J, Sikora A, Hardy M, Ouari O, Vasquez-Vivar J, Cheng G, Lopez M, Kalyanaraman B. Mitochondria-Targeted Triphenylphosphonium-Based Compounds: Syntheses, Mechanisms of Action, and Therapeutic and Diagnostic Applications. *Chem. Rev.* 117, **2017**, 10043–10120.

51. Lee JH, Yeo Y. Controlled Drug Release from Pharmaceutical Nanocarriers. *Chem. Eng. Sci.* 125, **2015**, 75–84.

52. Kamaly N, Yameen B, Wu J, Farokhzad OC. Degradable Controlled-Release Polymers and Polymeric Nanoparticles: Mechanisms of Controlling Drug Release. *Chem. Rev.* 116, **2016**, 2602–2663.

53. Langer R, Peppas N. Chemical and Physical Structure of Polymers as Carriers for Controlled Release of Bioactive Agents: A Review. *J. Macromol. Sci. Part C.* 23, **1983**, 61–126.

54. Uhrich KE, Cannizzaro SM, Langer RS, Shakesheff KM. Polymeric Systems for Controlled Drug Release. *Chem. Rev.* 99, **1999**, 3181–3198.

55. Peppas NA, Bures P, Leobandung W, Ichikawa H. Hydrogels in Pharmaceutical Formulations. *Eur. J. Pharm. Biopharm.* 50, **2000**, 27–46.

56. Korsmeyer RW, Peppas NA. Solute and Penetrant Diffusion in Swellable Polymers. III. Drug release from glassy poly5HEMA-*co*-NVP copolymers. *J. Control. Release.* 1, **1984**, 89–98.

57. Davoodi P, Lee LY, Xu Q, Sunil V, Sun V, Sun Y, Soh S, Wang C-H. Drug Delivery Systems for Programmed and On-Demand Release. *Adv. Drug Deliv. Rev.* 132, **2018**, 104–138.

58. Ringsdorf H. Structure and Properties of Pharmacologically Active Polymers. *J. Polym. Sci. C.* 51, **1975**, 135–153.

59. Ekladious I, Colson YL, Grinstaff MW. Polymer–Drug Conjugate Therapeutics: Advances, Insights and Prospects. *Nat. Rev. Drug Discov.* 18, **2019**, 273–294.

60. Mura S, Nicolas J, Couvreur P. Stimuli-responsive Nanocarriers for Drug delivery. *Nat. Mater.* 12, **2013**, 991–1003.

61. Fleige E, Quadir MA, Haag R. Stimuli-Responsive Polymeric Nanocarriers for the Controlled Transport of Active Compounds: Concepts and Applications. *Adv. Drug Deliv. Rev.* 64, **2012**, 866–884.

62. Lee Y, Kataoka K. Biosignal-Sensitive PolyIon Complex Micelles for the Delivery of Biopharmaceuticals. *Soft Matter.* 5, **2009**, 3810–3817.

63. Tian L, Bae YH. Cancer Nanomedicines Targeting Tumor Extracellular pH. *Colloids Surf. B Biointerfaces.* 99, **2012**, 116–126.

64. Kwon S, Ko H, You DG, Kataoka K, Park JH. Nanomedicines for Reactive Oxygen Species Mediated Approach: An Emerging Paradigm for Cancer Treatment. *Acc. Chem. Res.* 52, **2019**, 1771–1782.

65. de La Rica R, Aili D, Stevens MM. Enzyme-Responsive Nanoparticles for Drug Release and Diagnostics. *Adv. Drug Deliv. Rev.* 64, **2012**, 967–978.

66. Sun W, Gu Z. ATP-Responsive Drug Delivery Systems. *Expert Opin. Drug Deliv.* 13, **2016**, 311–314.

67. Li J, Dirisala A, Ge Z, Wang Y, Yin W, Ke W, Toh K, Xie J, Matsumoto Y, Anraku Y, Osada K, Kataoka K. Therapeutic Vesicular Nanoreactors with Tumor-Specific Activation and Self-Destruction for Synergistic Tumor Ablation. *Angew. Chem. Int. Ed. Engl.* 56, **2017**, 14025–14030.

68. Vemuri S, Rhodes CT. Preparation and Characterization of Liposomes as Therapeutic Delivery Systems: A Review. *Pharm. Acta Helv.* 70, **1995**, 95–111.

69. Kirby C, Clarke J, Gregoriadis G. Effect of the Cholesterol Content of Small Unilamellar Liposomes on Their Stability In Vivo and In Vitro. *Biochem. J.* 186, **1980**, 591–598.

70. Juliano RL, Stamp D. The effect of Particle Size and Charge on the Clearance Rates of Liposomes and Liposome Encapsulated Drugs. *Biochem. Biophys. Res. Commun.* 63, **1975**, 651–658.

71. Johnson SM. The Effect of Charge and Cholesterol on the Size and Thickness of Sonicated Phospholipid Vesicles. *Biochim. Biophys. Acta.* 307, **1973**, 27–41.

72. Miller CR, Bondurant B, McLean SD, McGovern KA, O'Brien DF. Liposome-Cell Interactions In Vitro: Effect of Liposome Surface Charge on the Binding and Endocytosis of Conventional and Sterically Stabilized Liposomes. *Biochemistry.* 37, **1998**, 12875–12883.

73. Campbell RB, Fukumura D, Brown EB, Mazzola LM, Izumi Y, Jain RK, Torchilin VP, Munn LL. Cationic Charge Determines the Distribution of Liposomes Between the Vascular and Extravascular Compartments of Tumors. *Cancer Res.* 62, **2002**, 6831– 6836.

74. Bangham A, De Gier J, Greville G. Osmotic Properties and Water Permeability of Phospholipid Liquid Crystals. *Chem. Phys. Lipids.* 1, **1967**, 225–246.

75. Reeves JP, Dowben RM. Formation and Properties of Thin-Walled Phospholipid Vesicles. *J. Cell. Physiol.* 73, **1969**, 49–60.

76. Biswas S, Dodwadkar NS, Deshpande PP, Torchilin VP. Liposomes Loaded with Paclitaxel and Modified with Novel Triphenylphosphonium-PEG-PE Conjugate Possess Low Toxicity, Target Mitochondria and Demonstrate Enhanced Antitumor Effects In Vitro and In Vivo. *J. Control. Release.* 159, **2012**, 393–402.

77. Hope MJ, Bally MB, Webb G, Cullis PR. Production of Large Unilamellar Vesicles by a Rapid Extrusion Procedure: Characterization of Size Distribution, Trapped Volume and Ability to Maintain a Membrane Potential. *Biochim. Biophys. Acta.* 812, **1985**, 55–65.

78. Popa R, Vranceanu M, Nikolaus S, Nirschl H, Leneweit, G. Entrance Effects at Nanopores of Nanocapsules Functionalized with Poly(ethylene glycol) and Their Flow Through Nanochannels. *Langmuir*. 24, **2008**, 13030–13036.

79. Jahn A, Vreeland WN, Gaitan M, Locascio LE. Controlled Vesicle Self-Assembly in Microfluidic Channels with Hydrodynamic Focusing. *J. Am. Chem. Soc.* 126, **2004**, 2674–2675.

80. Szoka F Jr, Papahadjopoulos D. Procedure for Preparation of Liposomes with Large Internal Aqueous Space and High Capture by Reverse-Phase Evaporation. *Proc. Natl. Acad. Sci. U.S.A.* 75, **1978**, 4194–4198.

81. Otake K, Shimomura T, Goto T, Imura T, Furuya T, Yoda S, Takebayashi Y, Sakai H, Abe M. Preparation of Liposomes Using an Improved Supercritical Reverse Phase Evaporation Method. *Langmuir*. 22, **2006**, 2543–2550.

82. Hwang SH, Maitani Y, Qi XR, Takayama K, Nagai T. Remote Loading of Diclofenac, Insulin and Fluorescein Isothiocyanate Labeled Insulin into Liposomes by pH and Acetate Gradient Methods. *Int. J. Pharm.* 179, **1999**, 85–95.

83. Dos Santos N, Cox KA, McKenzie CA, van Baarda F, Gallagher RC, Karlsson G, Edwards K, Mayer LD, Allen C, Bally MB. pH Gradient Loading of Anthracyclines into Cholesterol-Free Liposomes: Enhancing Drug Loading Rates Through Use of Ethanol. *Biochim. Biophys. Acta*. 1661, **2004**, 47–60.

84. Abraham SA, Edwards K, Karlsson G, Hudon N, Mayer LD, Bally MB. An Evaluation of Tansmembrane Ion Gradient-Mediated Encapsulation of Topotecan within lLiposomes. *J. Control. Release*. 96, **2004**, 449–461.

85. Han HD, Lee A, Song CK, Hwang T, Seong H, Lee CO, Shin BC. In Vivo Distribution and Antitumor Activity of Heparin-Stabilized Doxorubicin-Loaded Liposomes. *Int. J. Pharm.* 313, **2006**, 181–188.

86. Barenholz Y. Doxil–The first FDA-Approved Nano-drug: Lessons Learned. *J. Control. Release*. 160, **2012**, 117–134.

87. Safinya CR. Structures of Lipid-DNA Complexes: Supra-Molecular Assembly and Gene Delivery. *Curr. Opin. Struct. Biol.* 11, **2001**, 440–448.

88. Matsuura M, Yamazaki Y, Sugiyama M, Kondo M, Ori H, Nango M, Oku N. Polycation Liposome-Mediated Gene Transfer In Vivo. *Biochim. Biophys. Acta*. 1612, **2003**, 136–143.

89. Mohammed AR, Bramwell VW, Coombes AG, Perrie Y. Lyophilization and Sterilisation of Liposomal Vaccines to Produce Stable and Sterile products. *Methods.* 40, **2006**, 30–38.

90. Zuidam NJ, Lee SS, Crommelin DJ. Gamma-Irradiation of Non-Frozen, Frozen, and Freeze-Dried Liposomes. *Pharm. Res.* 12, **1995**, 1761–1768.

91. Toh MR, Chiu GN. Liposomes as Sterile Preparations and Limitations of Sterilization Techniques in Liposomal Manufacturing. *Asian J. Pharm. Sci.* 8, **2013**, 88–95.

92. Torchilin V. Tumor Delivery of Macromolecular Drugs Based on the EPR Effect. *Adv. Drug Deliv. Rev.* 63, **2011**, 131–135.

93. Yamada A, Taniguchi Y, Kawano K, Honda T, Hattori Y, Maitani Y. Design of Folate-Linked Liposomal Doxorubicin to its Antitumor Effect in Mice. *Clin. Cancer Res.* 14, **2008**, 8161–8168.

94. Gabizon A, Tzemach D, Gorin J, Mak L, Amitay Y, Shmeeda H, Zalipsky S. Improved Therapeutic Activity of Folate-Targeted Liposomal Doxorubicin in Folate Receptor-Expressing Tumor Models. Cancer Chemother. *Pharmacol.* 66, **2010**, 43–52.

95. Sofou S, Sgouros G. Antibody-Targeted Liposomes in Cancer Therapy and Imaging. *Expert Opin. Drug Deliv.* 5, **2008**, 189–204.

96. Maruyama K. PEG-Immunoliposome. *Biosci. Rep.* 22, **2002**, 251–266.

97. Bechara C, Sagan S. Cell-Penetrating Peptides: 20 years Later, Where do We Stand? *FEBS Lett.* 587, **2013**, 1693–1702.

98. Torchilin VP. Cell penetrating Peptide-Modified Pharmaceutical Nanocarriers for Intracellular Drug and Gene Delivery. *Biopolymers.* 90, **2008**, 604–610.

99. Copolovici DM, Langel K, Eriste E, Langel U. Cell-Penetrating Peptides: Design, Synthesis, and Applications. *ACS Nano.* 8, **2014**, 1972–1994.

100. Kloss A, Henklein P, Siele D, Schmolke M, Apcher S, Kuehn L, Sheppard PW, Dahlmann B. The Cell-Penetrating Peptide Octa-Arginine is a Potent Inhibitor of Proteasome Activities. *Eur. J. Pharm. Biopharm.* 72, **2009**, 219–225.

101. Biswas S, Dodwadkar NS, Deshpande PP, Parab S, Torchilin VP. Surface Functionalization of Doxorubicin-Loaded Liposomes with Octa-Arginine for Enhanced Anticancer Activity. *Eur. J. Pharm. Biopharm.* 84, **2013**, 517–525.

102. Biswas S, Deshpande PP, Perche F, Dodwadkar NS, Sane SD, Torchilin VP. Octa-Arginine-Modified Pegylated Liposomal Doxorubicin: An

Effective Teatment Strategy for Non-Small Cell Lung Cancer. *Cancer Lett.* 335, **2013**, 191–200.

103. Wust P, Hildebrandt B, Sreenivasa G, Rau B, Gellermann J, Riess H, Felix R, Schlag PM. Hyperthermia in Combined Treatment of Cancer. *Lancet Oncol.* 3, **2002**, 487–497.

104. Mills JK, Needham D. Lysolipid Incorporation in Dipalmitoylphosphatidylcholine Bilayer Membranes Enhances the Ion Permeability and Drug Release Rates at the Membrane Phase Transition. *Biochim. Biophys. Acta.* 1716, **2005**, 77–96.

105. Banno B, Ickenstein LM, Chiu GN, Bally MB, Thewalt J, Brief E, Wasan EK. The Functional Roles of Poly(ethylene glycol)-Lipid and Lysolipid in the Drug Retention and Release from Lysolipid-Containing Thermosensitive Liposomes In Vitro and In Vivo. *J. Pharm. Sci.* 99, **2010**, 2295–2308.

106. Tagami T, Ernsting MJ, Li SD. Efficient Tumor Regression by a Single and Low Dose Treatment with a Novel and Enhanced Formulation of Thermosensitive Liposomal Doxorubicin. *J. Control. Release.* 152, **2011**, 303–309.

107. Chen J, He CQ, Lin AH, Gu W, Chen ZP, Li W, Cai BC. Thermosensitive Liposomes with Higher Phase Transition Temperature for Targeted Drug Delivery to Tumor. *Int. J. Pharm.* 475, **2014**, 408–415.

108. Chen KJ, Chaung EY, Wey SP, Lin KJ, Cheng F, Lin CC, Liu HL, Tseng HW, Liu CP, Wei MC, Liu CM, Sung HW. Hyperthermia-Mediated Local Drug Delivery by a Bubble-Generating Liposomal System for Tumor-Specific Chemotherapy. *ACS Nano.* 8, **2014**, 5105–5115.

109. Kono K. Thermosensitive Polymer-Modified Liposomes. *Adv. Drug Deliv. Rev.* 53, **2001**, 307–319.

110. Yavlovich A, Singh A, Blumenthal R, Puri A. A Novel Class of Photo-Triggerable Liposomes Containing DPPC:DC8,9PC as Vehicles for Delivery of Doxorubicin to Cells. *Biochim. Biophys. Acta.* 1808, **2011**, 117–126.

111. Bressler N. Photodynamic Therapy of Sub Foveal Choroidal Neovascularization in Age-Related Macular Degeneration with Verteporfin: Two-Year Results of 2 Randomized Clinical Trials-Tap Report 2. *JAMA Ophthalmol.* 119, **2001**, 198–207.

112. Krishnamurthy S, Vaiyapuri R, g Zhang L, Chan. JM. Lipid-Coated Polymeric Nanoparticles for Cancer Drug Delivery. *Biomater. Sci.* 3, **2015**, 923–936.

113. Zhang L, Zhang L. Lipid–Polymer Hybrid Nanoparticles: Synthesis, Characterization and Applications. *Nano Life.* 1, **2010**, 163–173.

114. Zhang L, Chan JM, Gu FX, Rhee J-W, Wang AZ, Radovic-Moreno AF, Alexis F, Langer R, Farokhzad OC. Self-Assembled Lipid–Polymer Hybrid Nanoparticles: A Robust Drug Delivery Platform. *ACS Nano*. 2, **2008**, 1696–1702.

115. Kitajima H, Komizu Y, Ichihara H, Goto K, Ueoka R. Hybrid Liposomes Inhibit Tumor Growth and Lung Metastasis of Murine Osteosarcoma Cells. *Cancer Med*. 2, **2013**, 267–276.

116. Ichihara H, Nakagawa S, Matsuoka Y, Yoshida K, Matsumoto Y, Ueoka R. Nanotherapy with Hybrid Liposomes for Colorectal Cancer Along with Apoptosis In Vitro andIin Vivo. *Anticancer Res*. 34, **2014**, 4701–4708.

117. Matsumoto Y, Kato T, Iseki S, Suzuki H, Nakano K, Iwahara M, Ueoka R. Remarkably Enhanced Inhibitory Effects of Hybrid Liposomes on the Growth of Specific Tumor Cells. *Bioorg. Med. Chem. Lett*. 9, **1999**, 1937–1940.

118. Komizu Y, Matsumoto Y, Ueoka R. Membrane Targeted Chemotherapy with Hybrid Liposomes for Colon Tumor Cells Leading to Apoptosis. *Bioorg. Med. Chem. Lett*. 16, **2006**, 6131–6134.

119. Le Meins J-F, Schatz C, Lecommandoux S, Sandre O. Hybrid Polymer/Lipid Vesicles: State of the Art and Future Perspectives. *Mater. Today*. 16, **2013**, 397–402.

120. Lim SK, de Hoog H-P, Parikh AN, Nallani M, Liedberg B. Hybrid, Nanoscale Phospholipid/Block Copolymer Vesicles. *Polymers*. 5, **2013**, 1102–1114.

121. Quinteros D, Vicario-de-la-Torre M, Andres-Guerrero V, Palma S, Allemandi D, Herrero-Vanrell R, Molina-Martinez IT. Hybrid Formulations of Liposomes and Bioadhesive Polymers Improve the Hypotensive Effect of the Melatonin Analogue 5-MCA-NAT in Rabbit Eyes. *PLoS One*. 9, **2014**, e110344.

122. Nam J, Ha YS, Hwang S, Lee W, Song J, Yoo J, Kim S. pH-Responsive Gold Nanoparticles-In-Liposome Hybrid Nanostructures for Enhanced Systemic Tumor Delivery. *Nanoscale*. 5, **2013**, 10175–10178.

123. Matloob AH, Mourtas S, Klepetsanis P, Antimisiaris SG. Increasing the Stability of Curcumin in Serum with Liposomes or Hybrid Drug-In-Cyclodextrin-In-Liposome Systems: A comparative Study. *Int. J. Pharm*. 476, **2014**, 108–115.

124. Petralito S, Spera R, Pacelli S, Relucenti M, Familiari G, Vitalone A, Paolicelli P, Casadei MA. Design and Development of PEG-DMA Gel-In-Liposomes as a New Tool for Drug Delivery. *React. Funct. Polym*. 77, **2014**, 30–38.

125. Kaneda Y. Virosome: A Novel Vector to Enable Multi-Modal Strategies for Cancer Therapy. *Adv. Drug Deliv. Rev.* 64, **2012**, 730–738.

126. Kawaguchi Y, Miyamoto Y, Inoue T, Kaneda Y. Efficient Eradication of Hormone-Resistant Human Prostate Cancers by Inactivated Sendai Virus Particle. *Int. J. Cancer.* 124, **2009**, 2478–2487.

127. Kataoka K. Controlled Drug Delivery: The Next Generation, K. Park (ed.), ACS, Washington, DC, **1996**.

128. Daglar B, Ozgur E, Corman ME, Uzun L, Demirel GB. Polymeric Nanocarriers for Expected Nanomedicine: Current Challenges and Future Prospects. *RSC Adv.* 4, **2014**, 48639–48659.

129. Talelli M, Rijcken CJF, van Nostrum CF, Storm G, Hennink WE. Micelles Based on HPMA Copolymers. *Adv. Drug Deliv. Rev.* 62, **2010**, 231–239.

130. Cabral H, Miyata K, Osada K, Kataoka K. Block Copolymer Micelles in Nanomedicine Applications. *Chem. Rev.* 118, **2018**, 6844–6892.

131. Park JH, Saravanakumar G, Kim K, Kwon IC. Targeted Delivery of Low Molecular Drugs Using Chitosan and its Derivatives. *Adv. Drug Deliv. Rev.* 62, **2010**, 28–41.

132. Ossipov DA. Nanostructured Hyaluronic Acid-Based Materials for Active Delivery to Cancer. *Expert Opin. Drug Deliv.* 7, **2010**, 681–703.

133. Zhang X, Jackson JK, Burt HM. Development of Amphiphilic Diblock Copolymers as Micellar Carriers of Taxol. *Int. J. Pharm.* 132, **1996**, 195–206.

134. Wei Z, Hao J, Yuan S, Li Y, Juan W, Sha X, Fang X. Paclitaxel-Loaded Pluronic P123/F127 Mixed Polymeric Micelles: Formulation, Optimization and In Vitro Characterization. *Int. J. Pharm.* 376, **2009**, 176–185.

135. Kataoka K, Harada A, Nagasaki Y. Block Copolymer Micelles for Drug Delivery: Design, Characterization and Biological Significance. *Adv. Drug Deliv. Rev.* 47, **2001**, 113–131.

136. Itaka K, Kanayama N, Nishiyama N, Jang WD, Yamasaki Y, Nakamura K, Kawaguchi H, Kataoka K. Supra-Molecular Nanocarrier of siRNA from PEG-Based Block Catiomer Carrying Diamine Side Chain with Distinctive pKa Directed to Enhance Intracellular Gene Silencing. *J. Am. Chem. Soc.* 126, **2004**, 13612–13613.

137. Harada A, Kataoka K. Chain Length Recognition: Core-Shell Supramolecular Assembly from Oppositely Charged Block Copolymer. *Science.* 283, **1999**, 65–67.

138. Nakanishi T, Fukushima S, Okamoto K, Suzuki M, Matsumura Y, Yokoyama M, Okano T, Sakurai Y, Kataoka K. Development of the Polymer Micelle Carrier System for Doxorubicin. *J. Control. Release.* 74, **2001**, 295–302.

139. Hamaguchi T, Matsumura Y, Suzuki M, Shimizu K, Goda R, Nakamura I, Nakatomi I, Yokoyama M, Kataoka K, Kakizoe T. NK105, A Paclitaxel-Incorporating Micellar Nanoparticle Formulation, Can Extend In Vivo Antitumour Activity and Reduce the Neurotoxicity of Paclitaxel. *Br. J. Cancer.* 92, **2005**, 1240–1246.

140. van Nostrum CF. Polymeric Micelles to Deliver Photosensitizers for Photodynamic Therapy. *Adv. Drug Deliv. Rev.* 56, **2004**, 9–16.

141. Ideta R, Tasaka F, Jang W-D, Nishiyama N, Zhang G-D, Harada A, Yanagi Y, Tamaki, Y, Aida T, Kataoka K. Nanotechnology-Based Photodynamic Therapy for Neovascular Disease using a Supramolecular Nanocarrier Loaded with a Dendritic Photosensitizer. *Nano Lett.* 5, **2005**, 2426–2431.

142. Bae Y, Nishiyama N, Fukushima S, Koyama H, Matsumura Y, Kataoka K. Preparation and Biological Characterization of Polymeric Micelle Drug Carriers with Intracellular pH-Triggered Drug Release Property: Tumor Permeability, Controlled Subcellular Drug Distribution, and Enhanced In Vivo Antitumor Efficacy. *Bioconjugate Chem.* 16, **2005**, 122–130.

143. Takahashi A, Yamamoto Y, Yasunaga M, Koga Y, Kuroda J, Takigahira M, Harada M, Saito H, Hayashi T, Kato Y. Kinoshita T, Ohkohchi N, Hyodo I, Matsumura Y. NC-6300, An Epirubicin-Incorporating Micelle, Extends the Antitumor Effect and Reduces the Cardiotoxicity of Epirubicin. *Cancer Sci.* 104, **2013**, 920–925.

144. Mukai H, Kogawa T, Matsubara N, Naito Y, Sasaki M, Hosono A. A First-In-Human Phase 1 Study of Epirubicin-Conjugated Polymer Micelles (K-912/NC-6300) in Patients with Advanced or Recurrent Solid Tumors. *Invest. New Drugs.* 35, **2007**, 307–314.

145. Yen HC, Cabral H, Mi P, Toh K, Matsumoto Y, Liu X, Koori H, Kim A, Miyazaki K, Miura Y, Nishiyama N. Kataoka K. Light-induced cytosolic Activation of Reduction-Sensitive Camptothecin-Loaded Polymeric Micelles for Spatiotemporally Controlled In Vivo Chemotherapy. *ACS Nano.* 8, **2014**, 11591–11602.

146. Nishiyama N, Okazaki S, Cabral H, Miyamoto M, Kato Y, Sugiyama Y, Nishio K, Matsumura Y, Kataoka K. Novel Cisplatin-Incorporated Polymeric Micelles Can Eradicate Solid Tumors in Mice. *Cancer Res.* 63, **2003**, 8977–8983.

147. Quader S, Liu X, Chen Y, Mi P, Chida T, Ishii T, Miura Y, Nishiyama N, Cabral H, Kataoka K. cRGD Peptide-Installed Epirubicin-Loaded Polymeric Micelles for Effective Targeted Therapy Against Brain Tumors. *J. Control. Release.* 258, **2017**, 56–66.

148. Meng FH, Zhong ZY. Polymersomes Spanning from Nano- to Micro-scales: Advanced Vehicles for Controlled Drug Delivery and Robust Vesicles for Virus and Cell Mimicking. *J. Phys. Chem. Lett.* 2, **2011**, 1533–1539.

149. Ahmed F, Pakunlu RI, Brannan A, Bates F, Minko T, Discher DE. Biodegradable Polymersomes Loaded with Both Paclitaxel and Doxorubicin Permeate and Shrink Tumors, Inducing Apoptosis in Proportion to Accumulated Drug. *J. Control. Release.* 116, **2006**, 150–158.

150. Lee JS, Feijen J. Polymersomes for Drug Delivery: Design, Formation and Characterization. *J. Control. Release.* 161, **2012**, 473–483.

151. Chen W, Meng F, Cheng R, Zhong Z. pH-Sensitive Degradable Polymersomes for Triggered Release of Anticancer Drugs: A Comparative Study with Micelles. *J. Control. Release.* 142, **2010**, 40–46.

152. Khorsand B, Lapointe G, Brett C, Oh JK. Intracellular Drug Delivery Nanocarriers of Glutathione-Responsive Degradable Block Copolymers Having Pendant Disulfide Linkages. *Biomacromolecules.* 14, **2013**, 2103–2111.

153. Nam J, Vanderlick TK, Beales PA. Formation and Dissolution of Phospholipid Domains with Varying Textures in Hybrid Lipo-Polymersomes. *Soft Matter.* 8, **2012**, 7982–7988.

154. Liechty WB, Kryscio DR, Slaughter BV, Peppas NA. Polymers for Drug Delivery Systems. *Annu. Rev. Chem. Biomol. Eng.* 1, **2010**, 149–173.

155. Duncan R. The Dawning Era of Polymer Therapeutics. *Nat. Rev. Drug Discov.* 2, **2003**, 347–360.

156. Stirland DL, Nichols JW, Miura S, Bae YH. Mind the Gap: A Survey of How Cancer Drug Carriers Is Susceptible to the Gap Between Research and Practice. *J. Control. Release.* 172, **2013**, 1045–1064.

157. Seymour LW, Ferry DR, Kerr DJ, Rea D, Whitlock M, Poyner R, Boivin C, et al. Phase II Studies of Polymer-Doxorubicin (PK1, FCE28068) in the Treatment of Breast, Lung and Colorectal Cancer. *Int. J. Oncol.* 34, **2009**, 1629–1636.

158. Hoch U, Staschen CM, Johnson RK, Eldon MA. Nonclinical Pharmacokinetics and Activity of Etirinotecan Pegol (NKTR-102), A

Long-Acting Topoisomerase 1 Inhibitor, in Multiple Cancer Models. *Cancer Chemother. Pharmacol.* 74, **2014**, 1125–1137.

159. Chen GS, Jiang M. Cyclodextrin-Based Inclusion Complexation Bridging Supramolecular Chemistry and Macromolecular Self-Assembly. *Chem. Soc. Rev.* 40, **2011**, 2254–2266.

160. Wang Y, Wang H, Chen Y, Liu X, Jin Q, Ji J. Biomimetic Pseudopolyrotaxane Prodrug Micelles with High Drug Content for Intracellular Drug Delivery. *Chem. Commun.* 49, **2013**, 7123–7125.

161. Cheng J, Khin KT, Jensen GS, Liu A, Davis ME. Synthesis of Linear, β-Cyclodextrin-Based Polymers and Their Camptothecin Conjugates. *Bioconjugate Chem.* 14, **2003**, 1007–1017.

162. Paleos CM, Tsiourvas D, Sideratou Z, Tziveleka LA. Drug Delivery Using Multifunctional Dendrimers and Hyperbranched Polymers. *Expert Opin. Drug Deliv.* 7, **2010**, 1387–1398.

163. Sadekar S, Linares O, Noh G, Hubbard D, Ray A, Janát-Amsbury M, Peterson CM, Facelli J, Ghandehari H. Comparative Pharmacokinetics of PAMAM-OH Dendrimers and HPMA Copolymers in Ovarian Tumor-Bearing Mice. *Drug Deliv. Transl. Res.* 3, **2013**, 260–271.

164. Menjoge AR, Kannan RM Tomalia DA. Dendrimer Based Drug and Imaging Conjugates: Design Considerations for Nanomedical Applications. *Drug Discov. Today.* 15, **2010**, 171–185.

165. Stiribara SE, Frey H, Haag R. Dendritic Polymers in Biomedical Applications: From Potential to Clinical Use in Diagnostics and Therapy. *Angew. Chem., Int. Ed.* 41, **2002**, 1329–1334.

166. McNerny DQ, Leroueil PR, Baker JR. Understanding Specific and Nonspecific Toxicities: A Requirement for the Development of Dendrimer-Based Pharmaceuticals. *Nanomed. Nanobiotechnol.* 2, **2010**, 249–259.

167. Mintzer MA, Grinstaff MW. Biomedical Applications of Dendrimers: A Tutorial. *Chem. Soc. Rev.* 40, **2011**, 173–190.

168. Iezzi R, Guru BR, Glybina IV, Mishra MK, Kennedy A, Kannan RM. Dendrimer-Based Targeted Intravitreal Therapy for Sustained Attenuation of Neuroinflammation in Retinal Degeneration. *Biomaterials.* 33, **2012**, 979–988.

169. Zhou Z, Ma X, Murphy CJ, Jin E, Sun Q, Shen Y, Van Kirk EA, Murdoch WJ. Molecularly Precise Dendrimer-Drug Conjugates with Tunable Drug Release for Cancer Therapy. *Angew. Chem. Int. Ed.* 53, **2014**, 10949–10955.

170. Li H-J, Du J-Z, Du X-J, Xu C-F, Sun C-Y, Wang H-X, Cao Z-T, Yang X-Z, Zhu Y-H, Nie S, Wang J. Stimuli-Responsive Clustered Nanoparticles for Improved Tumor Penetration and Therapeutic Efficacy. *Proc. Natl. Acad. Sci. U.S.A.* 113, **2016**, 4164–4169.

171. Etrych T, Strohalm J, Chytil P, Černoch P. Starovoytova L, Pechar M, Ulbrich K. Biodegradable Star HPMA Polymer Conjugates of Doxorubicin for Passive Tumor Targeting. *Eur. J. Pharm. Sci.* 42, **2011**, 527–539.

172. Qiu L, Liu Q, Hong C-Y, Pan C-Y. Unimolecular Micelles of Camptothecin-Bonded Hyperbranched Star Copolymers via β-thiopropionate Linkage: Synthesis and Drug Delivery. *J. Mater. Chem. B.* 4, **2016**, 141–151.

173. Haxton KJ, Burt HM. Hyperbranched Polymers for Controlled Release of Cisplatin. *Dalton Trans.* 43, **2008**, 5872–5875.

174. Wu H, Yin T, Li K, Wang R, Chen Y, Jing L. Encapsulation Property of Hyperbranched Polyglycerols as Prospective Drug Delivery Systems. *Polym. Chem.* 9, **2018**, 300–306

175. Duro-Castano A, Movellan J, Vicent MJ. Smart Branched Polymer Drug Conjugates as Nano-sized Drug Delivery Systems. *Biomater. Sci.* 3, **2015**, 1321–1334.

176. Kolhe P, Khandare J, Pillai O, Kannan S, Lieh-Lai M, Kannan R. Hyperbranched Polymer-Drug Conjugates with High Drug Payload for Enhanced Cellular Delivery. *Pharm. Res.* 21, **2004**, 2185–2195.

177. Liu J, Pang Y, Chen J, Huang P, Huang W, Zhu X, Yan D. Hyperbranched Polydiselenide as a Self-Assembling Broad Spectrum Anticancer Agent. *Biomaterials.* 33, **2012**, 7765–7774.

178. Álvarez-Pérez M, Ali W, Marć MA, Handzlik J, Domínguez-Álvarez E. Selenides and Diselenides: A Review of Their Anticancer and Chemopreventive Activity. *Molecules.* 23, **2018**, 628.

179. Sharma K, Zolotarskaya OY, Wynne KJ, Yang H. Poly(ethylene glycol)-Armed Hyperbranched Polyoxetanes for Anticancer Drug Delivery. *J. Bioact. Compat. Polym.* 27, **2012**, 525–539.

180. Kavanda A, Anton N, Vandamme T, Serra CA, Chan-Seng D. Synthesis and Functionalization of Hyperbranched Polymers for Targeted Drug Delivery. *J. Control. Release.* 321, **2020**, 285–311.

181. Patra CR, Bhattacharya R, Mukhopadhyay D, Mukherjee P. Application of Gold Nanoparticles for Targeted Therapy in Cancer. *J. Biomed. Nanotechnol.* 4, **2008**, 99–132.

182. Sokolov K, Follen M, Aaron J, Pavlova I, Malpica A, Lotan R, et al. Real-Time Vital Optical Imaging of Precancer using Anti-

Epidermal Growth Factor Receptor Antibodies Conjugated to Gold Nanoparticles. *Cancer Res.* 63, **2003**, 1999–2004.

183. Mukherjee P, Bhattacharya R, Wang P, Wang L, Basu S, Nagy JA, et al. Antiangiogenic Properties of Gold Nanoparticles. *Clin. Cancer Res.* 11, **2005**, 3530–3534.

184. Moore CM, Pendse D, Emberton M. Photodynamic Therapy for Prostate Cancer—A Review of Current Status and Future Promise. *Nat. Clin. Pract. Urol.* 6, **2009**, 18–30.

185. El-Sayed IH, Huang X, El-Sayed MA. Surface Plasmon Resonance Scattering and Absorption of Anti-EGFR Antibody conjugated Gold Nanoparticles in Cancer Diagnostics: Applications in Oral Cancer. *Nano Lett.* 5, **2005**, 829–834.

186. Porcel E, Liehn S, Remita H, Usami N, Kobayashi K, Furusawa Y, et al. Platinum Nanoparticles: A Promising Material for Future Cancer Therapy? *Nanotechnology.* 21, **2010**, 85–103.

187. Roy I, Ohulchanskyy TY, Pudavar HE, Bergey EJ, Oseroff AR, Morgan J, et al. Ceramic-Based Nanoparticles Entrapping Water-insoluble Photosensitizing Anticancer Drugs: A Novel Drug-Ccarrier System for Photodynamic Therapy. *J. Am. Chem Soc.* 125, **2003**, 7860–7865.

188. Juzenas P, Chen W, Sun YP, et al. Quantum Dots and Nanoparticles for Photodynamic and Radiation Therapies of Cancer. *Adv. Drug Deliv. Rev.* 60, **2008**, 1600–1614.

189. Panwar N, Soehartono AM, Chan KK, Zeng S, Xu G, Qu J, et al. Nanocarbons for Biology and Medicine: Sensing, Imaging, and Drug Delivery. *Chem. Rev.* 119, **2019**, 9559–9656.

190. Miao W, Shim G, Lee S, Lee S, Choe YS, Oh Y-K. Safety and Tumor Tissue Accumulation of Pegylated Graphene Oxide Nanosheets for Co-Delivery of Anticancer Drug and Photosensitizer. *Biomaterials.* 34, **2013**, 3402–3410.

191. Wu H, Shi H, Wang Y, Jia X, Tang C, Zhang J, Yang S. Hyaluronic Acid Conjugated Graphene Oxide for Targeted Drug Delivery. *Carbon.* 69, **2014**, 379–389.

192. Song E, Han W, Li C, Cheng D, Li L, Liu L, Zhu G, Song Y, Tan W. Hyaluronic Acid-Decorated Graphene Oxide Nanohybrids as Nanocarriers for Targeted and pH-Responsive Anticancer Drug Delivery. *ACS Appl. Mater. Interfaces.* 6, **2014**, 11882–11890.

193. Foldvari M, Bagonluri M. Carbon Nanotubes as Functional Excipients for Nanomedicines: II. Drug Delivery and Biocompatibility Tissues. *Nanomedicine.* 4, **2008**, 183–200.

194. Bhatnagar I, Venkatesan J, Kim S-K. Polymer Functionalized Single Walled Carbon Nanotubes Mediated Drug Delivery of Gliotoxin in Cancer Cells. *J. Biomed. Nanotechnol.* 10, **2014**, 120–130.

195. Karousis N, Suarez-Martinez I, Ewels CP, Tagmatarchis N. Structure, Properties, Functionalization, and Applications of Carbon Nanohorns. *Chem. Rev.* 116, **2016**, 4850–4883.

196. Partha R, Mitchell LR, Lyon JL, Joshi PP, Conyers JL. Buckysomes: Fullerene-Based Nanocarriers for Hydrophobic Molecule Delivery. *ACS Nano.* 2, **2008**, 1950–1958.

197. Fan J, Fang G, Zeng F, Wang X, Wu S. Water-Dispersible Fullerene Aggregates as a Targeted Anticancer Prodrug with Both Chemo- and Photodynamic Therapeutic Actions. *Small.* 9, **2013**, 613–621.

198. Shi J, Zhang H, Wang L, Li L, Wang H, Wang Z, Li Z, Chen C, Hou L, Zhang C, Zhang Z. PEI-Derivatized Fullerene Drug Delivery Using Folate as a Homing Device Targeting to Tumor. *Biomaterials.* 34, **2013**, 251–261.

199. Zhang Y, Cui Z, Kong H, Xia K, Pan L, Li J, Sun Y, Shi J, Wang L, Zhu Y, Fan C. One-Shot Immunomodulatory Nanodiamond Agents for Cancer Immunotherapy. *Adv. Mater.* 28, **2016**, 2699–2708.

200. Li Y, Tong Y, Cao R, Tian Z, Yang, B, Yang P. In Vivo Enhancement of Anticancer Therapy Using Bare or Chemotherapeutic Drug-Bearing Nanodiamond Particles. *Int. J. Nanomed.* 9, **2014**, 1065–1082.

201. Zhao A, Chen Z, Zhao C, Gao N, Ren J, Qu X. Recent Advances in Bioapplications of C-Dots. *Carbon.* 85, **2015**, 309–327.

202. Della Rocca J, Liu D, Lin W. Nanoscale Metal-Organic Frameworks for Biomedical Imaging and Drug Delivery. *Acc. Chem. Res.* 44, **2011**, 957–68.

203. Rieter WJ, Pott KM, Taylor KML, Lin W. Nanoscale Coordination Polymers for Platinum-Based Anticancer Drug Delivery. *J. Am. Chem. Soc.* 130, **2008**, 11584–11585.

204. Liu DM, Poon C, Lu KD, He CB, Lin WB. Self-Assembled Nanoscale Coordination Polymers with Trigger Release Properties for Effective Anticancer Therapy. *Nat. Commun.* 5, **2014**, 5182.

205. Huxford RC, Dekrafft KE, Boyle WS, Liu D, Lin W. Lipid-Coated Nanoscale Coordination Polymers for Targeted Delivery of Antifolates to Cancer cells. *Chem. Sci.* 3, **2012**, 198–204.

206. He C, Liu D, Lin W. Nanomedicine Applications of Hybrid Nanomaterials Built from Metal–Ligand Coordination Bonds:

Nanoscale Metal–Organic Frameworks and Nanoscale Coordination Polymers. *Chem. Rev.* 115, **2015**, 11079–11108.

207. Lu K, He C, Lin W. Nanoscale Metal-Organic Framework for Highly Effective Photodynamic Therapy of Resistant Head and Neck Cancer. *J. Am. Chem. Soc.* 136, **2014**, 16712–16715.

208. He H, Liu L, Morin EE, Liu M, Schwendeman A. Survey of Clinical Translation of Cancer Nanomedicines-Lessons Learned from Successes and Failures. *Acc. Chem. Res.* 52, **2019**, 2445–2461.

209. Wolfram J, Ferrari M. Clinical Cancer Nanomedicine. *Nano Today.* 25, **2019**, 85–98.

210. van der Meel R, Sulheim E, Shi Y, Kiessling F, Mulder WJM, Lammers T. Smart cancer nanomedicine. *Nat. Nanotechnol.* 14, **2019**, 1007–1017.

211. Anselmo AC, Mitragotri S. Nanoparticles in the Clinic: An Update. *Bioeng. Transl. Med.* 4, **2019**, e10143.

212. Cabral H, Kataoka K. Progress of Drug-Loaded Polymeric Micelles into Clinical Studies. *J. Control. Release.* 190, **2014**, 465–476.

213. Varela-Moreira A, Shi Y, Fens MH, Lammers T, Hennink WE, Schiffelers RM. Clinical Application of Polymeric Micelles for the Treatment of Cancer. *Mater Chem. Front.* 1, **2017**, 1485–1501.

214. Germain M, Caputo F, Metcalfe S, Tosi G, Spring K, Aslund AKO, Pottier A, Schiffelers R, Ceccaldi A, Schmid R. Delivering the Power of Nanomedicine to Patients Today. *J. Control. Release.* 326, **2020**, 164–171.

215. Hashim P.K., Okuro K, Sasaki S, Hoashi Y, Aida T. Reductively Cleavable Nanocaplets for siRNA Delivery by Template-Assisted Oxidative polymerization. *J. Am. Chem. Soc.* 137, **2015**, 15608–15611.

216. Kohata A, Hashim P.K., Okuro K, Aida T. Transferrin appended Nanocaplet for Transcellular siRNA Delivery into Deep Tissues. *J. Am. Chem. Soc.* 141, **2019**, 2862–2866.

217. A Nano-Hospital in Everybody. Nanomedicine in Japan. *Nature.* https://www.nature.com/articles/d42473-018-00257-z (access date July 23, 2020).

Chapter 9

Magnetic Nanoparticles for Hyperthermia against Cancer

Sivakumar Balasubramanian and Allison J. Cowin

Future Industries Institute, University of South Australia, Mawson Lakes, Adelaide, SA 5095, Australia

siva.balasubramanian@unisa.edu.au; allison.cowin@unisa.edu.au

9.1 Introduction

The concept of administering heat to cure disease is very old.

"*Quae medicamenta non sanat; ferrum sanat. Quae ferrum non sanat; ignis sanat. Huae vero ignis non sanat; insanabilia reportari oportet*" (Those diseases which medicines do not cure, the knife cures; those, which the knife cannot cure, fire cures; and those which fire cannot cure, are to be reckoned wholly incurable)—from aphorism of Hippocrates of Kos, Western father of medicine (*ca.* 460 BC – *ca.* 370 BC), Hippocrates assumed that illnesses could be healed by slightly elevating patient's body temperature.

The term "hyperthermia" means the generation of heat and in literature is limited to the application of heat for therapeutic applications. Hyperthermia has been used alongside various

Bionanotechnology in Cancer: Diagnosis and Therapy
Edited by D. Sakthi Kumar and Aswathy Ravindran Girija

ISBN 978-981-4800-30-3 (Hardcover), 978-0-429-42291-1 (eBook)
www.jennystanford.com

cancer therapies such as chemotherapy, radiation therapy, immuno-therapy, gene therapy, and surgery. Hyperthermia is a normal or simulated process that raises the temperature of a body or body part, over threshold temperature set at a specific moment by thermoregulation pattern of an organism [1, 2]. When considering cancer therapies, hyperthermia is the artificial method of gradually raising body temperature by heat acquired from external sources to kill cancer cells or preventing the spread of cancer.

One of the major goals in oncology is the preferential killing of cancer cells without damaging healthy neighboring cells. Although there are various therapies, such as chemotherapy, radiotherapy, or surgery, available for the efficient treatment of cancers, none of them can treat the cancer without damaging normal cells. The destructive side effects of heavy doses of chemicals and radioactive agents adversely affect normal health of the subject during and after treatment. Existing therapies for cancer are extremely invasive, toxic, and painful accompanied with several side effects. Cancer recovery rates for several malignancies have not improved significantly over the last few decades. Hence, innovative treatment strategies, which are promising and less toxic, are highly recommended for improving the general health of cancer patients.

The prospective of hyperthermia as a successful cancer treatment strategy was first suggested following the finding that cancer cells are more sensitive to temperatures beyond 41 °C than normal cells [3, 4]. There are generally three sets of temperatures often employed in hyperthermia.

i. Temperature greater than 50 °C (up to 56 °C) is thermal ablation that is targeted to cause cells to undergo necrosis, coagulation, or carbonization. This is performed by direct heating techniques such as needle-type interstitial antenna, which induces hyperthermia locally. However, for larger areas of tissue, this technology is less effective.
ii. Moderate hyperthermia, which is usually termed as hyperthermia in general, employs temperatures approximately 41–46 °C and is related to anticancer effect both at cellular and tissue levels. Moderate hyperthermia results in activation of deterioration of several intra and

extracellular proteins, DNA cross-linking, initiation of apoptosis, multidrug resistance, and the expression of heat shock protein. However, the effectiveness of hyperthermia treatment significantly depends on the temperature generated, exposure, and features of particular cancer cells at the targeted sites of action.

iii. Diathermia, the third category refers to lower temperatures (<41 °C) for physiotherapy rather than cancer.

The National Cancer Institute distinguishes three types of hyperthermia based on the depth, site, and staging of the tumor such as (i) local hyperthermia, (ii) regional hyperthermia, and (iii) whole body hyperthermia (WBH).

i. **Local hyperthermia**: In this, heat is applied to a smaller area, such as a tumor, using several procedures that provide energy to heat up the tumor. For intracavital or intraluminal hyperthermia, heat is delivered through natural orifices, such as esophagus, prostate (via urethra), and rectum, to cancers. Whereas, interstitial hyperthermia is an invasive technique that is appropriate only for lesions (in the range of 5 cm).

ii. **Regional hyperthermia**: It is also known as body-part hyperthermia, and is often employed for larger areas, including organs, body cavity, and limb, which are heated with external applicators or regional perfusion. The three common approaches in this category are external applicator, which is used to heat deeply settled tumors; thermal perfusion, which is used to heat organs or limbs; and continuous hyperthermic peritoneal perfusion (CHPP) for cancers spread within abdominal cavity. CHPP is often performed along with chemotherapy perioperatively, distributing warm liquid with a temperature range of 41.0–42.5 °C with or without chemotherapeutic agents [5–8].

iii. **WBH**: It is used for the treatment of metastatic cancer such as melanoma and soft tissue sarcoma. Patients are subjected to general anesthesia during the treatment session and body is heated up to 39.5–41.0 °C around

3–4 h for moderate WBH or 42.0 °C for about 60 min in the case of extreme WBH. The advantage of WBH is the homogeneous thermal distribution compared to the other types of hyperthermia, but it is often associated with high risk of complications such as diarrhea, nausea, and vomiting. It also induces thermal stress to major organs like the heart, lungs, liver, or brain, resulting in unpredicted damage [9–13].

9.2 Biology of Hyperthermia and Cell Death

It has been observed in several studies that when cells or tissues are exposed to temperatures above 42 °C, it results in cell death. When cells are heated to temperature range of 41–47 °C that is mild to moderate hyperthermia, symbols of apoptosis are observed [13–15], whereas when the temperature is further increased to about 50 °C or above frank necrosis is observed accompanied with less apoptosis. When cells or tissues are exposed to lethal stimuli, such as extreme heat, cold, pressure, or any stress, it induces pro-apoptotic proteins to induce death-inducing signaling complexes such as caspase-8 or caspase-9. It can activate several apoptotic proteins, which initiate intracellular damages, most importantly mitochondrial damage that results in a cascade of events, eventually leading to cell death [15]. However, compared with apoptosis, necrosis is a faster mode to cause cell death. Necrosis is essentially based on denaturation of protein in the cells. Very high temperatures or contact with strong acids and bases mostly result in necrosis. It was reported in several research studies that high temperatures [16] can induce severe damage to cancer cells and sensitize the cells to other cancer treatments (radiotherapy, chemotherapy, and immunotherapy) with reduced or no damage to healthy tissues [12, 17–19]. As a result, hyperthermia generally employed as an adjuvant therapy, along with other cancer therapies [13, 20] for faster tumor regression.

A major characteristic of cancerous tissues is the poorly developed blood vessels and lower thermal resistance compared to normal tissue. Cancer cells are more vulnerable to heat than normal healthy cells, owing to their high metabolic rate. The blood vessels of normal tissue dilate when heated, which subsequently results in increased blood flow and decreased temperature. On the other hand, in tumors, the pathologic blood vessels do not dilate, resulting in the breakdown of few vessels. Hence, the induced heat is not dissipated by blood flow and the temperature inside the tumor tissue continues to make hyperthermia a promising cancer treatment modality. The overall result is that hyperthermia treatment makes cells sensitive to further radiotherapy or chemotherapy, so that the ionizing radiations induce re-oxygenation or enhance the effect of chemotherapeutics by boosting immune reactions, thereby destroying cancerous cells. The other cellular effects due to the elevated temperature include the denaturation of cellular proteins, disassembly of cytoskeleton and cell membrane, and destabilization of synthetic pathways of protein, nucleic acid, lipid metabolism, enzyme, signal transduction, and eventually to either necrosis or apoptosis of preheated cells. Also, it is stated in the literature that based on an Arrhenius plot where logarithm of the slope of cell survival curves as a function of temperature, it is explained that there is a cutoff point at about 43 °C. Rate of cell death will be doubled for an increase by 1 °C beyond 43 °C [17, 18].

9.3 Heat Generating Sources for Hyperthermia

The most commonly used external devices, which induce heat at tumor sites, use irradiation either through electromagnetic waves or with light. Currently, radiofrequencies that range between 100 kHz and 150 MHz, microwaves with the wavelengths from 433 MHz to 2450 MHz, and ultrasound are being used. Hot water perfusion, infrared radiators, resistive wire implants, ferromagnetic seeds, and application of nanoparticles are few other methods for inducing hyperthermia [12, 13]. Although hyperthermia treatments have been shown to be promising, there are a few drawbacks such as inadequate dissemination of heat into the body, heating of normal healthy neighboring tissue, and sometimes the target site may be suffering thermal under dosage, for example due to the thick scull or the bone of pelvis that protects deep tissues resulting in recurring tumor development. Therefore, hyperthermia is often employed as an adjunct along with other therapies for cancer.

9.4 Concepts of Nanotechnology and Hyperthermia: Nanothermotherapy

The effective and advantageous combination of concepts and features of nanotechnology in medicine has generated encouraging results that have led to significant improvements in cancer diagnosis and treatments. Owing to numerous advantageous factors such as great surface area to volume ratio and their tissue approachability, nanoparticles have been applied in in vitro diagnostic sector for four decades. Among the wide range of nanomaterials under exploration for cancer treatment, magnetic nanomaterials have gained substantial consideration owing to their exceptional properties that are not seen in other types of nanomaterials. Intense research has led to new developments in the fields of magnetic nanoparticles (MNPs) and engineered smart MNPs [21–23]. This has elevated their status as a front-line tool to attack cancer cells by exploiting properties for programmable heat, vibration, active drug release, and inducing mechanical stress and strain under an external magnetic field. Principles based on nanotechnology offer immense opportunities and innovative concepts for hyperthermia therapy. It is possible that nanotechnology can be used to engineer and precisely tailor the functional characteristics and properties of nanoparticles for biological applications and hence develop a comprehensive treatment strategy for cancer based on multifunctional nanoparticles.

Studies using nanotechnology and hyperthermia have led to the development of a new research area based on nano-thermotherapy that can offer significant advances in cancer therapy and treatment. MNPs have greater potential to revolutionize existing cancer diagnostics and therapeutic techniques. MNP-based hyperthermia offers several benefits over other conventional sources for inducing heat against tumors. It primarily includes the large surface to volume ratio, hence MNPs are able to escape the reticuloendothelial system (RES), circulate for longer time in the blood stream, and get taken up by cancer cells resulting in the distribution of therapeutic heat directly to the tumor. Active targeting can be achieved by specifically conjugating a targeting element, such as specific proteins, DNA or RNA aptamers, or carbohydrate to MNP, thus ensuring treatment is more specific and effective (Figure 9.1). Along with the delivery of heat for tumor tissue destruction, MNP-based hyperthermia can be used for effective delivery of drugs to the targeted site. MNPs are also able to cross the blood brain barrier and serve as treatment strategy for brain-related gliomas and other tumors, which are currently very difficult to treat. The oscillating magnetic field frequencies that are passed to the body are considered to be potentially harmless and highly specific leading to heat only being generated in the area of tissue with MNPs. This heat generation is therefore more effective and homogeneous than macroscopic implants. It is also possible to tailor MNPs with additional features for the fabrication of multifunctional nanoparticles for therapeutic applications.

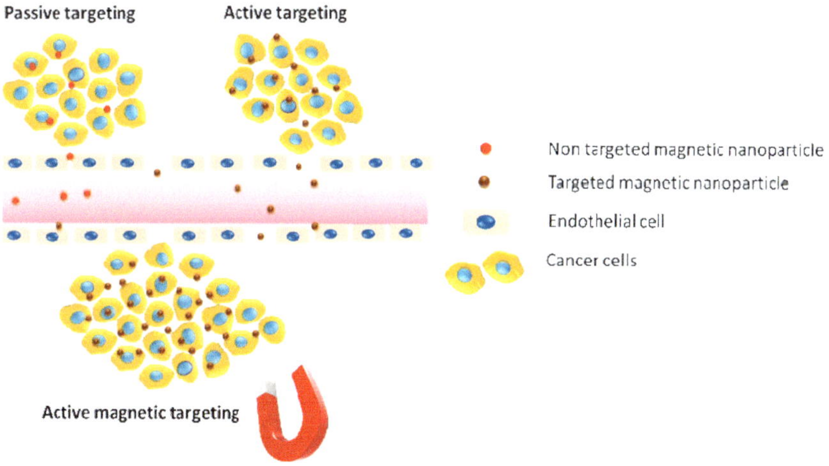

Figure 9.1 Schematic representation of passive, active, and active magnetic tumor targeting strategies.

9.5 Mechanism of Heat Generation

9.5.1 Metal Nanoparticle Heating

The MNP hyperthermia works on the concept that any metallic items when placed in an alternating magnetic field (AMF) develop an induced current flow. The current is directly proportional to the size of the magnetic field and the object. As the metal resists the flow of the current, it results in the development of heat, a process termed as inductive heating. The phenomenon is enhanced if the metal possesses magnetic property. Therefore, magnetic fluid becomes a potential heat source upon exposure to AMF and has the potential to kill the tumor cells [24]. Material's heating magnitude can be tuned by the Marie Curie temperature, the temperature where materials drop their magnetic properties and capability to heat. Several studies have been performed in the past two decades to advance hyperthermia treatment into clinical treatment settings.

9.5.2 MNP Heating

The heating capability of a magnetic material or electromagnetic device is quantified in terms of specific absorption rate (SAR) or specific loss power (SLP) that is defined as the amount of magnetic energy transformed into heat per time and mass [25]. SLP values can be determined by several parameters such as structure of MNPs, magnetic properties, applied magnetic field, and frequency of AMF. Thermal fluctuation that arises in the MNP due to initiation of rapid change of magnetic moment primarily governs the transformation of magnetic energy to heat energy due to an external AMF. Heat generation from MNPs is chiefly attributed to two mechanisms that are size dependent are relaxation and hysteresis loss. Single domain nanoparticles that are <100 nm exhibit two types of relaxations and they offer Neel and Brownian relaxations (Figure 9.2) [25, 26]. Brownian relaxation develops as a result of external dynamics that arises from physical rotation of MNPs. In an AMF medium, MNPs are experiencing the rotation and total particle oscillations, which result in friction between the MNPs and cause the generation of the heat. Heat transformation through Neel relaxation is due to the internal dynamics of the particle that arises from the rotation of the magnetic moment relative to crystal lattice with each field alternation. Brownian relaxation is hindered by the viscosity of the medium where the particles are placed [26] and Neel relaxation is stalled by anisotropic energy, which orients the magnetic domain relative to crystal lattice. Both sources of friction, Neel and Brownian, develop a phase lag among applied magnetic field, and magnetic moment direction results in the loss of thermal energy. When the MNPs enter the intracellular compartments of the cell, the Brownian movement of the particles is greatly restricted and hence the heat generates solely from Neel relaxation [27, 28]. Hergt et al. reported the most effective AMF parameters for hyperthermia that is 500 kHz (frequency) at 10 kA/m (amplitude) [29].

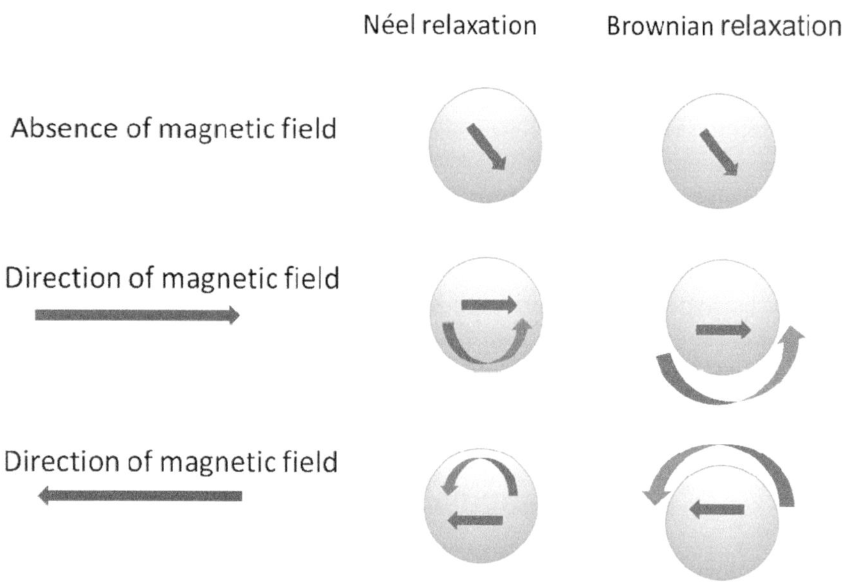

Figure 9.2 Mechanisms of heat generation in MNPs in response to an AMF. Gray solid circles represent magnetic iron oxide nanoparticles, red arrows depict direction of magnetic field, green curved arrows represent the movement, and blue curved arrows represent the change in magnetic moment direction.

9.6 Factors Influencing Design of MNPs for Hyperthermia

There are several factors, including physical and magnetic parameters, influence the design of MNPs for improving the hyperthermic potential for cancer therapy. The magnetic heating potential is greatly influenced by the size of nanoparticles. Monodisperse and highly crystalline iron oxide nanoparticles exhibited the highest SLP of 447 W/g for 14 nm particles, when these particles of different sizes ranging from 5–15 nm were screened with 24.5 kA/m at 400 kHz. It is remarkable to find the relationship between theoretical and experimental SLP values based on the size of particles and polydispersity indices. The size-dependent research reports evidently establish that monodispersed single domain iron oxide MNPs of 14–16 nm in size are highly efficient for hyperthermia [30, 31]. However, bacterial magnetosomes comprising ~30 nm γ-Fe_2O_3 exhibit maximum SLP value of 960 W/g (at 410 kHz and 10 kA/m) [32]. Carrier viscosity is yet another factor that can affect the magnetic heating rate [26]. SLP considerably reduces when viscosity of the medium is increased as was demonstrated by dispersing monodispersed maghemite and cobalt ferrite in water and glycerol [33]. The increase in the anisotropy of nanoparticles or increase in applied magnetic field strength influences MNP-based hyperthermia. Although MNPs with higher anisotropies are being developed based on Co and Co-based alloys, the inherent cytotoxicity associated with these particles limits their clinical application [34–38].

9.7 Magnetic Nanomaterials for Hyperthermia

MNPs have found to be effective in a variety of biological applications [39, 40], including imaging, cancer therapy, targeted drug delivery, non-invasive diagnosis, and hyperthermia for tumor therapy. There are a wide range of magnetic nanomaterials employed for hyperthermia in cancer treatment, mainly in the application of superparamagnetic iron oxide nanoparticles

(SPIONs), such as magnetite (Fe_3O_4) and maghemite (γ-Fe_2O_3), stabilized with ligands such as cationic liposome, hydrogel, lauric acid, dextran, polyvinyl alcohol, and polyethylene glycol [34–38, 41–45]. Several metallic nanoparticles and their oxides, such as Ni, Co, Fe, Mn, Mg, and Gd, have been explored for their use in magnetic heat treatments. The syntheses parameters of MNPs can be modified by doping of the guest atom in the structure resulting in the formation of doped ferromagnetic nanoparticles that are successfully investigated for hyperthermic potential [46]. Some examples include Fe doped Au, Mn–Zn–Gd doped iron oxide ($Mn_xZn_xGd_xFe_{(2-x)}O4$), and Zn–Mn doped iron oxides ($Zn_xMn_{(1-x)}$ Fe_3O_4) composites [30, 31]. Ferrites include manganese ferrite ($MnFe_2O_4$), nickel ferrite ($NiFe_2O_4$), cobalt ferrites ($CoFe_2O_4$), lithium ferrite ($Li_{0.5}Fe_{2.5}O_4$), ferrites of cobalt–nickel ferrite ($Co_xNi_{(1-x)}Fe_2O_4$), and nickel–zinc–copper ($Ni_{0.65}Zn_{0.35}Cu_{0.1}Fe_{1.9}O_4$) [34, 35, 37, 38, 44, 47]. Synthetic magnetic nanomaterials are being used as magnetic fluids (ferrofluids: colloidal suspensions associated to surface-modified MNPs of 2–20 nm in size dispersed in a fluid carrier). Novel magnetic materials, such as Mn–Zn–Fe, Co–Gd–Zn, and Zn–Gd–Fe, are reported to be better for cancer treatment than magnetic oxides, owing to their surface chemistry and ease of conjugation of functional groups. These nanoparticle systems have a Curie temperature, which is defined as the temperature above which a ferromagnetic material loses its strong magnetic property and become paramagnetic particle incidentally, which is just above the therapeutic temperature for killing cancer cells. These nanoparticles can be used to control heating via hyperthermia in the temperature range of 42–45 °C or above 46 °C by thermo-ablation phenomenon that creates permanent cellular injuries. The advantage of these particles is that once the desired therapeutic temperature is exceeded, heating can then be decreased significantly and hence no charring of the tissue occurs [48–53]. Several investigations have been reported to develop MNPs with a Curie temperature in the range 43–46 °C, and the reduction of the Curie temperature is mostly based on the size, shape, composition, and cationic distribution for the successful application for hyperthermia [54–63].

Core–shell nanomaterials are mostly hybrid materials that offer diverse properties and functionality, owing to the composition of the core and shell materials [64]. Several possibilities and prospects have been projected using this architecture. The shell over the core serves several functions such as it acts as a shielding layer for the core material, forms a platform for biofunctionalization for targeting, and also makes the nanoparticle biocompatible by reducing the toxicity of core material [48, 49, 65–68]. The fabrication of multifunctional nanoparticles also stems from this design, where detection moiety for targeting, imaging, and treatment modalities, by incorporating drugs or various treatment possibilities, can be integrated into core–shell nanoparticles [69, 70].

9.8 Significance of SPIONs in Hyperthermia

Although several novel MNPs with specific magnetic features are being developed, SPIONs displaying medium heating properties with slight control of in vivo temperature during hyperthermia continue to attract attention [71]. Advantages of using SPIONs include non-toxicity, excellent biocompatibility, low protein adsorption, and ability to escape RES that is highly essential for in vivo use [72]. Another advantage of iron-based nanoparticles is that they are metabolized to blood hemoglobin by heme oxygenase-1 effectively maintaining iron homeostasis in cells [73–76]. Most of the commercial MNPs for hyperthermia treatments are made up of SPIONs. The SPIONs are considered as smart nanomaterials, which have the potential to be used as theragnostic (used for diagnosis and therapy together) materials that aid in magnetic resonance imaging (MRI), multimodal imaging, magnetic hyperthermia, magnetically guided, and targeted delivery of therapeutic or biomolecules to cancer cells with negligible cytotoxicity to normal cells at the applied concentrations (Figure 9.3) [77–91]. Superparamagnetic MNPs are ideal candidates for in vivo applications compared to other types of MNPs. The main reason to use SPIONs is the disappearance of their magnetization when the external magnetic field is removed. Therefore, complications such as particle agglomeration and embolization of capillary vessels can be circumvented [92, 93].

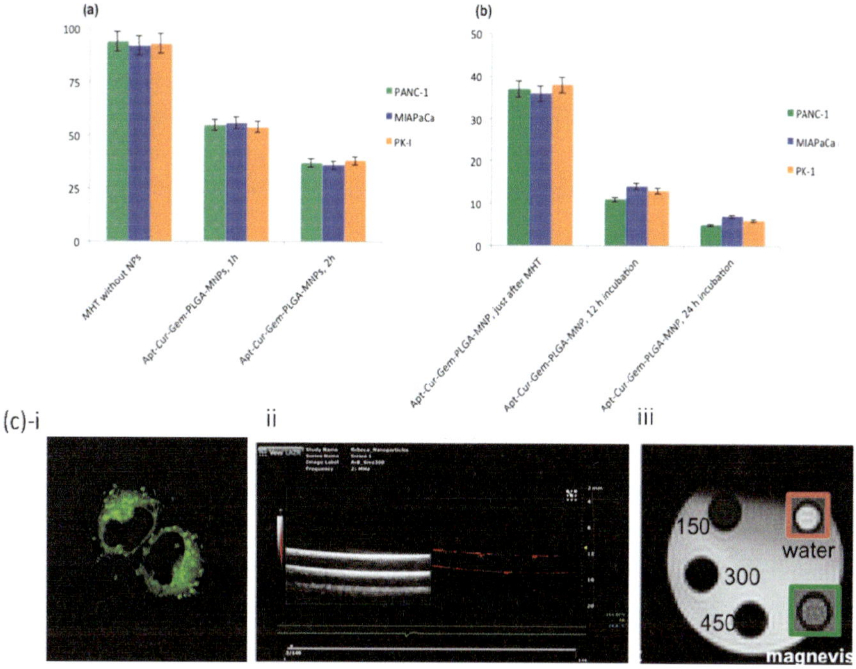

Figure 9.3 Application of SPIONs in magnetic hyperthermia. (a) Magnetic hyperthermia activity of Apt-Cur-Gem-PLGA-MNPs in pancreatic cancer cells (PANC-1, MIAPaCa, and PK-1 cells) treated with aptamer targeted curcumin, and gemcitabine loaded PLGA-MNPs (Apt-Cur-Gem-PLGA-MNPs) were exposed to AMF for hyperthermia treatment, and the viability was assessed by trypan blue staining. (b) Effect of magnetic hyperthermia on the viability of cancer cells at different time intervals. Reprinted with permission from Ref. [89]. Reproduced by permission of The Royal Society of Chemistry. (c) Multimodal imaging with aptamer conjugated curcumin and SPIONs encapsulated PLGA nanoparticles: (i) Curcumin encapsulation enabled optical imaging, (ii, iii) MNPs enabled photoacoustic and MRI. Reprinted with permission from Ref. [83]. Reproduced by permission of The Royal Society of Chemistry.

The biocompatibility is a crucial demand that any nanoparticles should possess and it can be attained via coating on SPIONs with several organic or inorganic coating materials, such as polysaccharide, alginate, gold, silica chitosan and poly-ethylenimine, polyethylene glycol, polyvinyl alcohol, polyethylene glycol fumarate, acrylate-based coatings, and synthetic polyesters [94–102]. Apart from the non-cytotoxicity, coating with cytocompatible materials also avoids the aggregation of SPIONs, conserves the functional groups in the nanoparticle, and attains preferred surface charge on SPIONs, which also avoids the opsonization that results in the rapid removal of nanoparticles from the blood through RES, hence leading to prolonged circulation within the body.

9.9 Engineered Smart Nanosystems: Drug Delivery via Hyperthermia

Engineered smart MNPs with magnetic cores can be encapsulated within multi-stimuli responsive shells, such as thermo- or pH-responsive polymer shells, for simultaneous heat generation by enhanced hyperthermia and releasing of drugs from the nanocomposites to the targeted site. Kumar et al. have elaborated two possible mechanisms of drug release based on MNP hyperthermia [103]. The first mechanism is the hyperthermia-based controlled drug delivery by bond breaking. In this type, the drug molecule is linked to MNPs via a linker molecule, which is broken due to the heating of the MNPs when an AMF is applied leading to the release of the drug molecule at specific sites [104, 105]. The method is often modified in complex drug release scenarios for the release of multiple drugs in a sequence or together by fluctuating the parameters controlling the electromagnetic field (EMF). The second method is described as the hyperthermia-based controlled drug delivery through enhanced permeability. Once the nanoparticle approaches the targeted site, based on physicochemical properties of the polymer shell, drug release can be attained by heat generated by magnetic cores and the drug is released at the tumor site [106–108]. Controlled drug release was observed by altering high frequency magnetic waves from folic acid (FA) and cyclodextrin (CD)-functionalized SPIONs, FA-CD-SPIONs; thus performing excellent platform for simultaneous drug delivery and hyperthermia (Figure 9.4) [109, 110]. A Fe_3O_4-CMCH-genistein nanosystem has also been developed that can efficiently release drugs by hyperthermia, thus enhancing SGC-7901 cancer cell

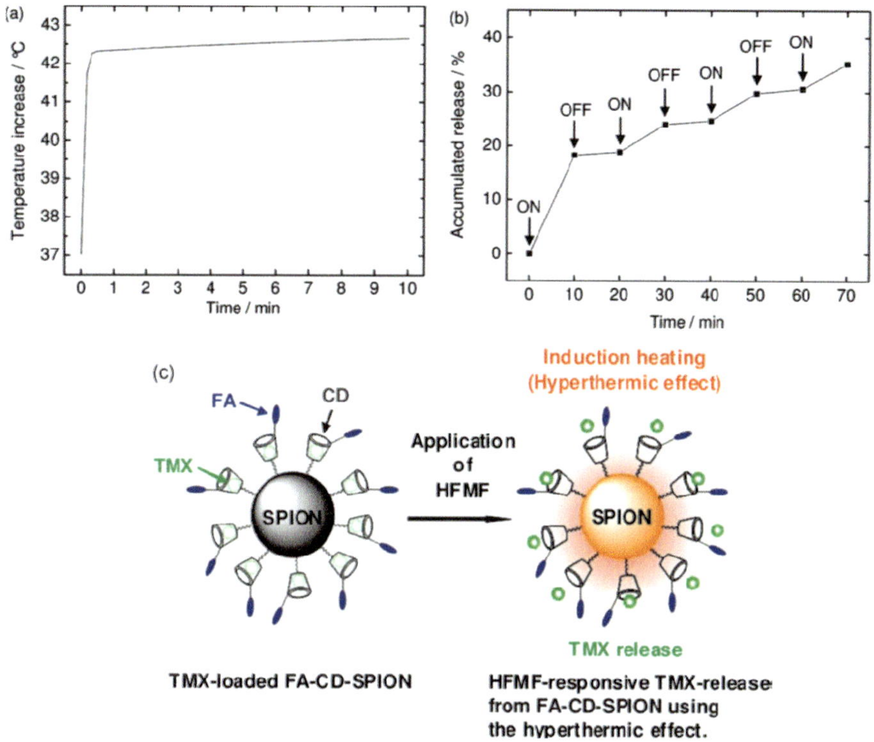

Figure 9.4 (a) The temperature increases of the water dispersed drug (tamoxifen)-loaded FA-CD-SPIONs under a high-frequency magnetic field (with 230 kHz in frequency and 100 Oe in amplitude. (b) Controlled release of tamoxifen (TMX) from FA-CD-SPIONs by switching a high-frequency magnetic field on and off. (c) Schematic illustration of release of TMX from FA-CD-SPIONs using the hyperthermic effect by applying a high-frequency magnetic field. Reprinted with permission from Ref. [109]. Copyright 2018, American Chemical Society.

apoptosis [111, 112]. The method has been extended to stimulate thermosenstive cation channel within the cell using magnetic field heating of nanoparticle and thus opens a new possibility to remotely operate cellular machinery for therapeutics [113]. Several parameters, such as magnetic property (strength and magnetic field), shape, size, composition, concentration, and depth of MNP in the tissue, must be optimized thoroughly for the application of smart drug delivery based on MNP hyperthermia. Apart from thermosenstive polymers, pH sensitive polymers and polymers offering cross-linking capability are also widely used for magnetically controlled drug delivery and release [102, 114, 115].

9.10 Examples of MNP Hyperthermia in Biology

There have been extensive studies on the application of MNPs hyperthermia for cancer treatments. Reports from both in vitro and in vivo experiments suggest that hyperthermia seems to be highly promising for cancer treatment. Table 9.1 explains and summarizes some works on magnetic hyperthermia in biological systems.

9.11 Conclusion and Future Prospects

Several factors remain to be optimized for the successful application of magnetic hyperthermia. The first and foremost is the biocompatibility of nanoparticles, thereby minimizing the toxic effect of nanoparticles. When MNPs are introduced into the tumor, the exact reason of the cell death is unclear, whether the cell death is due to the magnetic particle heating or due to the toxicity of nanoparticles or by both, and this issue must be properly addressed for the successful application of MNP hyperthermia. Also a comprehensive study of the physical properties of nanoparticles and their interaction with the biological system, e.g., cell, is highly essential. The choice and design of coating of a biomaterial on MNP to reduce the toxicity of nanoparticle should be a wise approach for the proper targeting for hyperthermia.

The fine-tuning of the physical property of MNP is critical for maximizing the SAR that can allow a lower dosage of magnetic material used in vivo. Clustering of nanoparticles in the tumor also affects the SAR gel-nanoparticle systems [139–141]. The arrangement of nanoparticles in cellular structures is also extremely important in magnetic hyperthermia. However, a few studies have been performed to explain this behavior so far.

Table 9.1 Magnetic hyperthermia studies in biological system

MNP	Experimental condition	Remark	Refs
Superparamagnetic dextran or aminosilane-coated magnetite nanoparticles of 3–13 nm crystal diameters	In vitro conditions	Comparable cytotoxic efficacy of water bath and MNP hyperthermia when used to heat cells	[116]
Superparamagnetic magnetite nanoparticles of 10–200 nm hydrodynamic diameter	In vivo tumor in mice (human breast adenocarcinoma)	Heterogenous tumor heating. Rise in temperature between 12 °C and 73 °C was registered at the tumor center and periphery	[117]
Magnetite cationic liposomes administered with Interleukin-2 (IL-2)	In vivo tumor in mice (B16 melanoma)	Investigated the combined effects of immunotherapy and magnetic hyperthermia on melanoma. Substantial improvement in mouse survival and tumor regression was observed	[118]
Magnetic cationic liposomes-based SPIONs	In vivo studies on hamster osteosarcoma	The tumor was heated over 42 °C, and complete regression of tumor was observed in all treated group hamsters and the average tumor volume of the hamsters treated with magnetic cationic liposomes was about 1/1000 of that of the control hamsters at day 12	[119]
Superparamagnetic aminosilane-coated iron oxide nanoparticles with 15 nm crystal diameter	In vivo glioblastoma-human brain tumors	Negligible cytotoxicity was observed in patients when brain tumor was heated to 42.4–49.5 °C with MNPs and concluded that deep cranial thermotherapy using MNPs can be safely applied for brain tumors	[120]
As_2O_3/Fe_3O_4 complex (Fe_3O_4 MNPs encapsulated in As_2O_3) of 50 nm in size	Xenograft HeLa cells (human cervical cancer cell line) in nude mice	Significant inhibiton on the xenograft cervical tumors mass (88.2%) and volume (91.57%) was observed and the expression of mRNA for CD44v6, VEGF-C, and MMP-9 related to cancer and/or metastasis were considerably inhibited	[121]

(*Continued*)

Table 9.1 (*Continued*)

MNP	Experimental condition	Remark	Refs
An implant entrapped with SPIONs embedded in silica microbeads	In vivo studies in nude mice with subcutaneous human colocarcinomas	45% of treated mice survived for 1 year without any tumor reappearance and hence a promising tumor therapy with magnetically induced moderate hyperthermia via injectable implants	[122]
Fe/Fe_3O_4 core/shell MNP	Subcutaneous mouse melanomas (B16-F10)	Micromolar concentrations of nanoparticles resulted in a substantial anti-tumor effect on murine B16-F10 melanoma with 10-minute AMF exposures for three times	[123]
Chitosan oligosaccharide-stabilized ferrimagnetic iron oxide nanocubes (Chito-FIONs)	In vitro studies	Nanoparticles exhibited enhanced magnetic heating capability with high SLP value (2614 W/g) and magnetically guided nanoparticles effectively destroyed cancer cells through caspase-mediated apoptosis	[124]
Iron oxide nanocrystals (IONCs)	In vitro studies	IONCs with an average diameter of 19 ±3 nm had significant SAR values in clinical conditions and in vitro trials studies on KB cancer cells exhibited effective hyperthermia performance. Cell mortality of about 50% was attained when an equilibrium temperature of 43 °C was touched after 1 h of treatment	[125]
Poly(lactic-co-glycolic acid) (PLGA) encapsulated MNPs (150 nm)	In vitro	Aptamer targeted drug loaded PLGA-MNPs exhibited excellent cellular cytotoxicity upon synergistic magnetic hyperthermia and drug mediated cytotoxicity in pancreatic cancer cell lines	[89]
PLGA encapsulated MNPs (150 nm)	In vitro	Folate and transferrin targeted dual drug (curcumin and 5-fluorouracil) loaded MNPs encapsulated in PLGA nanoparticles demonstrated excellent magnetic hyperthermia in cancer cells by inducing apoptosis	[90]

MNP	Experimental condition	Remark	Refs
PLGA encapsulated hybrid magnetic nanoparticles-Au-MNPs (100 nm)	In vitro	Hybrid MNPs have been integrated into PLGA with two chemotherapeutic agents (curcumin and gemcitabine) and decorated with specific targeting ligands (folate, transferrin, and AS1411 aptamer). The targeted nanoconjugate exerted its lethality by three distinctive modes (chemotherapeutic, photothermal, and magnetic hyperthermia) either independently or by the synergistic action of drug and hyperthermia	[86]
Manganese ferrite nanoassemblies (MNAs) (50 nm)	In vitro	Approximately 85% of cancer cells were killed by apoptosis owing to the heating of MNAs under external AMF	[126]
Ferromagnetic Fe0.6Mn0.4O nanoflowers (100 nm)	In vitro and in vivo	Nanoflowers displayed higher SAR values (535 W/g). Superior in vitro hyperthermia was exhibited at lower concentration of nanoflowers and total inhibition of tumor was observed without any side effects	[127]
PEG functionalized Fe_3O_4 multicore (30 nm)	In vitro	Multicore nanoparticles effectively killed cancer cells due to higher SAR value and enhanced cellular internalization	[128]
Superparamagnetic Fe_3O_4 nanoparticles coated with biocompatible hydroxyapatite (50–70 nm Fe_3O_4 nanosphere)	In vitro hyperthermia	MG-63 cancer cells were effectively killed at a temperature range of 42–47 °C after exposure to AMF for 30 min owing to the generation of reactive oxygen species	[129]
Bovine serum albumin coated Fe_3O_4 nanocubes (100 nm)	In vitro and in vivo	Nanoparticle exhibited enhanced magnetic hyperthermia owing to high tumor accumulation	[130]

(*Continued*)

Table 9.1 (*Continued*)

MNP	Experimental condition	Remark	Refs
Iron oxide nanoparticles integrated inside polymethylmethacrylate (PMMA-Fe_3O_4) (20–50 nm)	In vivo	Magnetic hyperthermia ablated MB-231 xenograft model of breast cancer using injectable nanomaterial at a lower dose of 0.1 mL under 180 s exposure of AMF. Magnetic material exhibiting liquid to solid phase transition property demonstrated magnetic hyperthermia-based ablation of tumors	[131]
Iron oxide	In vitro and in vivo	Nude mice with tumors demonstrated significant inhibiting effect of Fe_3O_4 nanometer magnetic fluid in glioma volume	[132]
Polymer coated Fe_3O_4 nanoparticles (10–50 nm)	In vitro and in vivo	Heat responsive contractible nanomaterial was used for tumor ablation through necrosis at temperature above 60 °C	[133]
Magnetic nanoemulsion hydrogel with zinc ferrite nanoparticles (55 nm)	In vivo	Gelation of thermoresponsive nanoemulsion inside tumor improved the heating efficiency resulting in necrosis and complete elimination of tumor without damaging normal tissues	[134]
Dextran-coated iron oxide nanoparticles (2.8 μm)	In vitro	Aptamer targeted dextran-coated iron oxide nanoparticles selectively eliminated human osteosarcoma cells U2OS under AMF	[135]
Antibody functionalized carmustine-loaded $ZnFe_2O_4$ Oleosome (100 nm)	In vitro and in vivo	Oleosome upon AMF exhibited significant anticancer effects owing to chemotherapy (by lipophilic carmustine) and magnetic hyperthermia under in vitro and in vivo conditions in a synergistic pattern	[136]
Temperature-sensitive polymer layered magnetic mesoporous silica nanoparticles (160 nm)	In vivo hyperthermia	Combined therapy effectively inhibited tumor within two days after treatment with nanoparticles	[137]
Magnesium doped γ-Fe_2O_3 nanoparticle (7 nm)	In vitro and in vivo	Nanocomposite exhibited higher of intrinsic loss power (100 times) than commercial Fe_3O_4. Complete destruction of the tumor was observed owing to high heating power under in vitro and in vivo studies	[138]

"Thermotolerance" and its variation in cells can be considered one of the major limitations of hyperthermia, and hence it is a major challenge to overcome for the smooth application of magnetic hyperthermia [14]. It is the ability to sustain viability with increasing temperature, frequency, or duration [17, 18]. The precise mechanisms and temporal interactions and the role of heat shock proteins are still to be explained, and determining the ideal temperature programming is a major task. Novel approaches to distinguish cellular heating (both intracellular and extracellular) based on in vivo studies and the extensive investigation on thermal inactivation in cancer and normal cells neighboring cells are required for mapping thermotolerance [142].

The strategy for curtailing the side effects of hyperthermia to healthy cells or tissues, charting or distribution of temperature at the targeted sites is an essential criterion. Although there are several techniques to map temperature, such as MRI thermometry, molecular diffusion by functional-MRI, and thermochromic fluorescent- and optical-films magnetically insensitive fiber-optic probes, the development of advanced noninvasive techniques for mapping local temperatures is highly challenging [143–146]. Studies based on modeling and fabrication of thermosensitive materials are promising approaches to overcome this challenge. There have been several efforts to monitor temperature distribution based on modeling. A few successful examples of software are: NanoPlan® developed by the Magforce Company and SigmaHyperPlan® developed by Dr. Sennewald Medizintechnik GmbH [147, 148]. More investigations based on modeling are required for the precise distribution of heat on tumor sites and neighboring healthy cells. Computer simulations based on modeling studies can guide physicians to control how, when, and where to apply the magnetic fluids during hyperthermia treatment. The development of thermosensitive materials for monitoring heat distribution can also contribute to mapping hyperthermia. Coupling of imaging moiety along with magnetic hyperthermia to the targeted site to monitor thermal distribution and drug delivery will be highly advantageous. Thus, simultaneous diagnosis and therapy, which is coined as "theranostics" or "theragnostics," can be attained and has immense potential in biomedical scenarios and in the near future for the design and application of personalized medicine. There are also challenges in targeting the design of stealth nanoparticles that have the capability to circulate in the blood for prolonged period as well as functionalization for bio recognition. The development of antitumor immunity along with reduction in tumor size was observed in several studies [149–153]. Hence, there are chances for hyperthermia to protect patients from further tumor recurrence. There are new opening for the development of vaccines based on immune and thermo-based therapies.

In the near future, it is highly anticipated that multifunctional nanomaterials with optimized pharmacokinetics and pharmacodynamics will provide effective therapies. Research on the development of "smart theranostics" can be highly beneficial for noninvasive, real-time monitoring for thermal distribution as well as delivery of drug to the targeted site, which could ultimately lead to a personalized medicine approach. In short, combination of nanotechnology, bioengineering, biomedical, toxicology, and pharmacology disciplines will augment the improvement of appropriate strategies to design advanced MNPs and nano-devices with biocompatible properties to treat cancerous tumors.

References

1. Dewhirst MW, Gibbs FA Jr, Roemer RB, Samulski TV. Hyperthermia. Gunderson LL, Tepper JE (eds.). Clinical radiation oncology. 1st Ed. Chapter 14. New York, NY: Churchill Livingstone. **2000**, 256–282.
2. Chicheł A, Skowronek J, Kubaszewska M, Kanikowski M. Hyperthermia–Description of a method and a review of clinical applications. *Rep. Pract. Oncol. Radiother.* 5, **2007**, 267–275.
3. Jordan A, Wust P, Scholz R, Tesche B, Fahling H, Mitrovics T, Vogl T, Cervos-Navarro J, Felix R. Cellular uptake of magnetic fluid particles and their effects on human adenocarcinoma cells exposed to AC magnetic fields in vitro. *Int. J. Hyperthermia.* 12, **1996,** 705–722.
4. Nielsen OS, Horsman M, Overgaard J. A future for hyperthermia in cancer treatment? *Eur. J. Cancer.* 37, **2001**, 1587–1589.
5. Ceelen WP, Hesse U, De Hemptinne B, Pattyn P. Hyperthermic intraperitoneal chemoperfusion in the treatment of locally advanced intraabdominal cancer. *Br. J. Surg.* 87, **2000**, 1006–1015.

6. Chang E, Alexander HR, Libutti SK, Hurst R, Zhai S, Figg WD, Bartlett DL. Laparoscopic continuous hyperthermic peritoneal perfusion. *J. Am. Coll. Surg.* 193, **2001**, 225–229.
7. Feldman AL, Libutti SK, Pingpank JF, Bartlett DL, Beresnev TH, Mavroukakis SM, Steinberg SM, Liewehr DJ, Kleiner DE, Alexander HR. Analysis of factors associated with outcome in patients with malignant peritoneal mesothelioma undergoing surgical debulking and intraperitoneal chemotherapy. *J. Clin. Oncol.* 21, **2003**, 4560–4567.
8. Glehen O, Mithieux F, Osinsky D, Beaujard AC, Freyer G, Francois GY, Peyrat P, Panteix G, Vignal J, Gilly FN. Surgery combined with peritonectomy procedures and intraperitoneal chemohyperthermia in abdominal cancers with peritoneal carcinomatosis: A phase II study. *J. Clin. Oncol.* 21, **2003**, 799–806.
9. Seegenschmiedt MH, Fessenden P, Vernon CC. Thermoradiotherapy and thermochemotherapy. Vol. 1: Biology, Physiology and Physics. Berlin: Springer-Verlag. **1995.**
10. Wehner H, von Ardenne A, Kaltofen S. Whole body hyperthermia with water-filtered infrared radiation: Technical-physical aspects and clinical experiences. *Int. J. Hyperthermia.* 17, **2001**, 19–30.
11. Repasky E, Issels R. Introduction: Physiological consequences of hyperthermia: heat, heat shock proteins and the immune response. *Int. J. Hyperthermia.* 18, **2002**, 486–489.
12. van der Zee J. Heating the patient: A promising approach? *Ann. Oncol.* 13, **2002**, 1173–1184.
13. Wust P, Hildebrandt B, Sreenivasa G, Rau B, Gellermann J, Riess H, Felix R, Schlag PM. Hyperthermia in combined treatment of cancer. *Lancet Oncol.* 3, **2002**, 487–497.
14. Hildebrandt B, Wust P, Ahlers O, Dieing A, Sreenivasa G, Kerner T, Felix R, Riess H. The cellular and molecular basis of hyperthermia. *Crit. Rev. Oncol. Hematol.* 43, **2002**, 33–56.
15. Milleron RS, Bratton SB. 'Heated' debates in apoptosis. *Cell. Mol. Life Sci.* 64, **2007**, 2329–2333.
16. Falk MH, Issels RD. Hyperthermia in oncology. *Int. J. Hyperthermia.* 17, **2001**, 1–18.
17. Overgaard J, Horsman MR. Hyperthermia. Steel GG. Basic clinical radiobiology. 2nd Ed. Edward Arnold. **1997**, 212–221.
18. Jones EL, Samulski TV, Vujaskovic Z, et al. Hyperthermia. Perez CA, Brady LW, Halperin EC, Schmidt-Ullrich RK (eds.). Principles and

practise of radiation oncology. 4th Ed. Chapter 24. Lippincott Williams & Wilkins. **2004**, 699–735.

19. Kapp DS, Hahn GM, Carlson RW: Principles of hyperthermia. Bast RC Jr, Kufe DW, Pollock RE, et al. (eds.). Cancer medicine e.5. 5th ed. Hamilton, Ontario: B.C. Decker Inc. **2000**.
20. Kim JH, Hahn EW, Ahmed SA. Combination hyperthermia and radiation therapy for malignant melanoma. *Cancer*. 50, **1982**, 478–482.
21. Bao Y, Wen T, Samia ACS, Khandhar A, and Krishnan KM. Magnetic nanoparticles: Material engineering and emerging applications in lithography and biomedicine. *J. Mater. Sci.* 51, **2016,** 513–553.
22. Li Z, Yang F, Yin Y. Smart materials by nanoscale magnetic assembly. *Adv. Funct. Mater.* 30, **2020**, 1903467.
23. Mai BT, Fernandes S, Balakrishnan PB, Pellegrino T. Nanosystems based on magnetic nanoparticles and thermo- or ph-responsive polymers: An update and future perspectives. *Acc. Chem. Res.* 51. **2018**, 999–1013.
24. Babincova M, Leszczynska D, Sourivong P, Babinec P. Selective treatment of neoplastic cells using ferritin-mediated electromagnetic hyperthermia. *Med. Hypoth.* 54, **2000**, 177–179.
25. Bornstein BA, Zouranjian PS, Hansen JL, Fraser SM, Gelwan LA, Teicher BA, Svensson GK. Local hyperthermia, radiation therapy, and chemotherapy in patients with local–regional recurrence of breast carcinoma. *J. Radiat. Oncol. Biol. Phys.* 25, **1993**, 79–85.
26. Jeyadevan B. Present status and prospects of magnetite nanoparticles-based hyperthermia. *J. Ceram. Soc. Jpn.* 118, **2010**, 391–401.
27. Riviere C, Wilhelm C, Cousin F, Dupuis V, Gazeau F, Perzynski R. Internal structure of magnetic endosomes. *Eur. Phys. J. E.* 22, **2007**, 1–10.
28. Zhao DL, Wang XX, Zeng XW, Xia QS, Tang JT. Preparation and inductive heating property of Fe_3O_4 chitosan composite nanoparticles in an AC magnetic field for localized hyperthermia. *J. Alloys Compd.* 477, **2009**, 739–743.
29. Hergt R, Dutz S. Magnetic particle hyperthermia: Biophysical limitations of a visionary tumour therapy. *J. Magn. Magn. Mater.* 311, **2007**, 187–192.
30. Wijaya A, Brown KA, Alper JD, Schifferli KH. Magnetic field heating study of Fe-doped Au nanoparticles. *J. Magn. Magn. Mater.* 309, **2007**, 15–19.

31. Sharma R, Chen CJ. Newer nanoparticles in hyperthermia treatment and thermometry. *J. Nanopart. Res.* 11, **2009**, 671–689.

32. Hergt R, Hiergeist R, Zeisberger M, Schüler D, Heyen U, Hilger I, Kaiser WA. Magnetic properties of bacterial magnetosomes as potential diagnostic and therapeutic tools. *J. Magn. Magn. Mater.* 293, **2005**, 80–86.

33. Fortin JP, Gazeau F, Wilhelm C. Internal heating of living cells with magnetic nanomediators. *Eur. Biophys. J.* 37, **2008**, 223–228.

34. Kaman O, Pollert E, Veverka P, Veverka M, Hadova E, Knýzek K, Marysko M, Kaspar P, Klementova M, Grunwaldova V, Vasseur S, Epherre R, Mornet S, Goglio G, Duguet E. Silica encapsulated manganese perovskite nanoparticles for magnetically induced hyperthermia without the risk of overheating. *Nanotechnology.* 20, **2009**, 275610.

35. Martina MS, Wilhelm C, Lesieur S. The effect of magnetic targeting on the uptake of magnetic-fluid-loaded liposomes by human prostatic adenocarcinoma cells. *Biomaterials.* 29, **2008**, 4137–4145.

36. Marcela GW, Matthias Z, Kannan MK. Size-dependent heating rates of iron oxide nanoparticles for magnetic fluid hyperthermia. *J. Magn. Magn. Mater.* 321, **2009**, 1947–1950.

37. Pradhan P, Giri J, SamantaG, Sarma HD, Mishra KP, Bellare J, Banerjee R, Bahadur D. Comparative evaluation of heating ability and biocompatibility of different ferrite-based magnetic fluids for hyperthermia application. *J. Biomed. Mater. Res. B Appl. Biomater.* 81, **2007**, 12–22.

38. Atsarkin VA, Levkin LV, Posvyanskiy VS, Melnikov OV, Markelova MN, Gorbenko OY, Kaul AR. Solution to the bioheat equation for hyperthermia with La_1–*x*AgyMn$O_{3-\delta}$ nanoparticles: The effect of temperature autostabilization. *Int. J. Hyperthermia.* 25, **2009**, 240–247.

39. Etienne D, Vasser S, Mornet S, Devoisselle J-M. Magnetic nanoparticles and their applications in medicine. *Nanomedicine.* 1, **2006**, 157–168.

40. Pankhurst QA, Connolly J, Jones SK, Dobson J. Applications of magnetic nanoparticles in biomedicine. *J. Phys. D. Appl. Phys.* 36, **2003**, R167.

41. Suto M, Hirota Y, Mamiya H, Fujita A, Kasuya R, Tohji K, Jeyadevan B. Heat dissipation mechanism of magnetite nanoparticles in magnetic fluid hyperthermia. *J. Magn. Magn. Mater.* 321, **2009**, 1493–1496.

42. Dennis CL, Jackson AJ, Borchers JA, Ivkov R, Foreman AR, PJ Hoopes, Strawbridge R, Pierce Z, Goerntiz E, Lau JW, Gruettner C. The influence of magnetic and physiological behaviour on the effectiveness of iron oxide nanoparticles for hyperthermia. *J. Phys. D Appl. Phys.* 41, **2008**, 134020.

43. Levy M, Wilhelm C, Siaugue J-M, Horner O, Bacri J-C, Gazeau F. Magnetically induced hyperthermia: Size-dependent heating power of γ-Fe_2O_3 nanoparticles. *J. Phys. Condens. Matter.* 20, **2008**, 204133.

44. Bae S, Lee SW, Takemura Y, Yamashita E, Kunisaki J, Zurn S, Kim CS. Dependence of frequency and magnetic field on self-heating characteristics of $NiFe_2O_4$ nanoparticles for hyperthermia. *IEEE Trans. Magn.* 42, **2006**, 3566–3568.

45. Kawai N, Futakuchi M, Yoshida T, Ito A, Sato S, Naiki T, Honda H, Shirai T, Kohri K. Effect of heat therapy using magnetic nanoparticles conjugated with cationic liposomes on prostate tumor in bone. *Prostate.* 68, **2008**, 784–792.

46. Duong GV, Turtelli RS, Nunes WC, Schafler E, Hanh N, Grössinger R, Knobel M. Ultrafine $Co_{1-x}Zn_xFe_2O_4$ particles synthesized by hydrolysis: Effect of thermal treatment and its relationship with magnetic properties. *J. Non Crystalline Solids.* 353, **2007**, 805–807.

47. Kim DH, Lee SH, Kim KN, Kim KM, Shim IB, Lee YK. Temperature change of various ferrite particles with alternating magnetic field for hyperthermic application. *J. Magn. Magn. Mater.* 293, **2005**, 320–327.

48. Han SK, Kim RS, Lee JH, Tae G, Cho SH, Yuk SH. Nanomaterials for medical diagnosis and therapy. Challa S.S.R. Kumar (ed.). Wiley-VCH. **2007**, 143–188.

49. Yamaguchi Y, Igarashi R. Nanomaterials for medical diagnosis and therapy. Challa S.S.R. Kumar (ed.). Wiley-VCH. **2007**, 310–341.

50. Matoussevitch N, Gorschinski A, Habicht W, Bolle J, Dinjus E, Boennemann H, Behrens S. Surface modification of metallic Co nanoparticles. *J. Magn. Magn. Mater.* 311, **2007**, 92–96.

51. Joo SH, Park JY, Tsung CK, Yamada Y, Yang P, Somorjai GA. Thermally stable Pt/mesoporous silica core.shell nanocatalysts for high-temperature reactions. *Nat. Mater.* 8, **2009**, 126–131.

52. Lisiecki I. Size, shape, and structural control of metallic nanocrystals. *J. Phys. Chem. B.* 109, **2005**, 12231–12244.

53. Zhang XB, Yan JM, Han S, Shioyama H, Xu Q. Magnetically recyclable Fe@Pt core shell nanoparticles and their use as electro catalysts

for ammonia borane oxidation: The role of crystallinity of the core. *J. Am. Chem. Soc.* 131, **2009** 2778–2779.

54. Akin Y, Obaidat IM, Issa B, Haik Y. Ni1-x Crx alloy for self-controlled magnetic hyperthermia. *Cryst. Res.Technol.* 44, **2009**, 386–390.
55. Park JH, Im KH, Lee AH, Kim DH, Lee DY, Lee YK, Kim KM, Kim KN. Preparation and characterization of magnetic chitosan particles for hyperthermia application. *J. Magn. Magn. Mater.* 293, **2005,** 328–333.
56. Aurich K, Schwalbe M, Clement JH, Weitschies W, Buske N. Polyaspartate coated magnetite nanoparticles for biomedical applications. *J. Magn. Magn. Mater.* 311, **2007,** 1–5.
57. Levy M, Wihlem C, Siaugue JM, Horner O, Bacri JC, Gazeau F. Magnetically induced hyperthermia: Size-dependent heating power of γ-Fe(2)O(3) nanoparticles. *J. Phys.: Condens. Matter.* 20, **2008**, 204133.
58. Kim DH, Nikles DE, Johnson DT, Brazel CS. Heat generation of aqueously dispersed $CoFe_2O_4$ nanoparticles as heating agents for magnetically activated drug delivery and hyperthermia. *J. Magn. Magn. Mater.* 320, **2008**, 2390–2396.
59. Drake P, Cho HJ, Shih PS, Kao CH, Lee KF, Kuo CH, Lin XZ, Lin YJ. Gd-doped iron-oxide nanoparticles for tumour therapy via magnetic field hyperthermia. *J. Mater. Chem.* 17, **2007**, 4914–4918.
60. Akin Y, Obaidat IM, Issa B, Haik Y. Ni1–xCrx alloy for self-controlled magnetic hyperthermia. *Cryst. Res. Technol.* 44, **2009**, 386–390.
61. Brusentsova TN, Brusentsov NA, Kuznesov VD, Nikiforov VN. Synthesis and investigation of magnetic properties of Gd-substituted Mn–Zn ferrite nanoparticles as a potential low-TC agent for magnetic fluid hyperthermia. *J. Magn. Magn. Mater.* 293, **2005**, 298–302.
62. Pollert E, Knizek K, Marysko M, Kaspar P, Vasseur S, Duguet E. New TC-tuned magnetic nanoparticles for self-controlled hyperthermia. *J. Magn. Magn. Mater.* 316, **2007**, 122–125.
63. Baldi G, Bonacchi D, Innocenti C, Lorenzi G, Sangregorio C. Cobalt ferrite nanoparticles: The control of the particle size and surface state and their effects on magnetic properties. *J. Magn. Magn. Mater.* 311, **2007**, 10–16.
64. Liu Z, Elbert D, Chien CL, Searson PC. FIB/TEM characterization of the composition and structure of core/shell Cu.Ni nanowires. *Nano Lett.* 8, **2008**, 2166–2170.

65. Silke B, Helmut B, Matoussevitch N, Eckhard D, Modrow H, Palina N, Frerichs M, Kempter V, Wolfgang MF, Heinemann A, Kammel M, Wiedenmann A, Pop L, Odenbach S, Uhlmann E, Bayat N, Hesselbach J, Guldbakke JM. Air-stable Co-, Fe-, and Fe/Co-nanoparticles and ferrofluids. *Z. Phys. Chem.* 220, **2006**, 3–40.

66. Wang L, Park HY, Lim SI, Schadt MJ, Mott D, Luo J, Wang X, Zhong CJ. Core@shell nanomaterials: Gold-coated magnetic oxide nanoparticles. *J. Mater. Chem.* 18, **2008**, 2629–2635.

67. De Villiers M, Lvov Y. Nanomaterials for medical diagnosis and therapy, Challa S.S.R. Kumar (ed.). Wiley-VCH. **2007**, pp. 527–556.

68. He X, Lin X, Wang K, Chen L, Wu P, Yuan Y, Tan W. Encyclopedia of nanoscience and nanotechnology, 1. H.S. Nalwa (ed.). American Scientific Publishers. **2004**, pp. 235–253.

69. Sengupta S, Eavarone D, Capila I, Zhao G, Watson N, Kiziltepe T, Sasisekharan R. Temporal targeting of tumour cells and neovasculature with a nanoscale delivery system. *Nature.* 436, **2005**, 568–572.

70. Schneider GF, Subr V, Ulbrich K, Decher G. Multifunctional cytotoxic stealth nanoparticles: A model approach with potential for cancer therapy. *Nano Lett.* 9, **2009**, 636–642.

71. Pollert E, Veverka P, Veverka M, Kaman O, Závěta K, Vasseur S, Epherre R, Goglio G, Duguet E. Search of new core materials for magnetic fluid hyperthermia: Preliminary chemical and physical issues. *Prog. Solid. State. Chem.* 37, **2009**, 1–14.

72. Mahmoudi M, Milani AS, Stroeve P. Surface architecture of superparamagnetic iron oxide nanoparticles for application in drug delivery and their biological response: A review. *Int. J. Biomed. Nanosci. Nanotechnol.* 1, **2010**, 164–201.

73. Gould P. Nanoparticles probe biosystems. *Mater. Today.* 7, **2004**, 36–43.

74. Corchero JL, Villaverde A. Biomedical applications of distally controlled magnetic nanoparticles. *Trends Biotechnol.* 27, **2009**, 468–476.

75. Huber DL. Synthesis, properties, and applications of iron nanoparticles. *Small.* 1, **2005**, 482–501.

76. Pradhan P, Giri J, Banerjee R, Bellare J, Bahadur D. Cellular interactions of lauric acid and dextran-coated magnetite nanoparticles. *J. Magn. Magn. Mater.* 311, **2007**, 282–287.

77. Mahmoudi M, Azadmanesh K, Shokrgozar MA, Journeay WS, Laurent S. Effect of nanoparticles on the cell life cycle. *Chem Rev.* 111, **2011**, 3407–3432.

78. Laurent S, Forge D, Port M, Roch A, Robic C, Vander Elst L, Muller RN. Magnetic iron oxide nanoparticles: Synthesis, stabilization, vectorization, physicochemical characterizations and biological applications. *Chem. Rev.* 108, **2008**, 2064–2110.

79. Mahmoudi M, Shokrgozar MA, Sardari S, Moghadam MK, Vali H, Laurent S, Stroeve P. Irreversible changes in protein conformation due to interaction with superparamagnetic iron oxide nanoparticles. *Nanoscale.* 3, **2011**, 1127–1138.

80. Mahmoudi M, Simchi A, Milani AS, Stroeve P. Cell toxicity of superparamagnetic iron oxide nanoparticles. *J. Colloid. Interface. Sci.* 336, **2009**, 510–518.

81. Mahmoudi M, Sant S, Wang B, Laurent S, Sen T. Superparamagnetic iron oxide nanoparticles (SPIONs): Development, surface modification and applications in chemotherapy. *Adv. Drug. Deliv. Rev.* 63, **2011**, 24–46.

82. Mahmoudi M, Hosseinkhani H, Hosseinkhani M, Boutry S, Simchi A, Journeay WS, Subramani K, Laurent S. Magnetic resonance imaging tracking of stem cells in vivo using iron oxide nanoparticles as a tool for the advancement of clinical regenerative medicine. *Chem. Rev.* 111, **2011**, 253–280.

83. Sivakumar B, Aswathy RG, Rebeca RA, Mitcham T, Mitchel KA, Nagaoka Y, Bouchard RR, Ajayan PM, Maekawa T, Sakthikumar DN. Highly versatile SPION encapsulated PLGA nanoparticles as photothermal ablators of cancer cells and as multimodal imaging agents. *Biomater. Sci.* 5, **2017**, 432–443.

84. Ankit KR, Sivakumar B, Aswathy RG, Sreejith R, Ankita B, Nagaoka Y, Nakajima Y, Maekawa T, Sakthikumar D. Dual mode of cancer cell destruction for pancreatic cancer therapy using Hsp90 inhibitor loaded polymeric nano magnetic formulation. *Int. J. Pharm.* 511, **2016**, 648–658.

85. Simrjit S, Neeraj K, Sivakumar B, Athulya A, Sakthikumar DN. Multifunctional CdS/$CoFe_2O_4$ fluorescent/magnetic core/shell nanocompo site structure for bio-applications. *Mater. Res. Express.* 3, **2016**, 045 024.

86. Sivakumar B, Aswathy RG, Nagaoka Y, Fukuda T, Iwai S, Venugopal K, Kato K, Maekawa T, Sakthikumar DN. An 'all in one' approach for simultaneous chemotherapeutic, photothermal and magnetic hyperthermia mediated by hybrid magnetic nanoparticles. *RSC Adv.* 5, **2015**, 25066–25078.

87. Sivakumar B, Aswathy RG, Nagaoka Y, Fukuda T, Iwai S, Hasumura T, Venugopal K, Kato K, Yoshida Y, Maekawa T, Sakthikumar DN. Augmented cellular uptake and antiproliferation against pancreatic cancer cells induced by targeted curcumin and SPION encapsulated PLGA nanoformulation. *Materials Express*, 4, **2014**, 183–195.

88. Sivakumar B, Aswathy RG, Sreejith R, Nagaoka Y, Iwai S, Suzuki M, Fukuda T, Hasumura T, Yoshida Y, Maekawa T, Sakthikumar DN. Bacterial exopolysaccharide based magnetic nanoparticles: A versatile nanotool for cancer cell imaging, targeted drug delivery and synergistic effect of drug and hyperthermia mediated cancer therapy. *J. Biomed. Nanotechnol.* 10, **2014**, 885–899.

89. Sivakumar B, Aswathy RG, Nagaoka Y, Iwai S, Venugopal K, Kato K, Yoshida Y, Maekawa T, Sakthikumar DN. Aptamer conjugated theragnostic multifunctional magnetic nanoparticles as a nanoplatform for pancreatic cancer therapy. *RSC Adv.* 3, **2013**, 20579–20598.

90. Sivakumar B, Aswathy RG, Nagaoka Y, Iwai S, Suzuki M, Venugopal K, Yoshida Y, Maekawa T, Sakthikumar DN. Curcumin and 5-fluorouracil loaded, folate and transferrin decorated polymeric magnetic nanoformulation: A synergistic cancer therapeutic approach, accelerated by magnetic hyperthermia. *Int. J. Nanomedicine.* 9, **2014**, 437–459.

91. Sivakumar B, Aswathy RG, Nagaoka Y, Suzuki M, Fukuda T, Yoshida Y, Maekawa T, Sakthikumar D. Multifunctional carboxymethyl cellulose-based magnetic nanovector as a theragnostic system for folate receptor targeted chemotherapy, imaging, and hyperthermia against cancer. *Langmuir.* 29, **2013**, 3453–3466.

92. Corr SA, Rakovich YP, Gun'ko YK. Multifunctional magnetic-fluorescent nanocomposites for biomedical applications. *Nanoscale. Res. Lett.* 3, **2008**, 87–104.

93. Mornet S, Vasseur S, Grasset F, Duguet E. Magnetic nanoparticle design for medical diagnosis and therapy. *J. Mater. Chem.* 14, **2004**, 2161–2164.

94. Chen KL, Mylon SE, Elimelech M. Aggregation kinetics of alginate-coated hematite nanoparticles in monovalent and divalent electrolytes. *Environ. Sci. Technol.* 40, **2006**, 1516–1523.

95. Chen TJ, Cheng TH, Chen CY, Hsu SCN, Cheng TL, Liu GC, Wang YM. Targeted herceptin dextran iron oxide nanoparticles for noninvasive imaging of HER2/neu receptors using MRI. *J. Biol. Inorg. Chem.* 14, **2009**, 253–260.

96. Lu Z, Prouty MD, Quo Z, Golub VO, Kumar CSSR, Lvov YM. Magnetic switch of permeability for polyelectrolyte microcapsules embedded with Co@Au nanoparticles. *Langmuir.* 21, **2005**, 2042–2050.

97. Tamer U, Gundogdu Y, Boyaci IH, Pekmez K. Synthesis of magnetic core.shell Fe_3O_4. Au nanoparticle for biomolecule immobilization and detection. *J. Nanopart. Res.* 12, **2010**, 1187–1196.

98. Tian J, Feng YK, Xu YS. Ring opening polymerization of D,L-lactide on magnetite nanoparticles. *Macromol. Res.* 14, **2006**, 209–213.

99. Li P, Zhu AM, Liu QL, Zhang QG. Fe_3O_4/poly(*N*-isopropylacrylamide)/chitosan composite microspheres with multiresponsive properties. *Ind. Eng. Chem. Res.* 47, **2008**, 7700–7706.

100. Lu CW, Hung Y, Hsiao JK, Yao M, Chung TH, Lin YS, Wu SH, Hsu SC, Liu HM, Mou CY, Yang CS, Huang DM, Chen YC. Bifunctional magnetic silica nanoparticles for highly efficient human stem cell labeling. *Nano. Lett.* 7, **2007**, 149–154.

101. Mahmoudi M, Simchi A, Imani M. Cytotoxicity of uncoated and polyvinyl alcohol coated superparamagnetic iron oxide nanoparticles. *J. Phys. Chem. C.* 113, **2009**, 9573–9580.

102. Mahmoudi M, Simchi A, Imani M, Hafeli UO. Superparamagnetic iron oxide nanoparticles with rigid cross-linked polyethylene glycol fumarate coating for application in imaging and drug delivery. *J. Phys. Chem. C.* 113, **2009**, 8124–8131.

103. Kumar CS, Mohammad F. Magnetic nanomaterials for hyperthermia-based therapy and controlled drug delivery. *Adv. Drug Deliv. Rev.* 63, **2011**, 789–808.

104. Derfus AM, Maltzahn GV, Harris TJ, Duza T, Vecchio KS, Ruoslahti E, Bhatia SN. Remotely triggered release from magnetic nanoparticles. *Adv. Mater.* 19, **2007**, 3932–3936.

105. McGill SL, Cuylear CL, Adolphi NL, Osiński M, Smyth HD. Magnetically responsive nanoparticles for drug delivery applications using low magnetic field strengths. *IEEE T. Nanobioscience.* 8, **2009**, 33–42.

106. Brazel CS. Magnetothermally-responsive nanomaterials: Combining magnetic nanostructures and thermally-sensitive polymers for triggered drug release. *Pharm. Res.* 26, **2009**, 644–656.

107. Chilkoti A, Dreher MR, Meyer DE, Raucher D. Targeted drug delivery by thermally responsive polymers. *Adv. Drug. Deliv. Rev.* 54, **2002**, 613–630.

108. Brazel CS, Peppas NA. Pulsatile local delivery of thrombolytic and antithrombotic agents using poly(*N*-isopropylacrylamide-*co*-methacrylic acid) hydrogels. *J. Control Release.* 39, **1996**, 57–64.

109. Hayashi K, Ono K, Suzuki H, Sawada M, Moriya M, Sakamoto W, Yogo T. High-frequency, magnetic-field-responsive drug release from magnetic nanoparticle/organic hybrid based on hyperthermic effect. *ACS Appl. Mater. Interfaces.* 2, **2010**, 1903–1911.

110. Mertoglu M, Garniera S, Laschewsky A, Skrabania K, Storsberg J. Stimuli responsive amphiphilic block copolymers for aqueous media synthesised via reversible addition fragmentation chain transfer polymerisation (RAFT). *Polymer.* 46, **2005**, 7726–7740.

111. Muqbil I, Masood A, Sarkar FH, Mohammad RM, Azmi AS. Progress in nanotechnology based approaches to enhance the potential of chemopreventive agents. *Cancers.* 3, **2011**, 428–445.

112. Si HY, Li DP, Wang TM, Zhang HL, Ren FY, Xu ZG, Zhao YY. Improving the anti-tumor effect of genistein with a biocompatible superparamagnetic drug delivery system. *J. Nanosci. Nanotechnol.* 10, **2010**, 2325–2331.

113. Huang H, Delikanli S, Zeng H, Ferkey DM, Pralle A. Remote control of ion channels and neurons through magnetic-field heating of nanoparticles. *Nat. Nanotechnol.* 5, **2010**, 602–606.

114. Dimitrov I, Trzebicka B, Muller AHE, Dworak A, Tsvetanov CB. Thermosensitive water-soluble copolymers with doubly responsive reversibly interacting entities. *Prog. Polym. Sci.* 32, **2007**, 1275–1343.

115. Meenach SA, Anderson KW, Hilt JZ. Synthesis and characterization of thermoresponsive poly(ethylene glycol)-based hydrogels and their magnetic nanocomposites. *J. Polym. Sci. A Polym. Chem.* 48, **2010**, 3229–3235.

116. Jordan A, Scholz R, Wust P, Schirra H, Schiestel T, Schmidt H, Felix R. Endocytosis of dextran and silan-coated magnetite nanoparticles and the effect of intracellular hyperthermia on human mammary carcinoma cells in vitro. *J. Magn. Magn. Mater.* 194, **1999**, 185–196.

117. Hilger I, Hiergeist R, Hergt R, Winnefeld K, Schubert H, Kaiser WA. Thermal ablation of tumors using magnetic nanoparticles: An in vivo feasibility study. *Investigative Radiol.* 37, **2002**, 580–586.

118. Ito A, Tanaka K, Kondo K, Shinkai M, Honda H, Matsumoto K, Saida T, Kobayashi T. Tumor regression by combined immunotherapy and hyperthermia using magnetic nanoparticles in an experimental subcutaneous murine melanoma. *Cancer Sci.* 94, **2003**, 308–313.

119. Matsuoka F, Shinkai M, Honda H, Kubo T, Sugita T, Kobayashi T. Hyperthermia using magnetite cationic liposomes for hamster osteosarcoma. *Biomagn. Res. Technol.* 25, **2004**, 3.

120. Maier-Hauff K, Rothe R, Scholz R, Gneveckow U, Wust P, Thiesen B, Feussner A, von Deimling A, Waldoefner N, Felix R, Jordan A. Intracranial thermotherapy using magnetic nanoparticles combined with external beam radiotherapy: Results of a feasibility study on patients with glioblastoma multiforme. *J. Neurooncol.* 81, **2007**, 53–60.

121. Du Y, Zhang D, Liu H, Lai R. Thermochemotherapy effect of nanosized As_2O_3/Fe_3O_4 complex on experimental mouse tumors and its influence on the expression of CD44v6, VEGF-C and MMP-9. *BMC Biotechnol.* 5, **2009**, 84.

122. Le Renard PE1, Buchegger F, Petri-Fink A, Bosman F, Rüfenacht D, Hofmann H, Doelker E, Jordan O. Local moderate magnetically induced hyperthermia using an implant formed in situ in a mouse tumor model. *Int. J. Hyperthermia.* 25, **2009**, 229–239.

123. Balivada S, Rachakatla RS, Wang H, Samarakoon TN, Dani RK, Pyle M, Kroh FO, Walker B, Leaym X, Koper OB, Tamura M, Chikan V, Bossmann SH, Troyer DL. A/C magnetic hyperthermia of melanoma mediated by iron(0)/iron oxide core/shell magnetic nanoparticles: A mouse study. *BMC Cancer.* 30, **2010,** 119.

124. Bae KH, Park M, Do MJ, Lee N, Ryu JH, Kim GW, Kim C, Park TG, Hyeon T. Chitosan oligosaccharide-stabilized ferrimagnetic iron oxide nanocubes for magnetically modulated cancer hyperthermia. *ACS Nano.* 6, **2012**, 5266–5273.

125. Guardia P, Di Corato R, Lartigue L, Wilhelm C, Espinosa A, Garcia-Hernandez M, Gazeau F, Manna L, Pellegrino T. Water-soluble iron oxide nanocubes with high values of specific absorption rate for cancer cell hyperthermia treatment. *ACS Nano.* 6, **2012**, 3080–3091.

126. Kumar S, Daverey A, Khalilzad-Sharghi V, Sahu NK, Kidambi S, Othman SF, Bahadur D. Theranostic fluorescent silica encapsulated magnetic nanoassemblies for in vitro MRI imaging and hyperthermia, *RSC Adv.* 5, **2015**, 53180–53188.

127. Liu XL, Ng CT, Chandrasekharan P, Yang HT, Zhao LY, Peng E, Lv YB, Xiao W, Fang J, Yi JB, Zhang H, Chuang KH, Bay BH, Ding J, Fan HM. Synthesis of ferromagnetic $Fe_{0.6}Mn_{0.4}O$ nanoflowers as a new class of magnetic theranostic platform for in vivo T_1-T_2 dual-mode magnetic resonance imaging and magnetic hyperthermia therapy. *Adv. Healthcare Mater.* 5, **2016**, 2092–2104.

128. Hemery G, Genevois C, Couillaud F, Lacomme S, Gontier E, Ibarboure E, Lecommandoux S, Garanger E, Sandre O. Monocore vs. multicore magnetic iron oxide nanoparticles: Uptake by glioblastoma cells and efficiency for magnetic hyperthermia. *Mol. Syst. Des. Eng.* 2, **2017**, 629–639.

129. Mondal S, Manivasagan P, Bharathiraja S, Santha Moorthy M, Nguyen V, Kim H, Nam S, Lee K, Oh J. Hydroxyapatite coated iron oxide nanoparticles: A promising nanomaterial for magnetic hyperthermia cancer treatment. *Nanomaterials.* 7, **2017**, 426.

130. Cho M, Cervadoro A, Ramirez M, Stigliano C, Brazdeikis A, Colvin V, Civera P, Key J, Decuzzi P. Assembly of iron oxide nanocubes for enhanced cancer hyperthermia and magnetic resonance imaging. *Nanomaterials.* 7, **2017**, 72.

131. Ling Y, Tang X, Wang F, Zhou X, Wang R, Deng L, Shang T, B Liang, Li P, Ran H, Wang Z, Hu B, Li C, Zuo G, Zheng Y. Highly efficient magnetic hyperthermia ablation of tumors using injectable polymethylmethacrylate-Fe_3O_4. *RSC Adv.* 7, **2017**, 2913–2918.

132. Xu H, Zong H, Ma C, Ming X, Shang M, Li K, He X, Cao L. Evaluation of nano-magnetic fluid on malignant glioma cells. *Oncol Lett.* 13, **2017**, 677–780.

133. Wang F, Yang Y, Ling Y, Liu J, Cai X, Zhou X, Tang X, Liang B, Chen Y, Chen H, Chen D, Li C, Wang Z, Hu B, Zheng Y. Injectable and thermally contractible hydroxypropyl methyl cellulose/Fe_3O_4 for magnetic hyperthermia ablation of tumors. *Biomaterials.* 128, **2017**, 84–93.

134. Wu H, Song L, Chen L, Huang Y, Wu Y, Zang F, An Y, Lyu H, Ma M, Chen J, Gu N, Zhang Y. Injectable thermosensitive magnetic nanoemulsion hydrogel for multimodal-imaging-guided accurate thermoablative cancer therapy, *Nanoscale.* 9, **2017**, 16175–16182.

135. Jurek PM, Zabłocki K, Waśko U, Mazurek MP, Otlewski J, Jeleń F. Anti-FGFR1 aptamer-tagged superparamagnetic conjugates for anticancer hyperthermia therapy. *Int. J. Nanomed.* 12, **2017**, 2941–2950.

136. Cho HY, Lee T, Yoon J, Han Z, Rabie H, Lee KB, Su WW, Choi JW. Magnetic oleosome as a functional lipophilic drug carrier for cancer therapy. *ACS Appl. Mater. Interfaces.* 10, **2018**, 9301–9309.

137. Guisasola E, Asín L, Beola L, De La Fuente JM, Baeza A, Vallet-Regí M. Beyond traditional hyperthermia: In vivo cancer treatment with magnetic-responsive mesoporous silica nanocarriers. *ACS Appl. Mater. Interfaces.* 10, **2018**, 12518–12525.

138. Jang JT, Lee J, Seon J, Ju E, Kim M, Il Kim Y, Kim MG, Takemura Y, Arbab AS, Kang KW, Park KH, Paek SH, Bae S. Giant magnetic heat induction of magnesium-doped γ-Fe_2O_3 superparamagnetic nanoparticles for completely killing tumors. *Adv. Mater.* 30, **2018**, 1704362.

139. Dennis CL, Jackson AJ, Borchers JA, Hoopes PJ, Strawbridge R, Foreman AR, van Lierop J, Grüttner C, Ivkov R. Nearly complete regression of tumors via collective behavior of magnetic nanoparticles in hyperthermia. *Nanotechnology.* 20, **2009**, 395103

140. AS Eggeman, SA Majetich, D Farrell, QA Pankhurst. Size and concentration effects on high frequency hysteresis of iron oxide nanoparticles. *IEEE Trans. Magnetics.* 43, **2007**, 2451–2453.

141. Giustini AJ, Ivkov R, Hoopes PJ. An in vivo transmission electron microscopy study of injected dextran-coated iron-oxide nanoparticle location in murine breast adenocarcinoma tumors versus time. *Energy-Based Treatment of Tissue and Assessment*, V 7181, **2009**, 71810.

142. Gordon RT, Hines JR, Gordon D. Intracellular hyperthermia a biophysical approach to cancer treatment via intracellular temperature and biophysical alterations. *Med. Hypothesis.* 5, **1979**, 83–102.

143. Maqueda RJ, Wurden GA, Terry JL, Stillerman JA. The new infrared imaging system on Alcator C-Mod. *Rev. Sci. Instrum.* 70, **2000**, 734–737.

144. Maswadi SM, Dodd SJ, Gao JH, Glickman RD. Temperature mapping of laser induced hyperthermia in an ocular phantom using magnetic resonance thermography. *J. Biomed. Opt.* 9, **2004**, 711 718.

145. Le Bihan DL, Delannoy J, Levin RL. Temperature mapping with MR imaging of molecular diffusion: Application to hyperthermia. *Radiology.* 171, **1989**, 853–857.

146. Amao Y, Okura I. Fullerene C_{60} immobilized in polymethylmethacrylate film as an optical temperature sensing material. *Analysis.* 28, **2000**, 847–849.

147. www.sennewald.de.

148. www.magforce.de.

149. Ito A, Shinkai M, Honda H, Yoshikawa K, Saga S, Wakabayashi T, Yoshida J, Kobayashi T. Heat shock protein 70 expression induces antitumor immunity during intracellular hyperthermia using magnetite nanoparticles. *Cancer Immunol. Immun.* 52, **2003**, 80–88.

150. Sincai M, Ganga D, Ganga M, Argherie D, Bica D. Antitumor effect of magnetite nanoparticles in cat mammary adenocarcinoma. *J. Magn. Magn. Mater.* 293, **2005**, 438–441.

151. Ito A, Nakahara Y, Fujioka M, Kobayashi T, Takeda T, Nakashima I, Honda H. Complete regression of hereditary melanoma in a mouse model by repeated hyperthermia using magnetite cationic liposomes. *Jpn. J. Hyperthermic Oncol.* 11, **2005**, 139–149.

152. Sivasai B, Rajashekar R, Hongwang W, Thilani SN, RajKumar D, Marla P, Franklin KO, Brandon W, Xiaoxuan L, Olga KB, Masaaki T, Viktor C, Stefan BH, Deryl TL. A/C magnetic hyperthermia of melanoma mediated by iron(0)/iron oxide core/shell magnetic nanoparticles: A mouse study. *BMC Cancer.* 10, **2010**, 1–9.

153. Suzuki M, Shinkai M, Honda H, Kobayashi T. Anti-cancer effect and immune induction by hyperthermia of malignant melanoma using magnetite cationic liposomes. *Melanoma Res.* 13, **2003**, 129–135.

Chapter 10

Phototherapy Using Nanomaterials

Resmi A. N.,[a] Resmi V. Nair,[a] Rekha C. R.,[a] Lakshmi V. Nair,[b] Shaiju S. Nazeer,[c] and Ramapurath S. Jayasree[a]

[a]*Division of Biophotonics and Imaging, Biomedical Technology Wing, Sree Chitra Tirunal Institute for Medical Sciences & Technology, Poojappura, Thiruvananthapuram, Kerala, India*
[b]*School of Materials Science and Engineering, NIT, Kozhikode, Kerala, India*
[c]*Department of Chemistry, Indian Institute of Space Sciences and Technology, Thiruvananthapuram, Kerala, India*

jayasree@sctimst.ac.in

10.1 Introduction

Cancer has been one of the leading causes of threat to human life for several years. Common therapeutic approaches currently used for its management include radiotherapies, chemotherapies, and surgeries. However, chemotherapies and radiotherapies are known for their poor selectivity toward cancer cells, thereby causing severe side effects to normal tissues. In the case of surgeries, complete removal of all cancer cells is hardly achieved in many cases. Newer, more efficient modes of therapies are always in demand and many techniques have emerged over the period of time. In view of this, there is a high demand for more efficient therapies, which can establish sufficient treatment

Bionanotechnology in Cancer: Diagnosis and Therapy
Edited by D. Sakthi Kumar and Aswathy Ravindran Girija

ISBN 978-981-4800-30-3 (Hardcover), 978-0-429-42291-1 (eBook)
www.jennystanford.com

outcomes. Phototherapy is an age-old technique, which was in use over many centuries and treatment with sunlight or "heliotherapy" was employed for curing skin diseases. More than 3500 years ago, ancient Egyptian and Indian healers adopted the technique of ingesting plant extracts or seeds, besides sunlight for treating "leucoderma." However, real modern phototherapy began with Nobel Prize winner Niels Finsen, the father of ultraviolet therapy. He developed a "chemical rays" lamp for treating patients with skin tuberculosis. The curative possibilities in oncology were widely explored by the introduction of extracorporeal photopheresis in 1987 for cutaneous T-cell lymphoma and of topical photodynamic therapy (PDT). However, there have been several limitations to these techniques though these were found to be very effective. Phototherapies based on nanomaterials have attracted extensive interest in recent years with the rapid development of nanoscience and nanotechnology. Nanomaterials exhibit unique physicochemical properties, due to their large surface-to-volume ratio. The biodistribution, cell uptake, pharmacokinetics, and targeting of these nanoparticles can be improved by modifying their surface with functional groups. They offer selectivity to tumor cells by means of light sources with superior tissue penetration ability in combination with phototherapeutic agents. This chapter will discuss about the phototherapy applications using nanomaterials.

10.2 Photothermal Therapy (PTT)

Cancer is one of the leading causes of human mortality worldwide, and different strategies have been employed for its treatment such as chemotherapy, radiotherapy, immunotherapy, and surgery. Nevertheless, there is a high demand for new treatment strategies, due to heterogeneous nature of tumor and undesired effects of conventional treatments on normal cells [1, 2]. PTT is one among the new approaches for cancer treatment and it employs light mediated heat therapy. This therapeutic method ablates cancer cells by utilizing heat generated from the electromagnetic radiation in the presence of phototherapeutic molecules [2]. Near-infrared (NIR) rays, visible light, radiofrequency, microwaves, and ultrasound are generally employed for the local heat generation [3]. Electromagnetic radiation excites photothermal sensitizer molecule and heat is released upon relaxation to the ground state. This heat energy is utilized for the treatment/therapeutic purpose. Antitumor activity of heat is owing to cell death, heat shock of proteins, inactivation of membrane proteins and mitochondria that include depolymerization of cytoskeletal filaments, membrane blebbing, and plasma membrane disruption due to cavitation effect [4]. PTT also induces programmed cell death (apoptosis) by activating the intrinsic pathway rather than following necrotic cell death [5], and it stimulates inflammatory responses that may usually compromise the antitumor activity.

Traditional photothermal therapy has been considered as a non-reliable technique for cancer treatment because of the strong extinction coefficients exhibited by human tissues in the visible range of the optical spectrum, limiting photothermal treatments to superficial tumors. In addition, non-selective absorption of light by both healthy and cancerous tissues reduces the efficacy of heat delivery within the tumor and an increment in the non-specific damage of the adjacent tissues. These limitations can be overcome by selecting the light source of wavelength lying in the specific wavelength such as biological windows where light absorption by healthy tissues is minimum. Biological windows are the spectral ranges where tissues become partially transparent, due to a simultaneous reduction in both absorption and scattering (Figure 10.1).

In a first order approximation, extinction of optical radiation into tissues is determined by the presence of absorption bands of constituents (such as water and hemoglobin) and by the scattering caused by natural tissue. The first biological window (I-BW) extends from 700 nm to 980 nm and corresponds to the spectral range defined between the visible absorption band of hemoglobin and the characteristic 980 nm absorption band of water. Here absorption cross-section vanishes, but shows scattering of light. The second biological window (II-BW) extends from 1000 nm to 1400 nm, both limits corresponding to water absorption bands. Here minimum optical scattering is observed with respect to first biological window.

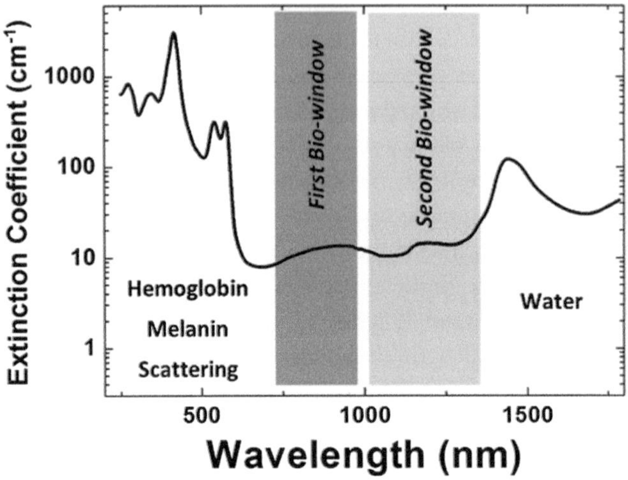

Figure 10.1 Extinction coefficient of a representative tissue. Two biological transparent windows are marked. Reproduced with permission from Ref. [10].

The development of nanomaterials having unique optical and physicochemical properties marks a remarkable step forward in PTT of cancer cells. Nanotechnology is the major driving force behind the technological revolution in many areas such as physical and chemical sciences, biological sciences, health sciences, and other interdisciplinary fields of science and technology [6, 7]. PTT-based cancer therapy has received significant attention with the development of nanotechnology-based photothermal sensitizing agents. In recent years, the remarkable developments in nanotechnology have provided a variety of nanostructures, which exhibit strong optical properties with higher absorption efficiency. The smaller size of nanoparticles aids easy excretion from bodies, which reduces undesired toxicity of nanoparticles left in bodies. Photothermal properties of nanoparticles are owing to their strong and tunable optical absorption in the visible-NIR region. The absorption is due to small particle effect known as surface plasmon resonance (SPR), which is the collective oscillation of conduction band electrons upon interaction with the electromagnetic field. This optical response of nanoparticles can be tuned to biological window, where the absorption by biological materials will be very low. From the perspective of cancer therapeutics, these nanoparticles act as PTT agents due to their larger absorption cross-sections, which ensure effective lower energy laser therapy. On excitation with external laser source, electron-electron scattering occurs in nanoparticles, which leads to electronic relaxation. This will be followed by electron-phonon coupling with the transfer of thermalized electron to nanoparticle lattice. This is a quasi-equilibrium state for nanoparticles and results in the formation of hot phonons [8, 9]. In the case of carbon-based nanomaterials, light to heat conversion is due to de-excitation processes, which involve luminescence and/or non-radiative relaxation [10]. By using these principles, a variety of nanoparticles are employed for PTT, which include gold nanoparticles [11], gold nanorods [12], carbon nanotubes (CNTs) [13], and graphene [14, 15] (Figure 10.2). Photothermal therapeutic approach using gold and carbon-based nanomaterials will be discussed in this section.

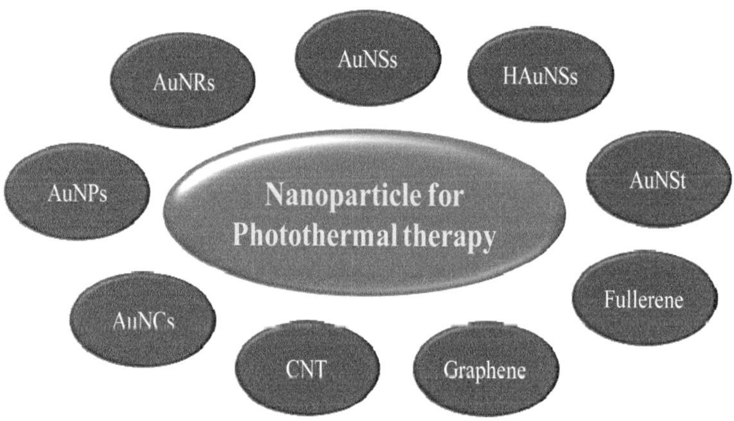

Figure 10.2 Commonly used nanoparticles for PTT: AuNPs – gold nanoparticles, AuNRs – gold nanorods, AuNSs – gold nanoshells, HAuNSs – hollow gold nanoshells, AuNSts – gold nanostars, fullerene, graphene, AuNCs – gold nanocages, and CNTs – carbon nanotubes.

10.2.1 Metal-Based Nanomaterial (Gold)

Gold is one of the unique noble metals, which has been known for its resistance against corrosion and oxidation. Gold nanomaterials have attracted considerable attention due to their interesting optical and physical properties, which are different from their bulk state. These properties have been known for centuries for use in medicinal applications [16]. Bulk gold is yellow in color due to the reduction in reflectivity for light toward the end of the electromagnetic spectrum [17]. When the size of gold is reduced to nanoparticle dimension, the ratio of radius of the particle and wavelength of light becomes important. Metallic nanoparticles exhibit attractive colors, due to the collective oscillation of the electrons in the conduction band, known as the surface plasmon oscillation. The oscillation frequency of these materials falls in the visible region, which results in strong SPR absorption. Free electrons in metals known as plasmons, move freely and can be easily excited by an electromagnetic wave. This collective oscillation of conduction electrons in metal is referred as SPR [6]. In other words, when the wavelength of electromagnetic radiation is much larger than the nanoparticle size, it can set up standing resonance conditions, which lead to the localized surface plasmon resonance (LSPR) in metal nanoparticles [7].

Photothermal property of gold nanoparticles depends on the SPR. Gold nanoparticles exhibit LSPR in the visible region of the electromagnetic spectrum, which indicates that they can absorb light in the visible region. Colloidal gold nanoparticle with 20 nm size exhibits a maximum absorption spectrum at 520 nm. Applying Mie's theory, for small spheres embedded in a medium, the plasmon resonances appear at certain frequencies of the excitation light, depending on the dielectric constants of the metal and its surrounding medium and on the diameter of the spherical particles [18]. However, when the size of gold nanoparticles changes to 99 nm, a significant red shift is observed in the absorption spectrum [19]. Absorption maxima can be tuned to NIR region for efficient phototherapy by controlling the-morphology of nanoparticles. On absorption of light, nanoparticles convert the energy into heat, which, in turn, can be utilized for PTT, which mostly suited for the destruction of cancerous cells. Gold nanomaterials are suitable for photothermal treatment of cancer because of lower cytotoxicity, efficient tumor uptake, activation using NIR laser light, and efficacy to multifaceted cancer therapy and drug delivery [20].

It has been observed that increasing the size of gold nanoparticles does not allow tuning of surface plasmon absorption beyond 600 nm. So, to achieve the required excitation matching, the biological window for efficient PTT and metallic nanoparticles with more complex geometries such AuNSs, AuNRs, HAuNSs, AuNCs, and AuNSts have been proposed. When the shape of the metallic nanoparticle deviates from the highly symmetric spherical shape, the dynamics of plasmonic oscillations is strongly modified. In order to tune gold nanomaterials for photothermal application, several design criteria are considered such as simple surface functionalization or encapsulation chemistry, plasmon resonance tunability, and high photothermal conversion efficiency [21, 22]. Rastinehad et al. published the initial results of a clinical trial of gold nanomaterial for using photothermal treatment for prostate cancer [23]. PTT based on some of these popular nanostructures will be discussed in the following sections.

10.2.1.1 Gold nanoshells (AuNSs)

Nanoshells are special class of nanocomposites materials. These are spherical nanoparticles having a dielectric core surrounded by thin gold shell. Generally, this core will be either hollow or silica/liposome core [24]. The uniqueness of AuNSs is that their SPR can be easily tunable from visible to NIR region especially in the therapeutic window (Figure 10.3). The light scattering is high and absorption is low in the therapeutic window. Due to this property, AuNSs have a wide range of applications in the field of photoacoustic imaging [25], PTT [26, 27], and optical coherence tomography (OCT) [28, 29].

The plasmon resonance frequency of AuNS can be tuned by changing the aspect ratio of shell thickness to the overall diameter [30]. AuNSs are excellent candidates for applications in the biomedical field, due to their inert nature and non-toxicity in liver, spleen, or kidney in mice [31]. The size of these nanoshells helps them to preferentially accumulate at tumor sites and they are efficient enough to convert the absorbed radiations into heat [32]. AuNSs with higher surface area can be fabricated to adopt flexible properties such as enhanced cellular uptake and drug loading efficiency. AuNSs can be engineered to absorb radiation at various frequencies and specific binding sites on the shell allow the target specificity to the materials.

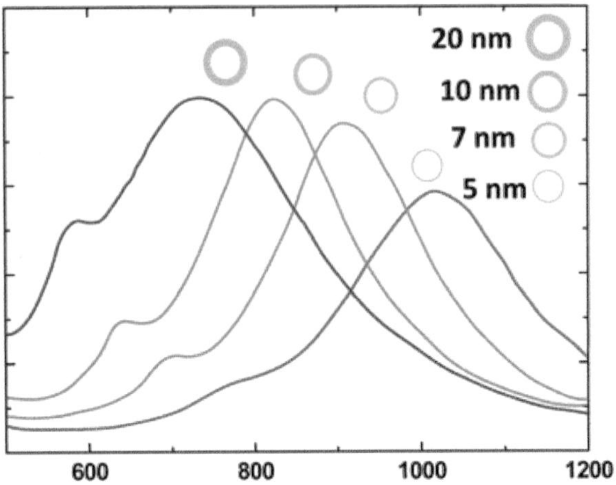

Figure 10.3 Extinction cross-section spectra of core (silica)-shell (gold) nanoparticles with different shell thicknesses. Reproduced with permission from Ref. [10].

When these nanoparticles are entrapped by cancer cells, laser source of suitable wavelength can excite them. When laser light passes through the body and reaches the nanoshell and upon absorption, light energy is converted into heat energy destroying the cells [33]. The underlying mechanism for this effect can be explained in the following way. The energy of NIR light is low and living cells do not absorb this radiation. The wave like excitation of electrons on the surface of AuNSs can convert the absorbed electromagnetic light into heat within 4–6 min of irradiation. PEGylated gold-silica nanoshells with 810 nm absorption were used after intravenous injection along with 808 nm laser (800 mW) at 4 W/cm^2 for 3 min, for PTT in tumor bearing immune-competent mice [27]. Accumulation of nanoshells in the tumor with 3–5.5 mm diameter was observed after 6 h. The temperature measured at the tumor site showed significant rise (~50 °C) by 30 s with the mean temperature of the laser and nanoshell-treated tumors than NIR-treated controls without nanoshell injection. Recovery from cancer was observed among the test animal group within 10 days. After 90 days of treatment, all mice remained healthy and free of tumors. The study highlights the high accumulation of AuNS in the tumor site by the enhanced permeability and retention (EPR) effect and photothermal efficacy of the material.

Rod-like mesoporous silica was coated with AuNSs (MSNR@Au hybrid) and this was modified with ultra-small gadolinium (Gd)-chelated supramolecular photosensitizer 5,10,15,20-Tetrakis (4-sulfonatophenyl)-porphyrin (TPPS), which was further used for therapies (PTT [using NIR laser] and PDT), quad modal imaging (using NIR fluorescence), multispectral opto-acoustic tomography (MSOT), computed tomography (CT), and magnetic resonance (MR). This material gives fluorescence signal within 4 h and get intense after 12 h due to the presence of TPPS in the system. It retains after 24 h indicating strong binding of AuNS construct in the tumor region. The biodistribution using fluorescence showed intense signal from liver other than tumor. MSOT imaging after 12 h of intravenous injection showed 2.8 times signal compared with pre-injected signal. The photoelectric property of gold in the nanoshell was used for CT imaging and an increase in contrast value was observed in the tumor region from 125 HU to 290 HU. The presence of Gd in the system gave T1-weighted MR imaging, and after 12 h, it showed 689.3% rise of the T1-weighted signal. These high intense signals were the clear evidence for the long-term accumulation of rod-shaped AuNS in the tumor region. The rod-shaped system generates 55.9 °C on irradiation with 808 nm laser irradiation (1.5 W/cm^2, 5 min) in vivo. The presence of photosensitizer provides PDT effect to the system. The combined effect of PDT and PTT using 660 nm and 808 nm lasers showed significant decrease in tumor volume and high treatment efficacy [34].

The anti-HER 2 antibody (over expressed in breast cancer) conjugated with PEG-AuNS has been accumulated in the cancer cells actively. The scattering of light by AuNS impart molecular-specific image contrast and the absorption of light induces cancer treatment based on NIR PTT. The active absorption helps AuNSs to bind to cancer cells more strongly, which increases the scattering efficacy compared with passive absorption. The darkfield microscope sensitive to scattered light was used for the imaging of AuNS. Immediately after imaging the system was used for PTT with 820 nm laser of 0.008 W/m^2 for 7 min. Silver staining, darkfield images, and Calcein AM staining showed the high efficiency of AuNS giving high uptake, imaging capacity along with photothermal property. The high cellular uptake efficacy of AuNS was increased with the help of rod-shaped mesoporous silica by endocytosys. This rod-shaped morphology made the system more suitable in the drug loading [35]. Similarly, modification using human serum albumin (HSA), indocyanin green, and folic acid to a branched nanostructure AuNSs with PLGA/DOXO-cores resulted in enhanced targeting efficacy and photothermal property [36]. It showed a virus capsid with a size of ca. 85 ± 15 nm and was grown on the doxorubicin-loaded polymeric matrix (PLGA/DOXO) with absorption of 770 nm; whereas in the absence of AuNS, polymeric nanomatrix showed only 490 nm absorption. This combined structure was again functionalized with human serum albumin/indocyanine green/folic acid complex (HSA-ICG-FA) for the incorporation of fluorescence imaging, PDT, and target specificity. This combination increases the absorption of AuNS to absorption peak maximum of 780 nm. For the therapy, laser of 808 nm and 2 W/cm^2 was irradiated and it showed an increase in temperature more above glass transition temperature of poly lactic-coglycolic acid (PLGA), which triggered drug release and also it initiated PDT in addition to the normal PTT.

A clinical trial for PTT using gold-silica nanoshells for prostate cancer upon laser excitation in combination with MR–ultrasound fusion imaging has been conducted [23]. The feasibility and safety data from 16 cases of patients identified with low or intermediate-risk localized prostate cancer was reported by the initial results of this device study. A superior efficacy was observed when a NIR light (810 ± 10 nm) was delivered continuously for 3 min with a median concentration of 8.28 μg/g (range: 1.15–33.12 μg/g) gold in the tumors. This study demonstrated that AuNS-based phototherapy is a safe and technically feasible procedure for the targeted destruction of prostate tumors. Even though these results showed nanoshells as promising PTT candidates, more studies in this field are still required to explore their photothermal efficacies in clinical use.

10.2.1.2 Gold nanorods (AuNRs)

AuNRs are gold nanostructures with narrow size distribution and are elongated along one direction. They have tunable absorption wavelength from visible to NIR region of the electromagnetic spectrum (Figure 10.4) upon changing the length to width ratio. This ratio of AuNRs is known as the aspect ratio. AuNRs possess unique optical properties compared to spherical nanoparticles due to LSPR [37–39]. Independent collective electron vibrations of AuNRs generate two type of oscillations viz. transverse (diameter wise vibration) surface plasmon resonance absorption (TSPR) in the visible range and longitudinal surface plasmon resonance absorption (lengthwise vibration). The vibration along the longitudinal axis results in the scattering of the incident light by the rod, which varies widely from the visible region to the NIR region [40]. The aspect ratio of AuNRs plays a significant role in tuning the optical properties of these materials for desired application.

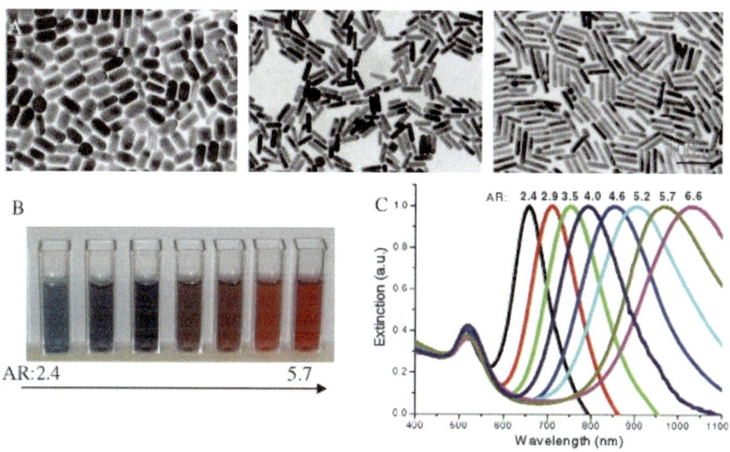

Figure 10.4 Tunable optical properties of AuNRs by changing the aspect ratios. AuNRs of different aspect ratios exhibit different dimensions as seen by TEM (A), in different color (B), and different SPR wavelength (C). Reproduced with permission from Ref. [39].

AuNRs have gained significant attention in PTT due to their unique NIR absorption properties [41]. El Sayed and co-workers have extensively studied AuNRs mediated plasmonic imaging and PTT by conjugating with anti-epidermal growth factor receptor (EGFR) antibody to the surfactant conjugated nanorods [42]. Darkfield imaging of cancer cells was conducted using scattered light of AuNRs and a NIR low-energy continuous-wave (CW) laser of 800 nm with 1 mm spot diameter was irradiated for 4 min for PTT. The laser power of above 160 mW (20 W/cm^2) showed complete destruction of all cells, but power of above 80 mW, which corresponds to 10 W/cm^2, showed photothermal effect. The high NIR absorption and target specificity of AuNRs lead to photothermal efficiency with low-energy harmless NIR laser. AuNRs along with defective titanium dioxide hybrid nanoparticles are fabricated for combined photodynamic and photothermal therapies. The combined system showed absorption from 700 nm to 1000 nm, which increases the photothermal property other than absorption at 525 nm. AuNRs-TiO_2 nanocomposite showed a temperature generation of 45 °C upon irradiation with 808 nm laser for 5 min. To evaluate the phototherapy, the authors did the same in vitro experiment dark without laser irradiation and showed high cell viability like untreated cells. It suggests that AuNR-TiO_2 nanocomposites induce phototherapies with the help of visible and NIR light. The reactive oxygen species (ROSs) generation from TiO_2 on irradiation with visible light increases with the help of LSPR-induced hot electrons and heat generation from AuNRs make the system more effective [43].

AuNRs, surface modified with different organic molecules, have been used for PTT by utilizing their NIR absorption property [44, 45]. In order to get high efficacy in PTT, loading of AuNRs to tumor region was increased by incorporation of AuNRs in pluripotent stem cell (AuNR-iPS). AuNR-iPS has been demonstrated for high targeting and accumulation in human gastric cancer cells. Heat is generated upon irradiation with 808 nm laser to the AuNR-iPS targeted gastric cancer cell. As a result, it showed PTT induces apoptosis on irradiation with 808 nm laser [46]. AuNRs in combination with other therapeutic agents have shown multimodal therapeutic effect. Doxorubicin loaded with AuNRs functionalized squaraine showed a turn on-turn off PTT and chemotherapy. Here AuNRs were used for PTT doxorubicin for chemotherapy and squaraine dye (SQ) as Raman reporter and using surface enhanced Raman scattering (SERS) property of AuNRs. Metalloproteinases (MMP) are class of zinc-dependent proteins highly expressed in many human cancer cells and recognition of MMP could provide valuable information in cancer progression. In order to detect MMP expressing cancer cells, a targeted theranostic nanoprobe designated as MMP-SQ@GNR@LAH-DOX was developed and cell studies were done in human fibrosarcoma cell line HT1080, which over-expresses MMPs. MMP-SQ@GNR@LAH-DOX was irradiated with laser (808 nm with power of 0.1 W/cm^2 for 2 min), which showed 63% apoptosis compared to 9% laser alone [47]. AuNRs were conjugated with protoporphyrin IX (PPIX) photothermal and PDT [48]. In this, the weak fluorescence of AuNRs was utilized for imaging of cancer. PPIX acts as a photodynamic therapeutic agent and electron acceptor in imaging process. Irradiation with 808 nm laser for 50 s showed considerable cell death due to photothermal property of AuNRs. Mitoxantrone-loaded polyethylene glycol (PEG)-functionalized AuNRs were used for photo-chemo therapy [49]. The loaded drug was delivered on irradiation with 808 nm laser of power 1 W for 1.30 min, which increased the heat generation and resulted in PTT. Here, SERS property of AuNRs was used for the evaluation of treatment efficacy. In another approach, AuNR and doxorubicin loaded in Halloysite nanotubes have been used for PTT and it was conjugated with folic acid and bovine serum albumin (BSA). This combined system showed a temperature generation of 26.8 °C after 8 min of 808 nm laser irradiation under 0.8 W/cm^2 [50]. This temperature increase triggered the chemotherapy and gave a PTT/chemotherapy effect and showed survival rate of 7.4% in MCF7 cells. Even though different approaches have been demonstrated to increase the photothermal therapeutic properties of AuNRs, more studies are needed in this direction as this modality has not been still used in clinical trials.

10.2.1.3 Hollow gold nanoshells (HAuNSs)

Hollow gold nanostructures have great potential applications in the biomedical field, owing to their significant LSPR feature, excellent biocompatibility, and photothermal conversion properties. These structures got popularized due to their low scattering efficacy such as in HAuNSs. They are considered as the second-generation gold nanostructures. HAuNS possess hollow interior with spherical shape of outer diameter of 30–50 nm. Due to the uniform structure, they showed strong tunable absorption band at NIR wavelength region. As described in the introduction, heat transfer from nanoparticle to its surroundings creates a rise in the local cell temperature, which, in turn, causes cell death.

The SPR frequency of HAuNSs is red-shifted by either an increase in outer diameter or a decrease in shell thickness (Figure 10.5) [51]. Sarah A. Lindley and Jin Z. Zhang demonstrated that by modifying the surface roughness of HAuNS, the absorption window can be tuned to NIR region. The roughness mainly depends on the hydroxyl ions present in the solution [52].

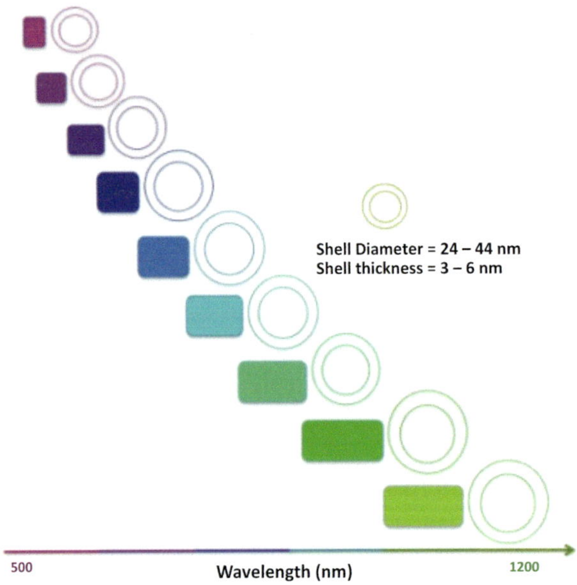

Figure 10.5 Absorption spectral range of AuNSs with varied shell diameter and thickness. Reproduced with permission from Ref. [24].

HAuNSs were synthesized using cobalt nanoparticle, which was prepared by nucleation process with galvanic exchange with chloroauric acid ($HAuCl_4$). These reactions are carried out in inert atmosphere. Cobalt nanoparticles are acting as a template for the synthesis of HAuNSs [53]. Li et al. demonstrated the PTT using HAuNS targeted with anti-EGFR monoclonal antibody. Higher photothermal efficacy in A431 cells and corresponding in vivo model was shown here [54]. Molecular specific photothermal coupling agents having EGFR conjugated hollow 30 nm AuNSs show excellent colloidal stability and efficient photothermal effect in the NIR region. A NIR 808 nm (output power of 40 W/cm^2) laser irradiation on A431 cells treated with anti-EGFR-HAuNS shows selective cell destruction compared to those cells treated with anti-EGFR-HAuNS alone, laser alone, or IgG-HAuNS. Using in-labeled HAuNS anti-EGFR-HAuNS could be delivered to EGFR-positive tumors at 6.8% of injected dose per gram of tissue. A preferential delivery to A431 tumor of anti-EGFR-HAuNS compared with IgG-HAuNS was confirmed using microscopic image of excised tumor with scattering signal from nanoshells. HAuNSs in combination with other nanoparticles have been used for imaging and photothermal application. HAuNSs in combination with iron oxide nanoparticles were used for magnetic imaging and PTT using folic acid targeting in HeLa cells. These ultrathin gold shells were synthesized, with an average outer diameter of 52.11 ± 3.54 nm and a shell thickness of 4.27 ± 0.62 nm. They were further coated with hydrophilic poly(vinylpyrrolidone) (PVP). It was further combined with iron oxide nanoparticle and the combined system showed NIR absorption shifted from 800 nm to 850 nm. Using the laser of wavelength 808 nm, the authors showed PTT in HeLa cells [55]. In another study, the HAuNSs were stabilized with a PEG coating and attached with a melanocyte-stimulating hormone (MSH) analog, [Nle4, D-Phe7] a-MSH (NDP-MSH), which is a potent agonist of melanocortin type-1 receptor (MC1R) over expressed in melanoma [56]. Here HAuNS showed absorption of 808 nm, which triggers the successful and selective photothermal ablation on irradiation with laser power of 16 W/cm^2 of 808 nm for 3 min, on B16/F10 melanoma. PTT was confirmed by histological evaluation and fluorodeoxyglucose positron emission tomography ([^{18}F] FDG-PET) at 24 h post NIR region laser irradiation at low-dose energy of 30 J/cm^2. These results showed the potential to mediate targeted PTT of melanoma. The same HAuNSs construct have shown to generate intense photoacoustic signals and used for imaging along with PTT in mice glioma [57].

Even though the hollow gold nanospheres (HGNSs) showed excellent results compared with other solid gold-based systems, more detailed investigation on PTT is still need to be explored. The effect of these nanoparticles in organs other than cancer cells has to be evaluated.

10.2.1.4 Gold nanocages (AuNCs)

AuNCs are hollow, porous structures with dimensions 10–150 nm. It is achieved through silver (Ag) template-engaged galvanic replacement reaction of $HAuCl_4$ [58]. LSPR of nanocages can be tuned from visible to NIR region by controlling the concentration of gold in the synthesis procedures (Figure 10.6). The LSPR of metal nanostructures results from incident light being scattered and absorbed at a resonant frequency due to the collective oscillation of conduction electrons [59]. By varying the size of AuNCs, the relative intensity of the scattering and absorption cross-sections can be tuned. Discrete dipole approximations (DDA) indicate that small edge AuNCs (edge length <45 nm) predominate in absorption of incident light, whereas larger prevails light scattering [60]. Therefore, in order to fabricate nanocages for specific applications, in addition to the position of LSPR, the size and the magnitude of their scattering and absorption cross-sections are also need to be considered.

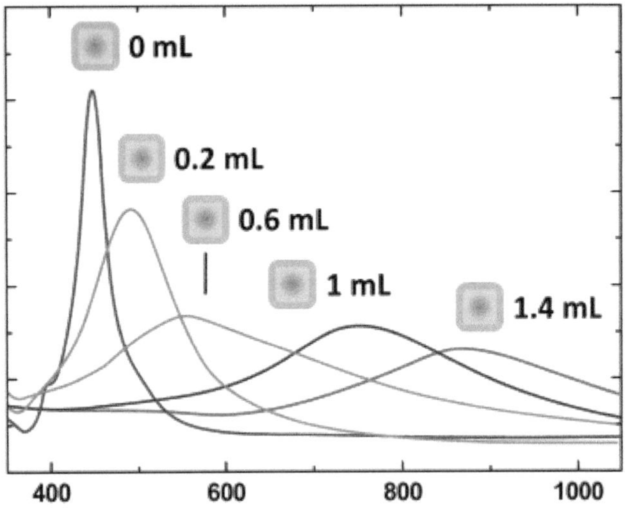

Figure 10.6 Absorption of AuNCs with different amount of gold contents. Reproduced with permission from Ref. [10].

AuNCs with large absorption cross-sections showed high photothermal effect by converting absorbed photons to phonons (lattice vibration). To study the thermal property of nanocages supported on copper transmission electron micrograph (TEM) grid (coated with carbon) exposed to a camera flashes after 10 exposures, the cages were completely melted and they became spherical in shape. In this experiment, the heat is not dissipated to surrounding and so it changes the shape [61]. These nanocages were further functionalized with polyethylene amine and HER2 antibody to target to SKBR3 cell lines. The functionalized nanocages showed absorption peak of 812 nm. Using 810 nm laser irradiation, AuNCs produced heat signals to kill cancer cells [62].

10.2.1.5 Gold nanostars (AuNSts)

AuNSts are another class of gold with protruding prolate tips, which exhibit hybridized plasmons due to a solid core and tips [63]. Depending on the number and length of their arms, the optical absorption varies from visible to NIR. The PTT using these nanostructures with plasmon peaks tunable in the infrared range has been popularized because of the ease of synthesis for large scale production, high surface-to-volume ratio useful for improving drug loading efficiency, and SPR in the "NIR transparent window" for deep tissue imaging [64–67]. Cross-section of AuNSts has played a significant role in photothermal property. Sharp branch tips of nanostars have shown high photothermal properties compared with smooth tips [67]. AuNSts possess superior light to heat conversion efficacy compared to AuNRs with similar plasmon absorption properties [10].

Nanostars with small core and multiple long thin branches exhibit high absorption cross-sections and low scattering making them photothermally efficient. AuNSts incubated with SKBR3 and irradiated with 980 nm laser with power of 15 W/cm^2 showed heat generation followed by cell death within 5 min of irradiation. The same AuNSts were covered with PEG and injected to mice model and subjected to thermal ablation therapy using 780 nm lasers (power of 1.1 W/cm^2) for 10 min. The result suggests high photothermal efficacy of AuNSts and PEG coating increases EPR in vivo system, thereby increasing therapy efficiency [68]. Another study described that when TAT peptide conjugated with AuNSts along with PEG showed high retention time in BT549 breast cancer cells. Photothermal effect was studied using 850 nm pulsed laser under 0.2 W/cm^2 irradiation and showed an increase in photothermal efficacy of AuNSts [69]. Nanostars can be combined with other therapeutic agents for PTT due to their peculiar characteristics. In one of the studies, PEG with different terminal groups (long-chain amine/carboxyl-terminated) was functionalized with AuNSts, so that it became pH response smart system. On the increase in surface functionality with amine and carboxylic groups, the cellular uptake increased even though surface charge is neutral. The pH mediated change in uptake and PTT using AuNts was seen in HeLa cells. In the normal cells with pH 7.4, the internalization is shown to be negligible and so the response is very less, and in case of cancer cells with pH 6.4, significant increase in uptake and efficient therapy is observed on irradiation with 808 nm laser at 2 W/cm^2 for 3 min [70]. AuNSts were functionalized with cyclic arginine-glycine-aspartic acid (RGD), MPA, and doxorubicin, for targeting, fluorescence imaging, and chemotherapy, respectively. This system demonstrated multimodal therapy both in vitro and in vivo. Photothermal therapy and chemotherapy were demonstrated in different cell lines as well as in in vivo using S180 tumor-bearing mouse models. It was observed that, after incubation with Au-cRGD-DOX and irradiation with NIR light (765 nm laser with power of 1.0 W/cm^2), the cell viability of MDA-MB-231 was only 40%. Advantage of the curative effect of the multi-functional nanoconjugates delivered by intratumoral injection over the tail vein injection was verified with the results obtained in this study [71].

10.2.2 Carbon Nanomaterials

Carbon is one of the versatile elements in the periodic table and can bonded together to form different allotropic forms. Carbon orbitals can hybridize to form sp, sp^2, and sp^3 forms. Various carbon nanomaterials, such as CNTs, carbon dots graphene, fullerenes, and nano horns, are used for different applications. The unique properties exhibited by these carbon nanomaterials are utilized in the field of energy storage [72], biomedical [73], and environmental science [74]. Carbon-based nanoparticles show high extinction (absorption plus scattering) along with the whole biological spectral range (i.e., for wavelengths above 700 nm). Therefore, carbon-based nanoparticles have the advantage of tunable PTT in a wide spectral range over plasmonic nanoparticle, where the heat generation occurs maximum on the specific spectral region [10].

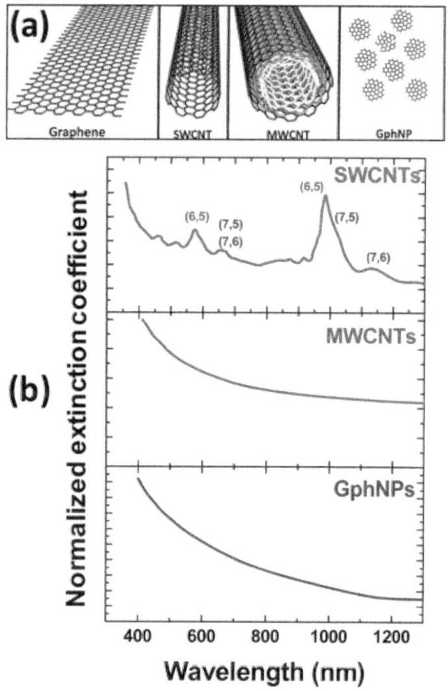

Figure 10.7 (a) Schematic drawing of the basic geometries of different carbon-based nanoparticles used for photothermal applications. (b) Room temperature extinction cross-section spectra of single-walled and multi-walled carbon nanotubes (SWCNTs and MWCNTs, respectively) and of graphene oxide nanoparticles (GphNPs). Reproduced with permission from Ref. [10].

This section mainly focuses on the PTT of CNT, graphene (Figure 10.7), and fullerenes. Choi et al. showed in vivo photothermal efficacy of PEGylated single-walled carbon nanotubes (SWCNTs) using 808 nm laser irradiation, which is considered as the first report in vivo PTT study using carbon nanomaterials [75, 76]. It showed complete destruction of tumor after 20 days of treatment and continued to 6 months without recurrence of the tumor.

10.2.2.1 CNTs

CNT is graphite sheet of one atom thickness rolled into tubular morphology in the nanometer dimension. It exhibits different properties depending on the way in which it is rolled [77]. Important properties of CNT include high modulus and tensile strength, high thermal, and electrical conductivity [78, 79]. CNTs are classified as SWCNTs and MWCNTs based on the number of graphite layers.

The major limitation of CNT in the biomedical field is its poor solubility in the biological environment. In order to overcome the above issue, functionalization with biomolecules or ligands was performed. Functionalized CNTs have many biomedical applications such as biosensing, disease diagnosis, and treatment [80]. Such CNTs are used to detect various biological targets for disease diagnosis [81] and deliver therapeutic molecules, including drugs and genes [82]. They can also help to monitor in vivo therapy, pharmacodynamical behaviour, and drug delivery efficiency.

One of the important properties of CNT is its wide absorption in the NIR region. CNT shows broad absorption in covering the full spectra of both the NIR I and II windows. Further tuning the number of walls, diameter, and length, the photophysical properties of CNTs can be varied [10]. Biological tissues exhibit a deep penetration with very low absorption of NIR photons in the wavelength of 700–1100 nm. The absorption property of CNTs in these regions provides the opportunity for selective photothermal treatment for cancer. Folic acid was conjugated to SWCNT with absorption peak of 980 nm (NIR) (FA-SWCNT) for the target binding of the tumor with high folate receptor [83]. FA-SWCNT targeted tumor cells were irradiated by a 980 nm laser (0.5 to 1 W/cm^2 for 2 min) in vitro and resulted in 85% cell death. Mammary tumors modeled with EMT6 cells in the female Balb/c mice were treated with FA-SWCNT. The mouse tumors with or without FA-SWCNT were irradiated with a 980-nm laser (1 W/cm^2 for 5 min) and showed a temperature rise of 54 °C and tumor death rate of 88% in folic acid conjugated system. In order to increase the imaging property along with PTT, quantum dot (QD) was incorporated in folic acid conjugated SWCNTs [84]. Photothermal therapeutic efficacy of Qd@CNT and FaQd@CNT was evaluated using 800 nm laser, after 3 h of incubation of the particles (1 mg/mL) with the cells. The change in temperature was continuously monitored using a thermal camera during laser exposure. It was observed that the temperature of FaQd@CNT reached 50–60 °C after 4 min, whereas the observed temperature rise was only up to 37 °C in the case of Qd@CNT, which remained unchanged even after 10 min of laser irradiation. MWCNTs were also used as photothermal agents. The oxidized MWCNT and coated with PEG for dispersing in the aqueous medium were used for melanoma treatment. The tumor bearing mice were irradiated with 808 nm laser with power 8 W/cm^2 for 10 min after injecting the MWCNT system. A sharp decrease in tumor indicated PTT efficacy compared to the control group, where only laser irradiation was employed [85].

10.2.2.2 Graphene

Graphene is sp^2 hybridized nano carbon material having a single or few-layered two-dimensional carbon sheet having numerous outstanding properties in physics and chemistry. Graphene has a poly-aromatic surface with an ultra-high surface area, which is available for efficient loading of aromatic drug molecules via π–π stacking for applications in drug delivery [86]. It possesses unique optical, electrical, chemical, and extraordinary electronic transport properties [87, 88].

Among different graphenes, graphene oxide (GO) and reduced graphene oxide (rGO) are the effective candidates for biocompatible application. GO is oxygenated counterpart of one atom thick graphene sheets [89]. It is widely used in biosensors [90], cellular imaging [91], drug delivery [92], and many other fields [93]. Compared to rGO, GO has low electrical conductivity because of the collective inhibition of functional groups present on the edges and the surface [94]. Thus, owing to its small size, intrinsic optical properties, large specific surface area, low cost, and useful non-covalent interactions, GO is a promising material for biomedical applications.

The presence of GO in GO–poly(allylamine hydrochloride) (PAH) makes suitable release of drug molecules upon excitation with NIR light source [95]. It happens due to the dual roles played by GO, such as it serves as a structural component of capsules and a strong NIR-light absorbing agent, which efficiently converts absorbed light into heat on irradiation with 1064 nm laser. Upon laser irradiation, the capsules were opened in a "pointwise fashion" due to local heating. Photothermal property and photoacoustic imaging property of GO were improved by the π–π stacking interaction of it with indocyanine green (ICG) [96]. The combined system showed 20 times higher absorption in 808 nm in comparison with GO, which increased the photothermal property of the system. Folic acid was functionalized to this system for target specificity. On irradiation with 808 nm, 1 W for 10 min, the combined system showed above 60 °C temperature generation.

The rGO prepared by ultra-sonication using BSA nano construct was directly used as a photoacoustic/ultrasonic imaging, and PTT showed excellent effects in living mice [97]. The PTT was done using an 808-nm laser with power of 0.6 W/cm^2 for 5 min irradiation. Clinical potential of pancreatic cancer treatment was demonstrated using rGO on mouse pancreatic cancer cells upon irradiating with 980 nm laser light [98]. The rGO has strong light absorption capability between 600 nm and 1100 nm. Different concentrations of rGO under 980 nm laser irradiation at different powers demonstrated that the enhancement of the photothermal conversion effect of rGO depended on concentration and light dose. In vivo experiments showed that higher laser dose (0.75 W/cm^2) combined with higher rGO concentration (2 mg/kg) could achieve higher treatment temperature and slower tumor growth. NIR absorbing graphene QD showed enhanced NIR imaging of tumor in living mice having a greater probability for in vivo NIR imaging-guided PTT in the NIR-II window [99]. Graphene QDs absorbing 1070 nm wavelength can efficiently ablate tumor cells and inhibit the tumor growth under NIR-II irradiation. In this study, 10 days grown tumor-bearing mice were randomly divided into four groups, two groups were treated with PBS solution and the other two groups were treated with graphene QDs. 24 h after injection, the tumors were irradiated or not irradiated with NIR-II laser (1064 nm, 1.0 W/cm^2) for 5 min. Tumor sizes in these groups were collected every 2 days until 14 days after treatment and the results suggested the greater possibility of these QDs for NIR imaging-guided PTT.

10.2.2.3 Fullerenes

Fullerenes (C60) are one of the allotropes of carbon [100]. The empty interior cavity of the fullerene molecule aids the metal-binding to form metallofullerene. As a result, it shows novel electronic and photophysical property, and hence it is used as MRI [101] agent for PDT in biological systems [102]. Functionalized fullerenes can be utilized for tumor diagnosis and therapy. Localized cell destruction was induced by heating polyhydroxy fullerenes (PHF) and carboxy fullerenes (CF) up to their ignition temperature by irradiating with a continuous wave laser of low intensity (<10^2 W/cm^2) [103]. Fullerene functionalized with cyanuric chloride at room temperature shows broad UV absorption in the wavelength range of 200–800 nm. This exhibited excellent photothermal conversion and fluorescence with a high quantum yield [104]. Photoacoustic effect of NIR absorbing fullerene derivatives with click reagents like 7,7,8,8-tetracyanoquiodimethane, 2,3,5,6-tetrafluoro-7,7,8,8-tetracyanoquiodimethane (TCNQ), and F4-TCNQ was studied. The F4-TCNQ derived fullerenes showed excellent triple efficacy (PA, PT, and PDT effect) compared with other four derivatives [105]. Studies in the photothermal property of fullerenes are still need to be studied in detail. In combination with CNT, alkylated fullerenes act as temperature sensors, which indicate the thermal sensing nature of fullerenes [106].

10.3 Photodynamic Therapy (PDT) – Introduction

PDT is a photochemistry-based diseased cell-ablative interventional therapeutic methodology. PDT uses a photo-activatable chemical, named as photosensitizer (PS), and light of a suitable wavelength to impart cytotoxic damage to the diseased cells through generating reactive molecular species [107–109]. Therapeutic activation of the PS is attained by exciting it using an appropriate light (typically 600–800 nm). The activated PS generates cytotoxic ROSs on cells as a result of transfer of energy to molecular oxygen such as singlet oxygen ($^1O_2$). This $^1O_2$ oxidizes cellular macromolecules, leading to diseased cell ablation [110].

People used to practice light therapy for diseases like psoriasis, combined with some chemicals even during ancient days. Approximately, 3000 years ago in Egypt and India, photo chemotherapy by sunlight activation of psoralens was used for the repigmentation of vitiligo [111]. Over 5000 years ago, sunlight exposure without any exogenous chemical agents had been utilized as therapeutic method in India, Egypt, and China. Later, this methodology paved the ground for the development of practice of heliotherapy, associated with the famous Greek physician Herodotus [112]. During the last century, revolutionary development has happened in the area of PDT for various diseases, especially for cancer treatment. Effective treatment of lung cancer, head and neck cancers, skin cancer, bladder cancer, and Barrett's esophagus has been achieved by PDT [113–117]. In order to reduce residual tumor burden after surgical resection, PDT is also being used as an adjunctive therapy [118]. Over the last two decades, PDT has gained immense progress due to the advances in the research and development in the area of bionanotechnology, addressing many of the challenges existed before. Accordingly, engineered multifunctional nanoparticles have been identified as delivery vehicles of PSs, down-converting PSs, and as energy transducers for PDT [110].

10.3.1 Challenges to Clinical Adoption of PDT

Major limitations of PDT are connected with the dark toxicity of PS due to its systemic distribution mainly affecting the skin. Advanced stages of infected and neoplastic diseases cannot be treated with this method, due to limitations connected with the difficulty in irradiation of large areas. Moreover, treatment of deep-seated lesions is also difficult due to the limited penetration efficacy of visible light. The majority of PDT PSs with heterocyclic ring structure are insoluble in polar solvents and body fluids, making challenging for their systemic administration. Furthermore, a balance between properties like high extinction coefficient and high quantum yield of $^1O_2$, and clinical requirements like low toxicity and high selectivity is important. Additional desired qualities are NIR wavelength absorption for better light-tissue penetration and rapid removal from the body. PSs must also have clinically useful half-life in tissues, good selectivity toward tumor tissue, and rapid elimination from non-target tissue. PSs have to be non-degradable on light activation to prevent production of toxic by-product [107, 109, 110, 114, 119, 120]. Although some photosensitizers satisfy all or some of these criteria, there are currently only a few PDT photosensitizers that have received official approval for clinical use around the world.

10.3.1.1 Photosensitizers

Photosensitizers absorb a photon of light and get excited from ground state to a short-lived singlet state. Molecular structure of most of the PSs has characteristic conjugated double bonds with a delocalized system of π-electrons. These electrons are spin paired in low energy orbitals in the ground state. Absorption of light results in the excitation of electron from the highest occupied molecular orbital (HOMO) of the PS to the lowest unoccupied molecular orbital (LUMO), an unstable and short-lived excited singlet state. The molecule undergoes several rapidly occurring processes in order to obtain stability [102].

The excited PS may either decay to the ground state (emitting fluorescence can be clinically utilized for imaging and photodetection) or can go through intersystem crossing to relatively long-lived triplet state. The most important process in PDT is intersystem crossing [121]. Reversals of the excited electrons spin result the electrons to be at triplet state of the PS. This state has less energy and a much longer lifetime (microseconds as opposed to nanoseconds) than the excited singlet state. Electron at the triplet state has a spin parallel to its former paired electron. So, the electron may not immediately revert to a lower energy level according to the Pauli Exclusion Principle. As a result, spin orientation of the excited electron in the PS triplet state may change (a relatively slow process) and the emission of energy as phosphorescence will occur. As the selection rules that specify triplet–singlet interactions are spin-forbidden, while triplet–triplet interactions are spin-allowed, the PS triplet can react with molecular oxygen. This molecular oxygen is one of the few molecules that is a triplet in the ground state (Figure 10.8). Radical cation or anion is formed from electron or proton as a result of this interaction. These charged ions then react with oxygen and produce oxygenated products such as superoxide anion radicals, hydroxyl radicals, and hydrogen peroxides (Type I reaction). Alternatively, the triplet PS may interact with cell membrane or molecule in its immediate environment. Singlet oxygen can be formed alternatively by transferring the energy of the excited PS to molecular oxygen (type II reaction) [122]. By-products of these two types of reactions are responsible for the therapeutic effect in PDT. Both of these reactions can occur simultaneously, but as indicated by most of the studies type II reactions are prominent, and therefore $^1O_2$ plays a significant role in PDT [110, 122, 123].

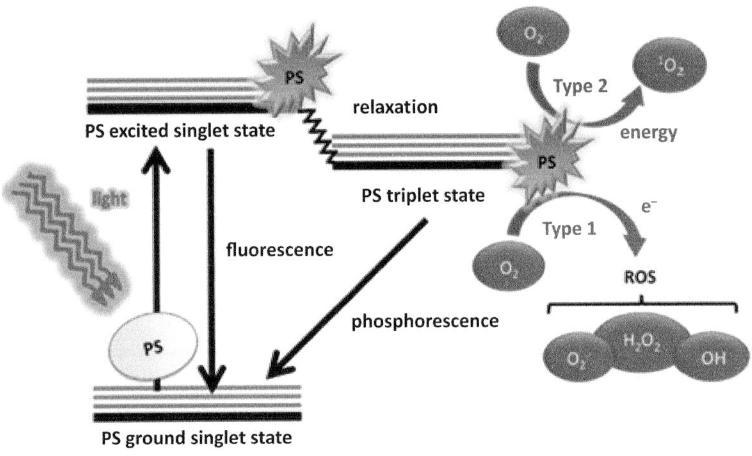

Figure 10.8 Jablonski Diagram illustrating electronic transitions as a resultant of absorption of light by a photosensitizer, and $^1O_2$ production via energy transfer to an oxygen molecule. Reproduced from reference with permission from Ref. [121].

Accumulation in tumor tissue and rapid clearance from the normal tissue is preferred for an ideal PS. Clinically successful PS must also possess amphiphilic nature. After systemic administration, some degrees of hydrophilicity and lipophilicity are required for unhindered travel to the target tissue as well as to bind to the targeted cells. A PS must also have high quantum yield of triplet state formation and minimal dark-toxicity. In order to generate sufficient amount of ROS via interaction with ground state oxygen or other substrates PS must also have appropriate triplet lifetime. PS with an activation light wavelength of above 700 nm is considered as an ideal system. Light below this wavelength may not reach the target tissues due to the limitation of penetration depth as a resultant of light absorption by various endogenous fluorophores and chromophores (Figure 10.9). Many clinically used PSs over the year do not satisfy all these requirements. Over the period of last five decades, PSs acquired all these capabilities during the course of the development of this treatment modality. Based on the time of development and improvement in photochemical properties, PSs are characterized as the first-, second-, and third-generation [110].

The first-generation PSs, developed in the 1970s and early 1980s, are prominently porphyrin-based PSs. These were the most useful and earliest PSs used in clinical trials. These PSs include hematoporphyrin derivative (HpD) containing a proprietary mix of porphyrin monomers, dimers, and oligomers, or porfimer sodium, as active materials. The PSs with negligible dark toxicity and effective tumor destruction capability can easily formulate in a water-soluble form to use for intravenous delivery. In the red portion of the spectrum where tissue penetration of light is optimal, these PSs have relatively weak absorption. This substantial decrease in light penetration ability is affecting the efficiency of the treatment [119, 124].

The second-generation PSs were developed in the late 1980s, to overcome the limitation of the first-generation PSs. The second-generation PSs comprising porphyrinoid compounds, including porphyrin or porphyrin-based macrocyclic structures, and non-porphyrinoid compounds, such as anthraquinones, curcuminoids, cyanines, phenothiazines, and xanthenes, with known chemical structures. These PSs usually show higher quantum yield $^1O_2$, higher tumor to normal tissue concentration compared to first-generation PSs, and improved anti-tumor effect [125–127].

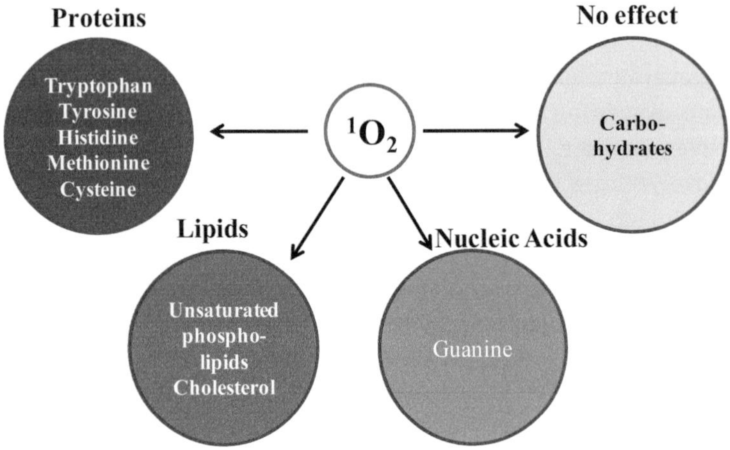

Figure 10.9 Singlet oxygen reacts readily with many biological substrates including certain amino acids in proteins, mainly tryptophan, tyrosine, histidine, cysteine, and methionine. It also reacts with guanine bases of DNA and RNA, as well as a variety of unsaturated lipids, including cholesterol and unsaturated fatty acids. It does not affect carbohydrates appreciably.

Current studies on PSs are mainly concentrated on the development of the third-generation PSs. Researchers are trying to improve PSs properties such as light activation wavelength more toward NIR region, shorter photosensitivity, and most importantly, enhanced tumor specificity. Better tumor specificity can be attained either by modifying PSs with tumor specific, targeting biological conjugates such as peptides-/antibody-/antisense-conjugates or encapsulation/chemical conjugation of PS in carriers or delivery vehicles that can efficiently transport the PS through intravenous-administration to the target tissue [107, 128–131].

10.3.1.2 Light wavelength

Activation of PS using appropriate light is an important procedure in PDT. Wavelength of the light used must be sufficient enough to activate the PS to produce $^1O_2$ and destroy the diseased cells. Selection of light wavelength mainly depends on criteria such as location of the diseased organ and distance from the affected tissue to the periphery of that organ. For diseases associated with skin, oral cavity, and cervix, light illumination is carried out directly on the affected site; whereas, for diseases associated with internal organs, endoscope assisted/post incision light illumination using the help of a fiber optic probe can be carried out [111, 114, 115, 132–134].

Penetration depth of typical red light using for PDT is only 1–3 mm. Surface of the target area is only illuminated evenly using this type of light source. Sufficient light cannot reach further down the surface of the tissue and the energy of the light reaching the required tissue site will be low. Endogenous fluorophores and chromophores within biological tissues, such as collagen, flavin, hemoglobin, and protoporphyrin, are causing this hindrance [135,136]. These tissue components have strong absorbance in the wavelength region of 300–600 nm. To avoid tissue absorbance problem and to allow deeper penetration of light, PS must be activated by wavelengths in the NIR region 700–100 nm, known as the 'optical window' of biological tissue [110].

An ideal PS must have an absorbance wavelength of 700 nm or above. But it is very difficult to develop any drug that fulfills so many disparate requirements, including absorbance wavelength, to be eligible as an ideal PS. The PSs largely belong to the first- and second-generation, which are already approved/ being tested for clinical use, are not satisfying this requirement. Most of these PSs are having absorbance in the wavelength range of 375–690 nm. Current studies on PDT are largely concentrated to tackle this issue by developing new types of PSs that absorb in the NIR range to improve the tissue penetrability [137, 138]. Due to low energy, NIR light at single-photon level lacks to fulfill the energy requirement to excite PSs as required to generate sufficient ROS. This emphasizes the requirement of a capable energy or light transducers to activate the nearby PS via absorbing the light in the NIR region and emitting in the visible region.

10.3.1.3 Selective drug delivery

Topical application PSs is the preferred way of applying drug for PDT on diseases associated with periphery of external organs like skin and oral cavity [114, 115]. But for the diseases associated with internal organs and deeper tissue area of external organs, drug delivery for PDT is performed via oral, intravenous, intratumoral, and inhalational administration roots. Drug targeting and delivery capability of PSs depends upon factors like the selectivity toward diseased tissue, rapid elimination from the normal tissue, $^1O_2$ generation capability, and good solubility in bodily fluids [107, 110].

Lack of tumor selectivity, poor bioavailability, unfavorable biodistribution, and prolonged photosensitivity are the major drawbacks with the first-generation PSs. As these PSs having low extinction coefficient value, a large amount of the drug is required to obtain adequate therapeutic response. The time between PS administration and light irradiation is typically around 48–72 h. Patient must be confined into a dark place for long time due to this reason. Apart from diseased tissues, these PSs also accumulate and retain in the normal tissue and skin for around 4–6 weeks after treatment that leads to severe photosensitivity [119].

Compared to the first-generation PSs, the second-generation PSs showed improvement in tissue accumulation time. Therapy becomes more patient friendly and can be conducted as an out-patient procedure, as the time between PS administration and light irradiation is reduced. Therapy can be carried out on the same day of administration. Photosensitivity window of the second-generation PSs also reduces as they stay less than 2 weeks only in the normal tissue [126]. These PSs with porphyrin ring structures are highly hydrophobic in nature. The degree of PS hydrophobicity has negative influence in the route of administration along with biodistribution/pharmacokinetic profile [122, 139]. Tumor to normal ratio is high for hydrophobic PSs (7:1–8:1) compared to hydrophilic PS (2:1) [140]. Even though hydrophobicity facilitatives the ability to penetrate the cell membrane and locate in the photosensitive sub-cellular compartment, formation of aggregates in aqueous solution is an issue that adversely affect $^1O_2$ generation and photo killing properties. Hydrophobicity can also lead to blood incompatibility and limit the clinical application. These hydrophobic PSs are not soluble in physiological solvents and body fluids that lead to blood incompatibility and limitation in clinical application [141, 142].

The third-generation PSs are currently at the developmental stage. Compared to the second-generation PSs, the third-generation PSs are showing improvements in terms of drug targeting and delivery abilities. Progress in areas like bionanotechnology as well as peptides-/antibody-/antisense-conjugate chemistry is providing immense improvement in drug targeting and delivery capability of PSs [110, 143].

10.4 Nanoparticles in PDT

As discussed in previous sections, PDT is a minimally invasive therapeutic tool with tremendous potential for cancer treatment [118, 144]. This process involves the administration of tumor localizing photosensitizer followed by irradiation with visible light of wavelength, usually in the NIR region of 700–1000 nm, known as 'optical window' of biological tissues [80]. Furthermore, an ideal photosensitizer must have an absorbance wavelength of 700 nm or above. The photosensitizer will be activated by light and forms ROSs that destroy cancer cells. Thus, the critical role of PDT is to induce efficient damage to tumor cells without causing any harm to the surrounding cells [116, 117]. Nevertheless, most photosensitizers are hydrophobic and tend to accumulate easily under physiological conditions, thereby causing a decrease in quantum yield. Moreover, poor selectivity of target tissues and absorption spectra at shorter wavelengths limit the application of these photosensitizers in clinical trials. The application of nanoparticles has emerged as a promising approach for resolving some of these challenges associated with PDT [144–146]. Nanoparticles possess a large surface-to-volume ratio, which can be modified with functional groups for altering their biological or physical properties, thereby enhancing the biodistribution, cell uptake, pharmacokinetics, and targeting. Moreover, nanoparticles help to prevent early release of PS and possible inactivation of drugs, thereby reducing the overall photosensitivity. They can improve the solubility of hydrophobic drugs and provide the advantages of hydrophilicity and proper size to aggregate in tumor tissue through the EPR effect. They are efficiently taken up by cells due to greater distribution volumes. Moreover, nanoparticles help to prevent early release of PS and possible inactivation of drugs, thereby reducing the overall photosensitivity [122, 147–152]. In PDT, nanotechnology is more precisely desirable for three major reasons: (i) potential targeting increases the concentration of PS at the desired site and decreases toxic effects to normal tissues/cells; (ii) nanoparticles can make hydrophobic PSs more dispersible in physiological media; and (iii) continuous PS delivery results in the zero-order release of kinetics, thereby retaining a continuous therapeutic dosage at the action site [153, 154]. Thus, in order to develop PDT drugs, the use of nanosized carriers for photosensitizers is a promising method that can increase photodynamic activity performance and can resolve several side effects associated with classic PDT [155]. An ideal drug delivery system must be biodegradable, small in size, and high load capacity, low immunogenicity, and should not induce side effects. Furthermore, it should selectively aggregate in therapeutic concentrations in the appropriate area [156]. The recent progress made in nanomedicine and the use of photosensitizers for PDT render the production of photoactive nanoparticles, an obvious option for current research. It also relates to the creation of anti-cancer drugs and sensors for tumor identification and imaging or to other PDT applications such as neovascular disorders [157]. A major downside of using nanoparticles is the rapid removal of these particles from circulation after administration of the reticuloendothelial system (RES) by macrophages. Through using functionalized lipids in their construction, the circulatory half-life of nanoparticles can be greatly enhanced. These nanoparticles have greater blood circulation and have less reactivity to serum proteins and are prone to RES absorption [158]. PS can be carried to the target by encapsulation in delivery agents such as liposomes, micelles, ceramic nanoparticles, gold nanoparticles, and polymer nanoparticles (Figure 10.10). For example, liposomes can encapsulate both hydrophobic and hydrophilic drugs. Micelles prevent elimination by the RES that improves their body circulation and the capacity to deliver the drug to specific cells.

Out of these strategies, nanoparticles that encapsulate photosensitive drugs that are biodegradable and non-biodegradable have a range of advantages. The RES system does not remove them from the body because of their small size, which leads to longer half-life times. Surface modification with active targeting ligands of PS-loaded nanoparticles also improves the targeted absorption of PS drugs into tumors.

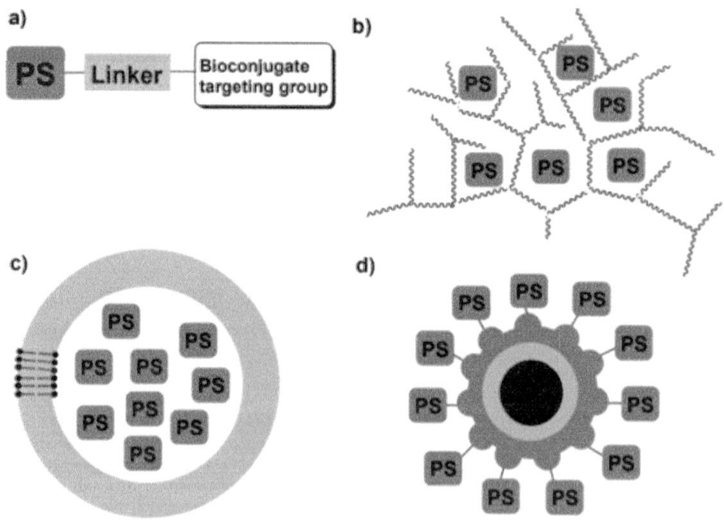

Figure 10.10 Nano strategies for PS modification. Reproduced with permission from Ref. [157].

The design of nanocarriers for effective PDT requires the administration of drugs to tumor sites with minimal loss to their volume and these drugs should only destroy tumor cells without harming normal cells. These needs can be facilitated by using two approaches viz. passive and active targeting using nanoparticles. Passive and active nano targets are classified based either on the absence or existence of any targeting molecules on the surface [159–163]. The strategy that allows parenteral administration without the involvement of specific targeting moieties are termed as 'passive'; whereas, the approach that follows specific delivery of PS to diseased tissues using target tissue receptors are termed as 'active' (Figure 10.11).

Usually, tumor vessels will be dilated with a number of pores, which enlarge the gap junctions between endothelial cells and compromized lymphatic drainage, and this leaky vascularization leads to EPR effect. Passive nanoparticles take advantage of this effect for enabling nanodrugs to accumulate in tumor tissues. Active targeting follows the strategy of attaching affinity ligands, such as antibodies, small molecules, peptides, and aptamers, which can bind only to specific receptors on the surface of the cancer cells. However, there are other functional formulations that describe an additional active intermediary role of carrier nanoparticles in the process of photodynamic activation, which will be detailed in the following sections. Through material composition, passive carriers may be sub-classified into (a) biodegradable polymer-based nanoparticles and (b) non-polymer-based nanoparticles. Active nanoparticles can be sub-classified by mechanism of activation into (a) photosensitizer nanoparticles, (b) self-lighting nanoparticles, and (c) upconversion nanoparticles (UCNPs). Details of these classifications are provided in Table 10.1.

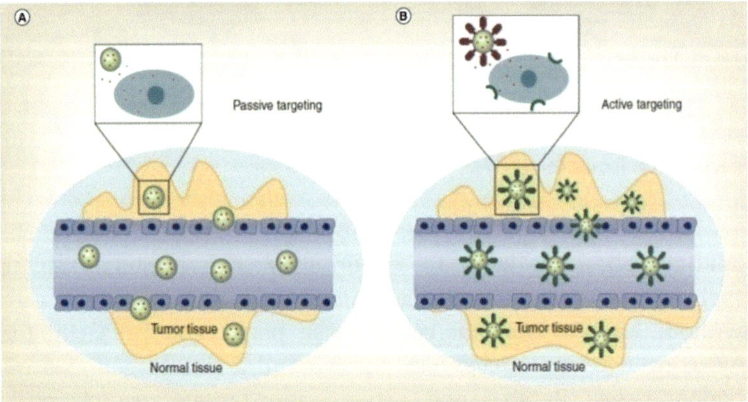

Figure 10.11 (A) Passive targeting: In this case, drugs may be released in the extracellular matrix and then diffuse through the tissue. (B) Active targeting: This can enhance the therapeutic efficacy of drugs by the increased accumulation and cellular uptake of nanoparticles through receptor-mediated endocytosis. Nanoparticles can be engineered to incorporate ligands that bind to endothelial cell surface receptors. In this case, the EPR effect does not pertain, and the presence of leaky vasculature is not required. Reproduced with permission from Ref. [163].

Table 10.1 Classification of nanoparticles used for PDT

Category	Material	Mechanism
Passive nanoparticles		
Biodegradable	PLA, PLGA, PVAL	Controlled release by biodegradation of encapsulated photosensitizer
	Alginate, cyclodextrin, albumin, chitosan	Micelles, dendrimers, liposomes, or polymeric nanoparticles are used to deliver the drug
Non-biodegradable	Silica	Photosensitizer is covalently adsorbed to porous shell
	Polyacrylamide	Micro emulsion encapsulates a two-photon dye
	Magnetic iron oxide	Drug is delivered directly or in micelle or polymeric nanoparticles External magnetic field achieves a targeted delivery
	Gold	Nanoparticles function as pure carriers, without any release
Active nanoparticles		
Photosensitizer	CdSe/CdS/ZnS	FRET mechanism from the nanoparticle to the photosensitizer, which causes indirect excitation of photosensitizer
	Fullerenes	Direct transfer of energy from incident light to the surrounding oxygen
	SWNTs	Regulation of $^1O_2$ generation through quenching action of CNTs
Self-lighting	$BaFBr:Eu^{2+}$, Mn^{2+}, $LaF_3:Tb^{3+}$	Scintillation or persistence luminescence
Up converting	$NaYF_4:Yb^{3+}$, $Er^{3+}NaYF_4:Yb^{3+}$, Tm^{3+}	Attached photosensitizer is activated by converting low-energy light to higher energy transmission

10.4.1 Passive Nanoparticles

Passive nanoparticles can penetrate and accumulate spontaneously in different biological sites characterized by abnormal vascular membranes, such as solid tumors or regions of infarction, through the EPR effect [164]. This effect is based on the assumption that under some circumstances, for example, inflammation characteristic of tumors and other pathological sites, the endothelial lining of the blood vessel wall is more permeable under physiological conditions, allowing particles in the interstitial space to accumulate between 200 nm and 800 nm in size [157, 165]. Tumor targeting on the basis of the EPR effect requires long-circulating drugs. Therefore, long-circulating pharmaceutical nanocarriers, such as liposomes, micelles, or polymeric nanoparticles, can accumulate via the EPR effect in the pathological region and are commonly used for the delivery of drugs to tumors [164]. Two specific approaches for producing PDT nanoparticle carriers are frequently adopted: (i) Biodegradable nanoparticles (alginate, chitosan, cyclodextrin, albumin, polymeric nanoparticles (PLA, PLGA, PVAL)) carrying the PS that could be enzymatically or hydrolytically degraded and thus excreted, reducing the long-term accumulation of carriers in the biological system. (ii) Non-biodegradable nanoparticles (polyacrylamide, silica, gold, and magnetic iron oxide) do not degrade easily in the biological system, but rather often give several functionalities in-built that make them useful as theranostic agents. In biodegradable nanoparticles, the composition of the substance may be limited to lipids or other polymers, while non-biodegradable nanoparticles stay in the body for long periods of time [166].

10.4.1.1 Biodegradable nanoparticle carriers

Biodegradable nanoparticles are made from polymers that degrade after PS release. Biodegradable polymer-based nanoparticles have gained a lot of attention because of their advantages in regulating drug release, flexibility in processes of material processing, less toxicity, and high drug loading [167]. The polymeric surface properties, morphologies, and structure can be designed to achieve the desired biocompatibility, regulate rate of degradation, and kinetics of drug release. Also, polymer's chemical composition and structure can be adapted to fit photosensitizers with varying degrees of hydrophobicity, molecular weight, charge, and pH. Synthetic polymers, such as aliphatic polylactide (PLA), polyglycolide (PGA), and their copolymer polymer poly(D,L-lactide-co-glycolide) (PLGA), are also used as drug delivery carriers because of their favorable properties such as strong biocompatibility, biodegradability, bioresorbability, and mechanical power [168, 169]. For example, a photosensitizer meso-tetra(hydroxyphenyl)porphyrin (*p*-THPP) was encapsulated with submicronic nanoparticles of poly(D,L-lactide-co-glycolide) (50:50 and 75:25 PLGA) and poly(D,L-lactide) (PLA) with drug loading of up to 7% (m/m) using emulsification-diffusion technique. The photodynamic activity was evaluated on EMT-6 mammary tumor cells as compared to free p-THPP. Irrespective of the nature of the delivery system, the cell viability of the drug was concentration dependent. The photo induced cytotoxicity of the dye was strongly affected by the PDT parameters, such as drug concentration, light dose, drug–light time interval, and copolymer molar ratio. It was observed that the copolymer molar ratios influence the rate of carrier uptake and consequently, the intracellular drug concentration [169].

In another work, the photosensitizer zinc (II) phthalocyanine (ZnPc) was encapsulated with biodegradable copolymer PLGA using solvent emulsion evaporation method. ZnPc photosensitizer is commonly used in PDT due to its high selectivity for tumor targets and enhanced cytotoxic efficiency due to $^1O_2$ photo generation properties. The photocytotoxicity of ZnPc loaded with PLGA nanoparticles was evaluated on P388-D1 cells by 6 h, exposed to red light (675 nm) for 120 s, and light dose of 30 J/cm^2. The results inferred that ZnPc loaded with PLGA nanoparticles is a promising drug delivery system for PDT [170]. Hypericin-loaded PLA nanoparticles have demonstrated greater trapping capacity and photoactivity than free drug on the NuTu-19 ovarian cancer cell model. This study provides an in vivo proof-of-concept for fluorescence photodetection and could improve the eradication of the disease by improving the surgical ablation of all detectable tumor nodules [171].

Meso-tetra (*p*-hydroxyphenyl) porphyrin (*m*-THPP) is a hydrophobic PS, incorporated into poly(D,L-lactide-co-glycolide) using an emulsification-diffusion method. Chick chorioallantoic membrane (CAM) of the chicken embryo was used to study the vivo photodynamic activity of m-THPP-loaded nanoparticles. The mTHPP photosensitizer-loaded PLGA nanoparticles were found to stay longer in the vascular compartment of CAM. Here, poly(vinyl alcohol) (PVAL) with molecular weight less than 30 kDa was used as a stabilizing agent, which helped to regulate the mTHPP release and the photodynamic activity in the CAM model [172].

Photodynamic activities of photosensitizers with different hydrophobicities were selected, which include porphyrins (meso-tetraphenylporphyrin (TPP) and meso-tetra-(carboxyphenyl) porphyrin (TCPP)) and chlorins (chlorin e6 (Ce6) and pheophorbide-a (PheoA)). These photosensitizers were entrapped in poly-D,L-lactic acid nanoparticles. The photothrombic efficiency was evaluated using the CAM of the chicken embryo as in vivo model. Analysis of the photodynamic activities of photosensitizers with different hydrophobicities showed that the nanoparticles formulated with the most hydrophobic PS (TPP) showed the highest entrapment efficiency and the highest degree of vascular damage [173].

Natural biodegradable polymers, including alginate, chitosan, dextran, albumin, gelatine, collagen, and agars, have been used in drug delivery systems. For example, aerosol OTalginate nanoparticles can be used as an efficient encapsulation and sustained delivery system for polar and weak base photosensitizer such as methylene blue (MB). Encapsulation of MB in aerosolOT-alginate nanoparticles significantly enhanced the intracellular ROS production and a higher incidence of necrosis was observed in breast cancer cell lines, MCF-7 and 4T1, following PDT. To study the effect of encapsulation in nanoparticles on ROS production, MB-encapsulated nanoparticles was exposed to laser irradiation 665 nm (1200 mJ/cm^2) [174].

In another work, aerosolOT-alginate nanoparticles were used for the simultaneous delivery of doxorubicin and MB. Balb/c mice bearing syngeneic JC tumors (mammary adenocarcinoma) were used as a drug-resistant tumor model and tumors were irradiated with 50 J/cm^2 dose of non-coherent light at

665 nm. Encapsulation of MB and doxorubicin in aerosolOT-alginate nanoparticles resulted in significantly enhanced intracellular and nuclear accumulation of the two drugs, which led to significant inhibition of tumor cell proliferation and increased induction of apoptosis [175].

Porphyrins and their metal-complexes have high triplet quantum yields and very long triplet lifetimes, which play a central role as suitable photosensitizers in PDT. The photodynamic activity of a carrier-sensitizer system consisting of heterotopic colloidal nanoparticles of a cationic amphiphilic cyclodextrin, and heptakis (2-o-amino-O-oligo(ethylene oxide)-6-hexylthio)-b-CD (SC6CDNH2) encapsulating the anionic 5,10,15,20-tetrakis(4-sulfonatophenyl)-21H,23H-porphyrin (TPPS) were investigated with in vitro studies on HeLa cells. Biological assays showed TPPS/SC6CDNH2, which acted as an efficient sensitizer of irradiation-induced cell death in HeLa cells at different molar ratios (1:10 and 1:50) [176].

Albumin is a plasma protein, which is responsible for blood colloidal osmotic pressure. This is non-antigenic and biodegradable and has been studied extensively as a drug carrier. The average HSA half-life is 19 days [177]. The photosensitizer pheophorbide was loaded on human serum albumin (Pheo-HSA) nanoparticles with different cross-linked ratios by non-covalent adsorption. Intracellular uptake and phototoxicity of both pheophorbide and Pheo-HSA nanoparticles were studied in Jurkat cells. For irradiation, a laser diode with emission at 668 nm and a light dose of 96 mJ/cm^2 were used as light source. Due to intramolecular interactions, $^1O_2$ quantum yield of pheophorbide-loaded HSA nanoparticles was very low and the final phototoxicity was at the same level as induced by free pheophorbide [178].

In another work, chitosan-based nanoparticles (CNPs) were used for PDT by self-assembling amphiphilic glycol chitosan–5β-cholanic acid conjugates in an aqueous environment and then encapsulating the water-insoluble photosensitizer (PpIX) with high drug-loading efficiency (>90%). As nanoscale drug carriers, PpIX-CNPs exhibited a sustained release profile in vitro and were non-toxic to tumor cells in the dark. In SCC7 tumor-bearing mice, PpIX-CNPs exhibited rapid cellular uptake and the released

PpIX from CNPs became highly phototoxic upon visible irradiation with 633 nm He–Ne laser (3 mW/cm^2) and enhanced tumor specificity and increased therapeutic efficacy compared to free PpIX [179].

10.4.1.2 Non-biodegradable nanoparticle carriers

Non-biodegradable nanoparticles play a different function in PDT due to their inability to degrade and release drugs in a controllable manner. Non-biodegradable nanoparticles have many benefits over biodegradable polymer nanoparticles because they can serve as multifunctional platforms for drug delivery. Other features include: (i) particle size, shape, porosity, and monodispersibility of these nanoparticles can be easily controlled during the synthesis process; (ii) some nanoparticles are made of inert materials, which are stable to environmental fluctuations; (iii) these are not affected by microbial attack; and (iv) control of pore size of these nanoparticles aids the diffusion of oxygen in and out of the particles, but it does not allow the drugs to escape [180]. Many non-degradable nanoparticles are either silica or metallic based. For example, mesoporous silica nanoparticles have been successfully encapsulated with photosensitizer PpIX for PDT treatment. The size (<50 nm) and surface charge of silica nanoparticles can help their uptake by HeLa cells through the EPR effect [181].

In another work, the photosensitize MB was encapsulated into three types of nanoparticles: polyacrylamide, sol–gel silica, and organically modified silicate (ORMOSIL). On light irradiation, the entrapped MB was able to generate $^1O_2$ and a significant amount diffused out of the matrix entered the adjacent cell membranes and destroyed tumor cells. Photocytotoxicity of these nanoparticles was evaluated on rat C6 glioma tumor cells in vitro [182]. In addition, nanoparticle based on polyacrylamide matrix provides protection for embedded MB against the reduction of diaphorase enzymes, thereby retaining the photoactive form of MB for efficient PDT treatment. It was observed that upon light irradiation, the MB-containing NP could induce photodynamic effects on rat C6 glioma cells. The encapsulated MB allows sufficient light-stimulated generation of $^1O_2$ as well as adequate diffusion out of the matrix to cause damage to the tumor cells [183].

Advances in nanotechnology allow researchers to develop nanoparticle-based MRI contract agents with higher magnetization and the required surface characteristics to satisfy the specific requirements for effective biodistribution [184]. There are two types of iron oxide that were specifically investigated for use in magnetic NP formulation: maghemite (α-Fe_2O_3) and magnetite (Fe_3O_4), both biocompatible, while the most promising candidate is magnetite. Typically, they are coated with dextran, phospholipids, or other compounds to inhibit aggregation and improve stability [185]. A nanocarrier made up of polymeric diacyl phospholipid–PEG micelles co-loaded with the photosensitizer 2-[1-hexyloxyethyl]-2-devinyl pyropheophorbide-a (HPPH) and magnetic Fe_3O_4 nanoparticles showed excellent stability and effective uptake by HeLa cells. The magnetic response of nanocarriers was demonstrated by their targeted delivery to tumor cells in vitro when exposed to an external magnetic field. The magnetophoretic regulation of the cellular uptake improved imaging and phototoxicity [186]. The magnetic core containing chitosan nanoparticles and photosensitizer carriers encapsulating photosensitizer 2,7,12,18-tetramethyl-3,8-di(1-propoxyethyl)-13,17-bis(3-hydroxypropyl) porphyrin (PHPP) was found to have excellent targeting and imaging ability. With these nanoparticles at the level of 0–100 mM, non-toxicity and high photodynamic efficacy on SW480 carcinoma cells were achieved, both in vitro and in vivo [187]

Metallic nanoparticles have been extensively used as diagnostic and therapeutic agents for the past few years. These nanoparticles possess the advantages of high loading capacity and slow degradation, which make them efficient drug delivery carriers in PDT. They assist in the targeted and controlled release of drugs, which enable them to operate as efficient nanocarriers in drug delivery [188, 189]. Gold nanoparticles are renowned for their chemical resistance and limited acute cytotoxicity. These nanoparticles possess unique chemical properties that give them promising applications in gene therapy and drug delivery to specific cells [190]. Gold nano capsules have been studied on various models of cancer, in vitro as well as in vivo [191]. A phthalocyanine derivative, which is used as the photosensitizer in monomeric form on gold nanoparticle surface incubated with HeLa cells, absorbs radiation at 695 nm and catalytically produces the cytotoxic species $^1O_2$ with high efficiency [192].

A highly efficient PEGylated gold nanoparticle conjugate was reported, which acted as a water-soluble and biocompatible cage and allowed delivery of a hydrophobic drug to the targeting site. PEGylated gold nanoparticles were mixed with silicon phthalocyanine (Pc4) and it was observed that the drug delivery time required for PDT had been greatly reduced to less than 2 h, compared to 2 days for the free drug [193].

In another work, 5-aminolevulinic acid (5-ALA), which is a PPIX precursor, was conjugated with gold nanoparticles and it offered selective and efficient destruction of tumor cells. It was inferred that PPIX accumulated in fibrosarcoma tumor cells treated with 5-ALA-conjugated gold nanoparticles yielded significantly higher ROS and 50% greater cytotoxicity than that with 5-ALA [194].

10.4.2 Active Nanoparticles

Strategy involving delivery of PS specifically to target tissue receptors is generally termed as 'active'. Nevertheless, there is an additional active intermediary role for nanoparticles in the process of photodynamic activation. Therefore, the former structural formulation can be replaced with a functional one, which defines nanoparticles as active participants in PS excitation. Active nanoparticles play the role of catalysts in PDT for the production of free radicals from dissolved oxygen, thereby serving as ROS modulators for the activation of photosensitizers. In order to act as an effective PDT light source, the emission spectrum of active nanoparticles should match the absorption spectrum of the photosensitizer. They must be easily linked with photosensitizers and should possess high luminescence efficiency. Moreover, active nanoparticles used in PDT should be hydrophilic, non-toxic, and highly stable in biological environments [180, 195]. They can be sub-classified into photosensitizer nanoparticles, self-lighting nanoparticles, and upconversion nanoparticles by means of their activation mechanism.

10.4.2.1 Photosensitizer nanoparticles

Photosensitizer nanoparticles are group of nanostructures where the actual nanoparticles themselves act as PS thereby absorbing light and producing ROS. Some types of QDs and fullerenes are examples for these kinds of nanoparticles. QDs are semiconductor nano constructs, which own numerous characteristics that make them potential candidates in PDT. The size of QDs, which is in the range of 1–6 nm, gives exceptionally high quantum yields, photo stability, and fluorescent emission properties. The optical properties of QDs can be tuned from the UV to the infrared region by changing their size and composition [196–199]. The emission properties of QDs can be tuned into the NIR region, which is different from the visible emission of most classic photosensitizers. Therefore, QDs allow accessibility to deep-rooted tumors as light scattering and absorption in the NIR region of the spectrum would be minimal. Moreover, QDs are strong absorbers, due to their large transition dipole moment, which make them ideal for PDT applications. Many efforts have been explored to investigate the potential of QDS as photosensitizers. QDs can be made water soluble and biocompatible by functionalizing their surface coatings, which facilitate systematic target delivery to specific pathologic areas. The first study describing the application of semiconductor QDs for PDT was done by linking CdSe QDs with a silicon phthalocyanine (Pc4) photosensitizer [197]. The spectral interaction between CdSe QDs and Pc4 was interrogated in this report and it was observed that Pc4 could be directly activated for excitation wavelengths in the range of 550–630 nm. Furthermore, it was confirmed that CdSe QD can act as primary energy donor in the excitation wavelength range of 400–500 nm through a mechanism known as fluorescence or Forster resonance energy transfer (FRET) [200]. In this system, when two chromophores are in close proximity, a part of the energy of the excited state of a fluorophore (termed as donor) can be transferred to the second chromophore (termed as acceptor) by means of direct dipole interaction. Thus, the conjugation of semiconductor QDs enabled the use of an excitation wavelength, where the photosensitizer alone cannot absorb. Moreover, the size of QDs can be adjusted so that their spectral properties match with those of any PDT photosensitizer. Upon further investigation of two-step energy transfer mechanism in this study, it was reported that Cd Se QDs, with an average diameter of 5 nm can generate $^1O_2$ in toluene. These QDs were linked to a silicon Pc photosensitizer (Pc4) through an alkyl amino group on the photosensitizer's axial substituent. Here, the intercalation of dissolved oxygen molecules with the TOPO layer at the QD surface accounts for the generation of $^1O_2$, without a mediating PS molecule in toluene. The possible application of these QDs as therapeutic agents for PDT was evidenced by the emission of $^1O_2$ at 1270 nm generated by the energy transfer from CdSe QDs alone in toluene under excitation at 488 nm. They suggested a similar interaction in water-soluble phospholipid-capped QDs, with lower solubility of oxygen in water compared to toluene. They further predicted that since the lowest excited state of CdSe QDs is a triplet state, often known as 'dark state', the triplet energy transfer (TET) can account for the generation of $^1O_2$ from $^3O_2$. However, the $^1O_2$ yield was only 5% for these CdSe QDs, which was too low comparing with 40–60% yield by conventional photosensitizers. This low efficiency may be due to the very small fractions of QD–$^3O_2$ pairs generated at any moment and the carrier trapping and non-radioactive carrier recombination taking place on the early picoseconds time scale. In order to overcome this inefficiency, many efforts have been made, which utilize organic bridges for the covalent conjugation of PS to CdSe/ZnS.

Another study reported a bifunctional system in which Ir-complex functionalized CdSe/ZnS QDs were designed. In this work, QD plays a significant role in imaging, whereas the Ir-complexes act as a sensitizer for the production of $^1O_2$ [201]. It was observed that the emission of CdSe/ZnS QDs could be located at the low-lying triplet state absorption of the transition metal complexes, by adjusting the size of CdSe core. The synthesis of water-soluble QDs, using meso-tetra (4-sulfonatophenyl) porphine dihydrochloride (TSPP) as a photosensitizer, which was bound to CdTe nanocrystals through electrostatic approach, was demonstrated [201]. It was suggested that excitation of the QD followed by a FRET-type mechanism produced $^1O_2$ in this case.

Another study reported the development of peptide-coated QD-photosensitizer conjugates [198]. In this work, covalent conjugation of rose bengal and chlorin e6 photosensitizers were done through phytochelatin-related peptides. These conjugates were further utilized to overcoat green and red-emitting CdSe/CdS/ZnS nanocrystals. Here, $^1O_2$ was produced by means of indirect excitation through FRET from nanocrystals to PSs or by direct excitation of PSs. In vitro applications of QDs for PDT have also been illustrated, but it was found that the toxic effect of QDs to healthy living cells can limit their in vivo application. This is due to the fact that Cd^{2+} generated protective shell deteriorates after prolonged circulation in the body. But these studies lacked the interrogation of toxic oxygen generation capacity of QDs as well as better water solubility.

Functionalized fullerenes, which are composed of 60 or 70 carbon atoms, have attracted wide interest in the research field of PDT. They can be very effective in photo inactivation of cancer cells, depending on the functional groups introduced into the molecule and can generate ROSs upon illumination [202]. The first in vitro study of capability of fullerenes to create ROSs and catalyze the phototoxic reaction upon photo excitation was demonstrated on thymocytes and Ehrlich ascites cells [203]. In this work, the aqueous solution of fullerenes C_{60} was irradiated with visible light to these cell suspensions for short interval of time and increased rate of ROS was demonstrated. One of the major drawbacks, which limit the applications of fullerenes in biological system, is its poor water solubility. This may be overcome by introducing substituents, which are water soluble, mixing with water-soluble polymers or lipid membranes, and reduction to water-soluble anions. The comparison of the photodynamic activity of six functionalized fullerenes with 1, 2, or 3 hydrophilic or 1, 2, or 3 cationic groups was detailed in another work. Here, they considered three mouse cancer cell lines viz. J774, LLC, and CT26, which were incubated for 24 h with fullerenes and illuminated with white light. It was found that the photodynamic activity of the functionalized fullerenes was inversely proportional to the degree of substitution of the fullerene nucleus for both the neutral and the cationic series. They further inferred that photo toxicity can be mediated both by superoxide and $^1O_2$ [202, 204].

The chemical conjugation of PEG to fullerene was studied and diethylenetriaminepentaacetic acid (C60-PEG-DTPA) was further introduced to its terminal group. This conjugate demonstrated selective accumulation as well as prolonged retention in the tumor tissue [205]. C60-PEG-DTPA was mixed with gadolinium acetate solution to obtain Gd^{3+}-chelated C60-PEG and intravenously injected into tumor-bearing mice, in order to investigate the PDT anti-tumor effect and MRI tumor imaging. The major drawback of fullerenes is their optical absorption properties. The highest absorption spectrum exhibited by fullerenes is in the UV and blue regions of the spectrum. The tissue penetration depth of illumination is the shortest in this region, owing to a combination of light absorption by cellular chromophores and light scattering by cellular structures. Therefore, the development of fullerenes that absorb red or NIR light may be extremely significant for clinical application in PDT.

There are other nano constructs made up of thin sheets of benzene ring carbons rolled up into the shape of a seamless tubular structure, which belong to the family of fullerenes, known as CNTs [206]. They can be classified into single-walled nanotubes (SWNTs) and multi-walled nanotubes (MWNTs) based on their structure. Molecules such as drugs, peptides, and nucleic acids can be incorporated into their walls and tips due to their ultra-high surface area. This enables them to cross the cell membrane by endocytosis or to identify cancer-specific receptors on the cell surface. SWNTs are capable of efficiently quenching fluorescence probes. A novel molecular complex of a photosensitizer (chlorin e6), an ssDNA aptamer, and SWNTs was constructed for regulating $^1O_2$ generation [207]. The restoration of $^1O_2$ generation was done by target binding with ssDNA aptamers to specific proteins, which could disturb the interaction between aptamers and SWNTs, thereby ensuring more selective PDT. Owing to their outstanding potential in molecular diagnosis and targeted therapy of tumors, many therapeutic agents can be incorporated into CNTs. A thorough understanding on the pharmacologic and toxicologic properties of CNTs is further required before recommending them for routine clinical use.

10.4.2.2 Self-lighting nanoparticles

X-rays can be used as efficient energy sources for initiating effective PDT. For this, scintillation nanoparticles (ScNps) with attached photosensitizers are employed, which serves the purpose of self-lighting photodynamic therapy (SLPDT) [208]. These materials can convert X-rays to UV-visible light and this concept improved the efficiency of PDT without employing an external light source. X-rays will be irradiated to the tumor site where the nano sensitizer systems have been accumulated. Upon irradiation, the encapsulated nanoparticles will emit light and this can be collected to activate the nearby PS. The scintillation process involves the conversion of incoming radiation into a large number of electron-hole pair and transfer of this electron-hole pair energy to the luminescent ions. Upon interaction with high energy photons with the lattice of ScNPs, many electron-hole pairs will be created and thermalized in the conduction and valence band respectively. Photoelectric effect and Compton Effect account for this interaction. Afterwards, these electron-hole pairs will migrate through the materials and repeated trapping at defects of these materials cause energy losses mainly due to non-radiative recombination. This will be followed by the final stage, which is luminescence, and here the energy from the SCNPs will excite the nearby PS causing $^1O_2$ generation [209, 210].

Scintillation nanoparticles, such as BaFBr: Eu^{2+}, Mn^{2+}, and $LaF_3:Tb^{3+}$, will emit luminescence for activating the photosensitizers following exposure to ionizing irradiation. This results in the production of $^1O_2$ which enhances the destruction of cancer cells. The efficiency of $LaF_3:Tb^{3+}$-meso-tetra (4-carboxyphenyl) porphine (MTCP) nanoparticle conjugated as photodynamic agents following X-ray irradiation was investigated [211]. They described the effective energy transfer and $^1O_2$ generation mechanism upon initiation by X-rays at low dose. These scintillation nanoparticles can be possibly utilized as a promising deep cancer treatment modality. Another study described SLPDT against ovarian cancer and they synthesized zinc oxide (ZnO) nanoparticles conjugated to porphyrin (ZnO-MTAP). The photoactivation of ZnO-MTAP conjugates led to ROS release, which could be effectively targeted to tumor cells [212]. In the case of persistent-luminescence nanoparticles, the luminescence continues even after the excitation source is ceased. This long decay lifetime will decrease the ionization dose and prolonged photosensitizer excitation required for cancer treatment. The application of these nanoparticles in biological systems needs to be further explored.

10.4.2.3 Upconversion nanoparticles (UCNPs)

One of the chief limitations in PDT is that even the longest wavelength used for exciting PS has relatively poor tissue penetration and also it has been observed that many PSs are better excited by blue light (400–430 nm) and green light (490–550 nm). In this view, UCNPs have been investigated, which ensure deeper-penetration of NIR light in PDT. UCNPs are modified nanocomposites for generating higher energy light from lower energy radiation by means of transition metal, lanthanide, or actinide ions doped into a host crystalline lattice. These ions possess long-lasting excited states, which provide longer time needed for multiple absorptions [213]. The advantage of this process is that it does not require coherent radiation and involves actual intermediate states. On the whole, an appropriate PS will be attached to these nanoparticles with an excitation band matching the emission of the nanoparticles, which will facilitate efficient $^1O_2$ species generation as shown in Figure 10.12. They can be modified as nanometer-sized composites for higher energy generation by incorporating transition metal, lanthanide, or actinide ions doped into a solid-state host. They are also referred as upconverting phosphors (UCPs), which are lanthanide-containing ceramic particles with size in the range of sub-micrometer. These nanoparticles can emit visible light by absorbing NIR or infrared light. Thus, PDT employing UCNPs can be utilized for deep-seated tumors [214].

The optical and morphological features of the desired nanocrystalline core should be appropriate for biomolecular conjugation. The surface properties and the growth dynamics of these nano constructs need to be properly transformed for high quality applications. $NaYF_4$ is one of the key materials used for UCPs. A versatile UCP $NaYF_4$:Yb^{3+}, Er^{3+} nanoparticles coated with a porous, thin layer of silica, which was doped with merocyanine-540 photosensitizer, was reported [215]. This was further functionalized with a tumor-targeting antibody. This work illustrated the potential of the proposed design to be an explicit PDT agent against MCF-7/AZ breast cancer cells.

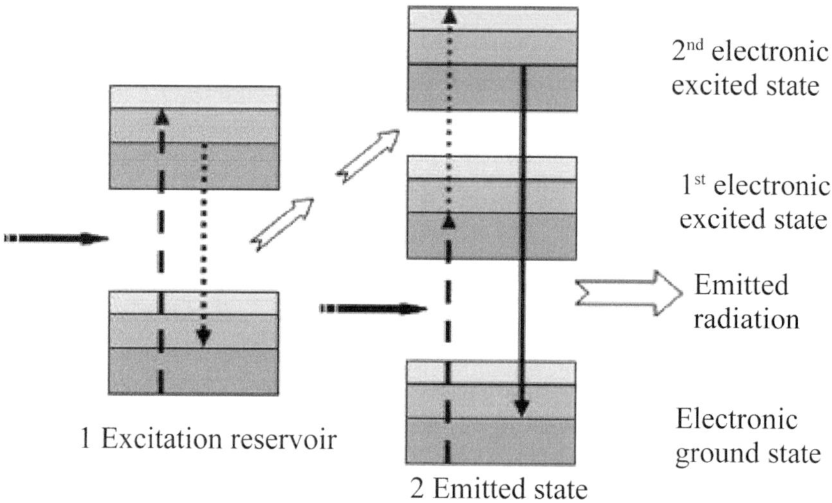

Figure 10.12 Mechanism upconversion: Two metastable entities are involved, the first serving as an excitation reservoir, and the second as the emitting state. Reproduced with permission from Ref. [180].

Another study reported UCP $NaYF_4$ nanocrystals coated with mesoporous silica, which was further incorporated with zinc phthalocyanine (ZnPc), acting as photosensitizer [216]. The nanocrystals convert NIR light to visible light upon excitation by a NIR laser, which activates the photosensitizer for releasing reactive $^1O_2$. Zinc phthalocyanine is secured from degradation even in the harsh biological environment. It has been demonstrated that even if the photosensitizers produce continuous $^1O_2$ upon excitation by a NIR laser, they were not released out of silica. A functional prototype, which co-localizes insoluble porphyrin with UCP crystals and efficiently generates $^1O_2$ under NIR illumination, was demonstrated [217]. In order to prevent clearance by the RES, a three-layer UCP composed of an

$NaYF_4$:Yb^{3+}, Er^{3+} nanoparticle interior, porphyrin photosensitizer coating, and a biocompatible PEG outer layer were designed and $^1O_2$ in mill molar amounts was generated at NIR intensities, which was far less than in two photon techniques.

Another work investigated the influence of different percentages of Er^{3+} and Yb^{3+} as doping ions, particle size, and host lattice on the upconversion process. They recorded the upconversion emission spectra in the visible region for efficiency and green/red emission relative intensity comparisons [218]. Lanthanide doped $NaYF_4$ nanocrystal exhibiting efficient multicolour upconversion emission in transparent colloidal solution was developed [212]. Another study designed a sandwich assay format for a nucleotide sensor using Er^{3+} doped $NaYF_4$ as energy donor and carboxytetramethylrhodamine (TAMRA) as energy acceptor. The key features of this sandwich design are high specificity, high sensitivity, and self-calibrating capability. The synthesis of water-soluble and biocompatible polyethyleneimine/NaYF4 nanoparticles doped with lanthanide ions using a one-pot synthesis method was reported. Strong upconversion fluorescences of different colors in aqueous solutions were shown by different lanthanide ions (Yb^{3+}, Er^{3+} and Tm^{3+}) doped into the nanoparticles, under excitation at 980 nm [219]. Another work reported the synthesis of PEI/$NaYF_4$:Yb^{3+}, Er^{3+} nanoparticles producing green/red emission on NIR excitation targeting to folate receptors on human colon cancer cells. It was observed that upon excitation, the nanoparticles modified with zinc phthalocyanine photosensitizer released $^1O_2$ after targeted binding to cancer cells, thereby causing significant cell destruction [214].

10.5 Conclusion and Future Prospects

Recent studies have shown promising results in the fields of PTT and PDT using a variety of nanomaterials, which can address major limitations of conventional therapeutic strategies. It has been well established that gold nanoparticles exhibiting LSPR have potential applications in PTT. The unique optical properties of these nanoparticles can be altered by modifying their structural dimensions, which facilitates strong absorption of tissue-penetrating NIR light. In addition, a variety of nanoparticles can be functionalized with various drugs and targeting molecules, thereby making them ideal multifunctional candidate for cancer therapeutics. It has also been demonstrated that nanotechnology is a promising approach where nanoparticles can be employed in the field of PDT as well and they exhibit many advantages compared to classic photosensitizers. Nanomaterials combined with PS drugs improve the water solubility of hydrophobic PS drugs. The target specificity of PS drugs to tumor tissues could be improved by the nanoparticles by means of EPR effect. Furthermore, PS-loaded nanoparticles can be surface modified with active targeting ligands, which enhances selective accumulation of PS drugs into tumors. Nevertheless, the long-term cytotoxicity studies of nanoparticles have not been fully explored. Even though promising credentials for the use of these nanoparticles in in vitro, preclinical and clinical studies exist, there are still concerns regarding clearance of nanoparticles from the body and long-term consequences resulting from their accumulation. In addition, the development of PS drugs that exhibit strong absorption at longer wavelength range along with high chemical and photo stability is also crucial. Therefore, it is worth interrogating future directions in these aspects, which will facilitate a further evolution of nanoparticles for cancer therapeutics.

Acknowledgement

The authors acknowledge Department of Biotechnology, Government of India (Project: BT/PR27222/NNT/28/1337/2017) for the financial support. Shaiju S Nazeer acknowledges Department of Biotechnology, Government of India for the Ramalingaswami Re-entry fellowship.

References

1. J.B. Vines, J.H. Yoon, N.E. Ryu, D.J. Lim, H. Park. Gold nanoparticles for photothermal cancer therapy. *Front. Chem.* 7 (2019), 1–16. https://doi.org/10.3389/fchem.2019.00167.

2. S. Gao, M. Zhang, X. Zhu, Z. Qu, T. Shan, X. Xie, Y. Wang, X. Feng. Apoptotic. *Int. J. Clin. Exp. Med.* 8 (2015), 15098. PMID: 26628993 PMCID: PMC4658882.

3. S. Roy, B. Tony L. Phototherapy in cancer prevention and treatment. *J. Cancer Prev. Curr. Res.* 7 (2017), 7–9. https://doi.org/10.15406/jcpcr.2017.07.00224.

4. J.L. Roti. Cellular responses to hyperthermia (40–46°C): Cell killing and molecular events. *Int. J. Hyperth.* 24 (2008), 3–15. https://doi.org/10.1080/02656730701769841.

5. M. Pérez-Hernández, P. Del Pino, S.G. Mitchell, M. Moros, G. Stepien, B. Pelaz, W.J. Parak, E.M. Gálvez, J. Pardo, J.M. De La Fuente. Dissecting the molecular mechanism of apoptosis during photothermal therapy using gold nanoprisms. *ACS Nano.* 9 (2015), 52–61. https://doi.org/10.1021/nn505468v.

6. S.K. Ghosh, T. Pal. Interparticle coupling effect on the surface plasmon resonance of gold nanoparticles: From theory to applications. *Chem. Rev.* 107 (2007), 4797–4862. https://doi.org/10.1021/cr0680282.

7. S. Eustis, M.A. El-Sayed. Why gold nanoparticles are more precious than pretty gold: Noble metal surface plasmon resonance and its enhancement of the radiative and nonradiative properties of nanocrystals of different shapes. *Chem. Soc. Rev.* 35 (2006), 209–217. https://doi.org/10.1039/b514191e.

8. S. Link, M.A. El-Sayed. Spectral properties and relaxation dynamics of surface plasmon electronic oscillations in gold and silver nanodots and nanorods. J. *Phys. Chem. B.* 103 (1999), 8410–8426. https://doi.org/10.1021/jp9917648.

9. X. Huang, P.K. Jain, I.H. El-Sayed, M.A. El-Sayed. Plasmonic photothermal therapy (PPTT) using gold nanoparticles. *Lasers Med. Sci.* 23 (2008), 217–228. https://doi.org/10.1007/s10103-007-0470-x.

10. D. Jaque, L. Martínez Maestro, B. Del Rosal, P. Haro-Gonzalez, A. Benayas, J.L. Plaza, E. Martín Rodríguez, J. García Solé. Nanoparticles for photothermal therapies. *Nanoscale.* 6 (2014), 9494–9530. https://doi.org/10.1039/c4nr00708e.

11. Y. Ma, V. Chechik. Aging of gold nanoparticles: Ligand exchange with disulfides. *Langmuir.* 27 (2011), 14432–14437. https://doi.org/10.1021/la202035x.

12. T. Niidome, M. Yamagata, Y. Okamoto, Y. Akiyama, H. Takahashi, T. Kawano, Y. Katayama, Y. Niidome. PEG-modified gold nanorods with

a stealth character for in vivo applications. *J. Control. Release.* 114 (2006), 343–347. https://doi.org/10.1016/j.jconrel.2006.06.017.

13. C. Liang, S. Diao, C. Wang, H. Gong, T. Liu, G. Hong, X. Shi, H. Dai, Z. Liu. Tumor metastasis inhibition by imaging-guided photothermal therapy with single-walled carbon nanotubes. *Adv. Mater.* 26 (2014), 5646–5652. https://doi.org/10.1002/adma.201401825.
14. H. Zhang, H. Wu, J. Wang, Y. Yang, D. Wu, Y. Zhang, Y. Zhang, Z. Zhou, S. Yang. Graphene oxide-BaGdF5 nanocomposites for multi-modal imaging and photothermal therapy. *Biomaterials.* 42 (2015), 66–77. https://doi.org/10.1016/j.biomaterials.2014.11.055.
15. K. Jibin, J.S. Prasad, G. Saranya, S.J. Shenoy, K.K. Maiti, R.S. Jayasree. Optically controlled hybrid metamaterial of plasmonic spiky gold inbuilt graphene sheets for bimodal imaging guided multimodal therapy. *Biomater. Sci.* 8 (2020), 3381–3391. https://doi.org/10.1039/d0bm00312c.
16. S.P. Pricker. Medical uses of gold compounds: Past, present and future. *Gold Bull.* 29, 53–60 (1996). https://doi.org/10.1007/BF03215464.
17. C. Cretu, E. Van Der Lingen. Coloured gold alloys. *Gold Bull.* 32 (1999), 115–126. https://doi.org/10.1007/BF03214796.
18. I.W. Sudiarta, P. Chýlek. Mie scattering by a spherical particle in an absorbing medium. *Appl. Opt.* 41 (2002), 3545. https://doi.org/10.1364/ao.41.003545.
19. S. Link, M.A. El-Sayed. Size and temperature dependence of the plasmon absorption of colloidal gold nanoparticles. *J. Phys. Chem. B.* 103 (1999), 4212–4217. https://doi.org/10.1021/jp984796o.
20. L.C. Kennedy, L.R. Bickford, N.A. Lewinski, A.J. Coughlin, Y. Hu, E.S. Day, J.L. West, R.A. Drezek. A new era for cancer treatment: Gold-nanoparticle-mediated thermal therapies. *Small.* 7 (2011), 169–183. https://doi.org/10.1002/smll.201000134.
21. J.R. Cole, N.A. Mirin, M.W. Knight, G.P. Goodrich, N.J. Halas. Photothermal efficiencies of nanoshells and nanorods for clinical therapeutic applications. J. *Phys. Chem. C.* 113 (2009), 12090–12094. https://doi.org/10.1021/jp9003592.
22. Y. Wang, K.C.L. Black, H. Luehmann, W. Li, Y. Zhang, X. Cai, D. Wan, S.Y. Liu, M. Li, P. Kim, Z.Y. Li, L. V. Wang, Y. Liu, Y. Xia. Comparison study of gold nanohexapods, nanorods, and nanocages for photothermal cancer treatment. *ACS Nano.* 7 (2013), 2068–2077. https://doi.org/10.1021/nn304332s.

23. A.R. Rastinehad, H. Anastos, E. Wajswol, J.S. Winoker, J.P. Sfakianos, S.K. Doppalapudi, M.R. Carrick, C.J. Knauer, B. Taouli, S.C. Lewis, A.K. Tewari, J.A. Schwartz, S.E. Canfield, A.K. George, J.L. West, N.J. Halas. Gold nanoshell-localized photothermal ablation of prostate tumors in a clinical pilot device study. *Proc. Natl. Acad. Sci. U.S.A.* 116 (2019), 18590–18596. https://doi.org/10.1073/pnas.1906929116.

24. Y.C. Wang, É. Rhéaume, F. Lesage, A. Kakkar. Synthetic methodologies to gold nanoshells: An overview. *Molecules.* 23 (2018), 1–28. https://doi.org/10.3390/molecules23112851.

25. V.W.K. Ng, R. Berti, F. Lesage, A. Kakkar. Gold: A versatile tool for in vivo imaging. *J. Mater. Chem. B.* 1 (2013), 9–25. https://doi.org/10.1039/c2tb00020b.

26. R. Vankayala, C.C. Lin, P. Kalluru, C.S. Chiang, K.C. Hwang. Gold nanoshells-mediated bimodal photodynamic and photothermal cancer treatment using ultra-low doses of near infra-red light. *Biomaterials.* 35 (2014), 5527–5538. https://doi.org/10.1016/j.biomaterials.2014.03.065.

27. D.P. O'Neal, L.R. Hirsch, N.J. Halas, J.D. Payne, J.L. West. Photothermal tumor ablation in mice using near infrared-absorbing nanoparticles. *Cancer Lett.* 209 (2004), 171–176. https://doi.org/10.1016/j.canlet.2004.02.004.

28. J. Hu, F. Sanz-Rodríguez, F. Rivero, E.M. Rodríguez, R.A. Torres, D.H. Ortgies, J.G. Solé, F. Alfonso, D. Jaque. Gold nanoshells: Contrast agents for cell imaging by cardiovascular optical coherence tomography. *Nano Res.* 11 (2018), 676–685. https://doi.org/10.1007/s12274-017-1674-4.

29. D.C. Adler, S.-W. Huang, R. Huber, J.G. Fujimoto. Photothermal detection of gold nanoparticles using phase-sensitive optical coherence tomography. *Opt. Express.* 16 (2008), 4376. https://doi.org/10.1364/oe.16.004376.

30. A.M. Schwartzberg, T.Y. Olson, C.E. Talley, J.Z. Zhang. Synthesis, characterization, and tunable optical properties of hollow gold nanospheres. *J. Phys. Chem. B.* 110 (2006), 19935–19944. https://doi.org/10.1021/jp062136a.

31. W. Lu, Q. Huang, G. Ku, X. Wen, M. Zhou, D. Guzatov, P. Brecht, R. Su, A. Oraevsky, L. V. Wang, C. Li. Photoacoustic imaging of living mouse brain vasculature using hollow gold nanospheres. *Biomaterials.* 31 (2010), 2617–2626. https://doi.org/10.1016/j.biomaterials.2009.12.007.

32. R.D. Averitt, S. L. Westcott, and N. J. Halas. Linear optical properties of gold nanoshells. *J. Opt. Soc. Am. B*, 16 (1999), 1824–1832. https://doi.org/10.1364/JOSAB.16.001824.

33. M.E. Khosroshahi, M.S. Nourbakhsh, L. Ghazanfari. Synthesis and biomedical application of SiO_2/Au nanofluid based on laser-induced surface plasmon resonance thermal effect. *J. Mod. Phys.* 2 (2011), 944–953. https://doi.org/10.4236/jmp.2011.29112.

34. S. Yang, Q. You, L. Yang, P. Li, Q. Lu, S. Wang, F. Tan, Y. Ji, N. Li. Rodlike MSN@Au nanohybrid-modified supermolecular photosensitizer for NIRF/MSOT/CT/MR quadmodal imaging-guided photothermal/photodynamic cancer therapy. *ACS Appl. Mater. Interfaces*. 11 (2019), 6777–6788. https://doi.org/10.1021/acsami.8b19565.

35. C. Loo, A. Lowery, N. Halas, J. West, R. Drezek. Immunotargeted nanoshells for integrated cancer imaging and therapy. *Nano Lett.* 5 (2005), 709–711. https://doi.org/10.1021/nl050127s.

36. A. Topete, M. Alatorre-Meda, P. Iglesias, E.M. Villar-Alvarez, S. Barbosa, J.A. Costoya, P. Taboada, V. Mosquera. Fluorescent drug-loaded, polymeric-based, branched gold nanoshells for localized multimodal therapy and imaging of tumoral cells. *ACS Nano*. 8 (2014), 2725–2738. https://doi.org/10.1021/nn406425h.

37. P.K. Jain, K.S. Lee, I.H. El-Sayed, M.A. El-Sayed. Calculated absorption and scattering properties of gold nanoparticles of different size, shape, and composition: Applications in biological imaging and biomedicine. *J. Phys. Chem. B*. 110 (2006), 7238–7248. https://doi.org/10.1021/jp057170o.

38. H.J. Moon, M. Ku, H. Lee, N. Yoon, J. Yang, K.W. Bong. Implantable photothermal agents based on gold nanorods-encapsulated microcube. *Sci. Rep*. 8 (2018), 1–12. https://doi.org/10.1038/s41598-018-31793-9.

39. X. Huang, M.A. El-Sayed. Gold nanoparticles: Optical properties and implementations in cancer diagnosis and photothermal therapy. *J. Adv. Res.* 1 (2010), 13–28. https://doi.org/10.1016/j.jare.2010.02.002.

40. L. Meng, J. Zhang, H. Li, W. Zhao, T. Zhao. Preparation and progress in application of gold nanorods. *J. Nanomater.* 2019 (2019), 1–11. https://doi.org/10.1155/2019/4925702.

41. S. Hwang, J. Nam, S. Jung, J. Song, H. Doh, S. Kim. Gold nanoparticle-mediated photothermal therapy: Current status and future perspective. *Nanomedicine*. 9 (2014), 2003–2022. https://doi.org/10.2217/nnm.14.147.

42. X. Huang, I.H. El-Sayed, W. Qian, M.A. El-Sayed. Cancer cell imaging and photothermal therapy in the near-infrared region by using gold nanorods. *J. Am. Chem. Soc.* 128 (2006), 2115–2120. https://doi.org/10.1021/ja057254a.
43. J. Lee, Y.H. Lee, C.B. Jeong, J.S. Choi, K.S. Chang, M. Yoon. Gold nanorods-conjugated TiO_2 nanoclusters for the synergistic combination of phototherapeutic treatments of cancer cells. *J. Nanobiotechnology.* 16 (2018), 1–12. https://doi.org/10.1186/s12951-018-0432-4.
44. S. Charan, K. Sanjiv, N. Singh, F.C. Chien, Y.F. Chen, N.N. Nergui, S.H. Huang, C.W. Kuo, T.C. Lee, P. Chen. Development of chitosan oligosaccharide-modified gold nanorods for in vivo targeted delivery and noninvasive imaging by NIR irradiation. *Bioconjug. Chem.* 23 (2012), 2173–2182. https://doi.org/10.1021/bc3001276.
45. X. Wang, H. Wang, Y. Wang, X. Yu, S. Zhang, Q. Zhang, Y. Cheng. A facile strategy to prepare dendrimer-stabilized gold nanorods with sub-10-nm size for efficient photothermal cancer therapy. *Sci. Rep.* 6 (2016), 2–11. https://doi.org/10.1038/srep22764.
46. Y. Liu, M. Yang, J. Zhang, X. Zhi, C. Li, C. Zhang, F. Pan, K. Wang, Y. Yang, J. Martinez De La Fuentea, D. Cui. Human induced pluripotent stem cells for tumor targeted delivery of gold nanorods and enhanced photothermal therapy. *ACS Nano.* 10 (2016), 2375–2385. https://doi.org/10.1021/acsnano.5b07172.
47. N. Narayanan, L.V. Nair, V. Karunakaran, M.M. Joseph, J.B. Nair, A.N. Ramya, R.S. Jayasree, K.K. Maiti. Investigation of apoptotic events at molecular level induced by SERS guided targeted theranostic nanoprobe. *Nanoscale.* 8 (2016), 11392–11397. https://doi.org/10.1039/c6nr03385g.
48. R. V. Nair, L. V. Nair, D.M. Govindachar, H. Santhakumar, S.S. Nazeer, C.R. Rekha, S.J. Shenoy, G. Periyasamy, R.S. Jayasree. Luminescent gold nanorods to enhance the near-infrared emission of a photosensitizer for targeted cancer imaging and dual therapy: experimental and theoretical approach. *Chem. - A Eur. J.* 26 (2020), 2826–2836. https://doi.org/10.1002/chem.201904952.
49. R.V. Nair, H. Santhakumar, R.S. Jayasree. Gold nanorods decorated with a cancer drug for multimodal imaging and therapy. *Faraday Discuss.* 207 (2018), 423–435. https://doi.org/10.1039/c7fd00185a.
50. J. Zhang, X. Luo, Y.P. Wu, F. Wu, Y.F. Li, R.R. He, M. Liu. Rod in tube: A novel nanoplatform for highly effective chemo-photothermal combination therapy toward breast cancer. *ACS Appl. Mater. Interfaces.* 11 (2019), 3690–3703. https://doi.org/10.1021/acsami.8b17533.

51. S. Adams, J.Z. Zhang. Unique optical properties and applications of hollow gold nanospheres (HGNS). *Coord. Chem. Rev.* 320 & 321 (2016), 18–37. https://doi.org/10.1016/j.ccr.2016.01.014.

52. S.A. Lindley, J.Z. Zhang. Bumpy hollow gold nanospheres for theranostic applications: Effect of surface morphology on photothermal conversion efficiency. *ACS Appl.* Nano Mater. 2 (2019), 1072–1081. https://doi.org/10.1021/acsanm.8b02331.

53. S.A. Lindley, J.K. Cooper, M.D. Rojas-Andrade, V. Fung, C.J. Leahy, S. Chen, J.Z. Zhang. Highly tunable hollow gold nanospheres: Gaining size control and uniform galvanic exchange of sacrificial cobalt boride scaffolds. *ACS Appl. Mater. Interfaces.* 10 (2018), 12992–13001. https://doi.org/10.1021/acsami.8b00726.

54. M.P. Melancon, W. Lu, Z. Yang, R. Zhang, Z. Cheng, A.M. Elliot, J. Stafford, T. Olson, J.Z. Zhang, C. Li. In vitro and in vivo targeting of hollow gold nanoshells directed at epidermal growth factor receptor for photothermal ablation therapy. *Mol. Cancer Ther.* 7 (2008), 1730–1739. https://doi.org/10.1158/1535-7163.MCT-08-0016.

55. L.Y. Bai, X.Q. Yang, J. An, L. Zhang, K. Zhao, M.Y. Qin, B.Y. Fang, C. Li, Y. Xuan, X.S. Zhang, Y. Di Zhao, Z.Y. Ma. Multifunctional magnetic-hollow gold nanospheres for bimodal cancer cell imaging and photothermal therapy. *Nanotechnology.* 26 (2015), 315701. https://doi.org/10.1088/0957-4484/26/31/315701.

56. W. Lu, M.P. Melancon, C. Xiong, Q. Huang, A. Elliott, S. Song, R. Zhang, L.G. Flores, J.G. Gelovani, L.V. Wang, G. Ku, R.J. Stafford, C. Li. Effects of photoacoustic imaging and photothermal ablation therapy mediated by targeted hollow gold nanospheres in an orthotopic mouse xenograft model of glioma. *Cancer Res.* 71 (2011), 6116–6121. https://doi.org/10.1158/0008-5472.CAN-10-4557.

57. C.L. Wei Lu, Chiyi Xiong, Guodong Zhang, Qian Huang, Rui Zhang, Jin Z. Zhang. Targeted photothermal ablation of murine melanomas with melanocyte-stimulating hormone analog-conjugated hollow gold nanospheres. *Clin Cancer Res.* 15 (2009), 876–886. https://doi.org/10.1038/jid.2014.371.

58. Y. Sun, Y. Xia. Mechanistic Study on the replacement reaction between silver nanostructures and chloroauric acid in aqueous medium. *J. Am. Chem. Soc.* 126 (2004), 3892–3901. https://doi.org/10.1021/ja039734c.

59. Y. Sun, B.T. Mayers, Y. Xia. Template-engaged replacement reaction: A one-step approach to the large-scale synthesis of metal

nanostructures with hollow interiors. *Nano Lett.* 2 (2002), 481–485. https://doi.org/10.1021/nl025531v.

60. S.E. Skrabalak, J. Chen, Y. Sun, X. Lu, L. Au, C.M. Cobley, Y. Xia. *Gold Acc. Chem. Res.* https://doi.org/10.1021/ar800018v.
61. S.E. Skrabalak, J. Chen, L. Au, X. Lu, X. Li, Y. Xia. Gold nanocages for biomedical applications. *Adv. Mater.* 19 (2007), 3177–3184. https://doi.org/10.1002/adma.200701972.
62. J. Chen, D. Wang, J. Xi, L. Au, A. Siekkinen, A. Warsen, Z.Y. Li, H. Zhang, Y. Xia, X. Li. Immuno gold nanocages with tailored optical properties for targeted photothermal destruction of cancer cells. *Nano Lett.* 7 (2007), 1318–1322. https://doi.org/10.1021/nl070345g.
63. F. Hao, C.L. Nehl, J.H. Hafner, P. Nordlander. Plasmon resonances of a gold nanostar. *Nano Lett.* 7 (2007), 729–732. https://doi.org/10.1021/nl062969c.
64. L. Sironi, S. Freddi, M. Caccia, P. Pozzi, L. Rossetti, P. Pallavicini, A. Donà, E. Cabrini, M. Gualtieri, I. Rivolta, A. Panariti, L. Dalfonso, M. Collini, G. Chirico. Gold branched nanoparticles for cellular treatments. *J. Phys. Chem. C.* 116 (2012), 18407–18418. https://doi.org/10.1021/jp305021k.
65. A.M. Smith, M.C. Mancini, S. Nie. NIH public access: *Second Nat. Nanotechnol.* 4 (2009), 710–711. https://doi.org/10.1038/nnano.-2009.326.
66. N.J. Halas. Plasmonics: An emerging field fostered by nano letters. *Nano Lett.* 10 (2010), 3816–3822. https://doi.org/10.1021/nl1032342.
67. H. Chatterjee, D.S. Rahman, M. Sengupta, S.K. Ghosh. Gold nanostars in plasmonic photothermal therapy: The role of tip heads in the thermoplasmonic landscape. *J. Phys. Chem. C.* 122 (2018), 13082–13094. https://doi.org/10.1021/acs.jpcc.8b00388.
68. H. Yuan, C.G. Khoury, C.M. Wilson, G.A. Grant, A.J. Bennett, T. Vo-Dinh. In vivo particle tracking and photothermal ablation using plasmon-resonant gold nanostars. *Nanomedicine.* 8 (2012), 1355–1363. https://doi.org/10.1016/j.nano.2012.02.005.
69. H. Yuan, A.M. Fales, T. Vo-Dinh. TAT peptide-functionalized gold nanostars: Enhanced intracellular delivery and efficient NIR photothermal therapy using ultralow irradiance. *J. Am. Chem. Soc.* 134 (2012), 11358–11361. https://doi.org/10.1021/ja304180y.
70. S. Wang, Z. Teng, P. Huang, D. Liu, Y. Liu, Y. Tian, J. Sun, Y. Li, H. Ju, X. Chen, G. Lu. Reversibly extracellular pH controlled cellular

uptake and photothermal therapy by PEGylated mixed-charge gold nanostars. *Small.* 11 (2015), 1801–1810. https://doi.org/10.1002/smll.201403248.

71. H. Chen, X. Zhang, S. Dai, Y. Ma, S. Cui, S. Achilefu, Y. Gu. Multifunctional gold nanostar conjugates for tumor imaging and combined photothermal and chemo-therapy. *Theranostics.* 3 (2013), 633–649. https://doi.org/10.7150/thno.6630.

72. W.A. Marmisollé, O. Azzaroni. Recent developments in the layer-by-layer assembly of polyaniline and carbon nanomaterials for energy storage and sensing applications from synthetic aspects to structural and functional characterization. *Nanoscale.* 8 (2016), 9890–9918. https://doi.org/10.1039/c5nr08326e.

73. L.D. Mei Zhang, R. Rajesh Naik. Carbon nanomaterials for biomedical applications. Springer Series in Biomaterials Science and Engineering 5 (2016). https://doi: 10.1007/978-3-319-22861-7.

74. F.M.M. Bergmann, P. Carlos. Carbon nanomaterials as adsorbents for environmental and biological applications. *Carbon Nanostructures,* 2015. https://doi. org/10.1007/978-3-319-18875-1.

75. H.K. Moon, S.H. Lee, H.C. Choi. In vivo near-infrared mediated tumor destruction by photothermal effect of carbon nanotubes. *ACS Nano.* 3 (2009), 3707–3713. https://doi.org/10.1021/nn900904h.

76. Y.Y. Hui, H.-C. Chang, H. Dong, X. Zhang, John Wiley & Sons: Hoboken, NJ, USA, (2019). ISBN: 978-1-119-37345-2.

77. P.G. Collins, K. Bradley, Mishigami, A. Zettl. Extreme oxygen sensitivity of electronic. *Science* (80). 287 (2000), 1801–1804. https://doi.org/10.1126/science.287.5459.1801.

78. S.Y. Yang, W.N. Lin, Y.L. Huang, H.W. Tien, J.Y. Wang, C.C.M. Ma, S.M. Li, Y.S. Wang. Synergetic effects of graphene platelets and carbon nanotubes on the mechanical and thermal properties of epoxy composites. *Carbon.* 49 (2011), 793–803. https://doi.org/10.1016/j.carbon.2010.10.014.

79. N. Hamada, S.I. Sawada, A. Oshiyama. New one-dimensional conductors: Graphitic microtubules. *Phys. Rev. Lett.* 68 (1992), 1579–1581. https://doi.org/10.1103/PhysRevLett.68.1579.

80. D. Tasis, N. Tagmatarchis, A. Bianco, M. Prato. Chemistry of carbon nanotubes. *Chem. Rev.* 106 (2006), 1105–1136. https://doi.org/10.1021/cr050569o.

81. Y. Zhang, Y. Bai, B. Yan. Functionalized carbon nanotubes for potential medicinal applications. *Drug Discov. Today.* 15 (2010), 428–435. https://doi.org/10.1016/j.drudis.2010.04.005.

82. S. Prakash, M. Malhotra, W. Shao, C. Tomaro-Duchesneau, S. Abbasi. Polymeric nanohybrids and functionalized carbon nanotubes as drug delivery carriers for cancer therapy. *Adv. Drug Deliv. Rev.* 63 (2011), 1340–1351. https://doi.org/10.1016/j.addr.2011.06.013.

83. F. Zhou, D. Xing, Z. Ou, B. Wu, D.E. Resasco, W.R. Chen. Cancer photothermal therapy in the near-infrared region by using single-walled carbon nanotubes. *J. Biomed. Opt.* 14 (2009), 021009. https://doi.org/10.1117/1.3078803.

84. L.V. Nair, Y. Nagaoka, T. Maekawa, D. Sakthikumar, R.S. Jayasree. Quantum dot tailored to single wall carbon nanotubes: A multifunctional hybrid nanoconstruct for cellular imaging and targeted photothermal therapy. *Small.* 10 (2014), 2771–2775. https://doi.org/10.1002/smll.201400418.

85. Z. Sobhani, M.A. Behnam, F. Emami, A. Dehghanian, I. Jamhiri. Photothermal therapy of melanoma tumor using multiwalled carbon nanotubes. *Int. J. Nanomedicine.* 12 (2017), 4509–4517. https://doi.org/10.2147/IJN.S134661.

86. Z. Liu, J.T. Robinson, X. Sun, H. Dai. PEGylated nanographene oxide for delivery of water-insoluble cancer drugs. *J. Am. Chem. Soc.* 130 (2008), 10876–10877. https://doi.org/10.1021/ja803688x.

87. K.D. Patel, R.K. Singh, H.W. Kim. Carbon-based nanomaterials as an emerging platform for theranostics. *Mater. Horizons.* 6 (2019), 434–469. https://doi.org/10.1039/c8mh00966j.

88. K.S. Novoselov, A.K. Geim, S.V. Morozov, D. Jiang, Y. Zhang, S.V. Dubonos, I.V. Grigorieva, A.A. Firsov. Electric field effect in atomically thin carbon films. *Science.* 306 (2004), 306, 666–669. https://doi.org/10.1126/science.1102896.

89. W.S. Hummers, R.E. Offeman. Preparation of graphitic oxide. *J. Am. Chem. Soc.* 80 (1958), 1339. https://doi.org/10.1021/ja01539a017.

90. Z. Wang, P. Huang, A. Bhirde, A. Jin, Y. Ma, G. Niu, N. Neamati, X. Chen. A nanoscale graphene oxide–peptide biosensor for real-time specific biomarker detection on the cell surface. *Chem. Commun.* 48 (2012), 9768–9770. https://doi.org/10.1039/c2cc31974h.

91. C. Guo, B. Book-Newell, J. Irudayaraj. Protein-directed reduction of graphene oxide and intracellular imaging. *Chem. Commun.* 47 (2011), 12658–12660. https://doi.org/10.1039/c1cc15052a.

92. X. Yang, Y. Wang, X. Huang, Y. Ma, Y. Huang, R. Yang, H. Duan, Y. Chen. Multi-functionalized graphene oxide based anticancer drug-carrier

with dual-targeting function and pH-sensitivity. *J. Mater. Chem.* 21 (2011), 3448–3454. https://doi.org/10.1039/c0jm02494e.

93. X. Yang, X. Zhang, Z. Liu, Y. Ma, Y. Huang, Y. Chen. High-efficiency loading and controlled release of doxorubicin hydrochloride on graphene oxide. *J. Phys. Chem. C.* 112 (2008), 17554–17558. https://doi.org/10.1021/jp806751k.

94. G. Gonçalves, M. Vila, I. Bdikin, A. De Andrés, N. Emami, R.A.S. Ferreira, L.D. Carlos, J. Grácio, P.A.A.P. Marques. Breakdown into nanoscale of graphene oxide: Confined hot spot atomic reduction and fragmentation. *Sci. Rep.* 4 (2014), 1–8. https://doi.org/10.1038/srep06735.

95. R. Kurapati, A.M. Raichur. Near-infrared light-responsive graphene oxide composite multilayer capsules: A novel route for remote controlled drug delivery. *Chem. Commun.* 49 (2013), 734–736. https://doi.org/10.1039/c2cc38417e.

96. Y.W. Wang, Y.Y. Fu, Q. Peng, S.S. Guo, G. Liu, J. Li, H.H. Yang, G.N. Chen. Dye-enhanced graphene oxide for photothermal therapy and photoacoustic imaging. *J. Mater. Chem. B.* 1 (2013), 5762–5767. https://doi.org/10.1039/c3tb20986e.

97. Z. Sheng, L. Song, J. Zheng, D. Hu, M. He, M. Zheng, G. Gao, P. Gong, P. Zhang, Y. Ma, L. Cai. Protein-assisted fabrication of nano-reduced graphene oxide for combined invivo photoacoustic imaging and photothermal therapy. *Biomaterials.* 34 (2013), 5236–5243. https://doi.org/10.1016/j.biomaterials.2013.03.090.

98. J. Wu, Z. Li, Y. Li, A. Pettitt, F. Zhou. Photothermal effects of reduced graphene oxide on pancreatic cancer. *Technol. Cancer Res. Treat.* 17 (2018), 1–7. https://doi.org/10.1177/1533034618768637.

99. H. Liu, C. Li, Y. Qian, L. Hu, J. Fang, W. Tong, R. Nie, Q. Chen, H. Wang. Magnetic-induced graphene quantum dots for imaging-guided photothermal therapy in the second near-infrared window. *Biomaterials.* 232 (2020). https://doi.org/10.1016/j.biomaterials.2019.119700.

100. H.W. Kroto, A.W. Allaf, S.P. Balm. C60: Buck minster fullerene. *Chem. Rev.* 91 (1991), 1213–1235. https://doi.org/10.1021/cr00006a005.

101. H. Kato, Y. Kanazawa, M. Okumura, A. Taninaka, T. Yokawa, H. Shinohara. Lanthanoid endohedral metallofullerenols for MRI contrast agents. *J. Am. Chem. Soc.* 125 (2003), 4391–4397. https://doi.org/10.1021/ja027555+.

102. J.W. Arbogast, A.P. Darmanyan, C.S. Foote, F.N. Diederich, R.L. Whetten, Y. Rubin, M.M. Alvarez, S J. Anz. Photophysical properties of sixty atom carbon molecule (C60). *J. Phys. Chem.* 95 (1991), 11–12. https://doi.org/10.1021/j100154a006.

103. V. Krishna, N. Stevens, B. Koopman, B. Moudgil. Optical heating and rapid transformation of functionalized fullerenes. *Nat. Nanotechnol.* 5 (2010), 330–334. https://doi.org/10.1038/nnano.2010.35.

104. Z. Beiranvand, A. Kakanejadifard, I.S. Donskyi, A. Faghani, Z. Tu, A. Lippitz, P. Sasanpour, F. Maschietto, B. Paulus, W.E.S. Unger, R. Haag, M. Adeli. Functionalization of fullerene at room temperature: Toward new carbon vectors with improved physicochemical properties. *RSC Adv.* 6 (2016), 112771–112775. https://doi.org/10.1039/c6ra23419d.

105. Z.H. Zhao, D. Wang, H. Gao, Z. Yang, H. Cao, W. He. Photoacoustic effect of near-infrared absorbing fullerene derivatives with click moieties. *Dye. Pigment.* 164 (2019), 182–187. https://doi.org/10.1016/j.dyepig.2019.01.022.

106. Y. Shen, A.G. Skirtach, T. Seki, S. Yagai, H. Li, H. Möhwald, T. Nakanishi. Assembly of fullerene-carbon nanotubes: Temperature indicator for photothermal conversion. *J. Am. Chem. Soc.* 132 (2010), 8566–8568. https://doi.org/10.1021/ja1026024.

107. Lakshmi V. Nair, Shaiju S. Nazeer, Ramapurath S. Jayasree and Ayyappanpillai Ajayaghosh. Fluorescence imaging assisted photodynamic therapy using photosensitizer-linked gold quantum clusters. *ACS Nano.* 9(6) 2015, 5825–5832. https://doi.org/10.1021/acsnano.5b00406.

108. T.J. Dougherty, G.B. Grindey, R. Fiel, K.R. Weishaupt, D.G. Boyle. Photoradiation therapy. II. Cure of animal tumors with hematoporphyrin and light. *J. Natl. Cancer Inst.* 55(1) (1975), 115–21. https://doi: 10.1093/jnci/55.1.115.

109. J.P. Celli, B.Q. Spring, I. Rizvi, C.L. Evans, K.S. Samkoe, S. Verma, B.W. Pogue, T. Hasan. Imaging and photodynamic therapy: Mechanisms, monitoring, and optimization. *Chem. Rev.* 110(5) 2010, 2795–2838. https://doi.org/10.1021/cr900300p.

110. S.S. Lucky, K.C. Soo, Y. Zhang. Nanoparticles in photodynamic therapy. *Chem. Rev.* 115(4) 2015, 1990–2042, https://doi.org/10.1021/cr5004198.

111. T.B. Fitzpatrick, M.A. Pathak. Basic considerations of the psoralens. *J. Invest. Dermatol.* 32 (1958), 229–231. https://doi.org/10.1038/jid.1959.40.

112. J.H. Epstein. Phototherapy and photochemotherapy. *N. Engl. J. Med.* 322 (1990), 1149–51.

113. K. Moghissi, K. Dixon. Is bronchoscopic photodynamic therapy a therapeutic option in lung cancer? *Eur. Respir. J.* (2003), 535–541. https://doi.org/10.1183/09031936.03.00005203.

114. R. Saini, N.V. Lee, K.Y.P. Liu, C.F. Poh. Prospects in the application of photodynamic therapy in oral cancer and premalignant lesions. *Cancers.* 8(9) (2016), 83. https://doi.org/10.3390/cancers8090083.

115. D.K. Cohen, P.K. Lee. Photodynamic therapy for non-melanoma skin cancers. *Cancers (Basel).* 8(10) (2016), 90. https://doi.org/10.3390/cancers8100090.

116. B.J. Qumseya, W. David, H.C. Wolfsen. Photodynamic therapy for Barrett's esophagus and esophageal carcinoma. *Clin. Endosc.* 46(1) (2013), 30–37. https://doi:10.5946/ce.2013.46.1.30.

117. B.F. Overholt, M. Panjehpour. Photodynamic therapy in Barrett's esophagus. *J. Clin. Laser. Med. Surg.* 14(5) (1996), 245–249. https://doi:10.1089/clm.1996.14.245.

118. P. Agostinis, K. Berg, K.A. Cengel, T.H. Foster, A.W. Girotti, S.O. Gollnick, S.M. Hahn, M.R. Hamblin, A. Juzeniene, D. Kessel. Photodynamic therapy of cancer: An update. *CA Cancer J. Clin.* 61(4) (2011), 250–281. https://doi:10.3322/caac.20114.

119. R.R. Allison, G.H. Downie, R. Cuenca, X. Hu, C.J.H. Childs, C.H. Sibata. Photosensitizers in clinical PDT. *Photodiagnosis Photodyn. Ther.* 1(1) (2004), 27–42. https://doi:10.1016/S1572-1000(04)00007-9.

120. R.R. Allison. Photodynamic therapy: Oncologic horizons. *Future Oncol.* 10(1) (2014), 123–124. https://doi:10.2217/fon.13.176.

121. T. Dai, B.B. Fuchs, J.J. Coleman, R.A. Prates, C. Astrakas, T.G.S. Denis, M.S. Ribeiro, E. Mylonakis, M.R. Hamblin. Concepts and principles of photodynamic therapy as an alternative antifungal discovery platform. *Front. Microbiol.* 3 (2012), 1–16. https://doi.org/10.3389/fmicb.2012.00120.

122. A.P. Castano, T.N. Demidova, M.R. Hamblin. Mechanisms in photodynamic therapy: Part one – Photosensitizers, photochemistry and cellular localization. *Photodiagnosis Photodyn. Ther.* 1 (2004), 279–293. https://doi.org/10.1016/S1572-1000(05)00007-4.

123. B.C. Wilson, M.S. Patterson. The physics of photodynamic therapy. *Phys. Med. Biol.* 31(4) (1986), 327–360. https://doi:10.1088/0031-9155/31/4/001.

124. T.J. Dougherty, C.J. Gomer, B.W. Henderson, G. Jori, D. Kessel, M. Korbelik, J. Moan, Q. Peng. Photodynamic Therapy. *J. Natl. Cancer. Inst.* 90(12) (1998), 889–905. https://doi:10.1093/jnci/90.12.889.

125. L.B. Josefsen, R.W. Boyle. Photodynamic therapy and the development of metal-based photosensitizers. *Met. Based. Drugs.* 2008 (2008) 276109. https://doi.org/10.1155/2008/276109.

126. Q. Peng, J. Moan, J.M. Nesland. Correlation of subcellular and intratumoral photosensitizer localization with ultrastructural features after photodynamic therapy. *Ultrastruct. Pathol.* 20(2) (1996), 109–129. https://doi:10.3109/01913129609016306.

127. A.B. Ormond, H.S. Freeman. Dye sensitizers for photodynamic therapy. *Materials.* 6 (2013), 817–840. https://doi.org/10.3390/ma6030817.

128. H. Kataoka, H. Nishie, N. Hayashi, M. Tanaka, A. Nomoto, S. Yano, T. Joh. New photodynamic therapy with next-generation photosensitizers, *Ann. Transl. Med.* 5(8) (2017), 183. doi: https://10.21037/atm.2017.03.59.

129. J. Taquet, C. Frochot, V. Manneville, M. Barberi-heyob. Phthalocyanines covalently bound to biomolecules for a targeted photodynamic therapy. *Curr. Med. Chem.* 14(15) (2007), 1673–1687. https://doi:10.2174/092986707780830970.

130. D. Yang, G. Yang, S. Gai, F. He, G. An, Y. Dai, R. Lv, P. Yang. Au25 cluster functionalized metal–organic nanostructures for magnetically targeted photodynamic/photothermal therapy triggered by single wavelength 808 nm near-infrared light. *Nanoscale.* 7 (2015), 19568–19578. https://doi.org/10.1039/c5nr06192j.

131. X.S. Haozhe He, Xiaohua Zheng, Shi Liu, Min Zheng, Zhigang Xie, Yong Wang, B. Meng Yu. Diketopyrrolopyrrole-based carbon dots for photodynamic therapy. *Nanoscale.* 10 (2018) 10991–10998. https://doi.org/10.1039/C8NR02643B.

132. C.M. Lee, C.J. Engelbrecht, T.D. Soper, F. Helmchen, E.J. Seibel.Scanning fiber endoscopy with highly flexible, 1 mm catheterscopes for wide-field, full-color imaging. *J. Biophotonics.* (2010), 385–407. https://doi.org/10.1002/jbio.200900087.

133. J.P. Rovers, A.E. Saarnak, A. Molina, J.J. Schuitmaker, H. Sterenborg, O.T. Terpstra. Effective treatment of liver metastases with photodynamic therapy, using the second-generation photosensitizer meta-tetra (hydroxyphenyl) chlorin (mTHPC), in a rat model. *Br. J. Cancer.* 81(4) (1999), 600–608. https://doi:10.1038/sj.bjc.6690736.

134. J. Akimoto. Photodynamic therapy for malignant brain tumors. *Neurol. Med. Chir.* (Tokyo). 56(4) (2016), 151–157. doi:10.2176/nmc.ra.2015-0296.

135. J.V. Frangioni. In vivo near-infrared fluorescence imaging. *Curr. Opin. Chem. Biol.* 7(5) (2003), 626–634. https://doi:10.1016/j.cbpa.2003.08.007.

136. M.Y. Berezin, S. Achilefu. Fluorescence lifetime measurements and biological imaging. *Chem. Rev.* (2010), 2641–2684. https://doi.org/10.1021/cr900343z.

137. L. Luan, L. Ding, W. Zhang, J. Shi, X. Yu, W. Liu. a naphthalocyanine based near-infrared photosensitizer: Synthesis and in vitro photodynamic activities. *Bio. Med. Chem. Lett.* 23 (2013), 3775–3779. https://doi.org/10.1016/j.bmcl.2013.04.093.

138. T. Lu, P. Shao, I. Mathew, A. Sand, W. Sun, Synthesis and photophysics of benzotexaphyrin: A near-infrared emitter and photosensitizer. *J. Am. Chem. Soc.* (2008), 15782–15783. https://doi.org/10.1021/ja807021n.

139. J. Ian, Macdonald, T.J. Dougherty. Basic principles of photodynamic therapy. *J. Orp. Phths.* (2001) 105–129. https://doi.org/10.1002/jpp.328.

140. A. Ruck, R. Steiner. Basic reaction mechanisms of hydrophilic and lipophilic photosensitisers in photodynamic tumour treatment. *Min. Inv. Ther. All. Tech.* 7(6), 503–509, https://doi.org/10.3109/13645709809152895.

141. Y.N. Konan, R. Gurny, E. Allemann. State of the art in the delivery of photosensitizers for photodynamic therapy. *J Photochem and Photobio B: Bio.* 66 (2002) 89–106, https://doi.org/10.1016/S1011-1344(01)00267-6.

142. C. Lim, J. Heo, S. Shin, K. Jeong, Y. Hun, W. Jang, C. Rae, S. Young, S. Kim, I. Chan. Nanophotosensitizers toward advanced photo dynamic therapy of cancer. *Cancer. Lett.* 334 (2013), 176–187. https://doi.org/10.1016/j.canlet.2012.09.012.

143. M. Srivastava, N. Ahmad, S. Gupta, H. Mukhtar. Involvement of Bcl-2 and Bax in photodynamic therapy-mediated apoptosis. Antisense Bcl-2 oligonucleotide sensitizes RIF 1 cells to photodynamic therapy apoptosis. *J. Biol. Chem.* 276(18) (2001), 15481–15488. doi: 10.1074/jbc.M006920200.

150. R. Singh, J.W. Lillard. Nanoparticle-based targeted drug delivery. *Exp. Mol. Pathol.* 86 (2009), 215–223. https://doi.org/10.1016/j.yexmp.2008.12.004.

151. Y. Dang, J. Guan. Nanoparticle-based drug delivery systems for cancer therapy. *Smart Mater. Med.* 1 (2020), 10–19. https://doi.org/10.1016/j.smaim.2020.04.001.

152. H. Maeda. The enhanced permeability and retention (EPR) effect in tumor vasculature: The key role of tumor-selective macromolecular drug targeting. *Adv. Enzyme Regul.* 41 (2001), 189–207. https://doi.org/10.1016/S0065-2571(00)00013-3.

153. C.A. Kruger, H. Abrahamse. Utilisation of targeted nanoparticle photosensitiser drug delivery systems for the enhancement of photodynamic therapy. *Molecules.* 23 (2018). https://doi.org/10.3390/molecules23102628.

154. T.A. Debele, S. Peng, H.C. Tsai. Drug carrier for photodynamic cancer therapy. *Int. J. Mol. Sci.* 16(9) (2015), 22094–22136. https://doi.org/10.3390/ijms160922094.

155. M.Q. Mesquita, C.J. Dias, S. Gamelas, M. Fardilha, M.G.P.M.S. Neves, M.A.F. Faustino. An insight on the role of photosensitizer nanocarriers for photodynamic therapy. *An. Acad. Bras. Cienc.* 90 (2018), 1101–1130. https://doi.org/10.1590/0001-3765201720170800.

156. Z. Yang, Z. Sun, Y. Ren, X. Chen, W. Zhang, X. Zhu, Z. Mao, J. Shen, S. Nie. Advances in nanomaterials for use in photothermal and photodynamic therapeutics: A review. *Mol. Med. Rep.* 20 (2019), 5–15. https://doi.org/10.3892/mmr.2019.10218.

157. E. Paszko, C. Ehrhardt, M.O. Senge, D.P. Kelleher, J.V. Reynolds. Nanodrug applications in photodynamic therapy. *Photodiagnosis Photodyn. Ther.* 8 (2011), 14–29. https://doi.org/10.1016/j.pdpdt.2010.12.001.

158. J.K. Patra, G. Das, L.F. Fraceto et al. Nano based drug delivery systems: Recent developments and future prospects. *J. Nanobiotechnol.* 16 (2018), 71. https://doi.org/10.1186/s12951-018-0392-8.

159. N. Hoshyar, S. Gray, H. Han, G. Bao. The effect of nanoparticle size on in vivo pharmacokinetics and cellular interaction. *Nanomedicine.* 11 (2016), 673–692. https://doi.org/10.2217/nnm.16.5.

160. D.K. Newman. Photodynamic therapy: Current role in the treatment of chorioretinal conditions. *Eye.* 30 (2016), 202–210. https://doi.org/10.1038/eye.2015.251.

161. E.J. Hong, D.G. Choi, M.S. Shim. Targeted and effective photodynamic therapy for cancer using functionalized nanomaterials. *Acta Pharm. Sin. B.* 6 (2016), 297–307. https://doi.org/10.1016/j.apsb.2016.01.007.

162. G. Obaid, M. Broekgaarden, A.L. Bulin, H.C. Huang, J. Kuriakose, J. Liu, T. Hasan. Photonanomedicine: A convergence of photodynamic

therapy and nanotechnology. *Nanoscale.* 8 (2016), 12471–12503. https://doi.org/10.1039/c5nr08691d.

163. S. Bamrungsap, Z. Zhao, T. Chen, L. Wang, C. Li, T. Fu, W. Tan. Nanotechnology in therapeutics: A focus on nanoparticles as a drug delivery system. *Nanomedicine.* 7 (2012), 1253–1271. https://doi.org/10.2217/nnm.12.87.

164. S.K. Golombek, J.N. May, B. Theek, L. Appold, N. Drude, F. Kiessling, T. Lammers. Tumor targeting via EPR: Strategies to enhance patient responses. *Adv. Drug Deliv. Rev.* 130 (2018), 17–38. https://doi.org/10.1016/j.addr.2018.07.007.

165. S. Barua, S. Mitragotri. Challenges associated with penetration of nanoparticles across cell and tissue barriers: A review of current status and future prospects. *Nano Today.* 9 (2014), 223–243. https://doi.org/10.1016/j.nantod.2014.04.008.

166. I. Rodica-Mariana. Photodynamic nanomedicine strategies in cancer therapy and drug delivery. Advances in Bioengineering, Pier Andrea Serra, *IntechOpen*, DOI: 10.5772/59624.

167. N. Kamaly, B. Yameen, J. Wu, O.C. Farokhzad. Degradable controlled-release polymers and polymeric nanoparticles: Mechanisms of controlling drug release. *Chem. Rev.* 116 (2016), 2602–2663. https://doi.org/10.1021/acs.chemrev.5b00346.

168. A. Mahapatro, D.K. Singh. Biodegradable nanoparticles are excellent vehicle for site directed in-vivo delivery of drugs and vaccines. *J. Nanobiotechnology.* 9 (2011), 55. https://doi.org/10.1186/1477-3155-9-55.

169. S.A. Sibani, P.A. McCarron, A.D. Woolfson, R.F. Donnelly. Photosensitiser delivery for photodynamic therapy, Part 2: Systemic carrier platforms. *Expert Opin. Drug Deliv.* 5 (2008), 1241–1254. https://doi.org/10.1517/17425240802444673.

170. E. Ricci-Júnior, J.M. Marchetti. Zinc(II) phthalocyanine loaded PLGA nanoparticles for photodynamic therapy use. *Int. J. Pharm.* 310 (2006), 187–195. https://doi.org/10.1016/j.ijpharm.2005.10.048.

171. M. Zeisser-Labouèbe, F. Delie, R. Gurny, N. Lange. Benefits of nanoencapsulation for the hypercin-mediated photodetection of ovarian micrometastases. *Eur. J. Pharm. Biopharm.* 71 (2009), 207–213. https://doi.org/10.1016/j.ejpb.2008.10.005.

172. A. Vargas, N. Lange, T. Arvinte, R. Cerny, R. Gurny, F. Delie. Toward the understanding of the photodynamic activity of m-THPP encapsulated in PLGA nanoparticles: Correlation between

nanoparticle properties and in vivo activity. *J. Drug Target.* 17 (2009), 599–609. https://doi.org/10.1080/10611860903118914.

173. B. Pegaz, E. Debefve, F. Borle, J.P. Ballini, H. van den Bergh, Y.N. Kouakou-Konan. Encapsulation of porphyrins and chlorins in biodegradable nanoparticles: The effect of dye lipophilicity on the extravasation and the photothrombic activity–A comparative study. *J. Photochem. Photobiol. B Biol.* 80 (2005), 19–27. https://doi.org/10.1016/j.jphotobiol.2005.02.003.

174. A. Khdair, B. Gerard, H. Handa, G. Mao, M.P.V. Shekhar, J. Panyam. Surfactant–Polymer nanoparticles enhance the effectiveness of anticancer photodynamic therapy. *Mol. Pharm.* 5 (2008), 795–807. https://doi.org/10.1021/mp800026t.

175. A. Khdair, Di Chen, Y. Patil, L. Ma, Q.P. Dou, M.P.V. Shekhar, J. Panyam. Nanoparticle-mediated combination chemotherapy and photodynamic therapy overcomes tumor drug resistance. *J. Control. Release.* 141 (2010), 137–144. https://doi.org/10.1016/j.jconrel.2009.09.004.

176. S. Sortino, A. Mazzaglia, L. Monsù Scolaro, F. Marino Merlo, V. Valveri, M.T. Sciortino. Nanoparticles of cationic amphiphilic cyclodextrins entangling anionic porphyrins as carrier-sensitizer system in photodynamic cancer therapy. *Biomaterials.* 27 (2006), 4256–4265. https://doi.org/10.1016/j.biomaterials.2006.03.035.

177. F. Kratz. Albumin as a drug carrier: Design of prodrugs, drug conjugates and nanoparticles. *J. Control. Release.* 132 (2008), 171–183. https://doi.org/10.1016/j.jconrel.2008.05.010.

178. K. Chen, A. Preuß, S. Hackbarth, M. Wacker, K. Langer, B. Röder. Novel photosensitizer-protein nanoparticles for photodynamic therapy: Photophysical characterization and in vitro investigations. *J. Photochem. Photobiol. B Biol.* 96 (2009), 66–74. https://doi.org/10.1016/j.jphotobiol.2009.04.006.

179. S.J. Lee, K. Park, Y.K. Oh, S.H. Kwon, S. Her, I.S. Kim, K. Choi, S.J. Lee, H. Kim, S.G. Lee, K. Kim, I.C. Kwon. Tumor specificity and therapeutic efficacy of photosensitizer-encapsulated glycol chitosan-based nanoparticles in tumor-bearing mice. *Biomaterials.* 30 (2009), 2929–2939. https://doi.org/10.1016/j.biomaterials.2009.01.058.

180. D.K. Chatterjee, L.S. Fong, Y. Zhang. Nanoparticles in photodynamic therapy: An emerging paradigm. *Adv. Drug Deliv. Rev.* 60 (2008), 1627–1637. https://doi.org/10.1016/j.addr.2008.08.003.

181. J. Qian, A. Gharibi, S. He. Colloidal mesoporous silica nanoparticles with protoporphyrin IX encapsulated for photodynamic therapy. *J. Biomed. Opt.* 14 (2009), 014012. https://doi.org/10.1117/1.3083427.

182. W. Tang, H. Xu, R. Kopelman, M.A. Philbert. Photodynamic characterization and in vitro application of methylene blue-containing nanoparticle platforms. *Photochem. Photobiol.* 81 (2005), 242. https://doi.org/10.1562/2004-05-24-ra-176.1.

183. W. Tang, H. Xu, E.J. Park, M.A. Philbert, R. Kopelman. Encapsulation of methylene blue in polyacrylamide nanoparticle platforms protects its photodynamic effectiveness. *Biochem. Biophys. Res. Commun.* 369 (2008), 579–583. https://doi.org/10.1016/j.bbrc.2008.02.066.

184. X. Mao, J. Xu, H. Cui. Functional nanoparticles for magnetic resonance imaging. *Wiley Interdiscip. Rev. Nanomedicine Nanobiotechnology.* 8 (2016), 814–841. https://doi.org/10.1002/wnan.1400.

185. M. Khalkhali, S. Sadighian, K. Rostamizadeh, F. Khoeini, M. Naghibi, N. Bayat, M. Habibizadeh, M. Hamidi. Synthesis and characterization of dextran coated magnetite nanoparticles for diagnostics and therapy. *BioImpacts.* 5 (2015), 141–150. https://doi.org/10.15171/bi.2015.19.

186. L.O. Cinteza, T.Y. Ohulchanskyy, Y. Sahoo, E.J. Bergey, R.K. Pandey, P.N. Prasad. Diacyllipid micelle-based nanocarrier for magnetically guided delivery of drugs in photodynamic therapy. *Mol. Pharm.* 3 (2006), 415–423. https://doi.org/10.1021/mp060015p.

187. Y. Sun, Z.L. Chen, X.X. Yang, P. Huang, X.P. Zhou, X.X. Du. Magnetic chitosan nanoparticles as a drug delivery system for targeting photodynamic therapy. *Nanotechnology.* 20 (2009). https://doi.org/10.1088/0957-4484/20/13/135102.

188. V. Mody, R. Siwale, A. Singh, H. Mody. Introduction to metallic nanoparticles. *J. Pharm. Bioallied Sci.* 2 (2010), 282. https://doi.org/10.4103/0975-7406.72127.

189. S.K. Nune, P. Gunda, P.K. Thallapally, Y.Y. Lin, M. Laird Forrest, C.J. Berkland. Nanoparticles for biomedical imaging. *Expert Opin. Drug Deliv.* 6 (2009), 1175–1194. https://doi.org/10.1517/17425240903229031.

190. D.A. Giljohann, D.S. Seferos, W.L. Daniel, M.D. Massich, P.C. Patel, C.A. Mirkin. Gold nanoparticles for biology and medicine. *Angew. Chemie-Int. Ed.* 49 (2010), 3280–3294. https://doi.org/10.1002/anie.200904359.

191. E.E. Connor, J. Mwamuka, A. Gole, C.J. Murphy, M.D. Wyatt. Gold nanoparticles are taken up by human cells but do not cause acute cytotoxicity. *Small.* 1 (2005), 325–327. https://doi.org/10.1002/smll.200400093.

192. M.E. Wieder, D.C. Hone, M.J. Cook, M.M. Handsley, J. Gavrilovic, D.A. Russell. Intracellular photodynamic therapy with photosensitizer-nanoparticle conjugates: Cancer therapy using a 'Trojan horse'. *Photochem. Photobiol. Sci.* 5 (2006), 727–734. https://doi.org/10.1039/b602830f.

193. Y. Cheng, A.C. Samia, J.D. Meyers, I. Panagopoulos, B. Fei, C. Burda. Highly efficient drug delivery with gold nanoparticle vectors for in vivo photodynamic therapy of cancer. *J. Am. Chem. Soc.* 130 (2008), 10643–10647. https://doi.org/10.1021/ja801631c.

194. M.K. Khaing Oo, X. Yang, H. Du, H. Wang. 5-Aminolevulinic acid-conjugated gold nanoparticles for photodynamic therapy of cancer. *Nanomedicine.* 3 (2008), 777–786. https://doi.org/10.2217/17435889.3.6.777.

195. H. Huang and T. Hasan. The "Nano" world in photodynamic therapy. *Austin J. Nanomed. Nanotechnol.* 2(3) (2014), 1020. ISSN:2381-8956.

196. R. Bakalova, H. Ohba, Z. Zhelev, M. Ishikawa, Y. Baba. Quantum dots as photosensitizers? *Nat. Biotechnol.* 22 (2004), 1360–1361. https://doi.org/10.1038/nbt1104-1360.

197. A.C.S. Samia, X. Chen, C. Burda. Semiconductor quantum dots for photodynamic therapy. *J. Am. Chem. Soc.* 125 (2003), 15736–15737. https://doi.org/10.1021/ja0386905.

198. J.M. Tsay, M. Trzoss, L. Shi, X. Kong, M. Selke, M.E. Jung, S. Weiss. Singlet oxygen production by peptide-coated quantum dot–Photosensitizer conjugates. *J. Am. Chem. Soc.* 129(21) (2007), 6865–6871, DOI: https://doi.org/10.1021/ja070713i.

199. L. Shi, B. Hernandez, M. Selke. Singlet oxygen generation from water-soluble quantum dot-organic dye nanocomposites. *J. Am. Chem. Soc.* 128 (2006), 6278–6279. https://doi.org/10.1021/ja057959c.

200. B. Hochreiter, A.P. Garcia, J.A. Schmid. Fluorescent proteins as genetically encoded FRET biosensors in life sciences. *Sensors (Switzerland).* 15 (2015), 26281–26314. https://doi.org/10.3390/s151026281.

201. J.M. Hsieh, M.L. Ho, P.W. Wu, P.T. Chou, T.T. Tsai, Y. Chi. Iridium-complex modified CdSe/ZnS quantum dots: A conceptual design for bifunctionality toward imaging and photosensitization. *Chem. Commun.* 6 (2006), 615–617. https://doi.org/10.1039/b517368j.

202. P. Mroz, G.P. Tegos, H. Gali, T. Wharton, T. Sarna, M.R. Hamblin. Photodynamic therapy with fullerenes. *Photochem. Photobiol. Sci.* 6 (2007), 1139–1149. https://doi.org/10.1039/b711141j.

203. A.P. Burlaka, Y.P. Sidorik, S. V. Prylutska, O.P. Matyshevska, O.A. Golub, Y.I. Prylutskyy, P. Scharff. Catalytic system of the reactive oxygen species on the C60 fullerene basis. *Exp. Oncol.* 26 (2004), 326–327.

204. P. Mroz, A. Pawlak, M. Satti, H. Lee, T. Wharton, H. Gali, T. Sarna, M.R. Hamblin. Functionalized fullerenes mediate photodynamic killing of cancer cells: Type I versus Type II photochemical mechanism. *Free Radic. Biol. Med.* 43 (2007), 711–719. https://doi.org/10.1016/j.freeradbiomed.2007.05.005.

205. J. Liu, S. ichi Ohta, A. Sonoda, M. Yamada, M. Yamamoto, N. Nitta, K. Murata, Y. Tabata. Preparation of PEG-conjugated fullerene containing Gd3+ ions for photodynamic therapy. *J. Control. Release.* 117 (2007), 104–110. https://doi.org/10.1016/j.jconrel.2006.10.008.

206. S. rong Ji, C. Liu, B. Zhang, F. Yang, J. Xu, J. Long, C. Jin, D. liang Fu, Q. xing Ni, X. jun Yu. Carbon nanotubes in cancer diagnosis and therapy. *Biochim. Biophys. Acta–Rev. Cancer.* 1806 (2010), 29–35. https://doi.org/10.1016/j.bbcan.2010.02.004.

207. Z. Zhu, Z. Tang, J.A. Phillips, R. Yang, H. Wang, W. Tan. Regulation of singlet oxygen generation using single-walled carbon nanotubes. *J. Am. Chem. Soc.* 130 (2008), 10856–10857. https://doi.org/10.1021/ja802913f.

208. A. Kamkaew, F. Chen, Y. Zhan, R.L. Majewski, W. Cai. Scintillating nanoparticles as energy mediators for enhanced photodynamic therapy. *ACS Nano.* 10 (2016), 3918–3935. https://doi.org/10.1021/acsnano.6b01401.

209. G. Blasse. Reviews: Scintillator materials. *Chem. Mater.* 6 (1994), 1465–1475. https://doi.org/10.1021/cm00045a002.

210. M. Nikl. Scintillation detectors for X-rays. *Meas. Sci. Technol.* 17 (2006). https://doi.org/10.1088/0957-0233/17/4/R01.

211. Y. Liu, W. Chen, S. Wang, A.G. Joly. Investigation of water-soluble X-ray luminescence nanoparticles for photodynamic activation. *Appl. Phys. Lett.*, 92 (2008), 1–4. https://doi.org/10.1063/1.2835701.

212. Y. Zhang, W. Chen, S. Wang, Y. Liu, C. Pope. Phototoxicity of zinc oxide nanoparticle conjugates in human ovarian cancer NIH: OVCAR-3 cells. *J. Biomed. Nanotechnolog.* 4 (2008), 432–48. https://doi.org/10.1166/jbn.2008.006.

213. M.R. Hamblin. Upconversion in photodynamic therapy: Plumbing the depths. *Dalt. Trans.* 47 (2018), 8571–8580. https://doi.org/10.1039/c8dt00087e.

214. D.K. Chatterjee, Y. Zhang. Upconverting nanoparticles as nanotransducers for photodynamic therapy in cancer cells. *Nanomedicine (Lond).* 3(1) (2008), 73-82. doi:10.2217/17435889.3.1.73.

215. P. Zhang, W. Steelant, M. Kumar, M. Scholfield. Versatile photosensitizers for photodynamic therapy at infrared excitation. *J. Am. Chem. Soc.* 129 (2007), 4526–4527. https://doi.org/10.1021/ja0700707.

216. H.S. Qian, H.C. Guo, P.C.L. Ho, R. Mahendran, Y. Zhang. Mesoporous-silica-coated up-conversion fluorescent nanoparticles for photodynamic therapy. *Small.* 5 (2009), 2285–2290. https://doi.org/10.1002/smll.200900692.

217. B. Ungun, R.K. Prud'homme, S.J. Budijon, J. Shan, S.F. Lim, Y. Ju, R. Austin. Nanofabricated upconversion nanoparticles for photodynamic therapy. *Opt. Express.* 17 (2009), 80. https://doi.org/10.1364/oe.17.000080.

218. A.M. Pires, S. Heer, H.U. Güdel, O.A. Serra. Er, Yb doped yttrium based nanosized phosphors: Particle size, "host lattice" and doping ion concentration effects on upconversion efficiency. *J. Fluoresc.* 16 (2006), 461–468. https://doi.org/10.1007/s10895-006-0085-9.

219. F. Wang, D.K. Chatterjee, Z. Li, Y. Zhang, X. Fan, M. Wang. Synthesis of polyethylenimine/$NaYF_4$ nanoparticles with upconversion fluorescence. *Nanotechnology.* 17 (2006), 5786–5791. https://doi.org/10.1088/0957-4484/17/23/013.

Chapter 11

Nanotechnology-Mediated Radiation Therapy

Ankita Borah and D. Sakthi Kumar

Bio-Nano Electronics Research Centre,
Graduate School of Interdisciplinary New Science,
Toyo University, Kawagoe, Saitama 350-8585, Japan

ankitaborah24@gmail.com, sakthi@toyo.jp

11.1 Introduction

In recent years, fueled by technological advancements in many fields, radiation therapy has upgraded the conventional procedures to bring an enhanced therapeutic outcome in cancer treatment. The culmination of advanced alternative strategies such as the incorporation of nanoparticles (NPs) in radiation therapies in conjunction with a deeper understanding of the biological interactions and responses has also caused a paradigm shift in radiation treatment. However, more refined comprehensive research needs to be carried out in bringing futuristic innovations and eliminating caveats to implement a safer and improved radiation-based cancer therapy. To know more about different ways of successful applications of NPs in radiation therapy, we should have a better understanding of the principles of

Bionanotechnology in Cancer: Diagnosis and Therapy
Edited by D. Sakthi Kumar and Aswathy Ravindran Girija

ISBN 978-981-4800-30-3 (Hardcover), 978-0-429-42291-1 (eBook)
www.jennystanford.com

radiation therapy, existing conventional approaches, and possibility of the inclusivity of nanotechnology-mediated approach that can revolutionize the existing scenario in radiation therapy.

11.2 Radiotherapy: Principles and Examples in Various Cancers

The discovery of X-rays by Wilhelm Roentgen in 1895 has found its presence ubiquitously in the field of medicine especially in radiation therapy as the first-line treatment in oncology. The application of high energy X-rays or gamma-rays in the effort to kill cancer cells via DNA-damage or by the generation of free radicals in the cells is a sought-after cancer treatment modality. It has been documented that a majority of cancer patients, an approximated 40–60% receive radiation therapy as the part of their treatment plan [1]. The decisive factor while planning a treatment module for cancer patients to use radiation therapy in conjunction with various other conventional procedures like surgery and chemotherapy often lies with the clinical oncologist, radiologists, and surgeons, to achieve the best outcomes in terms of the overall patient survival. For instance, radiation therapy could precede surgery or as a follow-up with systemic therapies to reduce tumor burden, alleviate toxicities, and prevent nearby organ damage. Radiation therapy could also be administered as palliative care in certain metastatic cases. Parameters related to the radiation therapy techniques, duration, dose, expected clinical outcomes, and toxicities vary in terms of disease scenario and treatment site. Radiation therapy modalities employ two different techniques namely teletherapy and brachytherapy. The former relies on a machine that positioned about a meter away from the patient to deliver the X-rays (photons, protons) or the sub-atomic particle beams; whereas, the latter depends on the implanted temporary or permanent radioactive sources inside the patient's body emanating radiation to achieve the expected result. A radiation oncologist should have expertise in the five foundation disciplines, such as encompassing radiobiology, medical physics, clinical oncology, imaging, and computer science, for safe and effective delivery of radiation therapy [2].

Since cancer follows the traditional dogma of a highly dysregulated proliferative state of cells demonstrating a tumor milieu with tumor stroma-immune cell interactions, radiation therapy particularly aims to target the proliferative cells by DNA-damage arresting cell growth and finally leading to cell death. To understand radiation-induced cell death, it is imperative to delineate the phases of the normal cell cycle in brief. A normal cell cycle comprises five phases namely the G0, G1, S, G2, and mitotic (M) phase. The resting G0-phase is where the cells spend much of their lives lasting a few hours to many years. Upon the reception of the cellular signal, the cell moves forward to the growth (G1) phase to make proteins, DNA, and RNA molecules lasting about 18–30 h, while preparing the cell for later stages in the cell cycle. G1 progresses into the DNA replication (S) phase to make new cells having the same copies of DNA. S-phase lasts approximately 18–20 h. G2-phase is followed thereafter lasting about 2–10 h preparing the cells for mitosis. The M-phase lasts only 30–60 min, where the cell divides to produce identical copies of cells. Radiation therapy utilizes this cell cycle to target the G1-phase cells that are rapidly proliferating and decelerating their proliferative capacity leading to cell growth. Considering the fact that radiation is unspecific, it affects normal cells and tissues causing undesirable side effects. However, normal cells possess the capacity to repair the DNA-damage, due to the internal checkpoints in the cell cycle (end of G1 and G2/M transition) much more effectively than cancer cells, which may succumb to differential killing by radiation. Moreover, the effect of radiation therapy is not immediate and taking days to months for the cells to die off and to achieve the final result. Once the treatment ends, the patient has to undergo constant monitoring of the damage inflicted upon the normal tissues and organs. Hence, during the radiation therapy, a balancing act of minimizing damage to the normal cells and effective killing of cancer cells is indispensable [3]. Despite the phenomenal success and advancement in radiation technology, the limitation that hovers around radiation therapy is the side effects to skin, nearby organs surrounding the tumor, hair loss, fatigue, and reduced blood counts [4]. One approach to circumvent such issues would be to predict the radio-sensitivity, and physiological and genetic risks to design a patient-specific radiation therapy module. Nonetheless, radiation therapy is the immediate treatment for early-stage cancers and has been studied well in a variety of cancers, which is discussed briefly in the following sections.

11.2.1 Breast Cancer

Breast cancer radiation therapy includes external beam radiation therapy (EBRT) and internal beam radiation therapy (IBRT), where the patient receives several daily doses as fractions. EBRT utilizes X-rays to target the external tumor tissue surface for a few minutes, while IBRT also known as brachytherapy uses radioactive implants near the tumor site to provide partial radiation. An approximate dosage of 4500–5000 centigray (cGy) is divided into five-weeks of treatment for breast cancers and tumor cells present in lymph nodes with an additional booster dose of 1000–2000 cGy [5]. Common side effects with this strategy often include swelling in the breast, skin irritation, discoloration of the skin, and fatigue [6].

11.2.2 Liver Cancer

In liver cancer treatment, apart from the current treatment modalities, such as IBRT and EBRT, it includes three-dimensional conformal radiation therapy (3D-CRT) and stereotactic body radiation therapy (SBRT). In 3D-CRT, a 3D-shaped tumor is targeted with multiple small volumes of radiation beams; whereas, in SBRT, a designated target receives high energy radiation beams with minimizing side effects to nearby tissues. Vomiting, nausea, gastritis, upper abdominal pain, and bleeding are common side effects in this scenario [7, 8].

11.2.3 Ovarian Cancer

The application of radiation therapy for the treatment of ovarian cancer depends upon the disease stage. Early-stage ovarian cancer can be treated with adjuvant radiation therapy, an advanced stage with consolidated radiation therapy, salvage radiation therapy to treat relapsed cancer, and palliative radiation therapy for metastatic disease. The patient receives a total dosage of 22.5–33 Gy in 10–24 fractions for five-weeks with a booster dose of 40–45 Gy often accompanied by bowel discomfort and vaginal irritation [9].

11.2.4 Head and Neck Cancer

A sophisticated form of 3D-CRT also known as intensity-modulated radiotherapy (IMRT) is one of the predominant techniques to treat head and neck cancer at the targeted site to minimize surrounding tissue damage along with brachytherapy. Hair loss, tooth decay, and trouble in swallowing are the associated side effects [10, 11].

11.2.5 Prostate Cancer

Prostate cancer is often treated with positron emission tomography (PET) combined with prostate-specific membrane antigen (PSMA)-based EBRT to achieve sensitivity and specificity in lymph nodes and metastases. Early-stage prostate cancer patients receive a radiation dosage of 70.2–79.2 Gy followed by a booster dosage of 19.8–28.8 Gy after surgery and later stages [12].

11.3 Techniques of Radiation Therapy

11.3.1 Intensity Modulated Radiation Therapy (IMRT)

Improvements in computer and imaging technologies pushed into the advancement of 3D-CRT to give rise to its next-generation technology known as IMRT, which can be used in diverse tumor shapes and sizes by applying external beam radiation. IMRT has two key additional features: (i) non-uniform intensity of radiation beams, and (ii) computerized inverse planning compared to its earlier version of CRT [13, 14]. The beneficial feature of computerized inverse planning in IMRT provides the radiation oncologist with the plan designed by the computer having adjusted beam intensities for achieving clinical outcomes in terms of tumor dose and normal structure dose limits. This is followed by the dose prescription decided by the clinician to be delivered at the tumor site and maintain a dose limit for other organs at risk. Different dose targets could be treated by inhomogeneous dose distribution planning referred to as "dose painting" or also simultaneously boosting the dose to treat areas of higher risk in an attempt to reduce treatment time. Dose prescription is followed by the delivery of IMRT using two methods: (i) segmental MRT, and (ii) dynamic IMRT use multi-leaf collimators (MLC). Due to its precision, it is recommended that the patient remains still and restricts movement during treatment [13]. Table 11.1 lists the several advantages and disadvantages of IMRT [13].

Table 11.1 Advantages and disadvantages of IMRT

Advantages	Disadvantages
Improved target conformity in concave shaped volumes	Needs extra time for clinicians to outline target tumor and organ
Provision of dose painting, i.e., intentional dose inhomogeneity	Extensive quality assurance required
Exclude normal tissues from receiving radiation	Increase machine treatment time
Dose escalation could be achieved to minimize treatment time	Increased total body irradiation dose
Compensate for missing tissue	Increased planning time

IMRT represents one of the successful radiotherapy techniques in treating the prostate cancer [15], head and neck cancer [16], and gynecological cancer [17], though posits further research in terms of tumor imaging for localization, manage organ motion, and image guidance to enhance clinical outcomes and reduce unwanted side effects.

11.3.2 Image-Guided Radiation Therapy (IGRT)

IGRT has a malleable definition that can be broadly defined as the imaging technique for detection and diagnosis, target volume evaluation, and dose distribution design with the ultimate aim to achieve precision in radiation delivery in reducing tumor size and avoiding toxicities [18]. IGRT has the advantage to allow the precise monitoring of tumors as treatment margin and conformity become tighter due to the possibility of organ motion in the tumor vicinity, which might affect normal tissues inadvertently [19]. The precision in radiation delivery leads to the feasibility of dose escalation finally improving the therapeutic ratio at tumor sites, which comes under the basic hypothesis of IGRT [20]. IGRT is demarcated into radiation and non-radiation-based systems depending on the type of imaging methods being used. Radiation-based systems comprise static as well as real-time tracking or hybrid methods such as (i) electronic portal imaging devices (EPID), (ii) cone-beam CT (CBCT), (iii) fan-beam KV CT (CT-on rails), or MV CT (TomoTherapy Hi ART II), and (iv) hybrid systems for real-time 4D tracking. Non-radiation-based systems utilize (i) ultrasound, (ii) camera-based systems, (iii) MRI-guided systems, and (iv) electromagnetic tracking systems [21].

The implementation of IGRT is based on the following parameters [21]:

(a) The difficulty in visualizing and alignment of soft tissue targets are assisted by fiducial markers that can be tracked in real-time to obtain the target coordinates in 3D.
(b) Planning target volume (PTV) determination is dependent on the intrafraction and inter-fraction target motion, which, in turn, relies on expertise, degree, and type of motion.
(c) Accuracy in patient positioning and reproducible fractionated treatments should be ensured relative to the chosen IGRT device and treatment unit.
(d) Calibration in the IGRT system must be warranted to achieve high quality in imaging, uniformity in slice thickness, image contrast, spatial resolution, and software accuracy to identify and correct couch misalignments to name a few.
(e) Radiation oncologists should carry out image review at the first fraction and periodically corroborate the accuracy and reproducibility of the treatment.

(f) Relevant QA procedures throughout treatment, simulation, and periodic verifications should be documented and maintained to assist departmental actions and future modifications of the treatment process.

American Society for Radiation Oncology (ASTRO) Board of Directors as a part of ASTRO's Target Safely Campaign commissioned a full series of white papers summarizing on patient safety consideration for IGRT, where they addressed questions that pertained to IGRT modality/frequency, PTV expansions, image verification methods, and perceived utility/ value of IGRT [22]. The full report could be read online at www.practicalradonc.org. Patients treated with IMRT, CRT, and SBRT likely to benefit from IGRT. Prostate cancer, head and neck cancers, paraspinal and retroperitoneal sarcomas, and thoracic and upper abdominal targets are also predicted to derive maximum benefit in this scenario [21].

11.3.3 Particle Therapy

A majority of cancer radiotherapy utilizes the delivery of high ionizing photons (X-rays or gamma rays) via IMRT culminating in the DNA damage, activating diverse biochemical reactions that set up the cancer cells toward the path to cellular death. However, ionizing photons are also known to be harmful to the adjacent healthy tissues, which serve as one of the aims of the radiation oncologist to minimize healthy tissue lethality and maximize tumor morbidity [23]. An alternative approach to ionizing photon beam therapy is the use of particle radiotherapy that includes charged particles and non-charged particles. Charged particles include protons, carbon, and helium ions; whereas, non-charged particles comprise neutrons [24]. Proton and carbon ion beams possess the advantage of high linear energy transfer (LET) radiation, where the proton beams can stop at a certain depth of the target material or tumor tissue offering enhanced depth-dose distribution to create an energy surge. This energy surge is called the Bragg peak that has a sharp and localized peak destroying the tumor tissue selectively owing to its high energy deposition, leaving the healthy tissue from the detrimental effects of the radiation [24, 25]. The physical characteristics of the charged particles adjusting the energy and intensities of the beam will create multiple Bragg peaks that are adequate to irradiate a whole tumor mass. Both proton and carbon ion beams, though share similarity in their depth-dose profiles, their relative biological effectiveness (RBE) in killing the cancer cells is markedly different [25]. The carbon ion beam is demonstrated to induce enhanced cancer cell killing at RBE of 2.0–3.0 in comparison to the proton beam estimated at 1.1 [26]. Additionally, the carbon ion beam has extra-biological advantages in terms of radiosensitivity contrary to proton beam, for instance: (i) low oxygen enhancement ratio, (ii) suppression of metastasis, and (iii) therapeutic efficacy against cancer stem cells. Particle radiotherapy has shown positive results in cancers such as glioblastoma, melanomas, and sarcomas, to name a few [27–29].

Neutron beams are also recognized as high LET radiation generated inside neutron generators, but do not exhibit any high energy surge. Unlike proton and carbon ion beams, the applications of neutron beam in clinical settings is quite limited due to the difficulty in generating neutrons, and uniform distribution of radiation at any depth of the body that might contribute to unwanted side effects [3].

11.3.4 3D Conformal Radiotherapy (3DCRT) and Stereotactic Body Radiation Therapy (SBRT)

The irregular shape and size of tumors demand better computer and special imaging techniques to assist radiologists in specifically locating the tumor and its surrounding organs, and designing and delivering a radiotherapy treatment to achieve optimum results without compromising the healthier tissue. The advent of 3D- CRT over 2D-radiation therapy started gaining popularity in the 1990s, due to its advantageous features like accurate localization of the tumor and its critical surrounding organs with the help of CT imaging, which facilitated improved dose calculations on the acquired 3D image sets [30]. This further allowed to arrange and maximize multiple radiation beams on the gross tumor volume (GTV) by a method called forward planning, keeping a margin for the clinical target volume (CTV), and an additional margin that might arise from the uncertain organ motion and set-up variations during the procedure known as the PTV [3, 31]. During the 3D-CRT treatment, the patient's body is kept still by fitting it in a plastic mold. As the name suggests, MLC in the 3D-CRT instrument maps the precise shape and location of the tumor followed by image acquisition in 3D. This accuracy in tumor localization helps to shield the normal tissues from unwanted radiation harms, allow the effective radiation dose to reach the tumor tissue, and fight cancer [30].

Intracranial stereotactic radiosurgery (SRS) led to the derivation of SBRT [32], followed by the US Food and Drug Administration (USFDA) approval of SRS in 2001 to establish the first SBRT delivery system to treat different areas of the body [33]. SBRT possesses an edge over the conventional radiation treatment by delivering large dose of highly conformal radiation in fewer fractions to the tumor site, which results in a clinically relevant biological effective dose (BED). Immobilization devices, for instance, skull-anchored head frame in Gamma Knife or custom whole-body vacuum bags assist clinicians to minimize uncertainties in SBRT delivery and maximize dose fall-off outside the target [30]. Even though there has been limited understanding in the radiobiology of SBRT treatments, the classic fractionation principle of Wither's four R's of radiotherapy (repair, repopulation, redistribution, and reoxygenation) to understand the biological effects of high radiation cannot be overlooked [34]. Along with the advantage of high radiation dose delivered in fractions, additional features such as shorter treatment time and inhomogeneous dose distribution allow SBRT to attain impressive local tumor control rates and improve overall survival [35]. Non-small cell lung cancer (NSCLC) patients have been demonstrated to derive maximum benefits from SBRT, owing to its minimal invasiveness, treatment convenience, limited side effects, and reduced mortality rates. A dose-escalation study in inoperable NSCLC patients conducted by Martel and colleagues reported the progression-free survival in the patients after receiving high doses as 84 Gy [36]. In another study, the safety and efficacy of SBRT in local stage I and solitary metastatic NSCLC patients were evaluated. The overall survival and local control rates were 36% and 53%, respectively, for all the 31 patients enrolled in the study, with four of the patients showing no further recurrence for more than five years after the second SBRT [37]. SBRT has also been shown to be effective in other tumors such as head and neck, prostate [38], hepatic [39], and oligo metastases.

11.3.5 Internal Radiation Therapy (Brachytherapy)

Brachytherapy (*Brachy* in Greek means short) or internal radiation therapy is a specific form of short-distance anti-neoplastic treatment that applies to the placement of radioactive sources into the tumor site or the body cavity, thus delivering high radiation doses to a small area. There are two types of brachytherapy based on the radioactive source placement in the body that can be temporary or permanent [30]. The radioactive sources, such as pellets, seeds, wires, and tubes, when placed directly into or next to the tumor, are called interstitial radiation; whereas, a container of radioactive material when placed in the body cavity like chest, rectum, or uterus is called intracavitary radiation. Permanent radioactive sources like the pellets or seeds give off radiation for several weeks or months until the radioactive material is completely used up. Temporary radioactive sources are further divided into low-dose-rate (LDR) and high-dose-rate (HDR) where the source is put inside a container and placed inside the body for a short period and removed once the treatment is over. During the LDR therapy, a clinical radiologist typically places the low-active radioactive source inside the patient, and to prevent the implant mobilization the patient is often advised to stay in the hospital for a period of 7–10 days till the removal of the implant. While in HDR brachytherapy, a high-intense radioactive source on a wire-end is contained in a computerized after-loader. The after-loader is then remotely guided by a machine to place the radioactive source at specified locations. This entire operation can be carried out by the hospital staff, which generally takes 10–20 min with repetition of the treatment course twice a day for a few days or once in a week for a few weeks [30]. The key factor in ensuring the therapeutic index of brachytherapy is to consider implant quality. The selection of brachytherapy for particular cancer treatment also depends upon the ease of placement of the catheter or applicator as the dose from the radiation source drops off quickly as a function of distance [40]. Radioactive sources when placed close to the surface of the target site receive greater therapeutic radiation dose efficacy. Breast [41, 42] and prostate cancer [43] are the two well-studied examples of brachytherapy treatments due to their high incidence rates and ideal localization. Brachytherapy often supersedes the techniques of external beam radiation by providing dosimetric advantage, optimal tumor-to-normal tissue ratio radiation dose, reduced treatment time, and minimized CTV margins. Brachytherapy can be used alone or in combination with EBRT to treat locally advanced primary tumors like cervical cancers [40].

11.4 Radiation Induced Cell Death Mechanisms

There are four types of radiation-induced cell death mechanisms namely, mitotic cell death, apoptosis and necrosis, autophagy, and senescence [44], as shown in Figure 11.1. The following sections will provide detailed information regarding these mechanisms.

11.4.1 Mitotic Cell Death

The high-energy X-rays or gamma rays, and charged particles emitting from radiation therapy target the most important sub-cellular molecule DNA to cause double-strand breaks, followed by the initiation of a chain of events to promote cell death via mitotic catastrophe, apoptosis, necrosis, and autophagy, thus dictate the central dogma of radiobiology [45]. The cell cycle is a tightly regulated process that is divided into two phases: interphase comprising (G1, S, and G2 phases) and mitosis. Cells exposed to radiation, undergo DNA damage, which is sensed by the two protein kinases ataxia-telangiectasia mutated (ATM) and ataxia-telangiectasia and Rad3-related protein (ATR) causing the arrest of the cell cycle and repairing the DNA damage. The double-strand DNA breaks are mainly repaired by non-homologous end joining (NHEJ) and homologous recombination (HR) pathways [44]. Mitotic catastrophe takes place when a cell

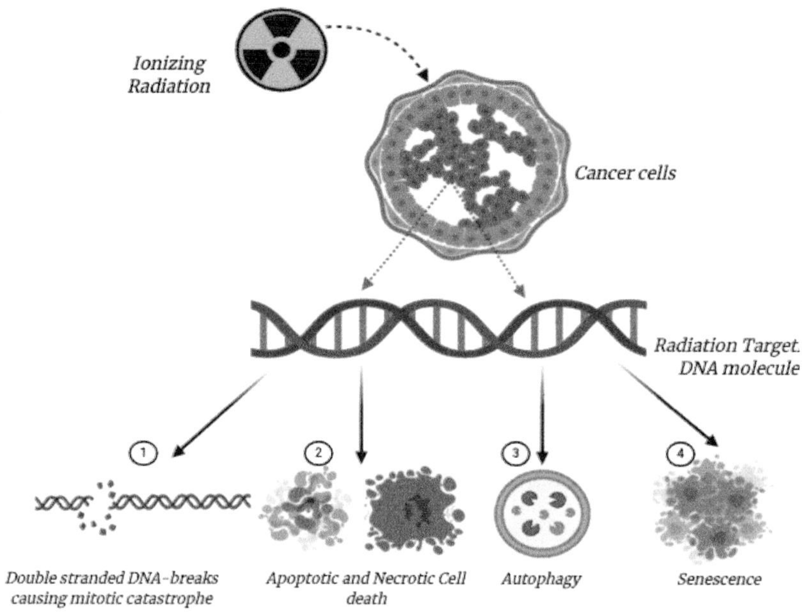

Figure 11.1 Schematic representation on the different mechanisms of cell death induced upon irradiation. The level of DNA damage in the irradiated cells dictates the type of cellular response to be initiated with apoptosis being one of the main cell death mechanisms (Image is made using Biorender.com).

loses its ability to complete mitosis, thus controlling the cells by triggering mitotic arrest and regulated cell death. Mitotic catastrophe in irradiated cells especially solid tumors is a delayed process taking around 2–6 days during which the cells might take several attempts to repair the damage and undergo cell divisions [46]. Though the exact mechanism of the initiation of mitotic catastrophe presents unclear answers requiring further investigation, nonetheless two different ideas have been proposed to shed light on this cellular event. The first idea proposed occurs as a result of DNA damage and dysfunctional cell cycle checkpoints. It is a well-established fact that tumors possess dysfunctional cell cycle checkpoints due to impaired p53 and aberrant apoptotic signaling pathways. The presence of mutated p53 leads to premature entry into the G2/M checkpoint that contains the unrepaired DNA, ultimately leading to cell death [47]. The second idea puts forward the concept of centrosome hyper amplification. Centrosomes are the microtubules organizing centers that facilitate the systematic chromosome segregation into daughter cells during mitosis. In the case of centrosome hyper amplification, it leads to the formation of multipolar mitotic spindle that promotes atypical chromosomal segregation, and the formation of abnormal giant cells with multiple nuclei culminating in cell death [48, 49].

11.4.2 Apoptosis and Necrosis

Apoptosis or programmed cell death has been regarded as a main cellular death mechanism following exposure to radiation and chemotherapy. Cell shrinkage, chromatin condensation, DNA fragmentation, and cell membrane blebbing are the characteristic features of apoptotic cells [50]. Earlier studies posit the importance of p53 as a prognostic marker in inducing early apoptosis in thymocytes, where wild-type p53 cells are highly radiosensitive and the mutated p53 cells appear to be radio resistant [51, 52]. Hence, tumor tissues susceptible to apoptosis undergo radiation-induced p53-dependent apoptotic process within a few hours of exposure, which helps to monitor the efficacy of radiotherapy. Radiation-induced apoptotic cell death usually takes place via the activation of the caspases cascade, which is either by intrinsic or extrinsic pathway. The intrinsic apoptotic pathway comes into its mode of action when the p53-mediated DNA damage repair machinery disrupts the harmonious balance between the pro- and anti-apoptotic factors. This causes the mitochondria to release cytochrome-c into the cytoplasm to activate caspase-9 [46, 53]. While the extrinsic pathway is triggered mainly by the binding of the tumor necrosis factor (TNF)-family of ligands to the death receptors, which, in turn, activates the extrinsic-pathway specific caspase-8 [54, 55]. An additional pathway known as the ceramide pathway triggers cell death utilizing the radiation-induced activation of acid sphingomyelinase present in the plasma membrane to hydrolyze sphingomyelin producing ceramide, finally initiating the apoptotic program [56, 57]. Conclusively, the three pathways converge to activate the caspase family of 3, 6, and 7 that sets in motion the apoptotic targets like structural proteins, DNA repair proteins, and cell cycle proteins to perform their actions in initiating cell death.

In the majority of the radiation-based cancer therapies, apoptosis takes the spotlight in causing cell death mechanisms; however, newer reports have started coming out to shift the focus toward whether cancer therapies can also induce other forms of cell death. In a study carried out by Lei G et al. demonstrated the link between ionizing radiation and a form of regulated cell death caused by lipid peroxidation called ferroptosis [58]. The authors analyzed that ionizing radiation in cancer cells led to the induction of reactive oxygen species (ROS) and upregulated expression of a lipid metabolizing enzyme ACSL4 to initiate ferroptosis. The study concluded that ferroptosis in cancer patients conferred better response and survival to radiotherapy.

Necrosis is a chaotic cell death mechanism elicited upon response to unfavorable pH changes, ionic imbalance, and energy loss after a cell experience irradiation. Necrosis is generally characterized by the gain in cell volume, organelle swelling, non-condensed chromatin, plasma membrane rupture, and cell lysis followed by the spilling of its intracellular contents. Radiation-induced cell death is specific to cell type and tumor type, and the radiation dose given to the tumor implicates the type of death induced in cells. High dose of radiation often leads to necrosis, whilst low radiation dose initiates apoptosis [59–61].

11.4.3 Autophagy

Autophagy has been observed as a regulatory cellular homeostatic process that protects genome integrity, and prevents the accumulation of misfolded proteins and action of damaged organelles through a lysosomal pathway [62, 63]. Autophagic cell death takes place upon induction of autophagy in cells treated with radiation. Literary evidences have shown that, in cancer cells, those receiving chemotherapy and radiotherapy, treatment response often results in tumor inhibition via the involvement of autophagy-related proteins (ATG) depending on factors like time and type of treatment by inhibiting autophagy [64, 65]. To reverse the pro-cell survival mechanisms exerted by autophagy, in certain cancer cells treatment with autophagy inhibitors combined with chemotherapy and radiotherapy will succumb the cancer cells toward cell death [66]. Cancer cells receiving ionizing radiation sometimes do not die via the usual programmed cell death apoptotic pathway, which could be due to the intrinsic resistance toward apoptosis on account of mutations in p53 and bax [67]. In this scenario, autophagy is activated as the type II programmed cell death pathway making it a pro-cell death mechanism. Administration of mammalian target of rapamycin (mTOR) inhibitors such as Everolimus induces autophagy and sensitizes cancer cells to ionizing radiation [68]. Consistent reports supporting the role of autophagic cell death in radiation-induced cancer cells, where it functions as pro-cell death phenomenon suggests that a blockade in the autophagy process might lead to a decrease in the cell death of the irradiated cells [69]. Autophagy is characterized by the formation of autophagosomes, a double membrane-bound vesicular structure that assists in the sequestration of old and damaged organelles by lysosomal degradation [53]. Though numerous signaling finds association in the initiation of the autophagy process, the PI3K-Akt-mTOR pathway is the crucial one, which is also studied to be connected with the induction of autophagy in radiation-induced cells [70]. It has become imperative to develop future cancer treatments in a way to modulate autophagic signaling pathways for improving radiotherapeutic outcomes in cancer patients to reap the maximum benefits of the treatment process.

11.4.4 Senescence

Hayflick and Moorhead coined the term "senescence" to define the lifetime capacity of cellular proliferation as a consequence of telomere shortening. Senescent cells though cease to proliferate and synthesize DNA, but remain viable, metabolically active, enlarged in size, and increased granularity with flattened shape [71, 72]. DNA damage is notably triggered in radiation-induced senescent cells followed by the p53 and pRb pathways activation to initiate cell-cycle arrest in addition to oxidative stress that drives the irradiated cells to die by apoptosis [73–75].

One of the prolific markers present in the perinuclear region of senescent cells is the senescence-associated β-galactosidase of lysosomal compartments along with surrogate markers involved in the DNA damage response (DDR) process. If the DNA damage is not repaired in the cell cycle process upon receiving a signal from the DDR, then the cell typically undergoes p53-dependent (apoptotic) or p53-independent (mitotic catastrophe) cell death. Sometimes the cells activate a chronic state of DDR signaling that leads to activation of senescence [46, 76]. These cells are also known to secrete senescence-associated secretory phenotype (SASP) pro-inflammatory factors known to modulate tumorigenic potential/inhibit tumorigenesis, proliferation, metastatic process, and immunosuppression. Senescent cells are often generated as a result of the therapy-induced consequences in a cancer patient, which in the long run might contribute to certain age-related side effects in treated cancer patients [77]. Henceforth, SASP inhibition or senolysis that targeted the elimination of senescent cells is a complementary strategy to mitigate the unwanted negative effects of radiation-based cancer therapies [78]. A few examples of inhibitors involved in the senolytic process that have been investigated in recent years are navitoclax (ABT-263) [79], obatoclax [80], and panobinostat [81], which demonstrate to kill senescence-like cells in breast cancer cells, triple-negative breast cancer cells, and NSCLC cells, respectively.

11.5 Risks Associated with Conventional Radiotherapy

Radiotherapy similar to chemotherapy comes with its share of unwanted side effects that have variations in the severity of toxicities from minor to extreme, which might range from weeks to a lifetime. Radiotherapy-associated toxicity can be graded as acute toxicity that develops rapidly or after the treatment, and late toxicity develops after months–years of completion of treatment leading to life-changing difficulties, which is greatly influenced by numerous factors like radiation dose, volume, patient medical history, genetics, and past treatments like chemotherapy. Acute toxicities include erythema, dermatitis, hair loss, diarrhea, and cystitis, which are often manageable and transient. Late toxicities, such as fibrosis, urinary incontinence, vascular damage, and atrophy, impair quality of life in several ways that can be permanent or life-threatening [82].

Chromosomal abnormalities, including dicentric chromosomes and translocated chromosomes, are one of the adverse effects caused by radiation therapy in cancer patients depending on the radiation dose received [83, 84]. Radiation mediates the production of free-radicals in cells that result in the bystander effect of DNA damage, chromosomal aberrations, and mutations finally causing genomic instability [85, 86]. Radiation-induced secondary malignancy (RISM) is accounted for one of the severe late-developing side effects in cancer patients, which occurs after the primary radiation therapy inducing DNA damage. RISM is characterized by three features: (i) there exist histological disparities in primary and secondary cancers, (ii) secondary cancers exist within the area of treatment receiving the primary radiation therapy, and (iii) secondary cancers develop after a latent period of five-years [87]. Approximately 17–19% of patients are reported to develop RISM due to genetic susceptibility, continued lifestyle, and modality of the primary radiation therapy; however, an accurate incidence rate is hard to predict due to multiple factors involved in the predisposition to secondary malignancies [88]. An established piece of evidence that concludes the case of secondary malignancies due to ionizing radiation are the survivors of the Hiroshima and Nagasaki A-bomb attack who went on to develop leukemia and solid tumors in the following years [89, 90]. Independent studies on the development of secondary malignancies after receiving primary radiation therapy have been studied extensively in recent years in breast cancer and rectal cancer patients. In the breast cancer study, patients receiving primary radiation therapy were reported to develop a greater rate of secondary malignancies than the endemic breast cancer patients [91] without any radiation therapy, while in the rectal cancer study, the incidence of developing pelvic and gynecological tumors was seen to be higher [92].

11.6 Nanotechnology-Mediated Radiotherapy Treatments

The immense popularity of nanotechnology in the 21st century has revolutionized the field of medicine in unprecedented ways from its applications in diagnostic, therapeutic to regenerative purposes. The best examples of nanomedicine could be seen is the development of rapid diagnostic tests for genetically predisposed disorders, infections caused by viruses, cancer treatments, cardiovascular and neurological disorders, and biocompatible tissue/organ implants, among the few [93]. The idea behind the incorporation of nanotechnology in medical research was to accomplish the objective to develop minimally invasive treatment procedures, early detection of life-threatening diseases by innovative imaging methods, personalized medicines, and prevent severe side effects in the long run. The foundation of nanomedicine is built on the three pillars of (i) nanomaterials and nano instruments that can be utilized for the delivery of active substances thus serving treatment purposes; (ii) wide-ranging knowledge on the subjects of molecular biology, genetics, proteomics, and microbiology; (iii) use of nanotechnology to develop rapid diagnostic tests, surgery and therapeutic procedures, repair and regenerative purposes, and to improve the physiological conditions [94]. Modern radiation therapy was founded on the wisdom of achieving greater tumor control by delivering fractionating treatments in smaller doses to minimize normal tissue toxicity. However, with the emergence of novel technologies and improved understanding of radiobiology, the testament of conventional radiotherapy procedures has been challenged and pushed to embrace sophisticated techniques. Apart from the developments in radiation techniques, medical oncology has also witnessed advancements in recent years, where we have observed crosstalk between physical sciences and the life sciences bridging the gap in multidisciplinary disciplines to arrive at a common objective of improving patient's lives [95]. In this context, the discernible convergence of radiation oncology with nanotechnology could be a step forward to surmount the challenges faced in current radiation therapy procedures to improve treatment, and patient quality of life.

Nanotechnology involves the study and manipulation of matter at a nanoscale range of 1–100 nm. The materials at the nanometer range are quite different from their bulk material concerning the physical and chemical properties. These nanomaterials are fabricated from different materials, such as organic and inorganic materials, by employing different techniques that can be harnessed based on their unique physical and chemical properties for diverse biological applications [96]. The facile synthesis and tuning of the size, shape, surface properties of the nanomaterials impart stability, biocompatibility, and delivery of active substances like chemotherapeutics and radiosensitizers. The large surface area of nanomaterials improves the interaction with cells, allows prolonged circulation in the physiological environments, and efficiently targets tissues [97].

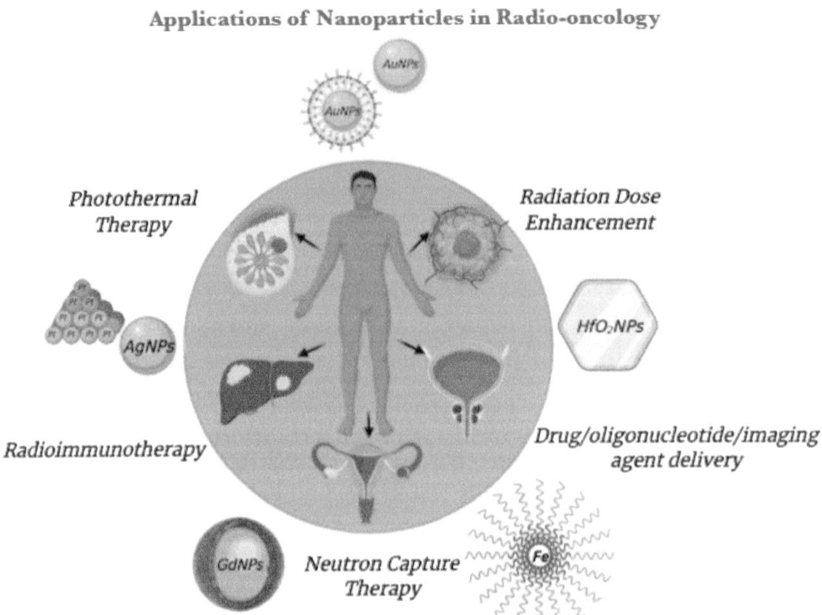

Figure 11.2 Schematic representation of the different NPs, for example, gold NPs (AuNPs), platinum NPs (PtNPs) and silver NPs (AgNPs), gadolinium NPs (GdNPs), superparamagnetic iron oxide NPs (SPIONs), and hafnium NPs (HfO_2NPs), which are used in myriad of radio-oncological applicationsdiscussed in the following sections (image is made using Biorender.com).

High atomic number NPs, such as gold, gadolinium, and iodine, have attracted attention in the context of radiotherapy, due to their ability to cause amplified radiation damage to tumor tissues specifically minimizing damage to healthy tissues. These NPs possess the specific property of releasing electrons in a medium due to the atomic excitation by incident radiation, and due to their nano size, these NPs are readily internalized by the tumor leaky vasculature allowing them to retain inside the tumor for an extended time, finally destroying the tumor tissue upon irradiation thus preventing off-target effects [97, 98]. Nanomaterials additionally offer the possibility to fabricate a multitasking therapeutic modality that will function as a drug delivery vehicle encapsulating a chemotherapeutic and radiosensitizer and function as a diagnostic medium by attaching fluorescent moieties on its surface. In this section, we discuss a few examples of NPs that have been extensively studied in radiotherapy applications as shown in Figure 11.2.

11.6.1 Gold Nanoparticles

Gold Nanoparticles (AuNPs) have demonstrated a myriad of applications in nanomedicine owing to the possession of unique properties such as biocompatibility, facile synthesis to produce the NPs in a variety of sizes, and surface functionalization with biological entities like antibodies, peptides, and oligonucleotides. The versatility of AuNP because of its physicochemical properties like conductivity, redox nature, and surface plasma resonance, and a high surface area to present multiple targeting agents such as fluorescent moieties and drugs make it possible to use in both diagnostic and therapeutic purposes, respectively [99]. Spherical AuNPs are quite popular in the arena of bionano-technological applications as they possess attributes like large surface-to-volume ratio, superior biocompatibility, and size- and shape-related optoelectrical properties. Spherical AuNPs, due to the changes in core size from 1 nm to 100 nm, exhibit a range of colors demonstrating a size-related absorption peak from 500 nm to 550 nm [100–102]. The first-generation of AuNPs was mainly designed to investigate the surface modifications to assess internalization by the cells, target accumulation, and analyze the in vitro cytotoxicity [103]. The second-generation of AuNPs primarily focused on the stealth modifications to facilitate the improved accumulation of the AuNPs in the target tissue [104, 105]. The present third-generation of AuNPs possesses superior biodistribution and disease responsive drug-release properties that are fabricated on the foundation of the challenges faced in the previous two generations [106, 107]. One example of the third-generation AuNPs platform is called the AuRad^{TM} designed specifically for radiation therapy applications considering the fundamental features of size and surface functionalization. The optimal size of the platform allows for increased uptake by the tumor tissue, prolonged blood circulation, and facilitated clearance. The PEGylation of the AuRad^{TM} platform imparts extended circulation when administered intravenously and the presence of multiple functional groups, such as carboxyl, amino, and methoxy ligand, on the PEG polymer allows to a conjugate variety of moieties like radiosensitizers, drugs, fluorophores, and radiolabels [106].

AuNPs-mediated radiation therapy modality is a burgeoning research topic because it improves the radiation therapy therapeutic efficacy by first targeting the tumor tissue with AuNPs followed by the irradiation of the AuNPs to further boost the radiotherapy process [108]. The intrinsic radio sensitization potential of AuNPs is due to its high atomic number making them interact with high-energy radiation (keV) via the photoelectric effect emitting photo-/Auger electrons in the micro meter range as evident in another high atomic number NP [109]. The improvement in radiotherapy efficiency by AuNPs is measured by the dose enhancement factor (DEF), which is the ratio of radiation dose absorbed by the tumor tissue in presence of AuNPs to the radiation dose absorbed by the cells in the absence of AuNPs. DEF may vary with the concentration of AuNPs localized within the tumor tissue [110, 111]. The applications of AuNPs in radiation therapy is attractive not only due to its intrinsic radio sensitization potential, but also due to excellent biocompatible features [112] in contrast to other high atomic number NPs, and AuNPs could be further utilized as an imaging contrast agent for enhancing the efficacy of radiation therapy [113]. The proof-of-concept of the therapeutic benefit of applying AuNPs during radiation therapy was proved by Hainfeld et al. in preclinical mouse models [114].

The mice when intravenously administered 7 mg/g of AuNPs with 250 kVp X-rays/photons were observed to have an 86% long-term survival rate compared to the mice that received radiation alone [114]. Cognizant of the fact that AuNPs have high X-ray absorption capacities along with exceptional chemical and optoelectronic properties, and they are frequently used as radiosensitizers to enhance radiation therapy. The first evidence of using Au as a radiosensitive material was demonstrated by Regulla et al. by adsorbing a thin layer of gold foil on to a poly (methylmethacrylate) slab surface. Mouse embryonic fibroblasts adhered to the gold surface was observed to have radiation dose enhancements by a factor of 50 upon irradiation [115]. These findings were further supported by the works of Herold et al. who used gold microspheres in a cell suspension of rodent and human tumors to achieve dose enhancement of 1.43 [116]. Chithrani et al. focused on the radiosensitive properties of the NPs in the size ranging from 14 nm to 74 nm and found that radiosensitivity dependent on the number of AuNPs internalized by the cells. The highest radio sensitization enhancement factor was observed for the 50 nm AuNPs that had superior uptake by the cells compared to the 14 nm and 74 nm AuNPs [117]. AuNPs are known to cause radiation enhancement in cells via physical, chemical, and biological pathways. The difference in energy-absorbing properties of Au and the soft tissues is taken into account for physical dose enhancement by producing photoelectrons, Auger electrons, and low energy electrons (LEEs) via the Compton effect or the photoelectric effect to cause cellular damage [118]. The physical effects of AuNPs are well demonstrated in many studies on plasmid DNA, where it revealed that the AuNPs release energy in the nanoscale range, whilst the LEEs are responsible for the increase in dose enhancement [119]. The chemical enhancement of AuNPs radiation exposure is theoretically proposed via two mechanisms: (i) chemically sensitizing DNA to IR-induced damage, which requires the nuclear translocation of the AuNPs [120], and (ii) activated AuNPs surface leads to increase the production of free radicals and catalysis [118, 121]. Previous studies reported three key biological pathways to elucidate the role of AuNPs in the biological enhancement: (i) oxidative stress and ROS production, (ii) inhibition of DNA repair mechanisms, and (iii) cell-cycle effects [118, 122]. PEGylated AuNPs functionalized with a known ROS sensor dihydro-rhodamine 123 (DHR-123) having a size of 20 nm were engineered to effectively monitor the ROS generation in tumor cells that led to apoptotic cell death following radiation exposure [123]. AuNPs with core sizes of 2 nm, 5 nm, and 19 nm when functionalized with prostate-specific membrane antigen-1 (PSMA-1) ligand reported the highest cellular uptake of the smallest size PSMA-1 targeted AuNPs in PSMA expressing tumors followed by improvement in radiation enhancement [124]. Over the years, significant progress has been made to exploit AuNPs not only as a versatile drug delivery platform and imaging modality, but also innovative radiosensitizers and obstacles in the form of in vivo efficacy/fate. However, for the full potential use of AuNPs, further study about safety concerns along with a comprehensive understanding of the radio sensitization mechanisms and radiation energy parameters needs to be conducted.

11.6.2 Platinum- and Silver-Based NPs

High atomic number metal-based NPs, commonly termed as nanoenhancers, display chemical inertness and produce photoelectrons and Auger electrons upon irradiation that contribute to physical dose enhancement, chemical contribution, and radiobiological effects. Due to their chemical inertness, metal-based radiosensitizers decrease potential side toxicities, an attribute required for intended clinical use [125]. Cisplatin, carboplatin, and oxaliplatin are popular platinum-based chemotherapeutic drugs and chemo-radiotherapy agents [126]. Due to the non-selective nature of the platinum-based chemotherapy drugs that result in severe side effects, platinum-based NPs are a safer answer to target the tumor tissues via the EPR effect with surface functionalization might leading to improved therapeutic efficacy [127]. Platinum NPs (PtNPs) have been studied to radiosensitize the tumor cells by inducing double-stranded breaks (DSB) in DNA in addition to producing hydroxyl radicals. The DNA damage mediated by the production of hydroxyl radicals originating from the inner shell of excited platinum atoms was reported by Kobayashi et al. and they found that the chloroterpyridine platinum, when bound to plasmid DNA, triggered X-ray induced DNA breaks in aqueous solutions [128]. Another study corroborating the presence of DSB in DNA was conducted by Porcel et al. to reveal that PtNPs comparatively enhanced the DNA breaks with the production of water radicals as a part of the radio sensitization process [129]. The advantages of biodegradable drug delivery systems, such as liposomes, polymer-based carriers include reduced systemic side effects and biocompatibility. These properties were harnessed to engineer a liposomal cisplatin formulation called Lipoplatin™ that displayed improved cellular uptake and enhanced radiosensitizing effect on glioma cells upon irradiation with γ-rays [130]. Ultra-small PtNPs with a size of 1.7 nm were duly investigated in a radioresistant organism *D. radiodurans* to observe the impact of the NPs in terms of its uptake, toxicity, and radiation effects. The ultra-small PtNPs were able to successfully penetrate the 150 nm cell wall thickness of the organism and amplified the gamma radiation responses by more than 40%, thus providing new avenues to utilize such NPs for efficient tumor targeting and overcome radio resistivity in certain cancers [131]. PtNPs possess remarkable thermos-plasmonic properties that can cause extreme heating when absorbed near-infrared (NIR) light to mediate the photothermal ablation of cancer cells. PtNPs of 50–70 nm size were shown to achieve stability in reaching surface temperatures up to 900 K when irradiated and resulting in the efficient killing of cancer cells [132]. Porous PtNPs were designed to have a combined effect of sufficient deposition of radiation energy in tumor and generation of oxygen molecules to induce cell death in tumor tissues. These porous PtNPs can induce DNA damage, ROS stress, and cell cycle arrest in the cancer cells, and increase the tumor oxygenation by the conversion of H_2O to O_2 molecules without any viable toxicities [133].

The therapeutic properties of silver (Ag) NPs have been investigated in a range of biological studies demonstrating their antimicrobial activity [134] and anti-cancer properties [135]. As anti-cancer therapeutic agents, AgNPs tend to kill cancer cells via apoptotic cell death, ROS generation, efflux activity inhibition, and reaction with glutathione molecules, and recent reports coming out in the direction supporting AgNPs as radiosensitizers [125]. Independent studies support that AgNPs have radiosensitizing effects on glioma cells both in vitro and in vivo conditions by enhancing the X-ray therapy [136]. In human hepatocellular carcinoma HepG2 cells, AgNPs enhanced the radiosensitivity by inducing increased DNA damage, triggering apoptotic pathway via elevated expression of Bax/caspase-3, and reduced levels of catalase and superoxide dismutase [137]. Novel AgNPs composites engineered by the PEGylation of a graphene-based quantum dot decorated with silver nano prisms (pGAgNPs) were analyzed to check the radiosensitivity in the radiation-sensitive (HCT116) and radiation-resistive (HT-29) colon cancer cells. In both the cell lines, the pGAgNPs were readily internalized and the in vivo biodistribution analysis of the NPs in nude mice followed by a single radiation dose of 10 Gy reported a decrease in the tumor growth contrary to the mice receiving radiation alone [138]. Another silver nanocomposite of 20 nm size fabricated by conjugating the AgNPs to an epidermal growth receptor-specific antibody (C225) was demonstrated to have anti-proliferative activity in nasopharyngeal carcinoma cells in a dose- and time-dependent manner. The radiosensitivity experiments were slightly more effective for the targeted Ag/C225 NPs than the non-targeted AgNPs as confirmed by the western blot analysis that revealed reduced expression levels of certain DNA damage/repair proteins including Ku-80, Ku-70, and Rad51 [139].

11.6.3 Gadolinium-Based NPs

Among the group of rare earth metals or the lanthanide series, gadolinium (Gd) is a popular choice for magnetic resonance imaging (MRI) T_1 contrast agents [140] and also an interesting material to pursue its application in neutron capture therapy (NCT) [141]. Gd-chelates provide precise irradiation when used in MRI-guided radiotherapy and their efficiency as radiosensitizer was reported by Young et al. while using a porphyrin complex of Gd (III) texaphyrin (Gd-tex^{2+}) [142]. The manipulation on the size of GdNP could be advantageous for its excretion by the kidney, which was proved by Mignot et al. who constructed

GdNPs (AGuIX) of 5 nm in diameter [143]. Since nonradioactive GdNPs have the capacity of high neutron capture, GdNPs-loaded chitosan NPs were demonstrated to be viable agents for Gd-NCT [144]. GdNPs promise to serve the dual purpose of diagnostic and therapeutic agents in clinical studies and combining the radiosensitizing effects of the GdNPs with X-rays, gamma rays, and charged particles contribute to enhanced killing effects [125, 145]. As discussed earlier regarding the small size of GdNP and its superiority as an MRI contrast agent, Hu et al. exploited these two properties of GdNPs to analyze their uptake and MRI-guided radio sensitization effects in hepatocellular carcinoma cells [142]. MRI imaging demonstrated that the ratio of AGuIX NPs in tumor/liver was the highest after 1 h of *i.v.* injection and a decrease in tumor size in presence of radiation therapy with AGuIX [146]. Commercially available gadolinium-based molecular agent DOTAREM® was compared with AGuIX NPs to evaluate the efficiency of MRI in healthy animals. AGuIX NPs and DOTAREM® when injected at the same concentrations, AGuIX NPs showed a better contrast than DOTAREM® due to their higher relaxivity and twice higher prolonged circulation in the blood [147]. Despite the availability of state-of-the-art therapeutic strategies, brain metastases have serious neurological and survival repercussions. A recent demonstration on the use of AGuIX NPs as effective theranostic and radiosensitizer agents with favorable toxicity levels has motivated its first use in human study to evaluate maximum tolerable dose in combination with whole-brain radiotherapy (WBRT) (trial registration number NCT02820454) [148]. Multifunctional bismuth gadolinium oxide (BiGdO) surface modified with PEG was constructed to perform a combination of functions like radiotherapy, computed tomography, and MRI. Bismuth and Gd provided CT contrast, while Gd alone is used as an MRI contrast agent, thus the NP served as a dual-modal imaging agent with contrast enhancement [149]. Another stealth GdNPs hyaluronic acid-functionalized Gd-oxide NPs were evaluated for their biocompatibility, radiosensitizing, and MRI agents in tumor cells displaying rapid intracellular uptake by the cancer cells via HA-receptor-mediated endocytosis, low cytotoxicity, high relaxivity as MRI agents, and enhancement in radiosensitivity [150].

11.6.4 Hafnium-Based NPs

Hafnium oxide (hafnia, HfO_2), due to their photo-luminescence, high atomic number, electron density, and chemical stability, potentiates its application as a radiosensitizing agent for radiation therapy, while exhibiting chemical inertness in cellular and subcellular systems [125]. Nanobiotix (http://www.nanobiotix.com/en/), a biotechnology company specializing in nanomedicine, developed the functionalized crystalline HfO_2 NPs, NBTXR3 with size of 50 nm. The surface negative charge of NBTXR3 allows for binding and facilitating rapid cellular internalization in the cancer cells. The aqueous suspension of NBTXR3 makes it feasible to administer both locally and intratumorally [151]. High-energy photon exposure of NBTXR3 resulted in significant radiation dose enhancement demonstrated by Monte Carlo simulation with tolerable toxicity levels [152]. Biological effects of NBTXR3 was studied in different cancer cell lines displaying differential uptake of the NPs forming clusters in the cytoplasm, where the DEF was increased depending on the concentration of NBTXR3 and radiation dose [153]. HfO_2 NPs in combination with radiation therapy was also studied in mice models to observe the anti-tumor immune response contrary to radiation therapy alone treatment revealing increased CD8+ cells in HfO_2 NPs plus radiation therapy treated and untreated tumors with no marked increase in radiation therapy alone treated tumors [154]. A clinical trial on the safety and efficacy of NBTXR3 activated by radiotherapy versus radiation therapy alone as a pre-operative treatment in soft tissue sarcoma patients is registered with ClinicalTrials.gov (NCT02379845) [155].

11.6.5 Superparamagnetic Iron Oxide Nanoparticles

Iron-based NPs that include iron-oxide NPs or superparamagnetic iron oxide nanoparticles (SPIONs) are ideal agents for theranostic applications especially cancer treatment due to their biocompatibility, biodegradability, and facile synthesis. Iron NPs have found their way in being useful as excellent MRI agents, photodynamic/photothermal (PDT/PTT) therapy, magnetic hyperthermia presented in varied investigations [156–158].

Iron NPs could be used as radiosensitizers even though iron seems to have a lower atomic number of (Z = 26) when combined with low LET kV and MV X-rays. Cells upon radiation exposure trigger the electrons from the electron transport chain with an increased production of superoxide anions to finally form hydrogen peroxide. Hydrogen peroxide could be catalyzed by iron oxide NPs to generate ROS that will aid in the killing of cancer cells [125]. Iron oxide NPs were synthesized and conjugated to a cell-penetrating peptide (TAT) with dextran coating to study its therapeutic and production of ROS in A549 cancer cells. The uncoated NPs were rapidly internalized in the lysosomes after cell uptake; however, TAT-functionalized coated NPs were present in the cytoplasm as well in the lysosomes having compromised lysosomal membranes. Loss of lysosomal membrane integrity is one of the upstream events seen in apoptosis, also proven in the study by lysosomal membrane permeability analysis revealing the specific targeting by TAT-functionalized NPs. TAT-functionalized NPs were able to increase the production of ROS levels in the cancer cells when combined with radiation therapy compared to the treatment with radiation therapy alone [159]. The efficiency of the radio sensitization potential of dextran-coated iron oxide NPs when irradiated with high-energy electron beams was studied in MCF-7 cells and HeLa cells. MCF-7 cells tend to show increased radio sensitization enhancement compared to HeLa cells, which could be due to the differential uptake of the NPs at a given concentration. The NPs did not show any extreme cytotoxicity to the cells at 10 μg/ml and 40 μg/ml, thus proving that iron oxide NPs assist in the improvement in electron therapy [160]. Overexpressing heat shock protein Hsp70 in high-grade gliomas was targeted using Hsp70 specific antibody-conjugated SPIONs to analyze the specificity of the NPs toward the tumor before and after ionizing radiation. The SPION-Hsp70 conjugates were seen to be selective toward a range of cancer cell lines, including C6 glioma cells, HeLa cervical cancer cells, and K562 leukemia cells, in a dose-dependent manner with the highest accumulation in the glioma cells after exposure to 10 Gy of ionizing radiation along with enhanced radio sensitization [161]. The enhancement of radio sensitization effect on glioma by SPIONs was also supported by another investigation, where polyglycerol-coated SPIONs (PG-SPIONs) were utilized to evaluate the cytotoxicity and radiosensitivity on the U87-MG cancer cells. The combinatorial effect of PG-SPIONs with 6 MV X-rays compromised the survival rate of the U87-MG cancer cells as demonstrated by the MTT and colony formation studies [162]. Iron oxide NPs are the go-to choice agents for cancer theranostics. However, a long way has to be covered for their intended application in clinical practice and the FDA approval due to certain unwanted toxicity issues reported through previous studies that could be overcome by various stealth modifications to improve their circulation time and evasion by the reticuloendothelial system [156]. A few examples of nanotechnology-based applications in radiotherapy, which are mostly used as radiosensitizers or radionuclides/radioisotopes delivered via nano-drug delivery systems, are mentioned in Table 11.2.

Table 11.2 Examples of nanotechnology based radiotherapy applications by using different NPs

Radiation modality	NP used	Observed effects
$^{111}In/^{177}Lu$; radionuclide	DTPA/PEG liposome	Efficient tumor targeting in CT-26 colon cancer cells by the use of radionuclide targeted liposomes [163]
^{64}Cu; radioisotope	Copper-sulfide NPs	Increased tumor accumulation, reduced undesired biodistribution of NPs, combined photodynamic therapy improved survival rates in 4T1 bearing mice [164]
Wortmannin (Wtmn); radiosensitizer	DSPE-PEG lipid shell/PLGA core NPs	Wtmn NPs are more effective as radiosensitizer compared to common radiosensitizer Cisplatin shown in colon, prostate, and glioma cells [165]
Au; radiosensitizer	Ultra-small glutathione-coated $Au_{29-43}(SG)_{27-37}$ nanoclusters	Improved uptake of gold nanoclusters at 24-h post-injection, inhibition of tumor upon irradiation, negligible side-toxicities [166]

Radiation modality	NP used	Observed effects
Curcumin; radiosensitizer	Poly (*N*-vinylpyrrolidone) (PVP)-polycaprolactone (PCL) NPs	PVP-PCL-Cur NPs time and dose-dependent cytotoxicity, radiosensitized A549 lung cancer cells, inhibited colony formation [167]
Gold; radiosensitizer	Chitosan-capped gold NPs encapsulating doxorubicin	Multifunctional nano chemo-radiosensitizer ensured efficient targeting of MCF-7 breast cancer cells, induced DNA damage vis DSB with a low radiation dose [168]
Iron (Fe II) complexes; radiotherapy $^{99m}Tc^{4+}$; radioisotope	Ultra-small gallic acid (GA)-Fe (II) nano-complexes and L-buthionine sulfoximine (BSO) co-encapsulated within liposomal nanocarrier chelated with $^{99m}Tc^{4+}$	$^{99m}Tc^{4+}$ chelated BSO/GA-Fe (II) liposomes could be imaged using SPECT, delivery of GA-Fe (II) generated OH radicals, and BSO mediate GSH depletion to generate oxidative stress in cancer cells [169]
Olaparib (Ola); radiosensitizer	mPEG-PCL NPs	Ola-NPs is a potent radiosensitizer compared to free Ola, which induces DSB repair inhibition, and promotes apoptosis in lung cancer cells [170]

11.7 Conclusion and Future Perspectives

Nanotechnology has found numerous applications in cancer diagnosis and treatment over the years and still has progressed by leaps and bounds in terms of research and innovation. Not lagging behind, radiation therapy continues to provide curative and palliative care to cancer patients with constantly evolving technological advances. Nanomaterials/NPs owing to their distinctive properties have modified tumor radiation response, enhanced radio sensitization, imaging, quantitative dosimetry, and IGRT. Despite the roaring success of nanotechnology in chemotherapeutic procedures, nanotechnological involvement in radiation oncology is still in its infancy requiring extensive studies to be carried out. Nevertheless, issues like toxicities caused to normal cells, tumor resistivity, and accuracy in radiotherapy delivery are readily overcome by nanotechnology. The successful transition of NPs into clinical settings can be pushed forward by navigating certain challenges in the pre-clinical stages such as (i) evaluation of the long-term safety of NPs by pre-clinical testing methods, along with documentation of the tolerable dose of the NPs, (ii) pharmacokinetic/pharmacodynamic analysis and clearance of the NPs from the physiological system should be investigated, and (iii) non-specific uptake of the NPs. The issue of non-specificity and to ensure proper delivery and targeting of the NPs to tumor tissue could be addressed by NPs size reduction, modulating the shape of NPs, surface charge of NPs (positively charged are cleared more rapidly than negatively charged), PEGylation of the NPs to evade opsonization, and inhibition of Kupffer cells activity in the liver [171]. Radiotherapy is known to increase the tumor antigen expression that leads to the production of inflammatory responses and immune cell infiltration. Checkpoint immunotherapy and radiotherapy could be together combined with nanotechnology to achieve synergistic effects for an ultimate annihilation of tumor and therapy response would be an interesting research arena [172]. Future perspectives also would be to perform more clinical trials of the chosen nano-radiotherapeutics and development of chemo-radiotherapy in combination with nanotechnology to surmount the challenges [173].

Acknowledgements

All the images in the chapter were created using Biorender.com online platform.

References

1. Ringborg U, Bergqvist D, Brorsson B, Cavallin-ståhl E, Ceberg J, Einhorn N, et al. The Swedish Council on Technology Assessment in Health Care (SBU) systematic overview of radiotherapy for cancer including a prospective survey of radiotherapy practice in Sweden 2001: Summary and Conclusions. *Acta Oncol (Madr)*. 2003; 42:357–65. https://doi.org/10.1080/02841860310010826.

2. Pieters R, Ding L, Bushe H, Aronowitz J. Principles of radiation oncology. Cancer concepts a guideb. Non-Oncologist, University of Massachusetts Medical School. 2014. https://doi.org/10.7191/cancer_concepts.1008.
3. Baskar R, Lee KA, Yeo R, Yeoh K-W. Cancer and radiation therapy: Current advances and future directions. *Int J Med Sci.* 2012; 9: 193–9. https://doi.org/10.7150/ijms.3635.
4. Mohan G, T P AH, A J J, K M SD, Narayanasamy A, Vellingiri B. Recent advances in radiotherapy and its associated side effects in cancer—a review. *J Basic Appl Zool.* 2019; 80:14. https://doi.org/10.1186/s41936-019-0083-5.
5. Dayes IS, Whelan TJ, Julian JA, Kuettel MR, Regmi D, Okawara GS, et al. Cross-border referral for early breast cancer: An analysis of radiation fractionation patterns. *Curr Oncol.* 2006. https://doi.org/10.3747/co.v13i4.96.
6. Harris JR, Lippman ME, Morrow M KOC. Diseases of the breast: Fifth edition. Wolters Kluwer Health Adis (ESP). 2014.
7. Dawson LA, McGinn CJ, Normolle D, Ten Haken RK, Walker S, Ensminger W, et al. Escalated focal liver radiation and concurrent hepatic artery fluorodeoxyuridine for unresectable intrahepatic malignancies. *J Clin Oncol.* 2000. https://doi.org/10.1200/JCO.2000.18.11.2210.
8. McGinn CJ, Ten Haken RK, Ensminger WD, Walker S, Wang S, Lawrence TS. Treatment of intrahepatic cancers with radiation doses based on a normal tissue complication probability model. *J Clin Oncol.* 1998. https://doi.org/10.1200/JCO.1998.16.6.2246.
9. Rai B, Bansal A, Patel FD, Sharma SC. Radiotherapy for ovarian cancers - Redefining the role. *Asian Pacific J Cancer Prev.* 2014. https://doi.org/10.7314/APJCP.2014.15.12.4759.
10. Eisbruch A, Ten Haken RK, Kim HM, Marsh LH, Ship JA. Dose, volume, and function relationships in parotid salivary glands following conformal and intensity-modulated irradiation of head and neck cancer. *Int J Radiat Oncol Biol Phys.* 1999. https://doi.org/10.1016/S0360-3016(99)00247-3.
11. Bjordal K, Kaasa S, Mastekaasa A. Quality of life in patients treated for head and neck cancer: A follow-up study 7 to 11 years after radiotherapy. *Int J Radiat Oncol Biol Phys.* 1994. https://doi.org/10.1016/0360-3016(94)90104-X.

12. Hoppe BS, Michalski JM, Mendenhall NP, Morris CG, Henderson RH, Nichols RC, et al. Comparative effectiveness study of patient-reported outcomes after proton therapy or intensity-modulated radiotherapy for prostate cancer. *Cancer*. 2014. https://doi.org/10.1002/cncr.28536.

13. Taylor A. Intensity-modulated radiotherapy—What is it? *Cancer Imaging*. 2004; 4:68–73. https://doi.org/10.1102/1470-7330.2004.0003.

14. Purdy JA. Intensity-modulated radiotherapy: Current status and issues of interest. *Int J Radiat Oncol Biol Phys*. 2001. https://doi.org/10.1016/S0360-3016(01)01749-7.

15. Wang-Chesebro A, Xia P, Coleman J, Akazawa C, Roach M. Intensity-modulated radiotherapy improves lymph node coverage and dose to critical structures compared with three-dimensional conformal radiation therapy in clinically localized prostate cancer. *Int J Radiat Oncol*. 2006; 66:654–62. https://doi.org/10.1016/j.ijrobp.2006.05.037.

16. Feng FY, Kim HM, Lyden TH, Haxer MJ, Feng M, Worden FP, et al. Intensity-modulated radiotherapy of head and neck cancer aiming to reduce dysphagia: Early dose–effect relationships for the swallowing structures. *Int J Radiat Oncol*. 2007; 68:1289–98. https://doi.org/10.1016/j.ijrobp.2007.02.049.

17. Mundt AJ, Lujan AE, Rotmensch J, Waggoner SE, Yamada SD, Fleming G, et al. Intensity-modulated whole pelvic radiotherapy in women with gynecologic malignancies. *Int J Radiat Oncol*. 2002; 52:1330–7. https://doi.org/10.1016/S0360-3016(01)02785-7.

18. Greco C, Clifton Ling C. Broadening the scope of image-guided radiotherapy (IGRT). *Acta Oncol (Madr)*. 2008; 47:1193–200. https://doi.org/10.1080/02841860802241956.

19. Langen K., Jones DT. Organ motion and its management. *Int J Radiat Oncol*. 2001; 50:265–78. https://doi.org/10.1016/S0360-3016(01)01453-5.

20. Michalski J, Purdy JA, Gaspar L, Souhami L, Ballow M, Bradley J, et al. Radiation therapy oncology group. Research Plan 2002–2006. Image-Guided Radiation Therapy Committee. *Int J Radiat Oncol Biol Phys*. 2001; 51:60–5.

21. Goyal S, Kataria T. Image guidance in radiation therapy: Techniques and applications. *Radiol Res Pract*. 2014; 2014:1–10. https://doi.org/10.1155/2014/705604.

22. Jaffray DA, Langen KM, Mageras G, Dawson LA, Yan D, EdD RA, et al. Safety considerations for IGRT: Executive summary. *Pract Radiat Oncol.* 2013; 3:167–70. https://doi.org/10.1016/j.prro.2013.01.004.

23. Trikalinos TA, Terasawa T, Ip S, et al. Particle beam radiation therapies for cancer. Rockville (MD): Agency for Healthcare Research and Quality (US) (Comparative Effectiveness Technical Briefs, No. 1.). 2009. Results available from: https://www.ncbi.nlm.nih.gov/books/NBK44543/.

24. Levin WP, Kooy H, Loeffler JS, DeLaney TF. Proton beam therapy. *Br J Cancer.* 2005;93:849–54. https://doi.org/10.1038/sj.bjc.6602754.

25. Shioyama Y, Tsuji H, Suefuji H, Sinoto M, Matsunobu A, Toyama S, et al. Particle radiotherapy for prostate cancer. *Int J Urol.* 2015; 22:33–9. https://doi.org/10.1111/iju.12640.

26. Kanai T, Matsufuji N, Miyamoto T, Mizoe J, Kamada T, Tsuji H, et al. Examination of GyE system for HIMAC carbon therapy. *Int J Radiat Oncol Biol Phys.* 2006. https://doi.org/10.1016/j.ijrobp.2005.09.043.

27. Ando K, Kase Y. Biological characteristics of carbon-ion therapy. *Int J Radiat Biol.* 2009. https://doi.org/10.1080/09553000903072470.

28. Ogata T, Teshima T, Kagawa K, Hishikawa Y, Takahashi Y, Kawaguchi A, et al. Particle irradiation suppresses metastatic potential of cancer cells. *Cancer Res.* 2005.

29. Cui X, Oonishi K, Tsujii H, Yasuda T, Matsumoto Y, Furusawa Y, et al. Effects of carbon ion beam on putative colon cancer stem cells and its comparison with X-rays. *Cancer Res.* 2011. https://doi.org/10.1158/0008-5472.CAN-10-2926.

30. Funk RK, Stockham AL, Laack NNI. Basics of radiation therapy. *Clin. Cardio-Oncology.* 2016. https://doi.org/10.1016/B978-0-323-44227-5.00003-X.

31. International Commission on Radiation Units and Measurements. ICRU Report 62. Prescribing, recording, and reporting photon beam therapy (Supplement to ICRU Report 50). *J ICRU.* 1999. https://doi.org/10.1259/bjr.74.879.740294.

32. Leksell L. Stereotactic radiosurgery. *J Neurol Neurosurg Psychiatry.* 1983. https://doi.org/10.1136/jnnp.46.9.797.

33. Amini A, Yeh N, Gaspar LE, Kavanagh B, Karam SD. Stereotactic body radiation therapy (SBRT) for lung cancer patients previously treated with conventional radiotherapy: *A review. Radiat Oncol.* 2014; 9:210. https://doi.org/10.1186/1748-717X-9-210.

34. Withers HR. The four R's of radiotherapy. *Advances in Radiation Biology*. 1975; 5:241–271. https://doi. org/10.1016/b978-0-12.
35. Macià i Garau M. Radiobiology of stereotactic body radiation therapy (SBRT). *Reports Pract Oncol Radiother*. 2017; 22:86–95. https://doi. org/10.1016/j.rpor.2017.02.010.
36. Martel M. Estimation of tumor control probability model parameters from 3D dose distributions of non-small cell lung cancer patients. *Lung Cancer*. 1999; 24:31–7. https://doi.org/10.1016/ S0169-5002(99)00019-7.
37. Ogawa Y, Shibamoto Y, Hashizume C, Kondo T, Iwata H, Tomita N, et al. Repeat stereotactic body radiotherapy (SBRT) for local recurrence of non-small cell lung cancer and lung metastasis after first SBRT. *Radiat Oncol*. 2018; 13:136. https://doi.org/10.1186/ s13014-018-1080-4.
38. Freeman DE, King CR. Stereotactic body radiotherapy for low-risk prostate cancer: Five-year outcomes. *Radiat Oncol*. 2011. https://doi. org/10.1186/1748-717X-6-3.
39. Wu QJ, Thongphiew D, Wang Z, Chankong V, Yin FF. The impact of respiratory motion and treatment technique on stereotactic body radiation therapy for liver cancer. *Med Phys*. 2008. https://doi. org/10.1118/1.2839095.
40. Chargari C, Deutsch E, Blanchard P, Gouy S, Martelli H, Guérin F, et al. Brachytherapy: An overview for clinicians. *CA Cancer J Clin*. 2019; 69:386–401. https://doi.org/10.3322/caac.21578.
41. Polgár C, Jánváry L, Major T, Somogyi A, Takácsi-Nagy Z, Fröhlich G, et al. The role of high-dose-rate brachytherapy boost in breast-conserving therapy: Long-term results of the Hungarian National Institute of Oncology. *Reports Pract Oncol Radiother*. 2010. https://doi.org/10.1016/j.rpor.2010.01.002.
42. Slampa, P., Ruzickova, J., Ondrova, B., Ticha, H., & Dolezelova, H. (2008). Sole conformal perioperative interstitial brachytherapy of early stage breast carcinoma using high-dose rate afterloading: longer-term results and toxicity. *Reports of Practical Oncology and Radiotherapy*, 13(2), 63–68.
43. Szlag M, Ślosarek K, Rembielak A, Białas B, Fijałkowski M, Bystrzycka J. Real-time brachytherapy for prostate cancer–Implant analysis. *Reports Pract Oncol Radiother*. 2008; 13:9–14. https://doi. org/10.1016/S1507-1367(10)60076-4.
44. Maier P, Hartmann L, Wenz F, Herskind C. Cellular pathways in response to ionizing radiation and their targetability for tumor

radiosensitization. *Int J Mol Sci.* 2016; 17:102. https://doi.org/10.3390/ijms17010102.

45. Golden EB, Apetoh L. Radiotherapy and immunogenic cell death. *Semin Radiat Oncol.* 2015; 25:11–7. https://doi.org/10.1016/j.semradonc.2014.07.005.
46. Eriksson D, Stigbrand T. Radiation-induced cell death mechanisms. *Tumor Biol.* 2010; 31:363–72. https://doi.org/10.1007/s13277-010-0042-8.
47. Ianzini F, Bertoldo A, Kosmacek EA, Phillips SL, Mackey MA. Lack of p53 function promotes radiation-induced mitotic catastrophe in mouse embryonic fibroblast cells. *Cancer Cell Int.* 2006. https://doi.org/10.1186/1475-2867-6-11.
48. Bourke E, Dodson H, Merdes A, Cuffe L, Zachos G, Walker M, et al. DNA damage induces Chk1-dependent centrosome amplification. *EMBO Rep.* 2007. https://doi.org/10.1038/sj.embor.7400962.
49. Eriksson D, Blomberg J, Lindgren T, Löfroth PO, Johansson L, Riklund K, et al. Iodine-131 induces mitotic catastrophes and activates apoptotic pathways in HeLa Hep2 cells. *Cancer Biother Radiopharm.* 2008. https://doi.org/10.1089/cbr.2008.0471.
50. Shinomiya N. New concepts in radiation-induced apoptosis: Premitotic apoptosis 1 and "postmitotic apoptosis." *J Cell Mol Med.* 2001. https://doi.org/10.1111/j.1582-4934.2001.tb00158.x.
51. Clarke AR, Purdie CA, Harrison DJ, Morris RG, Bird CC, Hooper ML, et al. Thymocyte apoptosis induced by p53-dependent and independent pathways. *Nature.* 1993. https://doi.org/10.1038/362849a0.
52. Lowe SW, Schmitt EM, Smith SW, Osborne BA, Jacks T. P53 is required for radiation-induced apoptosis in mouse thymocytes. *Nature.* 1993. https://doi.org/10.1038/362847a0.
53. Sia J, Szmyd R, Hau E, Gee HE. Molecular mechanisms of radiation-induced cancer cell death: A primer. *Front Cell Dev Biol.* 2020; 8. https://doi.org/10.3389/fcell.2020.00041.
54. Kobayashi T, Ruan S, Jabbur JR, Consoli U, Clodi K, Shiku H, et al. Differential p53 phosphorylation and activation of apoptosis-promoting genes Bax and Fas/APO-1 by irradiation and ara-C treatment. *Cell Death Differ.* 1998; 5:584–91. https://doi.org/10.1038/sj.cdd.4400382.
55. Embree-Ku M, Venturini D, Boekelheide K. Fas is involved in the p53-dependent apoptotic response to ionizing radiation in mouse testis. *Biol Reprod.* 2002. https://doi.org/10.1095/biolreprod66.5.1456.

56. Pettus BJ, Chalfant CE, Hannun YA. Ceramide in apoptosis: an overview and current perspectives. *Biochim Biophys Acta-Mol Cell Biol Lipids.* 2002; 1585:114–25. https://doi.org/10.1016/S1388-1981(02)00331-1.
57. Kolesnick R, Fuks Z. Radiation and ceramide-induced apoptosis. *Oncogene.* 2003; 22:5897–906. https://doi.org/10.1038/sj.onc.1206702.
58. Lei G, Zhang Y, Koppula P, Liu X, Zhang J, Lin SH, et al. The role of ferroptosis in ionizing radiation-induced cell death and tumor suppression. *Cell Res.* 2020; 30:146–62. https://doi.org/10.1038/s41422-019-0263-3.
59. Kroemer G, Galluzzi L, Vandenabeele P, Abrams J, Alnemri ES, Baehrecke EH, et al. Classification of cell death: Recommendations of the nomenclature committee on cell death 2009. *Cell Death Differ.* 2009; 16:3–11. https://doi.org/10.1038/cdd.2008.150.
60. Festjens N, Vanden Berghe T, Vandenabeele P. Necrosis, a well-orchestrated form of cell demise: Signalling cascades, important mediators and concomitant immune response. *Biochim Biophys Acta - Bioenerg.* 2006; 1757:1371–87. https://doi.org/10.1016/j.bbabio.2006.06.014.
61. Rainaldi G, Ferrante A, Luigi Indovina P, Santini MT. Induction of apoptosis or necrosis by ionizing radiation is dose-dependent in MG-63 osteosarcoma multicellular spheroids. *Anticancer Res.* 2003.
62. Dunn WA. Studies on the mechanisms of autophagy: Maturation of the autophagic vacuole. *J Cell Biol.* 1990; 110:1935–45. https://doi.org/10.1083/jcb.110.6.1935.
63. Carvalho H, Villar R. Radiotherapy and immune response: The systemic effects of a local treatment. *Clinics.* 2018; 73. https://doi.org/10.6061/clinics/2018/e557s.
64. Liu Y, Levine B. Autosis and autophagic cell death: The dark side of autophagy. *Cell Death Differ.* 2015; 22:367–76. https://doi.org/10.1038/cdd.2014.143.
65. Zeng X, Kinsella TJ. Impact of autophagy on chemotherapy and radiotherapy mediated tumor cytotoxicity: To live or not to live? *Front Oncol.* 2011; 1. https://doi.org/10.3389/fonc.2011.00030.
66. Bellodi C, Lidonnici MR, Hamilton A, Helgason GV, Soliera AR, Ronchetti M, et al. Targeting autophagy potentiates tyrosine kinase inhibitor-induced cell death in Philadelphia chromosome-positive cells, including primary CML stem cells. *J Clin Invest.* 2009. https://doi.org/10.1172/JCI35660.

67. Mrózek A, Petrowsky H, Sturm I, Kraus J, Hermann S, Hauptmann S, et al. Combined p53/Bax mutation results in extremely poor prognosis in gastric carcinoma with low microsatellite instability. *Cell Death Differ.* 2003; 10:461–7. https://doi.org/10.1038/sj.cdd.4401193.

68. Cao C, Subhawong T, Albert JM, Kim KW, Geng L, Sekhar KR, et al. Inhibition of mammalian target of rapamycin or apoptotic pathway induces autophagy and radiosensitizes PTEN null prostate cancer cells. *Cancer Res.* 2006. https://doi.org/10.1158/0008-5472.CAN-06-0802.

69. Jo GH, Bögler O, Chwae Y-J, Yoo H, Lee SH, Park JB, et al. Radiation-induced autophagy contributes to cell death and induces apoptosis partly in malignant glioma cells. *Cancer Res Treat.* 2014; 47:221–41. https://doi.org/10.4143/crt.2013.159.

70. Kondo Y, Kanzawa T, Sawaya R, Kondo S. The role of autophagy in cancer development and response to therapy. *Nat Rev Cancer.* 2005. https://doi.org/10.1038/nrc1692.

71. Hayflick L, Moorhead PS. The serial cultivation of human diploid cell strains. *Exp Cell Res.* 1961; 25:585–621. https://doi.org/10.1016/0014-4827(61)90192-6.

72. Roninson IB. Tumor cell senescence in cancer treatment. *Cancer Res.* 2003.

73. Rodier F, Coppé JP, Patil CK, Hoeijmakers WAM, Muñoz DP, Raza SR, et al. Persistent DNA damage signalling triggers senescence-associated inflammatory cytokine secretion. *Nat Cell Biol.* 2009. https://doi.org/10.1038/ncb1909.

74. Dai CY, Enders GH. p16(INK4a) can initiate an autonomous senescence program. *Oncogene.* 2000. https://doi.org/10.1038/sj.onc.1203438.

75. Steiner MS, Wang Y, Zhang Y, Zhang X, Lu Y. p16/MTS1/INK4A suppresses prostate cancer by both pRb dependent and independent pathways. *Oncogene.* 2000. https://doi.org/10.1038/sj.onc.1203428.

76. Collado M, Serrano M. Senescence in tumours: Evidence from mice and humans. *Nat Rev Cancer.* 2010. https://doi.org/10.1038/nrc2772.

77. Wang B, Kohli J, Demaria M. Senescent cells in cancer therapy: Friends or foes? *Trends in Cancer.* 2020. https://doi.org/10.1016/j.trecan.2020.05.004.

78. Sieben CJ, Sturmlechner I, van de Sluis B, van Deursen JM. Two-step senescence-focused cancer therapies. *Trends Cell Biol.* 2018. https://doi.org/10.1016/j.tcb.2018.04.006.

79. Chang J, Wang Y, Shao L, Laberge RM, Demaria M, Campisi J, et al. Clearance of senescent cells by ABT263 rejuvenates aged hematopoietic stem cells in mice. *Nat Med.* 2016. https://doi.org/10.1038/nm.4010.

80. Gayle SS, Sahni JM, Webb BM, Weber-Bonk KL, Shively MS, Spina R, et al. Targeting BCL-xL improves the efficacy of bromodomain and extra-terminal protein inhibitors in triple-negative breast cancer by eliciting the death of senescent cells. *J Biol Chem.* 2019. https://doi.org/10.1074/jbc.RA118.004712.

81. Samaraweera L, Adomako A, Rodriguez-Gabin A, McDaid HM. A novel indication for panobinostat as a senolytic drug in NSCLC and HNSCC. *Sci Rep.* 2017. https://doi.org/10.1038/s41598-017-01964-1.

82. West CM, Barnett GC. Genetics and genomics of radiotherapy toxicity: Towards prediction. *Genome Med.* 2011; 3:52. https://doi.org/10.1186/gm268.

83. Abe Y, Miura T, Yoshida MA, Ujiie R, Kurosu Y, Kato N, et al. Analysis of chromosome translocation frequency after a single CT scan in adults. *J Radiat Res.* 2016. https://doi.org/10.1093/jrr/rrv090.

84. Kanda R, Hayata I. Comparison of the yields of translocations and dicentrics measured using conventional Giemsa staining and chromosome painting. *Int J Radiat Biol.* 1996. https://doi.org/10.1080/095530096145445.

85. Kadhim M, Salomaa S, Wright E, Hildebrandt G, Belyakov O V., Prise KM, et al. Non-targeted effects of ionising radiation: Implications for low dose risk. *Mutat Res–Rev Mutat Res.* 2013. https://doi.org/10.1016/j.mrrev.2012.12.001.

86. Little JB. Genomic instability and bystander effects: A historical perspective. *Oncogene.* 2003. https://doi.org/10.1038/sj.onc.1206988.

87. De Gonzalez AB, Curtis RE, Kry SF, Gilbert E, Lamart S, Berg CD, et al. Proportion of second cancers attributable to radiotherapy treatment in adults: A cohort study in the US SEER cancer registries. *Lancet Oncol.* 2011. https://doi.org/10.1016/S1470-2045(11)70061-4.

88. Morton LM, Onel K, Curtis RE, Hungate EA, Armstrong GT. The rising incidence of second cancers: Patterns of occurrence and identification of risk factors for children and adults. *Am Soc Clin Oncol Educ B.* 2014. https://doi.org/10.14694/edbook_am.2014.34.e57.

89. Krueger SA, Joiner MC, Weinfeld M, Piasentin E, Marples B. Role of apoptosis in low-dose hyper-radiosensitivity. *Radiat Res.* 2007. https://doi.org/10.1667/RR0776.1.

90. National Research Council. 2006. Health Risks from Exposure to Low Levels of Ionizing Radiation: BEIR VII Phase 2. Washington, DC: The National Academies Press. https://doi.org/10.17226/11340.

91. Burt LM, Ying J, Poppe MM, Suneja G, Gaffney DK. Risk of secondary malignancies after radiation therapy for breast cancer: Comprehensive results. *The Breast.* 2017; 35:122–9. https://doi.org/10.1016/j.breast.2017.07.004.

92. Rombouts AJM, Hugen N, Elferink MAG, Feuth T, Poortmans PMP, Nagtegaal ID, et al. Incidence of second tumors after treatment with or without radiation for rectal cancer. *Ann Oncol.* 2017; 28: 535–40. https://doi.org/10.1093/annonc/mdw661.

93. Krukemeyer MG, Krenn V, Huebner F, Wagner W, Resch R (2015) History and Possible Uses of Nanomedicine Based on Nanoparticles and Nanotechnological Progress. *J Nanomed Nanotechnol* 6: 336. doi:10.4172/2157-7439.1000336.

94. Jain KK. The handbook of nanomedicine. Springer, Humana Press. 2008. https://doi.org/10.1007/978-1-60327-319-0.

95. Chatterjee DK, Wolfe T, Lee J, Brown AP, Singh PK, Bhattarai SR, et al. Convergence of nanotechnology with radiation therapy-insights and implications for clinical translation. *Transl Cancer Res.* 2013. https://doi.org/10.3978/j.issn.2218-676X.2013.08.10.

96. Mi Y, Shao Z, Vang J, Kaidar-Person O, Wang AZ. Application of nanotechnology to cancer radiotherapy. *Cancer Nanotechnol.* 2016; 7:11. https://doi.org/10.1186/s12645-016-0024-7.

97. Goel S, Ni D, Cai W. Harnessing the power of nanotechnology for enhanced radiation therapy. *ACS Nano.* 2017; 11:5233–7. https://doi.org/10.1021/acsnano.7b03675.

98. Kuncic Z, Lacombe S. Nanoparticle radio-enhancement: Principles, progress and application to cancer treatment. *Phys Med Biol.* 2018. https://doi.org/10.1088/1361-6560/aa99ce.

99. Yeh Y-C, Creran B, Rotello VM. Gold nanoparticles: Preparation, properties, and applications in bionanotechnology. *Nanoscale.* 2012; 4:1871–80. https://doi.org/10.1039/C1NR11188D.
100. Jain PK, Lee KS, El-Sayed IH, El-Sayed MA. Calculated absorption and scattering properties of gold nanoparticles of different size, shape, and composition: Applications in biological imaging and biomedicine. *J Phys Chem B.* 2006; 110:7238–48. https://doi.org/10.1021/jp057170o.
101. Eustis S, El-Sayed MA. Why gold nanoparticles are more precious than pretty gold: Noble metal surface plasmon resonance and its enhancement of the radiative and nonradiative properties of nanocrystals of different shapes. *Chem Soc Rev.* 2006; 35:209–17. https://doi.org/10.1039/B514191E.
102. Link S, El-Sayed MA. Size and temperature dependence of the plasmon absorption of colloidal gold nanoparticles. *J Phys Chem B.* 1999; 103:4212–7. https://doi.org/10.1021/jp984796o.
103. Hauck TS, Ghazani AA, Chan WCW. Assessing the effect of surface chemistry on gold nanorod uptake, toxicity, and gene expression in mammalian cells. *Small.* 2008. https://doi.org/10.1002/smll.200700217.
104. Zhang G, Yang Z, Lu W, Zhang R, Huang Q, Tian M, et al. Influence of anchoring ligands and particle size on the colloidal stability and in vivo biodistribution of polyethylene glycol-coated gold nanoparticles in tumor-xenografted mice. *Biomaterials.* 2009. https://doi.org/10.1016/j.biomaterials.2008.12.038.
105. Yao L, Danniels J, Moshnikova A, Kuznetsov S, Ahmed A, Engelman DM, et al. pHLIP peptide targets nanogold particles to tumors. *Proc Natl Acad Sci USA.* 2013. https://doi.org/10.1073/pnas.1219665110.
106. Kumar R, Korideck H, Ngwa W, Berbeco RI, Mike Makrigiorgos G, Sridhar S. Third generation gold nanoplatform optimized for radiation therapy. *Transl Cancer Res.* 2013. https://doi.org/10.3978/j.issn.2218-676X.2013.07.02.
107. Grieneisen ML, Zhang M. Nanoscience and nanotechnology: Evolving definitions and growing footprint on the scientific landscape. *Small.* 2011. https://doi.org/10.1002/smll.201100387.
108. Herold DM, Das IJ, Stobbe CC, Iyer RV, Chapman JD. Gold microspheres: A selective technique for producing biologically effective dose enhancement. *Int J Radiat Biol.* 2000. https://doi.org/10.1080/09553000050151637.

109. Cheung JYC, Tang F. The calculation of dose enhancement close to platinum implants for skull radiography. *Health Phys.* 2007; 93:267–72. https://doi.org/10.1097/01.HP.0000264450.81683.51.

110. Haume K, Rosa S, Grellet S, Śmiałek MA, Butterworth KT, Solov'yov A V., et al. Gold nanoparticles for cancer radiotherapy: A review. *Cancer Nanotechnol.* 2016; 7:8. https://doi.org/10.1186/s12645-016-0021-x.

111. Muddineti OS, Ghosh B, Biswas S. Current trends in using polymer coated gold nanoparticles for cancer therapy. *Int J Pharm.* 2015. https://doi.org/10.1016/j.ijpharm.2015.02.038.

112. Connor EE, Mwamuka J, Gole A, Murphy CJ, Wyatt MD. Gold nanoparticles are taken up by human cells but do not cause acute cytotoxicity. *Small.* 2005. https://doi.org/10.1002/smll.200400093.

113. Hainfeld JF, Smilowitz HM, O'connor MJ, Dilmanian FA, Slatkin DN. Gold nanoparticle imaging and radiotherapy of brain tumors in mice. *Nanomedicine.* 2013. https://doi.org/10.2217/nnm.12.165.

114. Hainfeld JF, Slatkin DN, Smilowitz HM. The use of gold nanoparticles to enhance radiotherapy in mice. *Phys Med Biol.* 2004; 49:N309–15. https://doi.org/10.1088/0031-9155/49/18/N03.

115. Regulla DF, Hieber LB, Seidenbusch M. Physical and biological interface dose effects in tissue due to X-ray-induced release of secondary radiation from metallic gold surfaces. *Radiat Res.* 1998; 150:92. https://doi.org/10.2307/3579649.

116. M. Herold, I. J. Das, C. C. Stobbe, D. Gold microspheres: A selective technique for producing biologically effective dose enhancement. *Int J Radiat Biol.* 2000; 76:1357–64. https://doi.org/10.1080/09553000050151637.

117. Chithrani DB, Jelveh S, Jalali F, van Prooijen M, Allen C, Bristow RG, et al. Gold nanoparticles as radiation sensitizers in cancer therapy. *Radiat Res.* 2010; 173:719. https://doi.org/10.1667/RR1984.1.

118. Her S, Jaffray DA, Allen C. Gold nanoparticles for applications in cancer radiotherapy: Mechanisms and recent advancements. *Adv Drug Deliv Rev.* 2017; 109:84–101. https://doi.org/10.1016/j.addr.2015.12.012.

119. Carter JD, Cheng NN, Qu Y, Suarez GD, Guo T. Nanoscale energy deposition by X-ray absorbing nanostructures. *J Phys Chem B.* 2007; 111:11622–5. https://doi.org/10.1021/jp075253u.

120. Zheng Y, Hunting DJ, Ayotte P, Sanche L. Radiosensitization of DNA by gold nanoparticles irradiated with high-energy electrons. *Radiat Res*. 2008; 169:19–27. https://doi.org/10.1667/RR1080.1.

121. Ionita P, Conte M, Gilbert BC, Chechik V. Gold nanoparticle-initiated free radical oxidations and halogen abstractions. *Org Biomol Chem*. 2007. https://doi.org/10.1039/b711573c.

122. Jain S, Coulter JA, Hounsell AR, Butterworth KT, McMahon SJ, Hyland WB, et al. Cell-specific radiosensitization by gold nanoparticles at megavoltage radiation energies. *Int J Radiat Oncol Biol Phys*. 2011. https://doi.org/10.1016/j.ijrobp.2010.08.044.

123. Choi BJ, Jung KO, Graves EE, Pratx G. A gold nanoparticle system for the enhancement of radiotherapy and simultaneous monitoring of reactive-oxygen-species formation. *Nanotechnology*. 2018. https://doi.org/10.1088/1361-6528/aae272.

124. Luo D, Wang X, Zeng S, Ramamurthy G, Burda C, Basilion JP. Prostate-specific membrane antigen targeted gold nanoparticles for prostate cancer radiotherapy: Does size matter for targeted particles? *Chem Sci*. 2019. https://doi.org/10.1039/c9sc02290b.

125. Liu Y, Zhang P, Li F, Jin X, Li J, Chen W, et al. Metal-based nanoenhancers for future radiotherapy: Radiosensitizing and synergistic effects on tumor cells. *Theranostics*. 2018. https://doi.org/10.7150/thno.22172.

126. De Vos D. Platinum complexes as anticancer agent. *Chem Weekbl*. 1978. https://doi.org/10.2174/157489206775246458.

127. Alshatwi AA, Athinarayanan J, Vaiyapuri Subbarayan P. Green synthesis of platinum nanoparticles that induce cell death and G2/M-phase cell cycle arrest in human cervical cancer cells. *J Mater Sci Mater Med*. 2015; 26:7. https://doi.org/10.1007/s10856-014-5330-1.

128. Kobayashi K, Frohlich H, Usami N, Takakura K, Le Sech C. Enhancement of X-ray-induced breaks in DNA bound to molecules containing platinum: A possible application to hadrontherapy. *Radiat Res*. 2002. https://doi.org/10.1667/0033-7587(2002)1570032:EOXRIB]2.0.CO;2.

129. Porcel E, Liehn S, Remita H, Usami N, Kobayashi K, Furusawa Y, et al. Platinum nanoparticles: A promising material for future cancer therapy? *Nanotechnology*. 2010; 21:085103. https://doi.org/10.1088/0957-4484/21/8/085103.

130. Charest G, Paquette B, Fortin D, Mathieu D, Sanche L. Concomitant treatment of F98 glioma cells with new liposomal platinum compounds and ionizing radiation. *J Neurooncol.* 2010. https://doi.org/10.1007/s11060-009-0011-5.

131. Li S, Porcel E, Remita H, Marco S, Réfrégiers M, Dutertre M, et al. Platinum nanoparticles: An exquisite tool to overcome radioresistance. *Cancer Nanotechnol.* 2017. https://doi.org/10.1186/s12645-017-0028-y.

132. Samadi A, Klingberg H, Jauffred L, Kjær A, Bendix PM, Oddershede LB. Platinum nanoparticles: A non-toxic, effective and thermally stable alternative plasmonic material for cancer therapy and bioengineering. *Nanoscale.* 2018. https://doi.org/10.1039/c8nr02275e.

133. Li Y, Yun K-H, Lee H, Goh S-H, Suh Y-G, Choi Y. Porous platinum nanoparticles as a high-Z and oxygen generating nanozyme for enhanced radiotherapy in vivo. *Biomaterials.* 2019; 197:12–9. https://doi.org/10.1016/j.biomaterials.2019.01.004.

134. Durán N, Durán M, de Jesus MB, Seabra AB, Fávaro WJ, Nakazato G. Silver nanoparticles: A new view on mechanistic aspects on antimicrobial activity. *Nanomedicine Nanotechnology Biol Med.* 2016; 12:789–99. https://doi.org/10.1016/j.nano.2015.11.016.

135. Asha Rani P V., Low Kah Mun G, Hande MP, Valiyaveettil S. Cytotoxicity and genotoxicity of silver nanoparticles in human cells. *ACS Nano.* 2009; 3:279–90. https://doi.org/10.1021/nn800596w.

136. Liu P, Jin H, Guo Z, Ma J, Zhao J, Li D, et al. Silver nanoparticles outperform gold nanoparticles in radiosensitizing U251 cells in vitro and in an intracranial mouse model of glioma. *Int J Nanomedicine.* 2016. https://doi.org/10.2147/IJN.S115473.

137. Zheng Q, Yang H, Wei J, Tong J long, Shu Y qian. The role and mechanisms of nanoparticles to enhance radiosensitivity in hepatocellular cell. *Biomed Pharmacother.* 2013. https://doi.org/10.1016/j.biopha.2013.04.003.

138. Habiba K, Aziz K, Sanders K, Santiago CM, Mahadevan LSK, Makarov V, et al. Enhancing colorectal cancer radiation therapy efficacy using silver nanoprisms decorated with graphene as radiosensitizers. *Sci Rep.* 2019. https://doi.org/10.1038/s41598-019-53706-0.

139. Yu D, Zhang Y, Lu H, Zhao D. Silver nanoparticles coupled to anti-EGFR antibodies sensitize nasopharyngeal carcinoma cells to

irradiation. *Mol Med Rep*. 2017; 16:9005–10. https://doi.org/10.3892/mmr.2017.7704.

140. Caravan P, Ellison JJ, McMurry TJ, Lauffer RB. Gadolinium(III) chelates as MRI contrast agents: Structure, dynamics, and applications. *Chem Rev*. 1999. https://doi.org/10.1021/cr980440x.
141. Franken NAP, Bergs JWJ, Kok TT, Kuperus RRN, Stecher-Rasmussen F, Haveman J, et al. Gadolinium enhances the sensitivity of SW-1573 cells for thermal neutron irradiation. *Oncol Rep*. 2006. https://doi.org/10.3892/or.15.3.715.
142. Young SW, Qing F, Harriman A, Sessler JL, Dow WC, Mody TD, et al. Gadolinium (III) texaphyrin: A tumor selective radiation sensitizer that is detectable by MRI. *Proc Natl Acad Sci USA*. 1996. https://doi.org/10.1073/pnas.93.13.6610.
143. Lux F, Mignot A, Mowat P, Louis C, Dufort S, Bernhard C, et al. Ultrasmall rigid particles as multimodal probes for medical applications. *Angew Chemie Int Ed*. 2011; 50:12299–303. https://doi.org/10.1002/anie.201104104.
144. Tokumitsu H, Hiratsuka J, Sakurai Y, Kobayashi T, Ichikawa H, Fukumori Y. Gadolinium neutron-capture therapy using novel gadopentetic acid-chitosan complex nanoparticles: In vivo growth suppression of experimental melanoma solid tumor. *Cancer Lett*. 2000. https://doi.org/10.1016/S0304-3835(99)00388-2.
145. Sancey L, Lux F, Kotb S, Roux S, Dufort S, Bianchi A, et al. The use of theranostic gadolinium-based nanoprobes to improve radiotherapy efficacy. *Br J Radiol*. 2014. https://doi.org/10.1259/bjr.20140134.
146. Hu P, Fu Z, Liu G, Tan H, Xiao J, Shi H, et al. Gadolinium-based nanoparticles for theranostic mri-guided radiosensitization in hepatocellular carcinoma. *Front Bioeng Biotechnol*. 2019; 7. https://doi.org/10.3389/fbioe.2019.00368.
147. Le Duc G, Roux S, Paruta-Tuarez A, Dufort S, Brauer E, Marais A, et al. Advantages of gadolinium based ultrasmall nanoparticles vs molecular gadolinium chelates for radiotherapy guided by MRI for glioma treatment. *Cancer Nanotechnol*. 2014. https://doi.org/10.1186/s12645-014-0004-8.
148. Verry C, Sancey L, Dufort S, Le Duc G, Mendoza C, Lux F, et al. Treatment of multiple brain metastases using gadolinium nanoparticles and radiotherapy: NANO-RAD, a phase I study protocol. *BMJ Open*. 2019; 9:e023591. https://doi.org/10.1136/bmjopen-2018-023591.

149. Rajaee A, Wang S, Zhao L, Wang D, Liu Y, Wang J, et al. Multifunction bismuth gadolinium oxide nanoparticles as radiosensitizer in radiation therapy and imaging. *Phys Med Biol.* 2019. https://doi.org/10.1088/1361-6560/ab2154.

150. Wu C, Cai R, Zhao T, Wu L, Zhang L, Jin J, et al. Hyaluronic acid-functionalized gadolinium oxide nanoparticles for magnetic resonance imaging-guided radiotherapy of tumors. *Nanoscale Res Lett.* 2020. https://doi.org/10.1186/s11671-020-03318-9.

151. Pottier A, Borghi E, Levy L. Metals as radio-enhancers in oncology: The industry perspective. *Biochem Biophys Res Commun.* 2015. https://doi.org/10.1016/j.bbrc.2015.09.027.

152. Laurence M, Darmon A, Vivet S, Polrot M, Zhang P, Deutsch E, et al. Abstract 2665: NBTXR3 hafnium oxide nanoparticle activated by ionizing radiation demonstrates marked radio-enhancement and antitumor effect via high energy deposit in human soft tissue sarcoma. 2011. https://doi.org/10.1158/1538-7445.am2011-2665.

153. Marill J, Anesary NM, Zhang P, Vivet S, Borghi E, Levy L, et al. Hafnium oxide nanoparticles: Toward an in vitro predictive biological effect? *Radiat Oncol.* 2014; 9:150. https://doi.org/10.1186/1748-717X-9-150.

154. Galon J, Laé M, Thariat JO, Carrere S, Papai Z, Delannes M, et al. Hafnium oxide nanoparticle activated by radiation therapy generates an anti-tumor immune response. *Int J Radiat Oncol.* 2018. https://doi.org/10.1016/j.ijrobp.2018.07.109.

155. Bonvalot S, Rutkowski PL, Thariat J, Carrère S, Ducassou A, Sunyach M-P, et al. NBTXR3, a first-in-class radioenhancer hafnium oxide nanoparticle, plus radiotherapy versus radiotherapy alone in patients with locally advanced soft-tissue sarcoma (Act.In.Sarc): A multicentre, phase 2–3, randomised, controlled trial. *Lancet Oncol.* 2019; 20:1148–59. https://doi.org/10.1016/S1470-2045(19)30326-2.

156. Revia RA, Zhang M. Magnetite nanoparticles for cancer diagnosis, treatment, and treatment monitoring: Recent advances. *Mater Today.* 2016. https://doi.org/10.1016/j.mattod.2015.08.022.

157. Di Corato R, Béalle G, Kolosnjaj-Tabi J, Espinosa A, Clément O, Silva AKA, et al. Combining magnetic hyperthermia and photodynamic therapy for tumor ablation with photoresponsive magnetic liposomes. *ACS Nano.* 2015. https://doi.org/10.1021/nn506949t.

158. Espinosa A, Di Corato R, Kolosnjaj-Tabi J, Flaud P, Pellegrino T, Wilhelm C. Duality of iron oxide nanoparticles in cancer therapy: Amplification of heating efficiency by magnetic hyperthermia and photothermal bimodal treatment. *ACS Nano*. 2016. https://doi.org/10.1021/acsnano.5b07249.

159. Hauser AK, Mitov MI, Daley EF, McGarry RC, Anderson KW, Hilt JZ. Targeted iron oxide nanoparticles for the enhancement of radiation therapy. *Biomaterials*. 2016. https://doi.org/10.1016/j.biomaterials.2016.07.032.

160. Rezaei MR, Khoshgard K, Hosseinzadeh L, Haghparast A, Eivazi MT. Application of dextran-coated iron oxide nanoparticles in enhancing the radiosensitivity of cancerous cells in radiotherapy with high-energy electron beams. *J Cancer Res Ther*. 2019. https://doi.org/10.4103/jcrt.JCRT_19_17.

161. Shevtsov MA, Nikolaev BP, Ryzhov VA, Yakovleva LY, Marchenko YY, Parr MA, et al. Ionizing radiation improves glioma-specific targeting of superparamagnetic iron oxide nanoparticles conjugated with cmHsp70.1 monoclonal antibodies (SPION-cmHsp70.1). *Nanoscale*. 2015. https://doi.org/10.1039/c5nr06521f.

162. Jafari S, Cheki M, Tavakoli MB, Zarrabi A, Ghazikhanlu SK, Afzalipour R. Investigation of combination effect between 6 MV X-ray radiation and polyglycerol coated superparamagnetic iron oxide nanoparticles on U87-MG cancer cells. *J Biomed Phys Eng*. 2020. https://doi.org/10.31661/jbpe.v0i0.929.

163. Wang H-E, Yu H-M, Lu Y-C, Heish N-N, Tseng Y-L, Huang K-L, et al. Internal radiotherapy and dosimetric study for ^{111}In/^{177}Lu-pegylated liposomes conjugates in tumor-bearing mice. Nucl Instruments Methods Phys Res Sect A Accel Spectrometers. *Detect Assoc Equip*. 2006; 569:533–7. https://doi.org/10.1016/j.nima.2006.08.124.

164. Zhou M, Zhao J, Tian M, Song S, Zhang R, Gupta S, et al. Radio-photothermal therapy mediated by a single compartment nanoplatform depletes tumor initiating cells and reduces lung metastasis in the orthotopic 4T1 breast tumor model. *Nanoscale*. 2015. https://doi.org/10.1039/c5nr04587h.

165. Karve S, Werner ME, Sukumar R, Cummings ND, Copp JA, Wang EC, et al. Revival of the abandoned therapeutic wortmannin by nanoparticle drug delivery. *Proc Natl Acad Sci USA*. 2012. https://doi.org/10.1073/pnas.1120508109.

166. Zhang XD, Luo Z, Chen J, Song S, Yuan X, Shen X, et al. Ultrasmall glutathione-protected gold nanoclusters as next generation

radiotherapy sensitizers with high tumor uptake and high renal clearance. *Sci Rep.* 2015. https://doi.org/10.1038/srep08669.

167. Wen C, Zhou Y, Zhou C, Zhang Y, Hu X, Li J, et al. Enhanced radiosensitization effect of curcumin delivered by PVP-PCL nanoparticle in lung cancer. *J Nanomater.* 2017. https://doi.org/10.1155/2017/9625909.

168. Fathy MM, Mohamed FS, Elbialy NS, Elshemey WM. Multifunctional chitosan-capped gold nanoparticles for enhanced cancer chemo-radiotherapy: An invitro study. *Phys Medica.* 2018. https://doi.org/10.1016/j.ejmp.2018.04.002.

169. Dong Z, Feng L, Chao Y, Hao Y, Chen M, Gong F, et al. Amplification of tumor oxidative stresses with liposomal fenton catalyst and glutathione inhibitor for enhanced cancer chemotherapy and radiotherapy. *Nano Lett.* 2019. https://doi.org/10.1021/acs.nanolett.8b03905.

170. Wu M, Liu J, Hu C, Li D, Yang J, Wu Z, et al. Olaparib nanoparticles potentiated radiosensitization effects on lung cancer. *Int J Nanomedicine.* 2018. https://doi.org/10.2147/IJN.S181546.

171. Chatterjee DK, Wolfe T, Lee J, Brown AP, Singh PK, Bhattarai SR, et al. Convergence of nanotechnology with radiation therapy-insights and implications for clinical translation. *Transl Cancer Res.* 2013; 2:256–68. https://doi.org/10.3978/j.issn.2218-676X.2013.08.10.

172. Mi Y, Shao Z, Vang J, Kaidar-Person O, Wang AZ. Application of nanotechnology to cancer radiotherapy. *Cancer Nanotechnol.* 2016. https://doi.org/10.1186/s12645-016-0024-7.

173. Denkova AG, Liu II, Men Y, Eelkema R. Enhanced cancer therapy by combining radiation and chemical effects mediated by nanocarriers. *Adv Ther.* 2020; 3:1900177. https://doi.org/10.1002/adtp.201900177.

Chapter 12

Role of Nanoparticles in Cancer Immunotherapy

Madhu Malinee[a] **and Alok Kumar**[b,#]

[a]*Department of Anatomy and Developmental Biology, Graduate School of Medicine, Kyoto University, Kyoto 606-8501, Japan*
[b]*Department of Immunology and Genomic Medicine, Graduate School of Medicine, Kyoto University, Kyoto 606-8501, Japan*
[#]*Department of Immunology, UPMC Hilman Cancer Center, 5117, Center Avenue, University of Pittsburgh, Pittsburgh, PA 15232, USA*

alokkumar.iitr2011@gmail.com

12.1 Introduction

Cancer immunotherapy deploys immune cells against cancer rather than direct killing of cancer cells. Recent developments in cancer immunology provide opportunities to harness the benefit of tumor-specific immunotherapy that can eradicate the cancer cells without harming the healthy cells. Immune checkpoint blockade-based cancer immunotherapy made a paradigm shift in cancer treatment because of impressive success rate, less side effects, recurrence-free survival, and applicability to a wide range of tumors. However, a significant fraction of patients show unresponsiveness to single agent therapy for immunoinhibitory receptors. These limitations should be addressed to improve the efficacy further. Nanoparticles (NPs), owing to their special

Bionanotechnology in Cancer: Diagnosis and Therapy
Edited by D. Sakthi Kumar and Aswathy Ravindran Girija

ISBN 978-981-4800-30-3 (Hardcover), 978-0-429-42291-1 (eBook)
www.jennystanford.com

features (e.g., small size and large surface-to-volume ratio), present immense potential that can be used as adjuvant or complementation to improve the antitumor immunity.

12.1.1 Innate and Adaptive Immunity

Immune system is capable of identifying and destroying non-self-pathogenic organisms. However, excessive and inappropriate activation of immune cells cause autoimmune diseases. Immune homeostasis is the balance between immune response by immune activation (accelerator) and immune tolerance by immune suppression (brake). Immune system comprises two arms, namely, innate immunity and adaptive immunity. The immunity reacts non-specifically (via innate immune response) or specifically (via adaptive immune response). Innate immunity provides immediate protection, while adaptive immunity provides long-term protection from the antigens. Innate immunity is non-specific, while adaptive immunity is antigen-specific and generates memory cells as well during the course of immune response. Innate immune cells like NK cells, macrophages, and dendritic cells recognize some conserved patterns (pathogen-associated molecular patterns (PAMP)) with their pattern-recognition receptors (PRRs), toll-like receptors (TLRs), and others (like cytoplasmic NOD-like receptor, RIG-I-like receptor, DNA sensor, mannose receptor, and asialoglycoprotein receptor) [1, 2]. Further, the adaptive immune response is categorized into humoral-based (where antibodies are the real effector player against antigens) and cell-based immunity (where killer $CD8^+$ T cells are the real effector player). Classical antigen presenting cells (APCs) like macrophages and dendritic cells digest the antigen and show the peptide fragments of antigen on their major histocompatibility complex (MHC) to the T cells, which recognize them through their T cell receptor (TCR). T lymphocytes are different with innate counterparts in the sense that they have randomized rearrangement of extensive arrays of gene cassettes that generates a huge repertoire of antigen receptors (TCR). T cells have different subsets based on different surface markers. $CD8^+$ T cells are the cytotoxic T cells that kill the antigens after primed by APC for a particular antigen.

12.1.2 Cancer and Treatment Methods

Cancer, defined as uncontrolled growth of cells with potency of invading from their primary origin to other parts of body, is the second topmost (14%) causes of deaths worldwide after cardiovascular diseases (30%) among non-communicable diseases [3]. In other words, every seventh person is died because of cancer. According to the WHO report on cancer statistics, in 2018, around 18 million new cases of cancer were reported, while about 9 million patients lost their life due to cancer [4]. The hallmarks of cancers are sustaining proliferative signaling, evading growth suppressors, resisting cell death, enabling replicative immortality, inducing angiogenesis, and activating invasion and metastasis [5, 6].

Different treatment strategies were proposed to curb cancer, e.g., surgery, radiation, and chemotherapy. In 1882, for the first time, Dr. William Halsted performed mastectomy to treat breast cancer and surgical procedure was one of the standard treatment modalities for the cancer up to 1950. In 1895, X-rays was discovered by Wilhelm Conrad Rontgen and the importance of ionizing radiation came into picture for the cancer treatment in 1896, for the first time, when Dr. Emil Herman Grubbe used X-rays to treat breast cancer patient [7]. In the later developments, German chemist Paul Ehrlich for the first time used the term 'chemotherapy' for the use of chemicals to treat diseases. Further, in 1947, the first chemotherapy practice was done for the cancer treatment where the doctors used anti-folate drugs (methotrexate) for cancer treatment [8]. These three modalities of cancer treatment, namely, surgery, radiation therapy, and chemotherapy had been considered as the first line of treatment for more than 100 years. While the surgery is limited only for solid cancer, the chemotherapy and radiotherapy are associated with severe side effects and high risk of recurrence.

The cancer immunotherapy is the latest in this arsenal and is considered as the fourth pillar of cancer treatment, which relies on enhancing host's immune system to fight back the cancer cells rather than targeting cancer cells in contrast to abovesaid approaches for cancer treatment. In the cancer immunology, research has been focused to study the 'cancer cell- immune cell interaction' and the factors that affect the tumor-specific T cells activation, expansion, and effector functions for effective treatment of the cancer.

12.2 Tumor Immune Surveillance and Immunoediting

The concept of tumor immune surveillance, i.e., the ability of immune cells to detect and kill the nascent tumor cells, was first of all proposed independently by M.F. Burnet and L. Thomas in 1950s [9, 10]. The idea of cancer immunosurveillance laid the foundation of tumor immunology and cancer immunotherapy. In later years, many clinical and animal studies supported the hypothesis of tumor immunosurveillance. For example, immunodeficient animals show high rate of tumor incidence and this trend is seen in immunodeficient human patients as well. However, immunosurveillance hypothesis could not clarify why the tumor cells still grow despite the ability of immune cells to recognize and eliminate them. To explain this observation, a more inclusive and updated hypothesis was proposed by Dr. Robert Schrieber in 2002, which is popularly known as 'cancer immunoediting' [11, 12] (Figure 12.1). The cancer immunoediting hypothesis incorporates both the aspects of immune system, namely, immunoprotective and immunoselective role. The immune system not only protects the host, but can also shape the immunogenicity of tumor. Cancer immunoediting encompasses in three stages defined by the three "Es" namely 'Elimination (equivalent to immune surveillance)', 'Equilibrium', and 'Escape'. Cancer cells evade the immune responses by employing or pirating the different immune suppression mechanism that were used to prevent overactivation of immune cells during immune homeostasis.

12.3 Cancer Immunotherapy

Cancer immunotherapy deals with the treating cancer by modulating the body's own immune system rather than directly killing of tumor cells unlike other treatment methods. Cancer immunotherapy revert back the cancer development process to immunosurveillance stage (Figure 12.1). It employs enhancing the immune activation (accelerator) and/or releases the immune brakes.

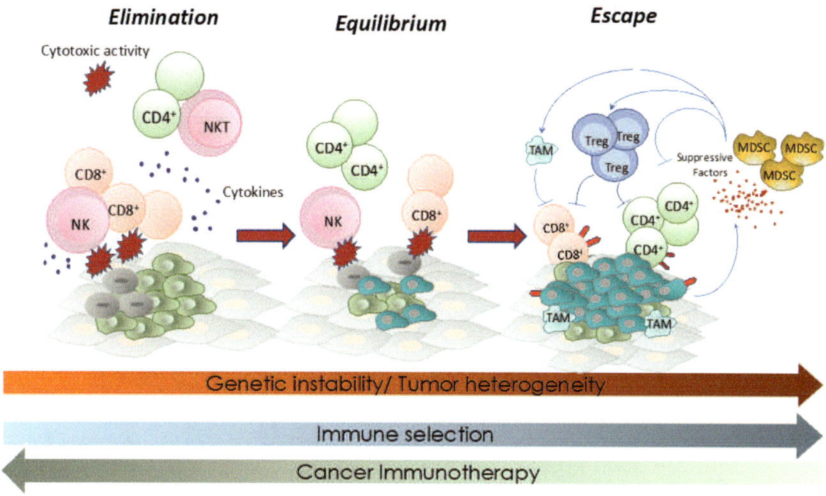

Figure 12.1 Cancer immunoediting. The three stages of cancer immunoediting: (i) Elimination, equivalent to immunosurveillance, (ii) equilibrium, where immune selection pressure promotes the generation of tumor cell variants with enhanced resistance to immune attack, and (iii) escape, where immune selected tumor cell variant grows uncontrollably. Immune selection, tumor heterogeneity, and genetic instability increase during the process of cancer immunoediting [11]. Cancer immunotherapy treatment reverses this process back to immunosurveillance stage [13–15].

Effective immune response needs (i) efficient delivery of tumor associated antigens (TAAs) and adjuvants to classical APCs, and (ii) efficient presentation of TAA-derived epitopes to cognate T cells (Figure 12.2). Multiple factors control the efficacy of immune responses like high-avidity tumor antigen specific T cells activation and expansion, memory T cells generation, and inhibition of T-cell intrinsic and immunosuppressive factors for T cells activation and function [16].

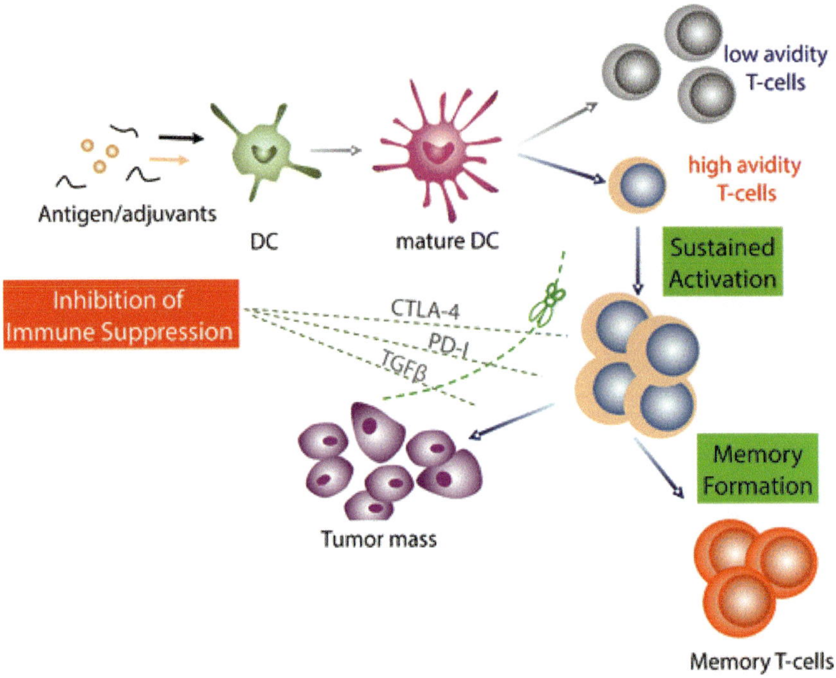

Figure 12.2 Antitumor immune response development: Effective immune response depends upon both factors, such as antigen presentation by APC to T cells and by tumor to T cells [16]. *Source*: Adapted from Ref. [16]. Available via license: CC BY-NC 3.0.

The fundamental strategy in cancer immunotherapy is to harness antitumor immune response. The immunosuppressive environment dampens the immune response and tumors escape the immune surveillance. The two possible strategies to counter the tumor growth are: (i) 'enhancement therapy' to enhance the immune responses nonspecifically, and (ii) release of immune brakes to normalize the immune responses, which are already dampened by the regulatory cells and immunosuppressive tumor microenvironment (TME). We have summarized the different strategies used under the umbrella of cancer immunotherapy for the cancer treatment in Figure 12.3. We will discuss these strategies in details in the below sections.

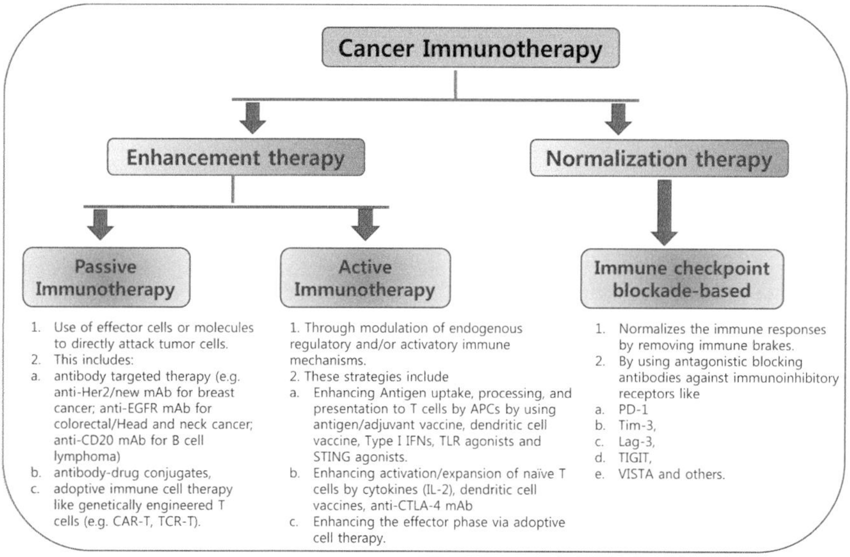

Figure 12.3 Cancer immunotherapy: Cancer immunotherapy broadly divided into enhancement therapy and normalization therapy. Enhancement therapy non-specifically enhances the immune responses (passively/actively) by using directly effector cells/molecule to target tumor cells or enhances the antigen presentation and effector phases of T cells. In normalization therapy, immune responses are being normalized by removing immune brakes using anti-PD-1/anti-CTLA-4/anti-Tim-3 monoclonal antibodies (mAb) and others [17].

12.3.1 Enhancement Immunotherapy (Tumor Nonspecific Immune Activation)

12.3.1.1 Passive immunotherapy

Since long time, it is believed that enhancing immune activation and functions have been important for a successful immune response. The enhancement strategy includes (i) passive immunotherapy and (ii) active immunotherapy. Here, we use effector cells/molecules to directly target tumor cells, e.g., antibody targeted therapy, CAR-T, and TCR-T, which come under this category (Figure 12.3).

Passive immunotherapy enhances the immune responses to higher levels and it sometimes causes even severe side effects.

The best examples of passive immunotherapy are the use of anti-Her2/neu mAb (for breast cancer), anti-EGFR mAb (for colorectal/ head & neck cancer), and anti-CD20 mAb (for B cell lymphoma). This includes genetically engineered T cells as well CAR-T and TCR-T.

12.3.1.2 Active immunotherapy

The active immunotherapy enhances the immune activation through the modulation of different endogenous regulatory and activation mechanisms by enhancing three steps of immune responses: (i) antigen uptake, processing, and presentation to T cells by APCs, (ii) activation and expansion of T cells, and (iii) exaggerate the effector phase of immune response (Figure 12.2).

Since long time, several reports have suggested the use of cytokines that promotes APC activity like IL-2, type I interferons (IFN-I). Agonists for TLR and STING enhance the antigen presentation activity by APCs. Antigen/adjuvant vaccine and dendritic cell vaccine also help in enhancing the antigen presentation.

Sometimes even after successful antigen presentation, there is not effective activation and expansion of T cells. Dendritic cell vaccine enhances the activation and expansion of T cells. Cytotoxic T lymphocyte antigen-4 (CTLA-4) is an early inhibitory receptor that gives negative signaling after binding with B7 molecules present in APCs or tumor cells. Treatment with anti-CTLA-4 mAb enhances the expansion of activated T cells.

Lastly, exaggerating the effector phase of the immune response through adoptive cell therapy also enhances the immune responses to fight back the cancer.

12.3.2 Normalization Immunotherapy (Tumor Specific Immune Activation)

Despite several attempts of enhancement therapy that enhances the immune responses systemically, the unfavorable response/ toxicity ratio made this approach unsafe for clinical use. The activation of immune responses systemically does not often correlate with better prognosis [18]. The discordance could be attributed to (i) local immunosuppression [19], and (ii) tumor immune escape mechanism for specific and non-specific immune responses against cancer [20]. The tumor immune escape mechanisms are mostly related with self-tolerance and this makes the enhancement therapy a very difficult task to achieve without immune-related adverse effects (irAEs).

Our immune system has developed many checkpoints for maintaining immune homeostasis. In 1990s, a great discovery of immune checkpoint in very late stages of T cell activation was witnessed. During the course of immune responses, immunoinhibitory receptors expressed on T cells hinder the process of elimination. CTLA-4 and programmed cell death protein 1 (PD-1) are two immune inhibitory receptors expressed on T cells, which were first discovered by Prof. James P. Allision and Prof. Tasuku Honjo in 1987 and 1992, respectively [21, 22]. APCs and cancer cells express ligands to these receptors, namely, B7-1/2 and PD-L1/2, which bind with the CTLA-4 and PD-1 on T cells, respectively. This interaction gives T cells negative signal, which results in inactivation of T cells [14]. Further, blocking the PD-1/PD-Ls interaction using antagonistic antibodies against the receptors or ligands releases T cells from inhibitory signaling and normalizes the immune response against the cancer cells.

Immunotherapy mediated by blockade of immunoinhibitory receptors programmed cell death 1 (PD-1) and CTLA-4 that reactivates T cells and clears the antigen/tumor load. Cancer immunotherapy surpassed the contemporary treatment methods in improved survival, efficacy, and applicability to a wide range of tumors irrespective of their origin and type. The beauty of cancer immunotherapy lies in the fact that it not only treats the primary cancer, but can also prevent metastasis and recurrence of the tumor post-treatment. Because of several advantages over other treatment methods, PD-1 blockade-based cancer immunotherapy has led to a paradigm shift in cancer treatment. In 2018, the most prestigious award 'Nobel Prize' in 'Medicine' for the discovery of PD-1 and CTLA-4 was awarded jointly to Prof. Tasuku Honjo and Prof. J.P. Allison [23]. Summary of the difference between immune checkpoint blockade-based cancer immunotherapy and other standard therapies is given in Table 12.1.

Table 12.1 Comparison of immune checkpoint blockade immunotherapy with other treatment methods

Particulars	Immune checkpoint blockade-based immunotherapy	Other methods (surgery, radiation, and chemotherapy)
Target for treatment	Immune cells	Cancer cells as well as other healthy cells
Agent for treatment	Antibody against inhibitory receptors (PD-1, CTLA-4) or their ligands (see Table 12.2 for details)	Surgery, radiation, chemicals
Applicability to all tumor types	Yes; Applicable to all tumors irrespective of their origin and type	Resectable tumors only (surgery)
Applicability to late stage cancers	Yes	No
Recurrence after cure	Less chance as memory T cells formed that can eliminate the similar cancer cell	Often recurred after treatment
Resistance developed	No	Yes (cancer cells develop resistance for chemotherapy drugs and radiations)
Efficacy	Most effective (40–50%) Overall survival significantly improved	Less effective
Side effects	Less side effects (manageable)	Severe side effects
FDA approval	Melanoma, Merkel-cell carcinoma, non-small cell lung cancer (NSCLC), small cell lung cancer, renal cell carcinoma (RCC), classical Hodgkin lymphoma, head and neck cancer, large B cell lymphoma, colorectal cancer, hepatocellular carcinoma, cervical cancer, gastric cancer, urothelial carcinoma, MSI-H and dMMR cancers, liver cancer	

MSI-H, Microsatellite instability-high; dMMR, DNA mismatch repair deficiency.

Table 12.2 US FDA approved mAbs for immune checkpoint blockade

Blockade	Drug (trade name)	Class	Pharma company	Cancer	Year
CTLA-4 blockade					
Anti-CTLA-4	Ipilumab (Yervoy)	Human IgG1	Medarex (later acquired by BMS)	Metastatic Melanoma	2011
PD-1 blockade					
Anti-PD-1 mAbs	Pembrolizumab (formerly MK-3475 or lambrolizumab, Keytruda)	Humanized IgG4	Merck	Melanoma, NSCLC, head and neck squamouse cell carcinoma (HNSCC)	2014
	Nivolumab (Opdivo)	Human IgG4	Bristol-Myers Squibb/Ono	Melanoma, squamous cell lung cancer, RCC, Hodgkin's lymphoma	2014
	Cemiplimab (Libtayo)	Human IgG4	Regeneron Pharma-ceuticals	Cutaneous squamous cell carcinoma (CSCC)	2018
Anti-PD-L1 mAbs	Atezolizumab (Tecentriq)	Humanized IgG1	Roche/ Genentech	Urothelial carcinoma, metastatic NSCLC	2016
	Durvalumab (Imfinzi)	Humanized IgG1	Astra Zeneca	Metastatic urothelial carcinoma, NSCLC after chemoradiation	2016
	Avelumab (Bavencio)	Humanized IgG1	Merck Serono and Pfizer	Metastatic Merkel-cell carcinoma	2017

These two important discoveries made a paradigm shift in the cancer treatment. The Food and Drug Administration (FDA), a federal agency of the United States Department of Health and Human Services, has approved antagonistic mAb against receptor PD-1, CTLA-4, or ligand PD-L1 for several types of cancers. Several recent reviews list the approved drugs and ongoing clinical trials of immune checkpoint blockade therapy for the cancer treatment. Summary of the US FDA approved drugs only for immune checkpoint blockade-based cancer therapy is given in Table 12.2. For more details, find the important discoveries and events in the time-scale of cancer immunotherapy (Figure 12.4).

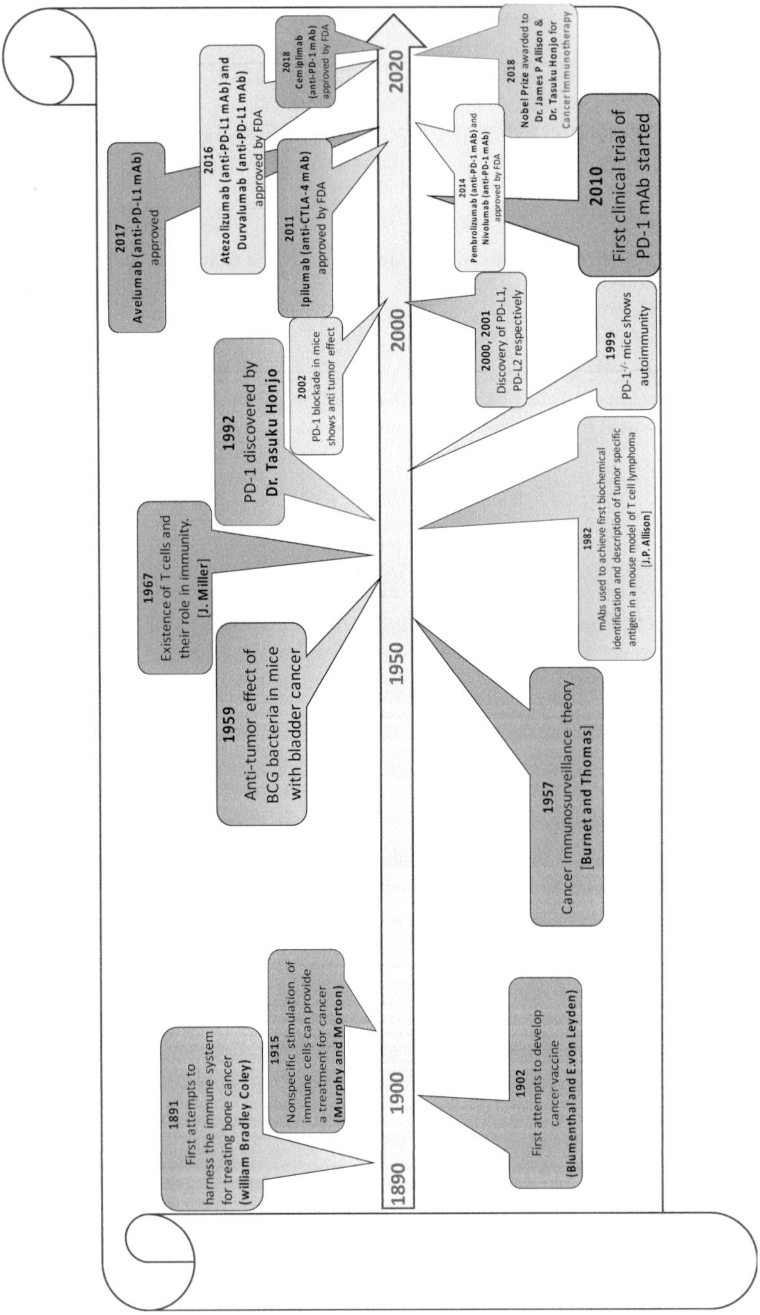

Figure 12.4 The history of cancer immunotherapy. Breakthrough discoveries to Nobel Prize (2018).

12.4 Immune Evasion Strategies by Tumor

Though tumor immunosurveillance hypothesis emphasizes the ability of immune cells to recognize and eliminate cancer cells; however, many patients develop cancer even in the presence of an apparently normal immune system. This indicates that tumor cells are able to escape immune surveillance. To avoid attack from the immune system, tumor cells develop different strategies to escape immune surveillance. Cancer immune evasion is a major limitation for developing an effective therapeutic approach to treat cancer. To have a successful anticancer immune response, the tumor-immunity cycle should be circulated uninterrupted [24]. Blockade at any step in the pathway leads to unresponsiveness. Cancer cells develop different strategies to interrupt the tumor-immunity cycle. In below sections, we review different escape mechanisms employed by cancer cells.

12.4.1 Downregulating MHC Class I Expression

Immune T cells recognize the antigens shown by MHC class I molecules present on the surface of cancer cells. Downregulating the antigen processing machinery is one of the fundamental mechanisms by which tumor cells evade immune responses. They downregulate major MHC I pathway, proteosome subunits latent membrane protein (LMP)2 and LMP7, transporter associated with antigen processing protein, and tapasin [25–27]. Kumar et al. have shown that some tumor lines do not respond to checkpoint blockade therapy, as they are deficient in MHC class I molecules. For examples, some sublines of B16 (murine melanoma cancer cell line) evade immune responses by downregulating MHC class I molecules from their surfaces [28]. The downregulation of antigen presentation machinery causes reduced tumor antigen that leads to enhanced tumor incidence and metastasis.

12.4.2 Developing Resistance to CTL-Mediated Killing Mechanisms

Tumor cells acquire apoptosis resistance mechanism as they progress and metastasize [29]. CTLs kill the cancer cells through

(i) perforin/granzyme pathway, and (ii) Fas pathway [30, 31]. Tumor cells evade from cytotoxic T lymphocytes (CTLs) lysis enhance the expression of anti-apoptotic genes Bcl2, and reduce the expression of death promoter BAX [32].

12.4.3 Turning Off Activated T-Cells via Direct Contact (Immune Inhibitory Pathways)

Tumor cells escape immune surveillance (attack) by utilizing a broad range of suppressive factors that are directly derived from the tumor and/or bystander tumor-related factors [33]. The direct factors include different ligands expressed on tumor cells surface that ligate with inhibitory receptors on immune effector $CD8^+$ T cells and inactivate them. Not only PD-L1, cancer cells develop ligands of other immune checkpoint receptors as well. For example, Galectin 9 (ligand of Tim-3), CD112, CD155 (ligand for necl-5), P-selectin glycoprotein ligand-1 (PSG-L1) ligand for selectin/VISTA, B7-1/B7-2 (ligand for CTLA-4), B7-H3 (ligand for TLT-2), and B7-H4 are expressed on the cancer cell surface [33, 34]. Summary of the different ligands that the cancer cells express to inhibit killer T cells by binding with their inhibitory receptors can be obtained from Table 12.3.

Table 12.3 Different ligands expressed by cancer cells to inhibit the T cells functions

Expressed ligand	Receptor on T cells	Reference
PD-L1 (CD274 or B7-H1)	PD-1 (CD279)	[35]
PD-L2 (CD273 or B7-DC)	PD-1 (CD279)	[36, 37]
Galectin 9	Tim-3	[38, 39]
CD112	CD112R, DNAM-1, TIGIT	[40]
CD155	DNAM-1, TIGIT, CD96	[40–42]
B7-1 (CD80), B7-2 (CD86)	CTLA-4 (CD152)	[22, 43]
B7-H3 (CD276)	TLT-2	[44, 45]
B7-H4 (VTCN1)		[44, 46]

DNAM-1, DNAX accessory molecule-1 (also known as CD226); VTCN1, V-set domain-containing T-cell activation inhibitor 1.

12.4.4 Releasing Soluble Factors to Inhibit Immune Cells (Paracrine Factors)

Cancer cells not only express the ligands of inhibitor receptors (expressed on T cells), but also release immunosuppressive cytokines and chemokines (e.g., IL-6, IL-10, TGF-β, PGE2, and others) that inhibit the activity of T cells indirectly [47]. In addition, they release metabolic products as well, which are inhibitory for the T cells functions [48–50].

In an unprecedented attempt to understand the mechanism of unresponsiveness, Kumar et al. devised a novel 'bilateral tumor model', where responsive and unresponsive tumors were inoculated on lateral sides of mouse, and immunosuppressive effects of unresponsive tumor on the immune response against responsive tumor were studied under PD-1 blockade treatment condition using anti-PD-L1 mAb. The finding was very intriguing in the sense that responsive tumor (MC38, a colon carcinoma) on right side was rejected by PD-1 blockade therapy when no tumor or B16 tumor was present on left side. On the contrary, MC38 was not rejected by PD-1 blockade therapy when LLC (a Lewis lung carcinoma) was present on the left side. These results indicate that LLC causes systemic immune suppression by releasing some soluble suppressive factors, while B16 does not. Further, it was shown that the secreted soluble factors from LLC were uncharacterized, non-proteinaceous, and small molecule (of size less than 3 kDa) [28].

12.4.5 Inhibiting T Cells through Bystander Effect

Unlike lymphomas, solid tumors make a close microenvironment that supports the tumor cells growth, while discourage the growth and function of immune effectors T cells. The TME consists of different cell populations and soluble factors.

Bystander factors indirectly suppress tumor-reactive T-cell activity. For examples, tumor cells produce cytokines and chemokines that induce regulatory T cells (Tregs), regulatory B cells (Bregs), myeloid-derived suppressive cells (MDSCs), tumor associated macrophages (TAMs), and some innate lymphoid cells (ILCs), which inhibit the effector $CD8^+$ T cells. In addition, cancer cells cause nutrient competition with T cells as well. In addition, cancer cells produce several metabolic products/byproducts (e.g., lactic acid, kynurenine, and adenosine) that inhibit the function of effector T cells [48–50].

Further, the chronic stimulation of T cells with the tumor antigen in the TME promotes the exhaustion states in T cells with enhanced expression of immunoinhibitory receptors like PD-1, Tim-3, Lag-3, and TIGIT.

In summary, tumor cells create a hostile environment for the T cells survival and functions.

12.5 Tumor Immunotherapy: Advantages, Drawbacks, and Need of Combination Approaches

The history of immunotherapy is almost 130 years old, when in 1991 for the first time William B Cooley used bacteria to harness the immune system to treat cancer through bystander antitumor inflammatory response. Further, about six decades later, Burnet and Thomas gave a hypothesis popularly known as 'Tumor Immunosurveillance' where they claimed that immune cells were capable of eliminating the cancer cells. This hypothesis was questionable up to very long time until 1990s when TAA were discovered. The discovery of TAA made a revolution in tumor-specific treatment based on peptide-, protein-, and cell-based vaccination therapies, which were capable of inducing pre-existing antitumor immunity. However, many classical vaccinations were unsuccessful because of inefficient antigen uptake, processing, and presentation of epitopes by APCs. Later, in late 1980s and early 1990s, CTLA-4 and PD-1, the two important negative regulators of T cells activation, were discovered and blocking these immunoinhibitory receptors reactivates T cells and eradicates tumors in animal model and human as well. The immunotherapy mediated by PD-1 or CTLA-4 blockade has made a paradigm shift in cancer treatment with improved overall survival and fewer side effects in contrast to contemporary methods of cancer treatments (surgery, radiation, and chemotherapy). However, almost 40–50% patients still show unresponsiveness to these immunotherapies. The following questions should be addressed in order to improve efficacy of cancer immunotherapy:

(i) How to identify predictive biomarker to recognize responsive and unresponsive cancer patients before start of therapy?
(ii) How to understand the mechanism of unresponsiveness by some patients?
(iii) What is the best combination therapy to rescue less responsive cancer patients?

12.6 Nano Immunotherapy

Small molecules (including aptamers, NPs, and others), owing to small size, cost-effective synthesis, and other several advantages over other large molecules (like antibody), are considered as the most suitable candidates for the diagnosis as well as therapeutic purposes in multidiscipline [51–53]. NPs have dimensions in the nano range (1–100 nm) and because of their small size, they have large surface-to-volume ratio that makes them important in many applications across different fields.

Nanoparticular immunotherapy represents novel approach for the cancer immunotherapy (Figure 12.5) [54].

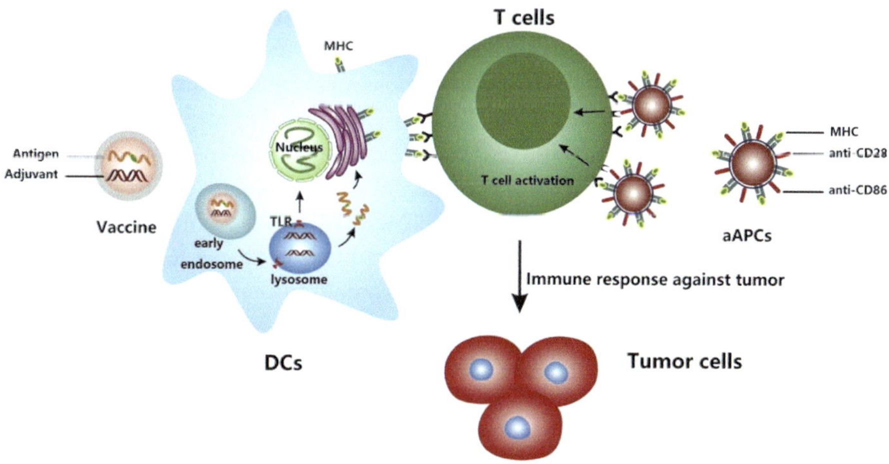

Figure 12.5 NP-based delivery system to activate classical APCs. *Source*: Adapted from Ref. [2] with permission from Elsevier.

NP has successfully been used as a vehicle for the delivery of drugs to the target tumor site in many chemotherapies [55]. NPs protect the cargo from degradation and so prolong the circulation in the blood and promote accumulation to the consumption site. These properties of NPs are beneficial for the delivery of tumor antigens/epitopes in vaccination strategies for cancer immunotherapy.

12.6.1 Delivery of TAA and Adjuvants to APCs

For a successful immune response, effective antigen presentation to the T cells is indispensable. NPs have shown the potential of delivering tumor antigen/epitopes to the classical APCs for better antigen presentations to elicit antitumor immune response during the vaccination.

Liposomes, polypropylene sulfide, gold nanoparticles (GNPs), PLGA NPs, and micelles have shown tremendous success in the delivery of TAA to the APCs. Summary of some recent important research works, which have shown their success story in the in vitro as well as in vivo mice models, can be obtained from Table 12.4.

Liposome-based nanocarriers have shown immense potential for delivering peptides to APCs. pH-sensitive liposomes are one of the good choices as they will be released into the cytosol by destabilizing or fusing with lysosomal membrane. Hyper-branched polyglycidol derivative liposomes confer escape from lysosomal degradation [69]. Adding adjuvant (pharmacological or immunological agent that improves the immune response of a vaccine) also along with antigen shows enhanced immune responses. Miyabe et al. have reported YSK05, a synthetic pH-sensitive liposome with optimal functionality at pH 6.5, as a suitable carrier (c-di-GMP/YSK05 liposomes) for cytosol transport of a cyclic dinucleotide cyclic-di-GMP (an adjuvant) to stimulate innate immune responses via STING-TBK1-IRF3 pathway [70]. However, despite several advantages, liposomes are not very suitable as they release the antigens quickly. Poly(lactic-co-glycolic acid) (PLGA) has advantage over liposome, as it has sustained release of antigens resulting in enhanced immune responses compared to fast release liposome [71]. Not only PLGA,

Table 12.4 Use of NPs-based delivery system in vaccination immunotherapy

	NPs	Composition	Payload	Tumor model	Reference
Delivery of TAAs to APCs	Polystyrene microspheres	$-NH_2$ group	Ovalbumin (Ova)	Ova-expressing melanoma	[56]
	Poly-gamma glutamic acid	–COOH group	Listerolysin-coated Ova		[57]
	Iron oxide–zinc oxide	Zinc oxide	Carcinoembryonic antigen (CEA)	MC38 tumor	[58]
	Liposomes		Ova	EG7.Ova	[59]
	PLGA		Poly I:C encapsulated Ova	Ova-expressing melanoma	[60]
	Artificial Exosomes micelles		hTERT-HSP70	S180/H22	[61]
Artificial APCs	Liposomes		MHC II peptide		[62]
	PLGA NPs		Anti-CD3, anti-CD28 and IL-2		[63]
	Nanoworms		Anti-CD3		[64]
	Artificial exosomes		MHC-I peptide		[65]
Direct activation of TAA-specific T cells	Microbeads		HLA-Ig and anti-CD28		[66]
	Iron-dextran particles	Dextran	MHC-Ig dimer and anti-CD28	Melanoma	[67]
	Polystyrene latex beads	Sulfate	Anti-His6 mAb, anti-CD28 and 4-1BB mAb	Melanoma	[68]

but poly(propylene) sulfide (PPS) NPs and GNPs as well show better antigen delivery to APCs [72, 73]. GNPs show low cytotoxicity, easily controllable size and shape, and designer surface chemistry, which make them a good choice for antigen delivery.

Since it is very difficult to identify tumor-specific antigens, some research groups have shown the inactivated tumor cells lysates as antigen to present to APCs for the induction of polyvalent immune response. Prasad et al. have shown the effectiveness of PLGA NPs incorporated with head and neck carcinoma cell lysates for better cancer immunotherapy [74]. Further, exosome, small 40–120 nm vesicles released from mammalian cells, have been shown as potential therapeutic to induce immune responses [75, 76].

12.6.2 Role of Artificial APCs

Though vaccination using nanovaccines approach has shown higher immunogenicity compared to traditional vaccines, it is still not satisfactory for an effective and complete tumor clearance. The reasons could be less efficient antigen delivery and some in-built inhibitory mechanism by natural DCs. To cope with this issue, it is required to develop artificial APCs (Figure 12.6). Artificial APCs provides two signals: (i) antigenic peptide to TCR, and (ii) costimulatory molecule or ligand to costimulatory receptor on T cells [77]. Perica et al. have synthesized artificial APCs by direct chemical coupling of soluble MHC-Ig dimer (signal 1) and B7.1-Ig (signal 2) in equimolar ratio to the surface of iron-oxide and dextran-coated particle (Figure 12.6) [77]. They have further shown that vaccination using cognate artificial APC and T cell has better immune response than T cell alone in vivo.

Some research groups have even provided third signal (some cytokines) necessary for T cell activation and memory development. In one such attempt, Steenblock et al. developed PLGA microparticles with adaptor elements for attaching CD3 and CD28, and a core for encapsulating IL-2. They have shown the sustained release of IL-2 in the T cell milieu that significantly improved T cell expansion [63].

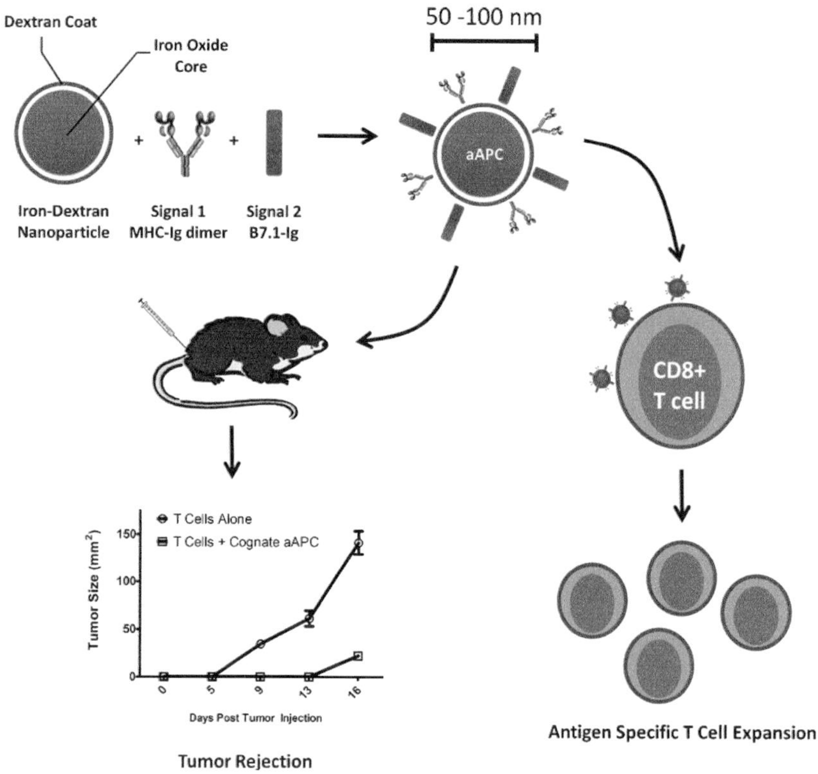

Figure 12.6 Artificial APC. Perica et al. have generated artificial APC by direct chemical coupling of soluble MHC-Ig dimer and B7.1-Ig to the surface of a paramagnetic iron-oxide and dextran-coated particle [77]. *Source*: Adapted from Ref. [77] with permission from Elsevier.

12.6.3 Direct Activation of TAA-Specific T Cells

In previous sections, we have seen the importance of NP-based delivery of antigens to APCs that present the antigens to T cells and induce immune response. However, the activated T cells express coinhibitory receptor CTLA-4 and PD-1, which may ligate with B7-1 and PD-L1, respectively, expressed on APC, and this interaction downregulates the T cells activity and immune responses. To cope with this issue, NP-based approaches can be developed that can directly trigger the activation of antigen-specific T-cells. In this process, the need of intermediate cells can be bypassed (Figure 12.7).

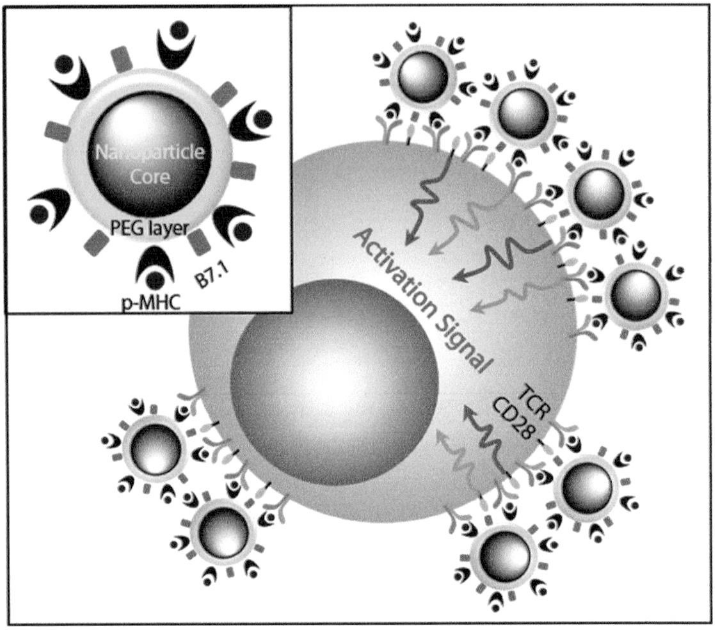

Figure 12.7 Direct activation of TAA-specific T cells. Direct T-cell-triggering micro/NP structures displaying TAA-relevant pMHC complexes and T-cell co-stimulatory molecules (B7.1) have the capacity to trigger T-cell activation. *Source*: Adapted from Ref. [16]. Available via license: CC BY-NC 3.0.

12.6.4 Role of NPs in Targeting Immunosuppressive TME

The antigen-primed T cells move to the periphery and reach to the target site (tumor). In the tumor site, T cells recognize the cancer cells with the peptide loaded on MHC of cancer cells and eliminate them by releasing cytotoxic granule granzyme or by other pathways. However, the TME is very hostile for the T cells functions. Many immune regulatory cells like TAM, MDSC, and Treg are present that inhibit the effector T cells. Presence of immunoinhibitory cells in the TME is one of the factors that limit the efficacy of cancer immunotherapy, including immune checkpoint blockade (PD-1 and CTLA-4).

MDSCs inhibit T cells-based immune responses. Lee et al. have shown the use of baccatin III (a precursor of paclitaxel) in reducing the infiltration of MDSC at tumor site [78]. Jeanbart et al. conjugated 6-thioguanine to polyethylene glycol(PEG)-PPS polymer micelles that efficiently depleted MDSCs and finally resulted in reduced tumor burden [79]. In another exciting report, Wang et al. have developed poly(amidoamine) dendrimers for 'all-trans retinoic acids (ATRA)' delivery that enhances the myeloid cells differentiation into mature DCs, macrophages, and granulocytes [80, 81].

Other major regulatory cells that inhibit the immune responses are Treg cells. Though Tregs are important to counter autoimmunity, the tumor residing Tregs are the major hurdles for effective antitumor immune responses. Depleting intratumoral Tregs without harming other useful Tregs could be one of the possibilities to enhance antitumor immunity. Intratumoral Tregs show higher expression of glucocorticoid-induced TNFR-related receptor (GITR). Sacchetti et al. designed PEG modified single-walled carbon nanotubes (PEG-SWCNTs) conjugated with anti-GITR mAb for targeting intratumoral Tregs [82]. TAMs, another subset of immunosuppressive cells that resides in TME and secretes immunosuppressive cytokine IL-10 that maintains the M2 phenotype in an autocrine manner. Some research groups have developed composite NPs for countering the infiltration of TAMs in the TME that ultimately enhance antitumor immunity [83, 84]. Summary of the use of NPs in targeting immunosuppressive cells in TME in cancer immunotherapy can be obtained from Table 12.5.

12.7 Conclusion and Future Perspectives

The 20th century witnessed a huge progress in the development of different treatment approaches to treat cancer. Cancer immunotherapy, the fourth pillar of cancer treatment, surpassed all the previous treatment modalities of cancer treatment because of several advantages over other treatment methods (surgery, radiation, and chemotherapy). Cancer immunotherapy has now become a standard modality for the cancer treatment and shown the potential to be the future of cancer treatment.

Table 12.5 Use of NPs in targeting immunosuppressive cells in the TME in cancer immunotherapy

	NPs	Payload	Tumor model	Refs
Targeting TAM	Polyion micelles	Poy(L-arginine) which	CT26 colon carcinoma tumor	[83]
	Polysaccharide-alendronate conjugation macrophage	Alendronate	S180	[85]
	PEG detachable active targeting NPs	CpG ODN anti-IL-10 ODN, and anti-IL-10 receptor ODN	Hepatome model	[84]
Targeting MDSC	PEG-PPS micelles	6-Thioguanine	E.G7-OVA thymoma cells and B16-F10 melanoma cells	[79]
	Liposome	Baccation III	4T1 breast carcinoma cells	[86]
Targeting Tregs	Poly(amidoamine) dendrimers	All-trans retinoic acids		[80]
	Ligand conjugated to PEGmodified single-walled CNT		B16 tumor model	[82]

Because of the great success of cancer immunotherapy (especially PD-1 blockade based), now it is believed that PD-1 is cancer equivalent of penicillin [87]. It is understandable that penicillin could not cure all infections; however, it prompted the discovery of many antibiotics that changed the treatment of infectious disease forever. In a similar way, though PD-1 blockade therapy alone could not cure cancer in all patients; however, it made a revolution in cancer treatment and in near future, hopefully, cancer would not be a serious illness as we have a solution like PD-1 blockade therapy. NPs, because of their special features, are very promising candidate to be used for therapeutic purposes. NPs can be used as delivery vehicles for peptides, as artificial APCs as well. It can be modulated to activate the antigen-specific T cells directly bypassing the intermediary cells. Moreover, NPs have been designed to target immune regulatory cells as well like MDSCs, Tregs, TAMs, etc. It is worth to perform combination therapy of PD-1 blockade with NP-based approaches for targeting TME residing immune regulatory cells to get better immune responses. PD-1 blockade therapy in combination with NPs for specific purposes gives us a ray of hope for better cancer treatment and improved longevity of cancer patients.

References

1. Kapadia CH, Perry JL, Tian S, Luft JC, DeSimone JM. Nanoparticulate immunotherapy for cancer. *Journal of Controlled Release*. 2015; 219:167–80.
2. Zang X, Zhao X, Hu H, Qiao M, Deng Y, Chen D. Nanoparticles for tumor immunotherapy. *Eur J Pharm Biopharm*. 2017; 115:243–56.
3. Nugent RA, Jamison DT. What can a UN health summit do? *Sci Transl Med*. 2011; 3(100):100–25.
4. Kumar A. Mitochondrial manipulation improves anti-tumor immunity. *Acta Scientific Cancer Biology*. 2019; 3(12):28–30.
5. Hanahan D, Weinberg RA. The hallmarks of cancer. *Cell*. 2000; 100(1):57–70.
6. Hanahan D, Weinberg Robert A. Hallmarks of Cancer: The next generation. *Cell*. 2011; 144(5):646–74.
7. Gianfaldoni S, Gianfaldoni R, Wollina U, Lotti J, Tchernev G, Lotti T. An Overview on radiotherapy: From its history to its current applications in dermatology. *Open Access Maced J Med Sci*. 2017; 5(4):521–5.
8. DeVita VT, Chu E. A history of cancer chemotherapy. *Cancer Research*. 2008; 68(21):8643.
9. Burnet M. Cancer: A biological approach. I. The processes of control. *Br Med J*. 1957; 1(5022):779–86.
10. Thomas L. Cellular and humoral aspects of the hypersensitive states. Lawrence H.S. (ed.), Hoeber-Harper, New York. 1959; 529–32.
11. Dunn GP, Bruce AT, Ikeda H, Old LJ, Schreiber RD. Cancer immunoediting: From immunosurveillance to tumor escape. *Nature Immunology*. 2002; 3(11):991–8.
12. Dunn GP, Old LJ, Schreiber RD. The immunobiology of cancer immunosurveillance and immunoediting. *Immunity*. 2004; 21(2):137–48.
13. Brahmer JR, Drake CG, Wollner I, Powderly JD, Picus J, Sharfman WH, et al. Phase I study of single-agent anti-programmed death-1 (MDX-1106) in refractory solid tumors: Safety, clinical activity,

pharmacodynamics, and immunologic correlates. *J Clin Oncol.* 2010; 28(19):3167–75.

14. Zou W, Wolchok JD, Chen L. PD-L1 (B7-H1) and PD-1 pathway blockade for cancer therapy: Mechanisms, response biomarkers, and combinations. *Sci Transl Med.* 2016; 8(328):328rv4.
15. Rotte A. Combination of CTLA-4 and PD-1 blockers for treatment of cancer. *Journal of Experimental & Clinical Cancer Research.* 2019; 38(1):255.
16. Shao K, Singha S, Clemente-Casares X, Tsai S, Yang Y, Santamaria P. Nanoparticle-based immunotherapy for cancer. *ACS Nano.* 2015; 9(1):16–30.
17. Sanmamed MF, Chen L. A paradigm shift in cancer immunotherapy: From enhancement to normalization. *Cell.* 2018; 175(2):313–26.
18. Rosenberg SA, Sherry RM, Morton KE, Scharfman WJ, Yang JC, Topalian SL, et al. Tumor progression can occur despite the induction of very high levels of self/tumor antigen-specific $CD8^+$ T cells in patients with melanoma. *The Journal of Immunology.* 2005; 175(9):6169.
19. Taube JM, Anders RA, Young GD, Xu H, Sharma R, McMiller TL, et al. Colocalization of inflammatory response with B7-H1 expression in human melanocytic lesions supports an adaptive resistance mechanism of immune escape. *Sci Transl Med.* 2012; 4(127):127ra37.
20. Vinay DS, Ryan EP, Pawelec G, Talib WH, Stagg J, Elkord E, et al. Immune evasion in cancer: Mechanistic basis and therapeutic strategies. *Seminars in Cancer Biology.* 2015; 35:S185–S98.
21. Ishida Y, Agata Y, Shibahara K, Honjo T. Induced expression of PD-1, a novel member of the immunoglobulin gene superfamily, upon programmed cell death. *Embo j.* 1992; 11(11):3887–95.
22. Brunet J-F, Denizot F, Luciani M-F, Roux-Dosseto M, Suzan M, Mattei M-G, et al. A new member of the immunoglobulin superfamily—CTLA-4. *Nature.* 1987; 328(6127):267–70.
23. Smyth MJ, Teng MW. 2018 Nobel Prize in physiology or medicine. *Clin Transl Immunology.* 2018; 7(10):e1041.
24. Chen DS, Mellman I. Oncology meets immunology: The cancer-immunity cycle. *Immunity.* 2013; 39(1):1–10.
25. Hicklin DJ, Marincola FM, Ferrone S. HLA class I antigen down-regulation in human cancers: T-cell immunotherapy revives an old story. *Mol Med Today.* 1999; 5(4):178–86.

26. Johnsen AK, Templeton DJ, Sy M, Harding CV. Deficiency of transporter for antigen presentation (TAP) in tumor cells allows evasion of immune surveillance and increases tumorigenesis. *J Immunol.* 1999; 163(8):4224–31.
27. Restifo NP, Esquivel F, Kawakami Y, Yewdell JW, Mulé JJ, Rosenberg SA, et al. Identification of human cancers deficient in antigen processing. *J Exp Med.* 1993; 177(2):265–72.
28. Kumar A, Chamoto K, Chowdhury PS, Honjo T. Tumors attenuating the mitochondrial activity in T cells escape from PD-1 blockade therapy. *Elife.* 2020; 9.
29. Lee H-M, Timme TL, Thompson TC. Resistance to lysis by cytotoxic T cells: A dominant effect in metastatic mouse prostate cancer cells. *Cancer Research.* 2000; 60(7):1927.
30. Shresta S, Pham CT, Thomas DA, Graubert TA, Ley TJ. How do cytotoxic lymphocytes kill their targets? *Curr Opin Immunol.* 1998; 10(5):581–7.
31. Rokhlin OW, Hostager BS, Bishop GA, Sidorenko SP, Glover RA, Gudkov AV, et al. Dominant nature of the resistance to Fas- and tumor necrosis factor-alpha-mediated apoptosis in human prostatic carcinoma cell lines. *Cancer Res.* 1997; 57(18):3941–3.
32. McConkey DJ, Greene G, Pettaway CA. Apoptosis resistance increases with metastatic potential in cells of the human LNCaP prostate carcinoma line. *Cancer Res.* 1996; 56(24):5594–9.
33. Kumar A, Chamoto K. Immune metabolism in PD-1 blockade-based cancer immunotherapy. *International Immunology.* 2020.
34. Andrews LP, Yano H, Vignali DAA. Inhibitory receptors and ligands beyond PD-1, PD-L1 and CTLA-4: Breakthroughs or backups. *Nat Immunol.* 2019; 20(11):1425–34.
35. Iwai Y, Ishida M, Tanaka Y, Okazaki T, Honjo T, Minato N. Involvement of PD-L1 on tumor cells in the escape from host immune system and tumor immunotherapy by PD-L1 blockade. *Proc Natl Acad Sci USA.* 2002; 99(19):12293–7.
36. Latchman Y, Wood CR, Chernova T, Chaudhary D, Borde M, Chernova I, et al. PD-L2 is a second ligand for PD-1 and inhibits T cell activation. *Nature Immunology.* 2001; 2(3):261–8.
37. Rozali EN, Hato SV, Robinson BW, Lake RA, Lesterhuis WJ. Programmed death ligand 2 in cancer-induced immune suppression. *Clinical and Developmental Immunology.* 2012; 2012:656340.

38. Li H, Wu K, Tao K, Chen L, Zheng Q, Lu X, et al. Tim-3/galectin-9 signaling pathway mediates T-cell dysfunction and predicts poor prognosis in patients with hepatitis B virus-associated hepatocellular carcinoma. *Hepatology*. 2012; 56(4):1342–51.
39. Chou F-C, Chen H-Y, Kuo C-C, Sytwu H-K. Role of galectins in tumors and in clinical immunotherapy. *Int J Mol Sci*. 2018; 19(2):430.
40. Nagumo Y, Iguchi-Manaka A, Yamashita-Kanemaru Y, Abe F, Bernhardt G, Shibuya A, et al. Increased CD112 expression in methylcholanthrene-induced tumors in CD155-deficient mice. *Plos One*. 2014; 9(11):e112415.
41. Molfetta R, Zitti B, Lecce M, Milito ND, Stabile H, Fionda C, et al. CD155: A multi-functional molecule in tumor progression. *Int J Mol Sci*. 2020; 21(3):922.
42. He W, Zhang H, Han F, Chen X, Lin R, Wang W, et al. CD155T/TIGIT signaling regulates CD8⁺ T-cell metabolism and promotes tumor progression in human gastric cancer. *Cancer Research*. 2017; 77(22):6375.
43. Fallarino F, Fields PE, Gajewski TF. B7-H1 engagement of cytotoxic T lymphocyte antigen 4 inhibits T cell activation in the absence of CD28. *The Journal of Experimental Medicine*. 1998; 188(1):205–10.
44. Sun Y, Wang Y, Zhao J, Gu M, Giscombe R, Lefvert AK, et al. B7-H3 and B7-H4 expression in non-small-cell lung cancer. *Lung Cancer*. 2006; 53(2):143–51.
45. Yonesaka K, Haratani K, Takamura S, Sakai H, Kato R, Takegawa N, et al. B7-H3 negatively modulates CTL-mediated cancer immunity. *Clinical Cancer Research*. 2018; 24(11):2653.
46. Sica GL, Choi I-H, Zhu G, Tamada K, Wang S-D, Tamura H, et al. B7-H4, a molecule of the B7 family, negatively regulates t cell immunity. *Immunity*. 2003; 18(6):849–61.
47. Kusmartsev S, Gabrilovich DI. Effect of tumor-derived cytokines and growth factors on differentiation and immune suppressive features of myeloid cells in cancer. *Cancer Metastasis Rev*. 2006; 25(3):323–31.
48. Brand A, Singer K, Koehl GE, Kolitzus M, Schoenhammer G, Thiel A, et al. LDHA-associated lactic acid production blunts tumor immunosurveillance by T and NK cells. *Cell Metab*. 2016; 24(5):657–71.

49. Katz JB, Muller AJ, Prendergast GC. Indoleamine 2,3-dioxygenase in T-cell tolerance and tumoral immune escape. *Immunol Rev.* 2008; 222:206–21.

50. Ohta A. A metabolic immune checkpoint: Adenosine in tumor microenvironment. *Frontiers in immunology.* 2016; 7:109.

51. Bansal M, Kumar A, Malinee M, Sharma TK. Nanomedicine: Diagnosis, treatment, and potential prospects. In: Daima HK, Pn N, Ranjan S, Dasgupta N, Lichtfouse E (eds.). Nanoscience in Medicine Vol 1. Cham: Springer International Publishing. 2020; 297–331.

52. Malinee M, Kumar A, Dhiman A, Sharma TK. Chapter 8: Aptamer-mediated nanobiosensing for health monitoring. In: Inamuddin, Khan R, Mohammad A, Asiri AM (eds.). Advanced Biosensors for Health Care Applications: Elsevier. 2019; 227–48.

53. Kumar A, Malinee M, Dhiman A, Kumar A, Sharma TK. Chapter 2: Aptamer technology for the detection of foodborne pathogens and toxins. In: Inamuddin, Khan R, Mohammad A, Asiri AM (eds.). Advanced Biosensors for Health Care Applications: Elsevier. 2019; 45–69.

54. Smith DM, Simon JK, Baker Jr JR. Applications of nanotechnology for immunology. *Nature Reviews Immunology.* 2013; 13(8):592–605.

55. Cho K, Wang X, Nie S, Chen ZG, Shin DM. Therapeutic nanoparticles for drug delivery in cancer. *Clin Cancer Res.* 2008; 14(5):1310–6.

56. Falo LD, Kovacsovics-Bankowski M, Thompson K, Rock KL. Targeting antigen into the phagocytic pathway in vivo induces protective tumour immunity. *Nature Medicine.* 1995; 1(7):649–53.

57. Uto T, Wang X, Sato K, Haraguchi M, Akagi T, Akashi M, et al. Targeting of antigen to dendritic cells with poly(γ-glutamic acid) nanoparticles induces antigen-specific humoral and cellular immunity. *The Journal of Immunology.* 2007; 178(5):2979.

58. Cho N-H, Cheong T-C, Min JH, Wu JH, Lee SJ, Kim D, et al. A multifunctional core–shell nanoparticle for dendritic cell-based cancer immunotherapy. *Nature Nanotechnology.* 2011; 6(10):675–82.

59. Yoshizaki Y, Yuba E, Komatsu T, Udaka K, Harada A, Kono K. Improvement of peptide-based tumor immunotherapy using pH-sensitive fusogenic polymer-modified liposomes. *Molecules.* 2016; 21(10):1284.

60. Rosalia RA, Cruz LJ, van Duikeren S, Tromp AT, Silva AL, Jiskoot W, et al. CD40-targeted dendritic cell delivery of PLGA-nanoparticle

vaccines induce potent anti-tumor responses. *Biomaterials*. 2015; 40:88–97.

61. Li K, Chang S, Wang Z, Zhao X, Chen D. A novel micro-emulsion and micelle assembling method to prepare DEC205 monoclonal antibody coupled cationic nanoliposomes for simulating exosomes to target dendritic cells. *International Journal of Pharmaceutics*. 2015; 491(1):105–12.
62. Giannoni F, Barnett J, Bi K, Samodal R, Lanza P, Marchese P, et al. Clustering of T cell ligands on artificial APC membranes influences T cell activation and protein kinase C θ translocation to the T cell plasma membrane. *The Journal of Immunology*. 2005; 174(6):3204.
63. Steenblock ER, Fahmy TM. A comprehensive platform for ex vivo T-cell expansion based on biodegradable polymeric artificial antigen-presenting cells. *Molecular Therapy*. 2008; 16(4):765–72.
64. Mandal S, Eksteen-Akeroyd ZH, Jacobs MJ, Hammink R, Koepf M, Lambeck AJA, et al. Therapeutic nanoworms: Towards novel synthetic dendritic cells for immunotherapy. *Chemical Science*. 2013; 4(11):4168–74.
65. De La Peña H, Madrigal JA, Rusakiewicz S, Bencsik M, Cave GWV, Selman A, et al. Artificial exosomes as tools for basic and clinical immunology. *Journal of Immunological Methods*. 2009; 344(2):121–32.
66. Perica K, De León Medero A, Durai M, Chiu YL, Bieler JG, Sibener L, et al. Nanoscale artificial antigen presenting cells for T cell immunotherapy. *Nanomedicine*. 2014; 10(1):119–29.
67. Kim J, Mooney DJ. In vivo modulation of dendritic cells by engineered materials: Towards new cancer vaccines. *Nano Today*. 2011; 6(5):466–77.
68. Manolova V, Flace A, Bauer M, Schwarz K, Saudan P, Bachmann MF. Nanoparticles target distinct dendritic cell populations according to their size. *Eur J Immunol*. 2008; 38(5):1404–13.
69. Yuba E, Harada A, Sakanishi Y, Kono K. Carboxylated hyperbranched poly(glycidol)s for preparation of pH-sensitive liposomes. *Journal of Controlled Release*. 2011; 149(1):72–80.
70. Miyabe H, Hyodo M, Nakamura T, Sato Y, Hayakawa Y, Harashima H. A new adjuvant delivery system 'cyclic di-GMP/YSK05 liposome' for cancer immunotherapy. *Journal of Controlled Release*. 2014; 184:20–7.

71. Irvine DJ, Hanson MC, Rakhra K, Tokatlian T. Synthetic nanoparticles for vaccines and immunotherapy. *Chemical Reviews.* 2015; 115(19):11109–46.

72. Zhou Q, Zhang Y, Du J, Li Y, Zhou Y, Fu Q, et al. Different-sized gold nanoparticle activator/antigen increases dendritic cells accumulation in liver-draining lymph nodes and CD8+ T cell responses. *ACS Nano.* 2016; 10(2):2678–92.

73. Niikura K, Matsunaga T, Suzuki T, Kobayashi S, Yamaguchi H, Orba Y, et al. Gold nanoparticles as a vaccine platform: Influence of size and shape on immunological responses in vitro and in vivo. *ACS Nano.* 2013; 7(5):3926–38.

74. Prasad S, Cody V, Saucier-Sawyer JK, Saltzman WM, Sasaki CT, Edelson RL, et al. Polymer nanoparticles containing tumor lysates as antigen delivery vehicles for dendritic cell-based antitumor immunotherapy. *Nanomedicine: Nanotechnology, Biology and Medicine.* 2011; 7(1):1–10.

75. Bell BM, Kirk ID, Hiltbrunner S, Gabrielsson S, Bultema JJ. Designer exosomes as next-generation cancer immunotherapy. *Nanomedicine: Nanotechnology, Biology and Medicine.* 2016; 12(1):163–9.

76. Batrakova EV, Kim MS. Using exosomes, naturally-equipped nanocarriers, for drug delivery. *Journal of Controlled Release.* 2015; 219:396–405.

77. Perica K, De León Medero A, Durai M, Chiu YL, Bieler JG, Sibener L, et al. Nanoscale artificial antigen presenting cells for T cell immunotherapy. *Nanomedicine: Nanotechnology, Biology and Medicine.* 2014; 10(1):119–29.

78. Lee Y-H, Lee Y-R, Park C-S, Im S-A, Song S, Hong JT, et al. Baccatin III, a precursor for the semisynthesis of paclitaxel, inhibits the accumulation and suppressive activity of myeloid-derived suppressor cells in tumor-bearing mice. *International Immunopharmacology.* 2014; 21(2):487–93.

79. Jeanbart L, Kourtis IC, van der Vlies AJ, Swartz MA, Hubbell JA. 6-Thioguanine-loaded polymeric micelles deplete myeloid-derived suppressor cells and enhance the efficacy of T cell immunotherapy in tumor-bearing mice. *Cancer Immunology, Immunotherapy.* 2015; 64(8):1033–46.

80. Wang Y, Wang H, Lv X, Liu C, Qi L, Song X, et al. Enhancement of all-trans retinoic acid-induced differentiation by pH-sensitive

nanoparticles for solid tumor cells. *Macromolecular Bioscience.* 2014; 14(3):369–79.

81. Mirza N, Fishman M, Fricke I, Dunn M, Neuger AM, Frost TJ, et al. All-trans-retinoic acid improves differentiation of myeloid cells and immune response in cancer patients. *Cancer Res.* 2006; 66(18):9299–307.
82. Sacchetti C, Rapini N, Magrini A, Cirelli E, Bellucci S, Mattei M, et al. In vivo targeting of intratumor regulatory T cells using PEG-modified single-walled carbon nanotubes. *Bioconjugate Chemistry.* 2013; 24(6):852–8.
83. Kudo S, Nagasaki Y. A novel nitric oxide-based anticancer therapeutics by macrophage-targeted poly(L-arginine)-based nanoparticles. *Journal of Controlled Release.* 2015; 217:256–62.
84. Huang Z, Zhang Z, Jiang Y, Zhang D, Chen J, Dong L, et al. Targeted delivery of oligonucleotides into tumor-associated macrophages for cancer immunotherapy. *Journal of Controlled Release.* 2012; 158(2):286–92.
85. Zhan X, Jia L, Niu Y, Qi H, Chen X, Zhang Q, et al. Targeted depletion of tumour-associated macrophages by an alendronate–glucomannan conjugate for cancer immunotherapy. *Biomaterials.* 2014; 35(38):10046–57.
86. Kullberg M, Martinson H, Mann K, Anchordoquy TJ. Complement C3 mediated targeting of liposomes to granulocytic myeloid derived suppressor cells. *Nanomedicine: Nanotechnology, Biology and Medicine.* 2015; 11(6):1355–63.
87. Coghlan A. Closing in on cancer. *New Scientist (1971).* 2016; 229(3063):34–7.

Chapter 13

Nanoparticle-Mediated Small RNA Deliveries for Molecular Therapies

Ramasamy Paulmurugan, Uday Kumar Sukumar, and Tarik F. Massoud

Molecular Imaging Program at Stanford, Department of Radiology, Stanford University School of Medicine, Stanford, California 94305, USA

paulmur8@stanford.edu

13.1 Introduction

Small RNAs and full-length genes are extensively used for therapeutic applications in the treatment of many diseases, including cancer. These therapeutic nucleic acid constructs work in different ways to achieve their intended goals, and therefore, their specific selection is mainly based on the end-point research or clinical application. Delivered genes code for full-length proteins, which provide or enhance the function of endogenous proteins that are either defective owing to mutation(s) or they are unavailable owing to deletion in cells. These types of genes are used to treat genetic, metabolic, and biochemical disorders during gene therapy. In contrast, there are genes that produce enzymes that convert delivered non-toxic prodrugs into toxic metabolites and kill the cells that express these enzymes. This

Bionanotechnology in Cancer: Diagnosis and Therapy
Edited by D. Sakthi Kumar and Aswathy Ravindran Girija

ISBN 978-981-4800-30-3 (Hardcover), 978-0-429-42291-1 (eBook)
www.jennystanford.com

strategy is normally applied in cancer therapy as suicide gene therapy. Accordingly, based on the respective applications, the delivered genes expressing full-length proteins can be used to treat genetic disorders during conventional gene therapy for biochemical disorders, or as suicide genes during cancer therapy (Figure 13.1). Similarly, small RNAs can be delivered in cells to alter the function of endogenous genes whose expression contributes to disease pathogenesis. Small RNAs are delivered as synthetic oligonucleotides (microRNA [miRNA] mimics, antisense miRNAs, and small interfering RNAs [siRNAs]), or they are expressed in cells by delivering mammalian expression vectors (short hairpin RNAs [shRNAs], siRNAs, miRNA mimics, and antisense miRNAs). Even though full-length genes and small RNAs are different in size, their delivery across cell membranes in vitro and in vivo is achievable by the adoption of common delivery systems with some minor technical modifications. Here, we will mainly consider small RNA delivery and its associated challenges.

13.1.1 Introduction to Small RNA Deliveries

Small RNAs, such as miRNAs, piwi-interacting RNAs [piRNAs], siRNAs, and shRNAs, are therapeutic molecules when delivered into the cells in intact forms (Figure 13.2). As stated earlier, although miRNAs and piRNAs are naturally expressed in cells, they can also be supplemented in functional forms by delivering synthetic RNA mimics. In some instance, the function of these endogenous molecules can also be blocked by delivering complementary RNA molecules (antisense-RNAs). Hence, any strategy that justifies the requirements for delivering small non-coding RNAs could be applied to delivery of all four types of RNA molecules mentioned earlier. Even though long non-coding RNAs (lncRNAs) are also clinically important, they have not been well studied as synthetic RNAs; rather they are transcriptionally expressed using expression vectors. While siRNAs, piRNAs, and miRNAs are delivered as small synthetic RNAs, these RNAs can also be transcriptionally expressed by constructing plasmid and viral vectors. Considering their potential clinical importance, several approaches have been

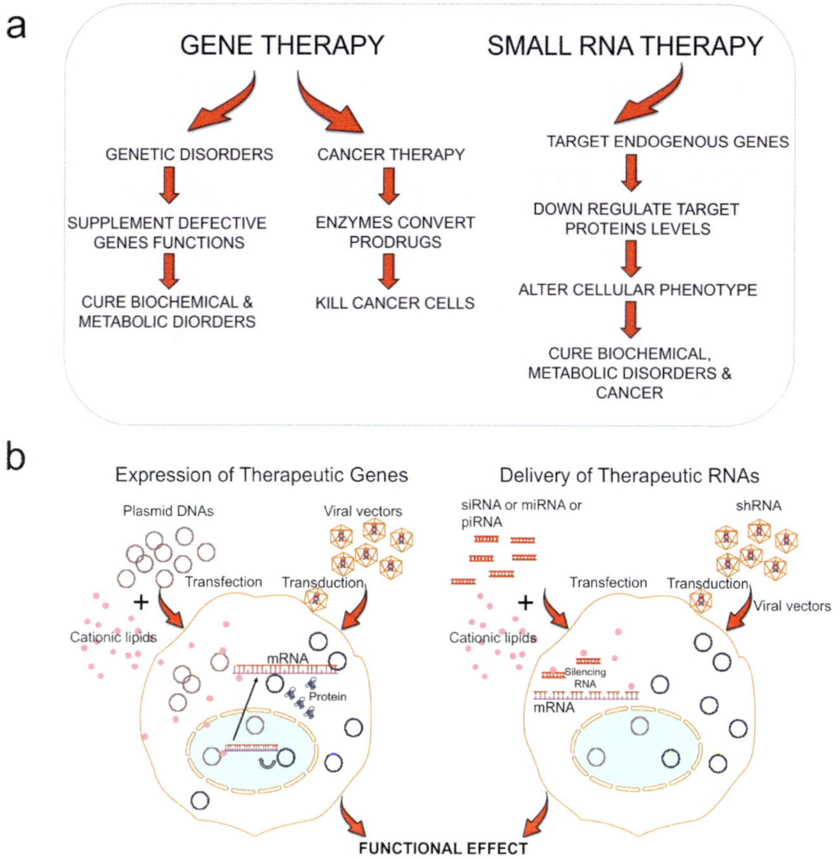

Figure 13.1 (a) Schematic and (b) pictorial illustrations highlight the use, delivery methods, and function of full-length genes and small RNAs for therapeutic applications in human diseases.

attempted to deliver these small RNAs using lipids and other nanodelivery vehicles. Although considerable research has been conducted on applications of various small RNAs for therapy, the epigenetic piRNA has not been well studied. Its role in cancer and other cellular diseases has been least explored, but such studies with promising results are increasing steadily [1, 2]. Recently, piRNA has received much attention, and hopefully soon this epigenetic RNA can also be used for therapeutic applications in various diseases, including cancer. Overall, we will mainly discuss lipid and other nanovectors that are used to deliver small RNAs for in vitro applications in cells, and in vivo pre-clinical animal model studies. We will also briefly cite some translational studies, which are in different stages of clinical trials.

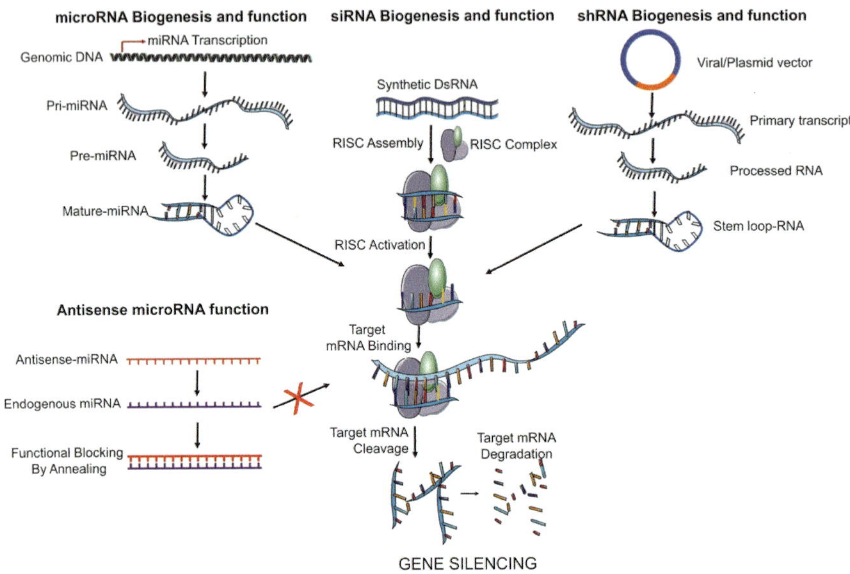

Figure 13.2 Schematic illustrations of the function of different small RNAs including antisense miRNA upon their delivery inside the cells by involving endogenous protein machineries to execute their functions on the target molecules.

Nucleic acids, especially RNAs are prone to digestion by highly stable RNases (RNase 1–8), which are predominantly present in cells and other extracellular spaces, including the circulating bloodpool [3]. Hence, naked RNAs cannot be delivered for therapeutic applications directly via systemic blood circulation. That is the case despite our knowledge that a major delivery route for therapeutic delivery for many drugs is indeed by systemic intravenous injection, because this route provides accessibility to most body tissues. Moreover, small RNAs also need to reach the cytoplasm of target cells to show their biological functions. Hence, nanoparticles of various types may be regarded as efficient delivery vehicles for small RNAs into target tissues. Importantly, nanoparticles may also protect RNAs from nucleases while facilitating their delivery. Identifying optimal nanoparticles for small RNA delivery is very important; nanoparticles of small size (less than 10 nm) are rapidly filtered through the kidneys and reticuloendothelial system (RES). Contrastingly, nanoparticles with sizes in the range of 100 nm to 150 nm have been shown to be efficiently taken-up by cells, while also providing sufficient space for loading RNAs within the nanoparticles to protect them from nucleases [4, 5].

13.2 Lipid-Based Nanovectors for Small RNA Deliveries

Nucleic acids (RNA and DNA) are negatively charged molecules of different sizes. They cannot cross biological barriers of cells, such as the cell and nuclear membranes. Hence, cationic lipids have been used as carriers for cellular delivery of nucleic acids. In particular, plasmid vectors designed to express full-length proteins need to be delivered across two biological barriers of cells, such as the plasma and nuclear membranes, to achieve functional transcription of delivered transgenes into mRNAs for protein synthesis. In contrast, small therapeutic RNAs (siRNAs and miRNAs) require only a cytoplasmic presence to establish their regulatory functions in targeting endogenous mRNAs. Hence, small RNA delivery is more efficient in achieving therapeutic and biological functions compared to the expression of full-length proteins from plasmid DNA vectors. The use of cationic lipids offers significant advantages compared to various other biomaterials for nucleic acid delivery. Biocompatibility, biodegradability, low cost, minimal toxicity, and limited immunogenicity are the primary advantages of lipids. Lipids are also available with varying cationic properties; hence lipid-based nanocarriers can be customized with the required properties to deliver nucleic acids at varying degrees, and can also be designed to lower the toxicity.

Liposomes are the widely used nanocarriers for drugs and nucleic acids deliveries. A highly diverse range of liposomes using different lipids has been applied to biological delivery of small molecule drugs, proteins, and nucleic acids of many sizes. Cationic lipids, such as phospholipids that form bilayers of lipids around aqueous solvents, have been frequently used to generate lipid nanocarriers. Lipids with different cationic properties have also been widely adopted for preparing liposomes. Liposomes can be prepared as a simple formulation, where plain lipids can be used for the delivery of hydrophobic drugs and negatively charged nucleic acids by a simple complexation. This type of liposome is useful for delivering drugs and nucleic acids to cells in culture. Liposomes are also prepared from PEGylated lipids, which provide reduced toxicity, while also facilitating extended in vivo bioavailability by extending their circulation time. The other forms of liposomes include targeted ligands (targeted liposomes) or targeted ligands in combination with imaging agents, such as fluorescent dyes for optical imaging or iron oxide/gadolinium for magnetic resonance imaging (MRI) contrast imaging (multifunctional liposomes) (Figure 13.3).

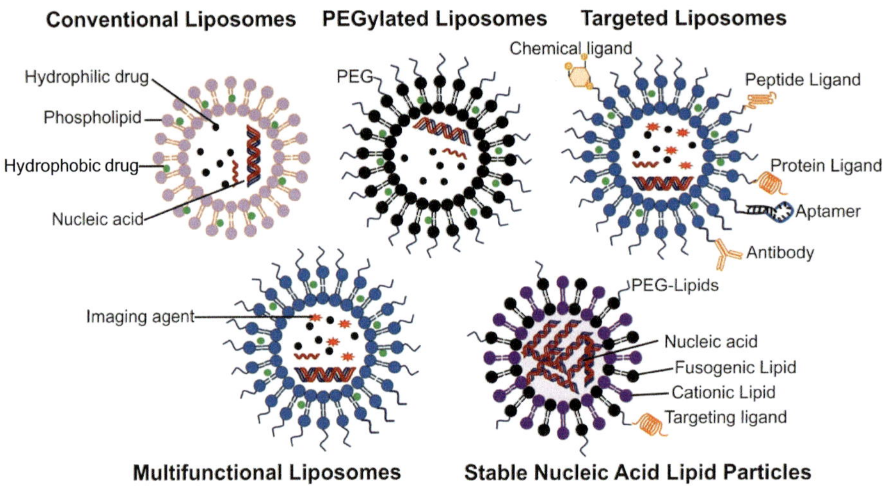

Figure 13.3 Schematic illustration of different types of lipid-based multifunctional nanoparticles currently used for non-targeted delivery and targeted delivery of drugs, imaging agents, and nucleic acids of various sizes for a variety of therapeutic and imaging applications.

13.2.1 Liposomes/Lipoplexes

Liposomes/lipoplexes are cationic lipids of a particular type used to deliver drugs and nucleic acids. Lipoplexes are complexed lipids with nucleic acids in certain ratios to achieve efficient transfection in cells. Lipoplexes release the loaded nucleic acids through endosomes, where they can use a flip-flop mechanism to release the loaded nucleic acids or can release the nucleic acids through pore formation on the cell membranes. The presence of a certain level of net cationic charge is needed for the liposomes to enter across the cell membranes (cytoplasmic and nuclear) to deliver the complexed nucleic acids. Lipoplexes are used to deliver full-length genes and small RNAs for cancer therapy. Since full-length genes are larger in sizes and with higher negative charge, they can be efficiently complexed to cationic liposomes for delivery across cell membranes. In contrast, this can negatively impact on their intracellular (cytoplasmic and nuclear) release for their functional expression [6]. At the same time, small RNAs, such as miRNAs and siRNAs, can show non-uniform complexation with liposomes, but can rapidly release in the cytoplasm to achieve target gene silencing. This requires optimization based on the cationic property of liposomes and the length of nucleic acids by calculating the nitrogen (liposome) to phosphate (nucleic acids) (N/P) ratio for effective functional delivery [7, 8].

In one study, a lipoplex for delivering siRNA targeting VE-cadherin in the endothelial cells was developed by using E-selectin targeted antibody conjugated cationic lipid (anti-E-selectin–SAINT lipoplex (SAINTarg)). It effectively transfected human umbilical vein endothelial cells (HUVECs) and showed biological function by downregulating E-cadherin expression [8]. Lipids such as 1,2-di-O-octadecenyl-3-trimethylammonium (DOTMA), 1,2-dioleoyl-3-trimethylammonium-propane (DOTAP), and dioleoylphosphatidylethanolamine (DOPE) are commonly used for making lipoplexes for transfection of nucleic acids, including siRNAs. Addition of cholesterol to lipoplexes makes these lipids deliver the loaded siRNAs into cells without encountering endosomes. This process is also called a fusogenic delivery mechanism [9]. PEGylated DC-Chol/DOPE liposomes loaded with siRNA targeting kinesin spindle protein (KSP) has shown an enhanced anticancer property against ovarian cancer in a preclinical mouse model without activating the immune system [10]. The use of polyethylene glycol (PEG) with lipids in lipoplex preparations has shown improved biodistribution of loaded siRNAs in the circulation, which facilitated improved tumor specific uptake in vivo. However, the use of PEG in lipoplexes can reduce the amount of siRNAs that can be complexed with the lipids [11].

13.2.2 Stable Nucleic Acid Lipid Particles

Stable nucleic acid lipid particles (SNALPs) are another type of nanoformulation in which nucleic acids are complexed with lipids to form stable nanoparticles for nucleic acid deliveries. SNALPs are approximately 100–150 nm in size, which facilitates efficient delivery of nucleic acids in vivo, while protecting them from nuclease-mediated degradation in the systemic circulation. SNALPs differ from lipoplexes in that the former forms physical nanostructures with certain sizes, whereas for lipoplexes, charge-based chemical composition helps them to form complexes and also facilitates transfection of loaded nucleic acids into the cells. Most SNALPs are made from cationic polymers using PEG as a base material. The bilayer lipid nanoparticles enter cells efficiently by endocytosis. Upon slow release of the loaded small RNA cargos inside the cells, efficient silencing of target genes may ensue. SNALPs were first formulated by Semple et al. for antisense oligonucleotide delivery [12]. In general, SNALPs are made from a combination of high transition temperature phospholipid, a PEGylated lipid, an ionizable cationic phospholipid (1,2-dioleyloxy-*N*,*N*-dimethyl-3-aminopropane (DODMA), and 1,2-dioleyl-3-dimethylammonium propane (DODAP)) [12]. Following the initial combination developed by Semple et al., several others have developed SNALPs by identifying various ionizable phospholipids of different saturation degrees to achieve effective fusogenic property to transfect cells efficiently [13, 14].

Since the discovery of silencing RNAs for modulating gene expression in plants [15], and later for its application in animal cells [16] and clinical translation for various therapies in humans [17], the development of clinically compatible delivery systems have become the prime focus of many research groups. Others have embarked on identifying new targets for RNA interference (RNAi) therapies. While lipoplexes are effective for complexing negatively charged nucleic acids and delivering them across cell membranes, the toxicity associated with lipoplexes, and challenges in identifying the optimal formulation needed to achieve a balanced charge ratio needed for in vivo delivery without toxicity, have both proven to be particular obstacles. In addition, the stability of these complexes in the systemic circulation and protecting the loaded nucleic acids from nucleases are other important challenges. Currently, SNALPs of various physicochemical properties have been used for siRNA and miRNA deliveries into cells and in vivo in living animals. Small RNAs are effective for the treatment of a range of human diseases, including cancer, but delivering intact forms of RNA to cells of target tissues in vivo after systemic administration remains a challenging task. In addition, it is crucial to deliver intact RNAs to the cytoplasm of cells without removal by kidney mediated filtration, RES-mediated cellular uptake and degradation, or aggregation with plasma proteins and blood cells. The existence of these constraints has led researchers to search for alternative delivery systems, as discussed later.

13.3 Structured Nanoparticles for Small RNA Deliveries

Cationic lipid serves as an efficient platform for delivery of nucleic acids independent of their sizes. However, cationic lipid offers limited abilities to functionalize nanoparticles for targeted delivery, especially for in vivo applications. In addition, formulating cationic nanoparticle with a correct N/P ratio for efficient nucleic acid complexation, while leaving some cationic charges for delivering the payload across cell membranes, remains a challenging task. Hence, other nanoparticles of various types and sizes derived from different materials with diverse chemical and physical properties offer alternative approaches to constructing nanodelivery vehicles for drugs and nucleic acids. Based on the materials used for nanoparticle synthesis, they may be broadly categorized into: (i) inorganic nanoparticles, (ii) organic nanoparticles, and (iii) hybrid nanoparticles. Although viruses and other cell membrane-derived vesicles can be considered as nanoparticles, considering the focus of this chapter, we will limit our discussion to various synthetic nanoparticles (Figure 13.4).

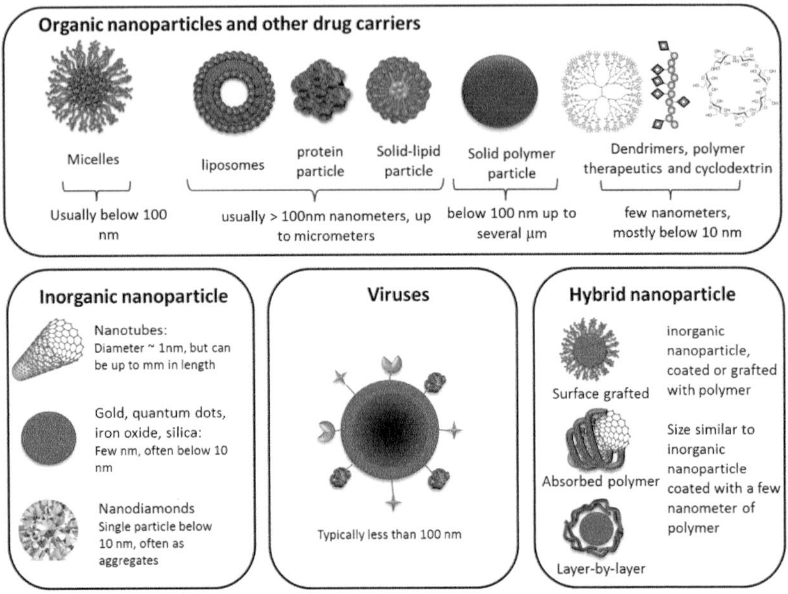

Figure 13.4 Schematic illustration of a series of images highlighting various types of nanoparticles that exist in soft and rigid categories, and belong to organic, inorganic, and hybrid chemical material types for the application of therapeutic delivery of drugs and nucleic acids. Reproduced with permission from Ref. [18].

Nanoparticles are clusters of atoms derived from one or more materials, assembled together to form particles in the nanometer size range (1–999 nm), with different shapes. Nanoparticles possess unique properties to deliver drug payloads into cells. Nanoparticle-mediated drug delivery is beneficial in several ways compared to free drugs. The use of nanoparticles allows for the possibility to functionalize their surface using specific targeting agents. They also can be loaded with imaging agents such as fluorescent dyes (for optical imaging) or contrast agents (for CT, MR, and X-ray imaging) to track them in cell culture during their trafficking to various cellular compartments, or in vivo to different organs including the intended target tissues. As stated earlier, based on the materials used for synthesis, nanoparticles can be categorized into inorganic nanoparticles, organic nanoparticles, and hybrid nanoparticles. While organic and inorganic nanoparticles are synthesized from respective materials, the hybrid nanoparticles are made from a combination of both, in most instances, inorganic materials are coated with organic chemicals. A wide range of organic materials are available for coating inorganic nanoparticles, including polymers, surfactants, and proteins with self-assembling properties. These hybrid nanostructures have been widely used for drugs and nucleic acids deliveries.

13.3.1 Inorganic Nanoparticles for Small RNAs Deliveries

Nanoparticles made from central core of pure inorganic materials are called inorganic nanoparticles. Nanoparticles prepared from different forms of carbon (graphene, graphene oxide, graphite, and fullerine), quantum dots (QDs, semiconductor nanocrystals), gold, iron, silica, etc., belong to this category. Inorganic nanoparticles have been widely used to deliver small molecule drugs and nucleic acids of different sizes to cells in vitro, and in vivo in living animals [19]. The size of these nanoparticles range from 1 nm to several nanometers. Nanoparticles made from these inorganic materials are not biodegradable and may even be toxic to animals and humans in some instances. While these nanoparticles can serve as efficient delivery vehicles (gold, carbon, and iron oxide), they also serve as contrast agents for imaging applications. QDs are predominantly used as imaging agents in many applications. Inorganic nanoparticles, such as gold, are also used as a therapeutic agent in the treatment of cancer in conjunction with photodynamic therapy (PDT) and photothermal therapy (PTT) [20, 21]. The use of inorganic nanoparticles for nucleic acid deliveries, including small RNAs, is limited compared to organic and hybrid nanoparticles. Hence, we briefly discuss later some of the inorganic nanoparticles used for small RNA deliveries, while providing a more extensive review of polymer nanoparticles.

13.3.1.1 Carbon nanoparticles for small RNA deliveries

Carbon in different forms has been successfully used to deliver siRNAs into cells in vitro, and in vivo in animal models. Carbon nanotubes and graphene nanosheets have been widely used for siRNA deliveries [22–25]. The needle-like structures of carbon nanotubes facilitate their ability to cross cell membranes into the cytoplasm by an endocytosis independent mechanism [26]. Since carbon nanomaterials are solid particles, they can be recognized by immune cells and macrophages to eliminate them from the blood circulation; and to overcome this, a variety of additional coatings using PEG and other materials have been employed. In addition, the electrical property of carbon allows it to be loaded with certain amounts of negatively charged nucleic acids, but it may not be sufficient for intracellular delivery of siRNAs to bring about effective silencing of target genes and to achieve a desired functional effect. In order to overcome this issue, PEGylation along with lipid coating of carbon nanotubes and nanosheets has been used to improve siRNA loading and deliveries [27, 28]. A few examples include siRNAs targeting the HDAC1 gene and the G12C mutant K-Ras gene delivered by PEGylated graphene oxide nanosheets effectively inhibited MIA PaCa-2 pancreatic adenocarcinoma cell growth, and a folic acid targeted version of the nanosheets effectively improved PTT in vivo [28]. Graphene is a two-dimensional (2D) material, basically a single layer of graphite with carbon atoms arranged in a hexagonal, honeycomb lattice. Graphene has been successfully adapted for the delivery of small RNAs in cells and living animals. Graphene oxide functionalized with anti-HER2 antibody successfully delivered siRNA targeting survivin mRNA and reduced breast carcinoma growth in vivo in animals [29]. To improve RNA loading efficiency, investigators have used polylysine (PLL), polyethyleneimine (PEI), and cationic peptides as functional molecules on graphene oxide. These functionalizations significantly improved the functional efficiency of loaded siRNAs.

Similar to siRNAs that are designed to target different cellular genes to achieve a certain functional effect, miRNAs are another group of small endogenous RNAs with similar structure and functional properties expressed in cells to regulate more than one cellular target (up to 80 genes) involved in multiple cellular pathways. MiRNAs have been used for therapeutic applications in treating various cellular diseases, including cancer. Carbon nanoparticles after different functionalizations have been used

for the delivery of miRNA mimics and antisense miRNAs. We have recently shown that fluorescently labeled antisense-miRNA (Cy5-antisense-miRNA-21) can be delivered using reduced graphene oxide, to measure its functional hybridization with intracellular miRNA-21 by using optical imaging [30].

13.3.1.2 QD for small RNAs deliveries

QDs are inorganic nanoparticles prepared from elements of the "classical" semiconductor groups III–V and nanocrystals of groups II–VI from the Periodic Table [31]. QDs are colloidal semiconductor nanocrystals widely used for drug delivery and imaging. QDs, with their superior optical and electronic properties, are excellent particles for effective delivery of small RNAs and drugs. Moreover, QDs have optical properties that render them useful in allowing their tracking, determination of their in vivo biodistribution, and target specific accumulation by using optical imaging. QDs have been successfully used to deliver siRNAs to different cellular targets for therapeutic applications. Some examples include the use of QDs to deliver siRNA targeting human epidermal growth factor receptor 2 (HER2) in breast cancer cells [32]. Similarly, QDs functionalized with hydrophobic N-palmitate and a small cationic peptide (VKIKK), were used to deliver siRNA targeting sphingomyelinase [33]. The cationic peptide allowed for efficient loading of siRNA to the QDs. In that study, the authors used a CdSe/ZnS core-shell QD with a light emission wavelength in the range of 520–625 nm. Even though their QDs showed effective deliveries of loaded siRNA, and facilitated imaging and tracking the cellular distributions of the nanoparticles, the toxicity of the QDs compromised their further extension to clinical applications. This applied to small RNAs deliveries, as well as for drug deliveries and imaging [34].

13.3.1.3 Gold nanoparticles/nanorods/nanostars for small RNAs deliveries

Gold is a non-toxic inorganic metal that is used in many medical applications. Gold nanoparticles (AuNPs) of various structures, sizes, and physicochemical properties have been used for biomedical applications in drugs and nucleic acid deliveries and imaging. AuNPs possess good biocompatibility, are easy to synthesize, display monodispersity, and can be functionalized. They serve as an attractive candidate to deliver various payloads (such as small molecule drugs), and large biomolecules (such as genes, siRNAs, and miRNAs). An advantage of using AuNPs is that the intracellular release could be controlled by pH, glutathione (GSH) responsive functionalization, or by external light stimuli [35]. Since AuNPs can be easily tuned for changing their optical, chemical, and physical properties, they have been used for multimodal deep tissue optical and photoacoustic imaging applications, and for laser-induced photothermal ablation of cancers. Recently, relative to lipid-based delivery agents, inorganic nanoparticles have emerged as synthetic nanovectors with several advantages, which include tunable sizes, surface properties, and multifunctional capabilities including their non-toxic nature. AuNPs, in particular, serve as attractive vehicles for nucleic acid delivery applications, owing to their structure, size, inert, non-toxic nature, and surface properties.

In contrast to the electrostatic interactions used for loading negatively charged nucleic acids onto cationic liposomes for transfection, in AuNPs, nucleic acids can be loaded on the surface of the nanoparticles through Au-S covalent chemistry or through supramolecular assemblies. For loading siRNAs and shRNAs, oligonucleotides in thiolated forms were commonly preferred for conjugation on the surface of AuNPs. In addition, nucleic acid hybridization has been used as a strategy for functionalizing antibodies or other targeting ligands on the surface of AuNPs for targeted deliveries [36]. In one study, AuNPs pre-conjugated with single stranded oligonucleotides through thiol-maleimide chemistry, were further hybridized with complementary strands of oligonucleotide conjugated targeting ligands and antibodies. In addition to direct conjugation, AuNPs have also been surface-coated with cationic chitosan for efficient loading of nucleic acids via electrostatic interactions. We have recently shown the use of Au nanostars surface coated with chitosan-cyclodextrin hybrid polymer for successful intranasal delivery of miRNAs (miR-100 and amiR-21) for glioblastoma (GBM) therapy (Figure 13.5) [37]. This process significantly improved the miRNA loading efficiency of AuNPs, while also facilitating intranasal mucosal membrane transport to the brain.

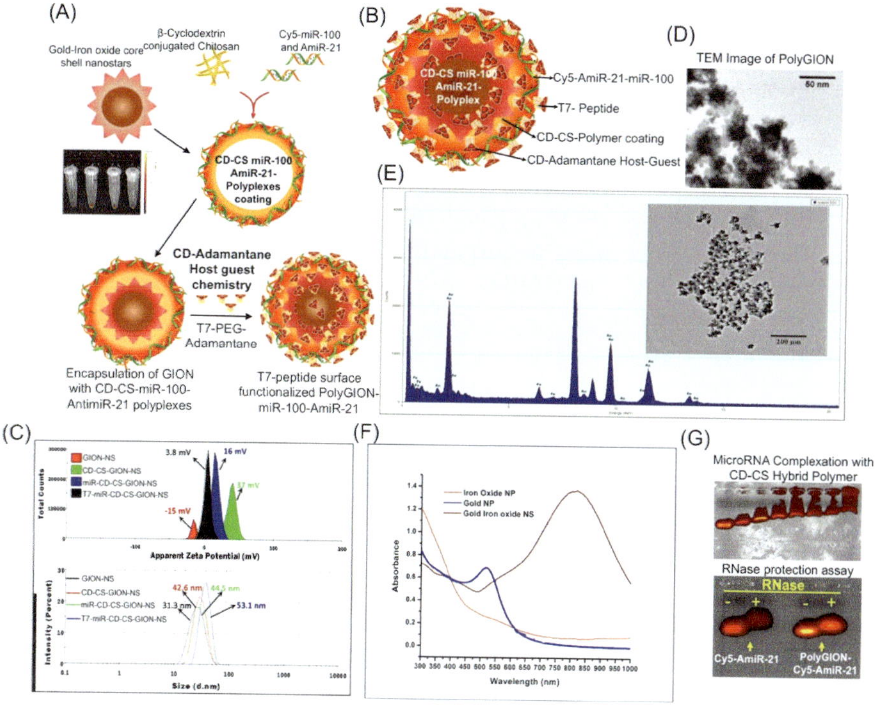

Figure 13.5 (A) Schematic illustration of synthesis of PolyGIONs and fluorescence images of Cy5 labeled miR-100 and antimiR-21 loaded CD-CS complexes. (B) Schematic representation of as-prepared polyGIONs structure and their various compositions. (C) DLS measuring zeta potential and size of nanoparticles after each step of GION surface modification. (D) TEM micrograph of GION. (E) GION coated with miR-100/antimiR-21 encapsulated CD-CS polymer layer, its EDS analysis, and corresponding images. (F) UV-Vis spectra of as-prepared GION nanoparticle and its precursor nanoparticle. (G) Gel retardation assay for Cy5-miR-100/antimiR-21 encapsulation in CD-CS hybrid polymer and RNase protection assay for measuring the stability of encapsulated miRNAs in polyGION-CD-CS. Reproduced with permission from Ref. [37].

13.3.1.4 Other inorganic nanoparticles in small RNA deliveries

Other than the above discussed inorganic nanoparticles, there are several other inorganic nanoparticles, such as particles made from silica and silver that have also been used to deliver siRNAs for a wide range of applications. PEI-complexed mesoporous silica nanoparticles showed successful release of loaded siRNA for nearly 30 days in vitro with efficient knock down of target gene expression in vivo [38]. Similarly, siRNA targeting the expression of multidrug resistance-associated protein 1 (MRP1) was delivered using a similar type of PEI-capped mesoporous silica nanoparticles for treating GBM in vitro and in vivo [39]. Silver nanoparticles conjugated to quercetin and siRNA (AgNP-Qe-siRNA) were used to target the antioxidant drug resistance mechanism and to inhibit drug-resistant *Bacillus subtilis* for effective gene-silencing-assisted antibacterial effect in vitro and in vivo [40].

13.3.2 Organic Nanoparticles for Small RNA Deliveries

Organic nanoparticles are small solid nanostructures of less than 1000 nm in size, fabricated from polymers, proteins, carbohydrates, lipids, and other organic compounds at a characteristic dimension for various applications. Over the last decade, the use of organic nanoparticles has expanded beyond the transitions from electronic to photonic, conducting materials to sensors, and medicine to biotechnology. The most frequently used organic nanoparticles are liposomes, dendrimers, carbon nanomaterials, and polymeric micelles. Carbon nanoparticles can be categorized under both organic (carbon nanotubes and graphene) and inorganic nanoparticles (graphite and diamond). We earlier discussed the use of carbon nanoparticles for small RNA deliveries, and we also reviewed the use of liposomes. In this section, we will focus mainly on organic polymer nanoparticles and micelles, which are used for small RNA deliveries.

13.3.2.1 Polymeric nanoparticles

Nanoparticles prepared from polymers of various types, lengths, and electrochemical properties have been used for drugs and small RNA deliveries in therapeutic applications. Polymers with biodegradable properties are useful to prepare nanoparticles for controlled drug delivery; this has shown significant improvements in cancer therapy [41]. It is also possible to load a wide range of drugs with different chemical properties, and small RNAs of various types (siRNA, shRNA, and miRNAs) in polymer nanoparticles. They can be used as delivery vehicles for both in vitro and in vivo applications. Since these drugs and RNAs-loaded nanoparticles show slow and sustainable release for a longer period in target cells, these nanoparticles provide enhanced antitumor efficacy, with reduced systemic side effects to non-target tissues, when compared to drugs delivered in their native forms. Moreover, these nanocarriers also protect drugs and RNAs from their rapid metabolism and nuclease-mediated degradation, respectively, in the systemic circulation, and clearance by the liver, kidneys, and RES, which further enhances the stability and target specificity of loaded drugs and nucleic acids [42, 43]. Several polymer-based nanoparticles such as poly(lactic-co-glycolic acid) or poly(lactide-co-glycolide) (PLGA), polylactic acid or polylactide (PLA), poly glycolic acid or polyglycolide (PGA), polycaprolactone (PCL), poly (D,L-lactide) (PDLLA), chitosan, and PLGA-PEG have been developed for drug deliveries, and are currently in various stages of clinical trials, and a few of them have been approved by the Food and Drug Administration (FDA) for clinical use. We reviewed these polymeric nanoparticles previously [44].

As stated earlier, polymer nanoparticles offer significant advantages to load drugs using a variety of methods, selected based on the properties of the drugs [45–47]. Most common methods used for preparation of polymer nanoparticles are emulsion-solvent evaporation/diffusion, double emulsion, nanoprecipitation, salting out, dialysis, and supercritical fluid techniques [47]. The use of a water-in-oil-in-water (W/O/W) double emulsion solvent evaporation method is common for encapsulation of hydrophilic drugs, while nanoprecipitation technique may be used for encapsulation of hydrophobic drugs in most of polymeric nanoparticles preparation. Other methods, such as salting out, dialysis, and supercritical fluid techniques, are commonly used for micelles, where critical micelle concentration is important for the preparation of nanoparticles. Polymer nanoparticles with solid structures are more preferable for in vivo drug deliveries where dilution of nanoparticles by the systemic blood volume cannot affect the particle structure, but, in contrast, this is an important consideration for micelles.

13.3.2.1.1 Polymer nanoparticles for therapeutic anticancer drug deliveries

Even though polymer nanoparticles are exceptional biocompatible vehicles for loading and delivering drugs of many types to treat diseases, they are extensively used for anticancer drug deliveries in cancer therapy, and some of the important formulations are currently in the clinic to treat cancer. We describe here a few of them that are currently in preclinical evaluation, followed by the formulations that have reached clinical applications.

13.3.2.1.1.1 Polymer nanoparticles loaded with therapeutic drugs in pre-clinical anticancer studies

PEG–PLGA conjugated polymer is a widely used biopolymer for preparing many important anticancer drugs loaded intoLnanoparticles. Paclitaxel (PTX) [48] loaded in PEG–PLGA nanoparticles using a simple emulsion and nanoprecipitation technique has been evaluated for its therapeutic effects in various cancers. This nanoparticle shows a biphasic drug release pattern for PTX, with an initial burst release followed by a slower and continuous release for a longer period (for several days). This property is quite common for many drugs and nanoparticle types. The treatment of HeLa cells with PEG–PLGA-loaded PTX shows significant levels of improvement in apoptotic induction in vitro, and greater levels of antitumor effects in vivo compared to respective controls treated with free PTX [48]. Cisplatin is another important and widely used anticancer agent for treating many cancers. However, its therapeutic applications are hampered by dose limiting toxicity causing many serious side effects, notably liver and kidney toxicity, when administered directly as a free drug. To overcome this, several cisplatin-entrapped nanoparticles have been prepared for slow delivery of drugs, without any loss of their activities [49, 54]. PLGA-*m*-PEG nanoparticle loaded with cisplatin using the double emulsion method is a very attractive approach for preparing this drug for anticancer therapy in vivo [49].

5-Fluoro uracil (5-FU) is a pyrimidine analog widely used as an antineoplastic cytotoxic drug in cancer therapy. 5-FU inhibits cellular thymidylate synthase to render its anticancer effect.

It also has non-specific effects when delivered as a free drug. 5-FU loaded in PLGA nanoparticles, using a double emulsion method, shows better cytotoxic effects against GBM (U87MG) and breast adenocarcinoma (MCF7) cell lines compared to free drug [55]. The formulation of 5-FU in galactosylated chitosan (GC) polymer nanoparticles shows an enhanced therapeutic effect in hepatocellular carcinoma (HCC) [56, 57]. In addition, other polymers, such as hyaluronic acid–poly(ethylene glycol)–poly(lactide-co-glycolide) (HA–PEG–PLGA) and poly(γ-benzyl-L-glutamate)–poly(ethylene glycol) (PBLG–PEG), have been also evaluated for 5-FU delivery in vitro and in vivo for anticancer therapy [58, 59]. Doxorubicin (DOX) is an anthracycline antibiotic that inhibits DNA synthesis, and has been widely used anticancer chemotherapy for many cancers [45, 60]. DOX has many adverse side effects when administered as a free drug. DOX-encapsulated PLGA nanoparticles using an oil-water nanoprecipitation technique with the resulting average size of 174 nm have shown enhanced therapeutic effects in breast cancer cells compared to free drug [61]. DOX entrapped in a nanoformulation prepared from biodegradable amphiphilic diblock methoxy poly(ethylene glycol)-b-poly(ε-caprolactone-co-γ-hydroxyl-ε-caprolactone) (mPEG-*b*-P(CL-co-HCL)) copolymer bearing pendant hydroxyl groups showed good cytotoxicity against HepG2 cells in vitro [62]. Poloxamer 188 coated lecithin-PLGA-human serum albumin (HSA) nanoparticles loaded with DOX (DOX-L-PLGA-HSA) enabled delivery of DOX across the blood-brain barrier at therapeutically effective concentrations in vivo in rat GBM model [63]. DOX-loaded cyclo-(1,12)-PenITDGEATDSGC (cLABL) peptide conjugated PLGA nanoparticles (DOX-cLABL-PLGA NPs) show more rapid cellular uptake by A549 lung epithelial cancer cells compared to non-peptide conjugated control nanoparticles by targeting encapsulated DOX to ICAM-1 expressing cells [64]. DOX-entrapped poly(lactic-co-glycolide)-gold half-shell nanoparticles (DOX-PLGA-Au-H-S NPs) can deliver both DOX and heat simultaneously to a selected tumorigenic region when irradiated by the near-infrared (NIR) light, resulting in higher cytotoxicity and more controlled delivery into HeLa cells [65]. DOX-loaded PLGA nanoparticles show sustained drug release rate in fibroblast cells in vitro without initial burst release [66]. DOX-loaded peptide-conjugated PLGA-PEG-NH_2

nanoparticles containing an amino end group display three-fold higher uptake than corresponding peptide-free PLGA-PEG-NH_2 nanoparticles in a human ovarian cancer SKOV3 cells with high expression of epidermal growth factor receptor (EGFR) [67].

In addition to the above mentioned drugs, several other drugs such as tamoxifen (TMX) and 4-hydroxytamoxifen (4-OHT)-loaded nanoparticles have been also used effectively to treat ER positive breast cancers [68]. Owing to relatively poor oral bioavailability and several side effects of these drugs, polymer nanoparticles have gained much interest for tamoxifen delivery [69–72]. Gemcitabine, mitomycin C, epirubicin, and curcumin are other frequently used antineoplastic agents that have been loaded in polymer-based nanoparticles for anticancer therapy and have been tested them in many cancers in vitro and in vivo [73–85]. Overall, these anticancer agents achieve a wide range of beneficial effects when loaded in nanoparticles compared to free drugs, which include: (i) reduction in toxicity, (ii) change in bioavailability, (iii) enhanced specificity, (iv) change in solubility, and (v) improved therapeutic effects. The use of biodegradable polymers has an added advantage to improve the efficiency of these drugs further for their utilizations in the clinic because of their non-toxic biocompatible nature.

13.3.2.1.1.2 Polymer nanoparticles loaded with therapeutic drugs in clinical applications

A number of nanoparticle-loaded anticancer drugs are currently in clinical trials and a few of them are approved for clinical treatment of specific types of cancers. Interestingly, most of the anticancer drug-loaded nanoparticles in clinical trials are based on biodegradable polymers. In particular, lipid polymer-based liposomes loaded with anticancer drugs are the type predominantly in clinical trials and have been successfully approved for human treatment. Liposomal formulations of a cytarabine and daunorubicin combination, such as VYXEOS and CPX-351, have been approved for the treatment of acute myeloid leukemia. Similarly, liposome nanoformulations independently loaded with DOX, vincristine, mifamurtide, and irinotecan, are other clinically approved nanoformulations for the treatment of a variety of cancers. Albumin-loaded anticancer PTX has also been approved for the treatment of advanced stage non-small cell lung carcinoma and metastatic breast and pancreatic cancers [86]. Li et al. have reviewed a list of drug-liposome nanoformulations that are currently in the clinic for treatment of many cancers [87].

13.3.2.1.2 Polymer nanoparticles for small silencing RNA (siRNA, shRNA, and miRNA) deliveries in cancer therapy

Small RNA-mediated gene silencing was first discovered in 1993 [88]. Since then, numerous small RNAs, including siRNAs, shRNAs, miRNAs, and piRNAs, have been identified and widely used for different biological applications, including cancer therapy [89]. These RNAs can be listed under two categories: (i) intrinsic and endogenously expressed RNAs, and (ii) synthetic and extrinsic RNAs. In contrast to their different origins, these RNAs follow similar endogenous pathways to achieve target gene silencing. The discovery of siRNA (or RNAi) was serendipitous. Even though siRNA mediated gene silencing was first discovered, siRNA mediated gene silencing and RNAi mediated gene silencing naturally use the endogenous miRNA mediated gene silencing pathway to execute their functions (Figure 13.2). SiRNAs are ~20–25 nucleotide long double-stranded RNA (dsRNA) molecules that, after binding to RNA-induced silencing complex, enable the cleavage and degradation of recognized mRNA. While siRNAs are delivered as synthetic dsRNAs, siRNA can be expressed as a synthetic vector-based expression system called shRNA. Even though the endpoint applications of siRNA and shRNA are similar, the step at which these RNAs target the intrinsic pathway to achieve functional effects is different [90]. The advantage of shRNA compared to siRNA is that, because of its low and continuous expression, shRNA shows long-term functional effects compared to synthetic siRNAs, which are delivered at biologically excess levels that rapidly degrade and lose their long-term functional effects [90].

MiRNAs are small single stranded RNAs of 18–23 nucleotides that are partially or fully complimentary to the 3' non-coding region of target mRNA sequence. MiRNAs achieve gene silencing through blocking mRNA translation or by degrading target mRNAs. The ability to achieve sequence-specific gene silencing using RNAi (siRNA, shRNA, and miRNA) has been used in a wide range of applications, including the treatment of numerous diseases, particularly for the cancer therapy [91]. Nevertheless, in vivo systemic administration and delivery of the small silencing RNA-based therapeutics to cells and tumors remain a challenge, owing to limitations such as poor cellular uptake, degradation by serum nucleases, and rapid renal clearance after administration [92]. To overcome these limitations, the development of effective and safe nanocarriers for systemic delivery of small silencing RNAs is required for efficient cancer therapy. Biodegradable polymer nanoparticles have been found to be safe and effective nanocarriers for the delivery of small silencing RNAs and show promising translational clinical applications.

13.3.2.1.2.1 Polymer nanoparticles for siRNA deliveries

SiRNA is dsRNA of 20–25 base pairs long designed against any region of the target mRNA (coding region or 3' UTR) to achieve gene silencing by mRNA degradation. Sequence-specific post transcriptional gene silencing can be achieved by delivering synthetic siRNAs via RNAi pathway in mammalian cells [93]. To achieve improved silencing effect, many of the currently available siRNAs are designed as pooled molecules, where more than one region of the target mRNAs is used as mixtures. The siRNA technology has rapidly evolved as a powerful tool in selectively silencing gene expression [93, 94]. Even though siRNAs are efficient for gene silencing, their intracellular delivery in intact form is challenging, and more so when considering this approach for in vivo gene silencing. Hence, while developing strategies to synthesize stable RNA molecules using modified bases (2-O-Me, 2'OMe-PS, 2'OMe-PO, 2'MOE, 2'MOE-PS, LNA, FANA, etc.,), enormous efforts have also been spent in developing nanodelivery systems for intact delivery of siRNAs by overcoming nuclease-mediated degradations in the systemic circulation. This can be achieved only by preventing nuclease access to RNAs. Nanoparticles synthesized from various polymers have served as excellent platforms that encapsulate RNAs within the nanoparticles and prevent them from exposure to RNases.

PLGA nanoparticle formulations with a higher siRNA loading capacity using the double-emulsion solvent evaporation technique were first developed for stable delivery of siRNA in vitro and in vivo [95]. In that study, the authors used topical administration of a single dose of PLGA-loaded siRNA nanoparticles, which showed efficient and sustained gene silencing throughout the female mouse reproductive tract when applied externally. The knockdown of gene expression was also observed to be proximal and distal to the site of topical delivery. These siRNA-loaded PLGA nanoparticles penetrated deep into the vaginal tissue after topical administration and demonstrated effective delivery capability for siRNA to the vaginal mucosa [95]. While PLGA nanoparticles are suitable for encapsulating small RNAs, in another study, PLGA with cationic chitosan (PLGA–CHS-NP) was used to load a codon-optimized plasmid-DNA (p-DNA) expressing eGFP and siRNA targeted against HBV-X gene. The results of that study showed a higher HBV-X gene-silencing efficiency with the expression of eGFP gene in HepG2.2.15 cells in vitro [96]. Similarly, monomethoxypoly (ethylene glycol)–poly(lactic-co-glycolide)–poly-L-lysine (mPEG–PLGA–PLL) nanoparticles loaded with siRNA targeting platelet-derived growth factor BB (PDGF-BB) in combination with microbubble (MB) and ultrasound (US) administration were safely used to increase the delivery of siRNA to rat retinal pigment epithelium (RPE-J) cells and achieved PDGF-BB gene silencing with a substantial decrease in mRNA and protein expression [97]. The mPEG–PLGA–PLL triblock copolymer nanoparticles were also efficiently used to transfect siRNA into human lung cancer (SPC-A1-GFP) cells steadily expressing green fluorescent protein (GFP), and produced much higher intracellular siRNA delivery efficiencies than cells transfected with siRNA without any transfection agent or with the lipofectamine-siRNA complex [98]. The mPEG–PLGA–PLL nanoparticles were also used for the cyclic Arg-Gly-Asp (cRGD) targeted delivery of human vascular endothelial growth factor (hVEGF)-siRNA using MB and US delivery approach to reduce VEGF expression in PC3 cells, which showed low cytotoxicity and high transfection efficiency. The results demonstrated that the targeted delivery of biodegradable mPEG–PLGA–PLL nanoparticles might prove useful as a new vector system for gene delivery [99]. Another approach consisted of Ala-Pro-Arg-Pro-Gly (APRPG) peptide-conjugated PEI PEG (siRNA–PEI–PEG–APRPG) nanoparticles was successfully used to silence VEGF expression in MCF7 breast cancer cells in vitro and in vivo [100]. This nanoformulation inhibited the microvessel formation, and suppressed tumor growth in vivo. The siRNA/PEI–PEG–APRPG nanoparticles improved siRNA delivery and provided a more efficient antiangiogenic therapy than free siRNA or siRNA/PEI [100]. In addition, several other polymer nanoformulations were successfully used to deliver siRNAs targeting different cellular genes to achieve intended biological functions.

13.3.2.1.2.2 Polymer nanoparticles for shRNA deliveries

The shRNAs are another class of small RNAs with a stem of 19–29 base pairs and a loop of 4–10 nucleotides that are expressed from plasmid or viral vectors to induce RNAi-mediated gene silencing. Even though both siRNA and shRNA mediate target gene silencing by similar RNAi mechanisms, the siRNA is delivered as a synthetic double stranded RNA, while shRNA is expressed as a dsRNA with a loop structure using gene expression vectors [101]. The expression of shRNA in cells can be achieved by delivery of plasmids or via viral vectors. The shRNA vectors commonly utilize an U6 promoter, a type III RNA polymerase III promoter, to achieve a primary transcript for its further processing using an endogenous RNAi mechanism [102–104]. shRNAs have been extensively used for various applications including cancer therapy and in drug screening for target specific identification of small molecule drugs.

The intravitreal delivery of a shRNA vector targeting hypoxia-inducible factor 1α (HIF-1α) and GFP co-expressing pDNA-loaded (pshHIF-1α NPs) PLGA nanoparticles showed its inhibitory efficiency on experimental choroidal neovascularization (CNV). The PLGA nanoparticles prepared for this study using water in the oil-in-water method resulted with a mean size of 303.7 nm and −8.9 zeta potential. The in vitro release behavior of pDNA-loaded (pshHIF-1α) nanoparticles showed that about 77% of total encapsulated pDNA was released within 10 days, followed by a continuous and sustained release for four weeks. CNV was induced by laser photocoagulation in rats and the results indicated that the expression of GFP preferentially occurred in the retinal pigment epithelium cell layer, lasting for four weeks. Thus, pshHIF-1α PLGA nanoparticles possibly will act as a novel therapeutic route of transferring specific pDNA and inhibiting the formation of experimental CNV [105].

PLGA nanoparticles prepared using the double emulsion-solvent evaporation method and loaded with a plasmid expressing shRNA targeting VEGF were able to regress corneal neovascularization in a more sustained and vigorous manner than naked plasmids alone, making them a good in vivo non-viral gene therapy tool for sustainable, non-immunogenic, and efficient regression of murine CNV [106]. Similarly, a construct expressing shRNA targeting androgen receptor (AR) and delivered by PLGA nanoparticle has shown an antiproliferative effect in prostate cancer. AR-shRNA construct-loaded and prostate-specific membrane antigen (PSMA) A10 aptamer-conjugated PLGA nanoparticles strongly improved cellular uptake of nanoparticles in both cell culture and xenograft-based models, and considerably enhanced tumor regression compared to non-conjugated nanoparticles in various prostate cancer cell lines (22RV1, LAPC-4, and LNCaP). The tissue-specific delivery of AR-shRNA using PLGA nanoparticles indicated an effective therapy for prostate cancer [107]. VEGF is a key driving force and regulator in tumor-associated angiogenesis and tumor development in HCC and many other cancers [108]. RNAi-based therapeutics have been shown to block this angiogenesis stimulator [109]. Due to low cellular uptake and poor stability of small RNAs, the clinically suitable, safe, and effective delivery systems are required for RNAi-based therapeutics [110]. Low molecular weight chitosan-shRNA (LMWC-shRNA) nanocomplexes directed against VEGF considerably inhibited VEGF expression of HCC cells and liver tumor tissues. Intravenous injection of LMWC-shRNA nanocomplexes displayed greater and prolonged deposition of shRNA at the tumor site in orthotopic allograft liver tumor-bearing mice [111].

Other polymer nanoparticles such as poly-β-amine esters nanoparticles loaded with vector expressing EGFP-shRNA (ShRNA-PAE NPs) have shown efficient cellular uptake and EGFP silencing in U87MG cells and HEK-293 cells. Intravenously injected shRNA-PAE NPs in U87MG-GFP tumor-bearing mice demonstrated high accumulation of nanoparticles in tumors, with effective silencing of intra-tumoral EGFP expression. The relationship between RNAi efficiency, cytotoxicity, and PAE polymer structure has indicated that the presence of disulfide bonds and nitrogen atom in the backbone of PAE nanoparticles has demonstrated a remarkable influence on in vitro and in vivo cytotoxicity and gene silencing efficiency [112].

13.3.2.1.2.3 Polymer nanoparticles for miRNA deliveries

The miRNAs are small, highly conserved, non-coding RNAs of 18–24 nucleotides in length endogenously expressed in cells that are engaged in the post transcriptional regulation of gene expression through the RNAi pathway [16, 89, 91–94, 113–118]. The miRNA expression is dysregulated in cellular diseases including cancer, where miRNA expressions are closely associated with cancer development, growth, invasion, and metastasis [16, 89, 91–94, 113–118]. Based on their functions, they are categorized as oncogenic miRNAs (oncomiRs) and tumor suppressor miRNAs. OncomiRs promote tumor growth by inhibiting tumor suppressor and apoptotic genes, whereas anti-oncomiRs block the function of proteins involved in cell cycle and apoptosis [93, 116]. There are numerous miRNAs that are reported to be tumor suppressors (miR-17-5p, miR-21, miR-29, miR-34, miR-127, miR-155, let-7, etc.) [115]. This list has been growing, and it is expected that there may be many more miRNAs with important functions yet to be discovered. Recently, we and several other groups have been investigating miRNAs as a new class of molecularly targeted anticancer therapeutics [4, 5].

Effective overexpression strategies are needed to examine the effects of miRNA-directed regulation of gene expression in vivo. AntagomiRs or anti-miRs are small chemically modified antisense oligonucleotides that are used to block the function of endogenous miRNAs. They inhibit miRNA activity by irreversibly binding to target miRNA [119]. Specific miRNAs can be silenced by anti-miRNAs complementary to the mature miRNA sequences [120]. Each miRNA controls the expression of numerous target genes by blocking translation of the target protein, and about 200 target genes may be regulated by a single miRNA [121]. AntagomiRs are effective silencers of miRNA function expression in mice. AntagomiR-122 has been used to specifically silence the function of miR-122 in the lung, liver, skin, intestine, heart, and bone marrow of normal mice. The effect of targeted gene silencing was found to last for more than a week after a single intravenous injection [119, 121]. We have previously shown that delivery of anti-sense miRNAs blocking the function of endogenous miR-21 and miR-10b by PLGA nanoparticles can prevent metastasis while showing significant antitumor effect in triple negative breast cancer in vivo in small animal models (Figure 13.6) [4].

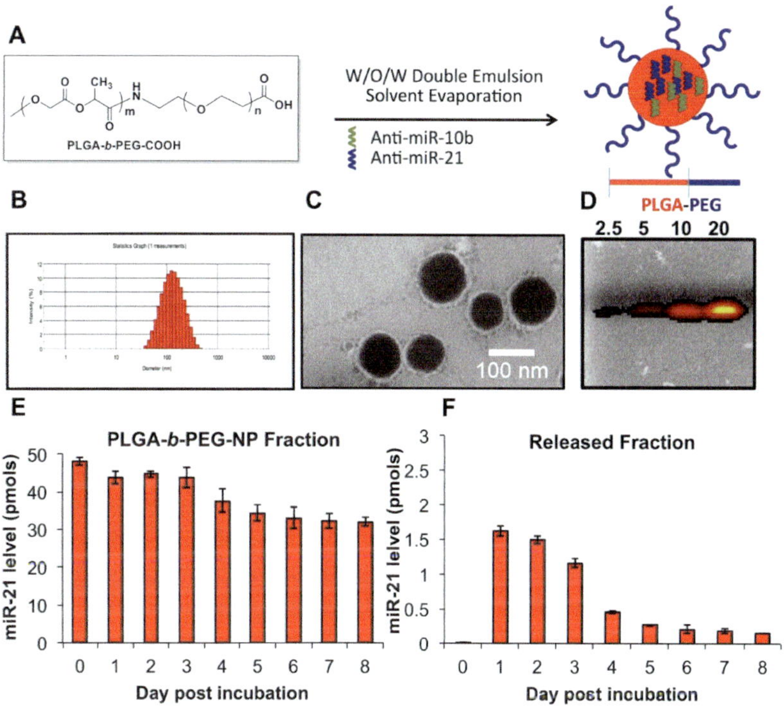

Figure 13.6 PLGA nanoparticles preparation and characterization. (A) Schematic illustration of nanoparticle formulation. (B) Hydrodynamic size of antisense-miRNA encapsulated PLGA-b-PEG nanoparticles measured by dynamic light scattering (DLS). (C) TEM image of antisense-miRNA encapsulated PLGA-b-PEG nanoparticles taken after staining with 1% phosphotungstic acid (Scale bar, 100 nm). (D) Evaluation of co-loaded Cy5-labeled antisense-miR-21 (10%) from the encapsulated PLGA-b-PEG nanoparticles after resolved in 3% agarose gel electrophoresis by optical CCD-camera imaging with the excitation of 570 nm and emission filter at 660 nm. (E & F) PLGA-b-PEG nanoparticles loaded miRNA-21 release profile evaluated after seeding the co-loaded nanoparticles in PBS at physiological pH for eight days of incubation at 37 °C by qRT-PCR analysis: (E) miR-21 fraction presents in nanoparticles in different time points after incubation. (F) miR-21 fraction released from nanoparticles in different time points after incubation. Reproduced with permission from [4].

Babar et al. reported the synthesis of anti-miR-155 encapsulated PLGA nanoparticles using a double emulsion solvent evaporation method [122]. PLGA nanoparticles' surfaces were modified with cell penetrating peptides (CPP) after loading with anti-miR-155, which upon delivery reduced the growth of pre-B-cell tumors in vivo in mice and appeared as an encouraging therapeutic option for lymphoma and leukemia [122]. The miRNA-10b positively controls breast cancer cell migration and invasion through inhibition of HOXD10 target synthesis, and directly inhibits breast cancer metastasis. Anti-miRNA-10b-loaded poly-L-lysine (PLL-anti-miR-10b) nanoparticles, when administered to MDA-MB-231 breast cancer cells, strongly inhibit the invasive property of cells [123]. The miR-21 is an important player in a majority of cancers. Anti-miR-21 and 5-FU-loaded poly(amidoamine) dendrimer nanoparticles significantly enhanced the cytotoxicity of 5-FU and strongly increased the apoptosis of U251 GBM cells, and the migration ability of tumor cells was significantly decreased [124].

Liang et al. reported a preparation of cationic PLGA-based nanoparticles of about 60 nm in size by coating PLGA core with a cationic PEI shell for gene delivery using the emulsion evaporation method [125]. These cationic PLGA–PEI nanoparticles were loaded with miRNA-26a expression vector (pDNA) and studied in HepG2 HCC cells. The results showed efficient delivery and expression of miRNA in HepG2 cells, and improved the effect of functional miRNA-26a delivery. Those results showed that these PLGA–PEI nanoparticles are promising non-viral vectors for gene delivery [125]. In another study, cyclic RGD peptide (cRGD) conjugated PEGylated liposome-polycation-hyaluronic acid (PEG–LPH) targeted nanoparticles efficiently delivered the anti-miR-296 oligonucleotides (AMOs) to cytoplasm and down-regulated target mRNA in HUVECs by blocking the function of endogenous miR-296 [101]. These targeted cRGD peptide-conjugated PEG–LPH nanoparticles also showed potential anti-angiogenesis activity with the delivery of anti-miR-296 oligonucleotides to the targeted site in vivo, as assessed using the matrigel plug assay [101].

13.3.2.1.3 Advantages of polymer nanoparticles for drug delivery

Nanobiotechnology allows the delivery of drugs through nanoformulations that provide desired release kinetics for prolonged periods of time. Polymer nanoparticles are biodegradable and surface modifiable. These nanoparticles protect and deliver the drugs and small RNAs. In addition, the polymer matrix is also subsequently degraded over time through normal metabolic process. Moreover, polymer nanoparticles possess unique capabilities to deliver multiple drugs with different properties in one carrier (drugs in combination with small RNAs). Polymer nanoparticles can both passively and actively target tumor cells, and can increase the effectiveness of drugs as compared to direct delivery [126–129]. In addition, PEGylation of polymer nanoparticles also increases the circulation time of nanoparticles in blood and passively enhances their concentration at tumor sites by the EPR effect [130, 131]. Moreover, polymer nanoparticles containing targeted ligands can actively target tumors through specific ligand–receptor interactions. The efficiency with which targeted nanoparticles selectively accumulate at the tumor site mainly depends on various factors such as (i) selective over expression of the receptor in the tumor cells in comparison to normal non-target cells, and (ii) receptor accessibility to the ligands [130, 131]. Although important progress has been made in tumor molecular targeting, significant developments are still required for therapists in practice to target drug delivery within solid tumors selectively.

An important advantage of the biodegradable polymer nanoparticles as drug delivery systems is that there is no post-treatment clearance required to remove the polymer nanoparticle waste after the drug administration period, owing to the degradation of the polymer. The use of biodegradable polymer nanoparticles offers a better alternative to conventional cancer treatments by capping the drug material inside the nanoparticles and shielding the drug from the immune system. However, further investigations are necessary regarding the possibility of enzymatic degradation of these nanoparticles, owing to biophysical resemblance, as well as further evaluation of their cytotoxicity [132].

13.4 Natural Polymers

While many synthetic biodegradable polymers have been used to synthesize nanoparticles for drugs and nucleic acid deliveries, a number of polymers of natural sources (natural polymers) have also been successfully used for nanoparticle synthesis in drug delivery applications. Carbohydrates, proteins, and muscle fibers are some natural polymers used in nanoparticle synthesis. However, most of these natural polymer materials are derived from plants (starch, hemicellulose, cellulose, glucomannan, pectin, agar, etc.), microbes (curdlan, gellan, and xanthan), algae (alginate and carrageenan), and fungi (pullulan, chitin, and scleroglucon). The gelling property of many of these natural polymers has been explored for drug delivery, especially for the slow release of drugs. Originally natural polymers were used as adjuvants in vaccine deliveries. Natural polymers are widely used in small molecule drug deliveries, but their use in small RNA delivery is very limited. Predominantly, polysaccharide-based nanoparticles were used for siRNA deliveries in cancer therapy. Similarly, chitosan and hyaluronic acid are the other two natural polymers used for siRNA delivery [133]. Since chitosan is a positively charged polymer, it is commonly used to deliver negatively charged nucleic acids. The nanoparticles prepared from other neutral and negatively charged polymers need additional coating for loading of negatively charged nucleic acids or direct conjugation of nucleic acids using cleavable linkers before they can be used for deliveries. Chitosan is combined with other synthetic polymers as hybrid polymers for nucleic acid deliveries. Chitosan has also been used as a coating material to functionalize metal nanoparticles for nucleic acid deliveries. We recently showed that gold–iron oxide nanoparticles coated with chitosan–cyclodextrin hybrid polymer can be used for efficient intranasal delivery of miRNAs as a GBM therapy [37].

13.5 Small RNA Deliveries in Clinical Trials

SiRNAs are currently in different clinical trial stages for the treatment of many human diseases. Predominantly, siRNA therapies are used for the treatment of metabolic and biochemical disorders, chronic viral infections, and for oncological diseases. Stable siRNA-liposome complexes have been well explored compounds for clinical siRNA therapies. Table 13.1 shows some of the siRNAs currently in clinical trials for different diseases.

Table 13.1 Nanoparticle-mediated siRNA deliveries currently in clinical trials [134]

Small RNA	Nanoparticles	Target pathway	Disease	Trial status	Sponsor
CALAA-01 (siRNA to RRM2)	Rondel® nanoparticles (CD)	M2 subunit of ribonucleotide reductase (RRM2)	Solid tumors	I	Calando Pharmaceuticals
Atu027	siRNA with 2'-O-Me and cationic lipid	Protein kinase N3 (PKN3)	Advanced solid tumors (metastatic pancreatic cancer)	I	Silence Therapeutics GmbH
ALN-RSV	Lipid nanoparticles	VEGF gene and KSP gene	Solid tumors (liver metastasis from colon cancer)	I	Alnylam Pharmaceuticals
DCR-MYC	Lipid nanoparticles	Myc	HCC	I	Dicerna Pharmaceuticals
siRNAEphA2-DOPC	DOPC liposomes	Ephrin type-A receptor2 (EphA2) gene	Advanced cancers	I	M.D. Anderson Cancer Center
siG12D-LODER	LODER® (Polymer)	KRAS (mutation G12D in KRAS oncogene)	Solid tumors (advanced pancreatic cancer)	II	Silenseed Ltd

When compared to siRNAs, research on miRNA applications has started much later, but miRNA-based therapeutics delivered using different nanocarriers is currently receiving much interest. Since miRNAs are endogenous and tend to target multiple genes, extensive preclinical validations are important compared to siRNAs, which are designed and intended for a one-RNA-one-gene therapeutic concept. Table 13.2 shows some of the nanoparticle-based miRNA deliveries currently in various stages of clinical trials for different diseases.

Table 13.2 Nanoparticle-mediated miRNA deliveries currently in clinical trials [135]

Companies	Product	Targeted miR	Disease type	Development phase	Reference
Roche/Santaris	Miravirsen	miR-122	HCV	II	roche.com
Regulus Therapeutics	RG-101	miR-122	HCV	II (hold)	regulusrx.com
	RG-012	miR-21	Alport syndrome	I	
	RG-125	miR-103/107	NASH	I	
MiRagen Therapeutics	MRG-201	miR-29b	Fibrosis	II	miragentherapeutics.com
	MRG-106	miR-155	Lymphoma and leukemia	I and II	
	MRG-107	miR-155	ALS	Pre-clinical	
	MRG-110	miR-92	Ischemia	I	
ENGeneIC	Mesomir	miR-16	Mesothelioma	I	engeneic.com
Abivax	ABX464	miR-124	IBD	II	abivax.com
Synlogic	Screening				synlogictx.com
Opko	Screening				opko.com
Alnylam Pharmaceuticals	Screening				alnylam.com
InteRNA Technologies	Screening				interna-technologies.com
Mello Biotech	Screening				mellobiotech.com

13.6 Conclusion and Future Perspectives

The current era of nanotechnology offers an unprecedented opportunity to deliver therapeutic molecules to treat a wide variety of cellular diseases, including cancer. In parallel, the availability of novel molecularly targeted therapeutic agents (miRNAs, siRNAs, and shRNAs) to reprogram cells (along with additional small molecule drugs), can add value to a molecular therapeutic strategy. Nanoparticles synthesized from biocompatible and biodegradable polymers of natural and synthetic sources, with properties that allow the loading or co-loading of drugs and small RNAs, facilitate further enhancement of available treatment options. Recent developments in bioengineering designs may also allow for locoregional selective delivery of therapeutics to target tissues using intravascular catheterization. Further, we recently developed US–MB-mediated drug delivery approach for efficient local delivery of miRNAs to target tumor tissues [10, 136–138]. This method showed effective treatment enhancement in mouse models of breast cancer and HCC. Combining catheter-based targeted delivery along with the US–MB mediated local delivery in future applications will likely further improve cancer therapy with minimal toxic side effects.

References

1. Cheng, Y.; Wang, Q.; Jiang, W., Bian, Y.; Zhou, Y.; Gou, A.; Zhang, W.; Fu, K.; Shi, W., Emerging roles of piRNAs in cancer: Challenges and prospects. *Aging (Albany NY)* **2019**, *11* (21), 9932–9946.
2. Chalbatani, G. M.; Dana, H.; Memari, F.; Gharagozlou, E.; Ashjaei, S.; Kheirandish, P.; Marmari, V.; Mahmoudzadeh, H.; Mozayani, F.; Maleki, A. R.; Sadeghian, E.; Nia, E. Z.; Miri, S. R.; Nia, N. Z.; Rezaeian, O.; Eskandary, A.; Razavi, N.; Shirkhoda, M.; Rouzbahani, F. N., Biological function and molecular mechanism of piRNA in cancer. *Pract Lab Med* **2019**, *13*, e00113.
3. Sorrentino, S., The eight human "canonical" ribonucleases: Molecular diversity, catalytic properties, and special biological actions of the enzyme proteins. *FEBS Lett* **2010**, *584* (11), 2194–200.
4. Devulapally, R.; Sekar, N. M.; Sekar, T. V.; Foygel, K.; Massoud, T. F.; Willmann, J. K.; Paulmurugan, R., Polymer nanoparticles mediated

codelivery of antimiR-10b and antimiR-21 for achieving triple negative breast cancer therapy. *ACS Nano* **2015**, *9* (3), 2290–302.

5. Ananta, J. S.; Paulmurugan, R.; Massoud, T. F., Tailored nanoparticle codelivery of antimiR-21 and antimiR-10b augments glioblastoma cell kill by temozolomide: Toward a "personalized" anti-microRNA therapy. *Mol Pharm* **2016**, *13* (9), 3164–75.
6. Wagner, E., Advances in cancer gene therapy: Tumor-targeted delivery of therapeutic pDNA, siRNA, and dsRNA nucleic acids. *J BUON* **2007**, *12 Suppl 1*, S77–82.
7. Cardoso, A.; Trabulo, S.; Moreira, J. N.; Duzgunes, N.; de Lima, M. C., Targeted lipoplexes for siRNA delivery. *Methods Enzymol* **2009**, *465*, 267–87.
8. Asgeirsdottir, S. A.; Talman, E. G.; de Graaf, I. A.; Kamps, J. A.; Satchell, S. C.; Mathieson, P. W.; Ruiters, M. H.; Molema, G., Targeted transfection increases siRNA uptake and gene silencing of primary endothelial cells in vitro: A quantitative study. *J Control Release* **2010**, *141* (2), 241–51.
9. Koynova, R.; Tarahovsky, Y. S.; Wang, L.; MacDonald, R. C., Lipoplex formulation of superior efficacy exhibits high surface activity and fusogenicity, and readily releases DNA. *Biochim Biophys Acta* **2007**, *1768* (2), 375–86.
10. Chowdhury, S. M.; Lee, T.; Bachawal, S. V.; Devulapally, R.; Abou-Elkacem, L.; Yeung, T. A.; Wischhusen, J.; Tian, L.; Dahl, J.; Paulmurugan, R.; Willmann, J. K., Longitudinal assessment of ultrasound-guided complementary microRNA therapy of hepatocellular carcinoma. *J Control Release* **2018**, *281*, 19–28.
11. Hattori, Y.; Nakamura, M.; Takeuchi, N.; Tamaki, K.; Shimizu, S.; Yoshiike, Y.; Taguchi, M.; Ohno, H.; Ozaki, K. I.; Onishi, H., Effect of cationic lipid in cationic liposomes on siRNA delivery into the lung by intravenous injection of cationic lipoplex. *J Drug Target* **2019**, *27* (2), 217–227.
12. Semple, S. C.; Klimuk, S. K.; Harasym, T. O.; Dos Santos, N.; Ansell, S. M.; Wong, K. F.; Maurer, N.; Stark, H.; Cullis, P. R.; Hope, M. J.; Scherrer, P., Efficient encapsulation of antisense oligonucleotides in lipid vesicles using ionizable aminolipids: Formation of novel small multilamellar vesicle structures. *Biochim Biophys Acta* **2001**, *1510* (1 & 2), 152–66.
13. Hayes, M. E.; Drummond, D. C.; Kirpotin, D. B.; Zheng, W. W.; Noble, C. O.; Park, J. W.; Marks, J. D.; Benz, C. C.; Hong, K., Genospheres:

Self-assembling nucleic acid-lipid nanoparticles suitable for targeted gene delivery. *Gene Ther* **2006**, *13* (7), 646–51.

14. Conceicao, M.; Mendonca, L.; Nobrega, C.; Gomes, C.; Costa, P.; Hirai, H.; Moreira, J. N.; Lima, M. C.; Manjunath, N.; Pereira de Almeida, L., Intravenous administration of brain-targeted stable nucleic acid lipid particles alleviates Machado-Joseph disease neurological phenotype. *Biomaterials* **2016**, *82*, 124–37.
15. Matzke, M. A.; Primig, M.; Trnovsky, J.; Matzke, A. J., Reversible methylation and inactivation of marker genes in sequentially transformed tobacco plants. *EMBO J* **1989**, *8* (3), 643–9.
16. Fire, A.; Xu, S.; Montgomery, M. K.; Kostas, S. A.; Driver, S. E.; Mello, C. C., Potent and specific genetic interference by double-stranded RNA in Caenorhabditis elegans. *Nature* **1998**, *391* (6669), 806–11.
17. Davis, M. E.; Zuckerman, J. E.; Choi, C. H.; Seligson, D.; Tolcher, A.; Alabi, C. A.; Yen, Y.; Heidel, J. D.; Ribas, A., Evidence of RNAi in humans from systemically administered siRNA via targeted nanoparticles. *Nature* **2010**, *464* (7291), 1067–70.
18. Young, S. W.; Stenzel, M.; Yang, J. L., Nanoparticle-siRNA: A potential cancer therapy? *Crit Rev Oncol Hematol* **2016**, *98*, 159–69.
19. Jiao, M.; Zhang, P.; Meng, J.; Li, Y.; Liu, C.; Luo, X.; Gao, M., Recent advancements in biocompatible inorganic nanoparticles towards biomedical applications. *Biomater Sci* **2018**, *6* (4), 726–745.
20. Kim, H. S.; Lee, D. Y., Near-infrared-responsive cancer photothermal and photodynamic therapy using gold nanoparticles. *Polymers (Basel)* **2018**, *10* (9).
21. Vines, J. B.; Yoon, J. H.; Ryu, N. E.; Lim, D. J.; Park, H., Gold nanoparticles for photothermal cancer therapy. *Front Chem* **2019**, *7*, 167.
22. Siu, K. S.; Zhang, Y.; Zheng, X.; Koropatnick, J.; Min, W. P., Non-covalently functionalized of single-walled carbon nanotubes by DSPE-PEG-PEI for siRNA delivery. *Methods Mol Biol* **2016**, *1364*, 151–63.
23. Cao, Y.; Huang, H. Y.; Chen, L. Q.; Du, H. H.; Cui, J. H.; Zhang, L. W.; Lee, B. J.; Cao, Q. R., Enhanced lysosomal escape of pH-responsive polyethylenimine-betaine functionalized carbon nanotube for the codelivery of survivin small interfering RNA and doxorubicin. *ACS Appl Mater Interfaces* **2019**, *11* (10), 9763–9776.
24. Li, J.; Ge, X.; Cui, C.; Zhang, Y.; Wang, Y.; Wang, X.; Sun, Q., Preparation and characterization of functionalized graphene oxide carrier for siRNA delivery. *Int J Mol Sci* **2018**, *19* (10).

25. Imani, R.; Shao, W.; Taherkhani, S.; Emami, S. H.; Prakash, S.; Faghihi, S., Dual-functionalized graphene oxide for enhanced siRNA delivery to breast cancer cells. *Colloids Surf B Biointerfaces* **2016**, *147*, 315–325.

26. Varkouhi, A. K.; Foillard, S.; Lammers, T.; Schiffelers, R. M.; Doris, E.; Hennink, W. E.; Storm, G., SiRNA delivery with functionalized carbon nanotubes. *Int J Pharm* **2011**, *416* (2), 419–25.

27. Siu, K. S.; Zheng, X.; Liu, Y.; Zhang, Y.; Zhang, X.; Chen, D.; Yuan, K.; Gillies, E. R.; Koropatnick, J.; Min, W. P., Single-walled carbon nanotubes noncovalently functionalized with lipid modified polyethylenimine for siRNA delivery in vitro and in vivo. *Bioconjug Chem* **2014**, *25* (10), 1744–51.

28. Yin, F.; Hu, K.; Chen, Y.; Yu, M.; Wang, D.; Wang, Q.; Yong, K. T.; Lu, F.; Liang, Y.; Li, Z., SiRNA delivery with PEGylated graphene oxide nanosheets for combined photothermal and genetherapy for pancreatic cancer. *Theranostics* **2017**, *7* (5), 1133–1148.

29. Wang, X.; Sun, Q.; Cui, C.; Li, J.; Wang, Y., Anti-HER2 functionalized graphene oxide as survivin-siRNA delivery carrier inhibits breast carcinoma growth in vitro and in vivo. *Drug Des Devel Ther* **2018**, *12*, 2841–2855.

30. Paulmurugan, R.; Ajayan, P. M.; Liepmann, D.; Renugopalakrishnan, V., Intracellular microRNA quantification in intact cells: A novel strategy based on reduced graphene oxide based fluorescence quenching. *MRS Commun* **2018**, *8* (3), 642–651.

31. Volkov, Y., Quantum dots in nanomedicine: Recent trends, advances and unresolved issues. *Biochem Biophys Res Commun* **2015**, *468* (3), 419–27.

32. Tan, W. B.; Jiang, S.; Zhang, Y., Quantum-dot based nanoparticles for targeted silencing of HER2/neu gene via RNA interference. *Biomaterials* **2007**, *28* (8), 1565–71.

33. Getz, T.; Qin, J.; Medintz, I. L.; Delehanty, J. B.; Susumu, K.; Dawson, P. E.; Dawson, G., Quantum dot-mediated delivery of siRNA to inhibit sphingomyelinase activities in brain-derived cells. *J Neurochem* **2016**, *139* (5), 872–885.

34. Ghaderi, S.; Ramesh, B.; Seifalian, A. M., Fluorescence nanoparticles "quantum dots" as drug delivery system and their toxicity: A review. *J Drug Target* **2011**, *19* (7), 475–86.

35. Guo, S.; Huang, Y.; Jiang, Q.; Sun, Y.; Deng, L.; Liang, Z.; Du, Q.; Xing, J.; Zhao, Y.; Wang, P. C.; Dong, A.; Liang, X. J., Enhanced gene delivery and

siRNA silencing by gold nanoparticles coated with charge-reversal polyelectrolyte. *ACS Nano* **2010**, *4* (9), 5505–11.

36. Zhang, K.; Hao, L.; Hurst, S. J.; Mirkin, C. A., Antibody-linked spherical nucleic acids for cellular targeting. *J Am Chem Soc* **2012**, *134* (40), 16488–91.
37. Sukumar, U. K.; Bose, R. J. C.; Malhotra, M.; Babikir, H. A.; Afjei, R.; Robinson, E.; Zeng, Y.; Chang, E.; Habte, F.; Sinclair, R.; Gambhir, S. S.; Massoud, T. F.; Paulmurugan, R., Intranasal delivery of targeted polyfunctional gold-iron oxide nanoparticles loaded with therapeutic microRNAs for combined theranostic multimodality imaging and presensitization of glioblastoma to temozolomide. *Biomaterials* **2019**, *218*, 119342.
38. Pinese, C.; Lin, J.; Milbreta, U.; Li, M.; Wang, Y.; Leong, K. W.; Chew, S. Y., Sustained delivery of siRNA/mesoporous silica nanoparticle complexes from nanofiber scaffolds for long-term gene silencing. *Acta Biomater* **2018**, *76*, 164–177.
39. Tong, W. Y.; Alnakhli, M.; Bhardwaj, R.; Apostolou, S.; Sinha, S.; Fraser, C.; Kuchel, T.; Kuss, B.; Voelcker, N. H., Delivery of siRNA in vitro and in vivo using PEI-capped porous silicon nanoparticles to silence MRP1 and inhibit proliferation in glioblastoma. *J Nanobiotechnology* **2018**, *16* (1), 38.
40. Sun, D.; Zhang, W.; Li, N.; Zhao, Z.; Mou, Z.; Yang, E.; Wang, W., Silver nanoparticles-quercetin conjugation to siRNA against drug-resistant Bacillus subtilis for effective gene silencing: In vitro and in vivo. *Mater Sci Eng C Mater Biol Appl* **2016**, *63*, 522–34.
41. Chan, J. M.; Valencia, P. M.; Zhang, L.; Langer, R.; Farokhzad, O. C., Polymeric nanoparticles for drug delivery. *Methods in Molecular Biology* **2010**, *624*, 163–75.
42. Chen, K. J.; Tang, L.; Garcia, M. A.; Wang, H.; Lu, H.; Lin, W. Y.; Hou, S.; Yin, Q.; Shen, C. K.; Cheng, J.; Tseng, H. R., The therapeutic efficacy of camptothecin-encapsulated supramolecular nanoparticles. *Biomaterials* **2012**, *33* (4), 1162–9.
43. Liu, Y.; Zhang, N., Gadolinium loaded nanoparticles in theranostic magnetic resonance imaging. *Biomaterials* **2012**, *33* (21), 5363–75.
44. Devulapally, R.; Paulmurugan, R., Polymer nanoparticles for drug and small silencing RNA delivery to treat cancers of different phenotypes. *Wiley Interdiscip Rev Nanomed Nanobiotechnol* **2014**, *6* (1), 40–60.

45. Kumari, A.; Yadav, S. K.; Yadav, S. C., Biodegradable polymeric nanoparticles based drug delivery systems. *Colloids and Surfaces. B, Biointerfaces* **2010**, *75* (1), 1–18.

46. Soppimath, K. S.; Aminabhavi, T. M.; Kulkarni, A. R.; Rudzinski, W. E., Biodegradable polymeric nanoparticles as drug delivery devices. *Journal of Controlled Release* **2001**, *70* (1 & 2), 1–20.

47. Rao, J. P.; Geckeler, K. E., Polymer nanoparticles: Preparation techniques and size-control parameters. *Progress in Polymer Science* **2011**, *36* (7), 887–913.

48. Danhier, F.; Lecouturier, N.; Vroman, B.; Jerome, C.; Marchand-Brynaert, J.; Feron, O.; Preat, V., Paclitaxel-loaded PEGylated PLGA-based nanoparticles: In vitro and in vivo evaluation. *Journal of Controlled Release: Official Journal of the Controlled Release Society* **2009**, *133* (1), 11–17.

49. Gryparis, E. C.; Mattheolabakis, G.; Bikiaris, D.; Avgoustakis, K., Effect of conditions of preparation on the size and encapsulation properties of PLGA-mPEG nanoparticles of cisplatin. *Drug Deliv* **2007**, *14* (6), 371–80.

50. Gryparis, E. C.; Hatziapostolou, M.; Papadimitriou, E.; Avgoustakis, K., Anticancer activity of cisplatin-loaded PLGA-mPEG nanoparticles on LNCaP prostate cancer cells. *Eur J Pharm Biopharm* **2007**, *67* (1), 1–8.

51. Cheng, L.; Jin, C.; Lv, W.; Ding, Q.; Han, X., Developing a highly stable PLGA-mPEG nanoparticle loaded with cisplatin for chemotherapy of ovarian cancer. *PLoS One* **2011**, *6* (9), e25433.

52. Dhar, S.; Gu, F. X.; Langer, R.; Farokhzad, O. C.; Lippard, S. J., Targeted delivery of cisplatin to prostate cancer cells by aptamer functionalized Pt(IV) prodrug-PLGA-PEG nanoparticles. *Proc Natl Acad Sci USA* **2008**, *105* (45), 17356–61.

53. Dhar, S.; Kolishetti, N.; Lippard, S. J.; Farokhzad, O. C., Targeted delivery of a cisplatin prodrug for safer and more effective prostate cancer therapy in vivo. *Proc Natl Acad Sci USA* **2011**, *108* (5), 1850–5.

54. Graf, N.; Bielenberg, D. R.; Kolishetti, N.; Muus, C.; Banyard, J.; Farokhzad, O. C.; Lippard, S. J., alpha(V)beta(3) integrin-targeted PLGA-PEG nanoparticles for enhanced anti-tumor efficacy of a Pt(IV) prodrug. *ACS Nano* **2012**, *6* (5), 4530–9.

55. Nair, K. L.; Jagadeeshan, S.; Nair, S. A.; Kumar, G. S., Biological evaluation of 5-fluorouracil nanoparticles for cancer chemotherapy and its dependence on the carrier, PLGA. *International Journal of Nanomedicine* **2011**, *6*, 1685–97.

56. Cheng, M.; He, B.; Wan, T.; Zhu, W.; Han, J.; Zha, B.; Chen, H.; Yang, F.; Li, Q.; Wang, W.; Xu, H.; Ye, T., 5-Fluorouracil nanoparticles inhibit hepatocellular carcinoma via activation of the p53 pathway in the orthotopic transplant mouse model. *PLoS One* **2012**, *7* (10), e47115.

57. Cheng, M. R.; Li, Q.; Wan, T.; He, B.; Han, J.; Chen, H. X.; Yang, F. X.; Wang, W.; Xu, H. Z.; Ye, T.; Zha, B. B., Galactosylated chitosan/5-fluorouracil nanoparticles inhibit mouse hepatic cancer growth and its side effects. *World J Gastroenterol* **2012**, *18* (42), 6076–87.

58. Yadav, A. K.; Agarwal, A.; Rai, G.; Mishra, P.; Jain, S.; Mishra, A. K.; Agrawal, H.; Agrawal, G. P., Development and characterization of hyaluronic acid decorated PLGA nanoparticles for delivery of 5-fluorouracil. *Drug Deliv* **2010**, *17* (8), 561–72.

59. Li, S.; Wang, A.; Jiang, W.; Guan, Z., Pharmacokinetic characteristics and anticancer effects of 5-fluorouracil loaded nanoparticles. *BMC Cancer* **2008**, *8*, 103.

60. Danhier, F.; Ansorena, E.; Silva, J. M.; Coco, R.; Le Breton, A.; Preat, V., PLGA-based nanoparticles: An overview of biomedical applications. *Journal of Controlled Release: Official Journal of the Controlled Release Society* **2012**, *161* (2), 505–22.

61. Betancourt, T.; Brown, B.; Brannon-Peppas, L., Doxorubicin-loaded PLGA nanoparticles by nanoprecipitation: Preparation, characterization and in vitro evaluation. *Nanomedicine* **2007**, *2* (2), 219–32.

62. Chang, L.; Deng, L.; Wang, W.; Lv, Z.; Hu, F.; Dong, A.; Zhang, J., Poly(ethyleneglycol)-b-poly(epsilon-caprolactone-co-gamma-hydroxyl-epsilon-caprolactone) bearing pendant hydroxyl groups as nanocarriers for doxorubicin delivery. *Biomacromolecules* **2012**, *13* (10), 3301–10.

63. Wohlfart, S.; Khalansky, A. S.; Gelperina, S.; Maksimenko, O.; Bernreuther, C.; Glatzel, M.; Kreuter, J., Efficient chemotherapy of rat glioblastoma using doxorubicin-loaded PLGA nanoparticles with different stabilizers. *PLoS One* **2011**, *6* (5), e19121.

64. Chittasupho, C.; Xie, S. X.; Baoum, A.; Yakovleva, T.; Siahaan, T. J.; Berkland, C. J., ICAM-1 targeting of doxorubicin-loaded PLGA nanoparticles to lung epithelial cells. *Eur J Pharm Sci* **2009**, *37* (2), 141–50.

65. Park, H.; Yang, J.; Lee, J.; Haam, S.; Choi, I. H.; Yoo, K. H., Multifunctional nanoparticles for combined doxorubicin and photothermal treatments. *ACS Nano* **2009**, *3* (10), 2919–26.

66. Amjadi, I.; Rabiee, M.; Hosseini, M. S.; Mozafari, M., Synthesis and characterization of doxorubicin-loaded poly(lactide-co-glycolide) nanoparticles as a sustained-release anticancer drug delivery system. *Appl Biochem Biotechnol* **2012**, *168* (6), 1434–47.

67. Liu, C. W.; Lin, W. J., Polymeric nanoparticles conjugate a novel heptapeptide as an epidermal growth factor receptor-active targeting ligand for doxorubicin. *International Journal of Nanomedicine* **2012**, *7*, 4749–67.

68. Wolf, D. M.; Langan-Fahey, S. M.; Parker, C. J.; McCague, R.; Jordan, V. C., Investigation of the mechanism of tamoxifen-stimulated breast tumor growth with nonisomerizable analogues of tamoxifen and metabolites. *J Natl Cancer Inst* **1993**, *85* (10), 806–12.

69. Perez, E.; Benito, M.; Teijon, C.; Olmo, R.; Teijon, J. M.; Blanco, M. D., Tamoxifen-loaded nanoparticles based on a novel mixture of biodegradable polyesters: Characterization and in vitro evaluation as sustained release systems. *J Microencapsul* **2012**, *29* (4), 309–22.

70. Cavalli, R.; Bisazza, A.; Bussano, R.; Trotta, M.; Civra, A.; Lembo, D.; Ranucci, E.; Ferruti, P., Poly(amidoamine)-cholesterol conjugate nanoparticles obtained by electrospraying as novel tamoxifen delivery system. *J Drug Deliv* **2011**, *2011*, 587604.

71. Jain, A. K.; Swarnakar, N. K.; Godugu, C.; Singh, R. P.; Jain, S., The effect of the oral administration of polymeric nanoparticles on the efficacy and toxicity of tamoxifen. *Biomaterials* **2011**, *32* (2), 503–15.

72. Renoir, J. M.; Stella, B.; Ameller, T.; Connault, E.; Opolon, P.; Marsaud, V., Improved anti-tumoral capacity of mixed and pure anti-oestrogens in breast cancer cell xenografts after their administration by entrapment in colloidal nanosystems. *J Steroid Biochem Mol Biol* **2006**, *102* (1–5), 114–27.

73. Arya, G.; Vandana, M.; Acharya, S.; Sahoo, S. K., Enhanced antiproliferative activity of Herceptin (HER2)-conjugated gemcitabine-loaded chitosan nanoparticle in pancreatic cancer therapy. *Nanomedicine: Nanotechnology, Biology, and Medicine* **2011**, *7* (6), 859–70.

74. Arias, J. L.; Reddy, L. H.; Couvreur, P., Superior preclinical efficacy of gemcitabine developed as chitosan nanoparticulate system. *Biomacromolecules* **2011**, *12* (1), 97–104.

75. Aggarwal, S.; Yadav, S.; Gupta, S., EGFR targeted PLGA nanoparticles using gemcitabine for treatment of pancreatic cancer. *J Biomed Nanotechnol* **2011**, *7* (1), 137–8.

76. Papa, A. L.; Basu, S.; Sengupta, P.; Banerjee, D.; Sengupta, S.; Harfouche, R., Mechanistic studies of Gemcitabine-loaded nanoplatforms in resistant pancreatic cancer cells. *BMC Cancer* **2012**, *12*, 419.

77. Jia, L.; Zheng, J. J.; Jiang, S. M.; Huang, K. H., Preparation, physicochemical characterization and cytotoxicity in vitro of gemcitabine-loaded PEG-PDLLA nanovesicles. *World J Gastroenterol* **2010**, *16* (8), 1008–13.

78. Erdogar, N.; Iskit, A. B.; Mungan, N. A.; Bilensoy, E., Prolonged retention and in vivo evaluation of cationic nanoparticles loaded with Mitomycin C designed for intravesical chemotherapy of bladder tumours. *J Microencapsul* **2012**, *29* (6), 576–82.

79. Erdogar, N.; Mungan, A.; Bilensoy, E., Preparation and characterization of cationic nanoparticles loaded with Mitomycin C by double emulsion and ionotropic gelation techniques. *Journal of Controlled Release: Official Journal of the Controlled Release Society* **2010**, *148* (1), e78–9.

80. Hou, Z.; Zhan, C.; Jiang, Q.; Hu, Q.; Li, L.; Chang, D.; Yang, X.; Wang, Y.; Li, Y.; Ye, S.; Xie, L.; Yi, Y.; Zhang, Q., Both FA- and mPEG-conjugated chitosan nanoparticles for targeted cellular uptake and enhanced tumor tissue distribution. *Nanoscale Res Lett* **2011**, *6* (1), 563.

81. Yordanov, G.; Skrobanska, R.; Evangelatov, A., Entrapment of epirubicin in poly(butyl cyanoacrylate) colloidal nanospheres by nanoprecipitation: Formulation development and in vitro studies on cancer cell lines. *Colloids and Surfaces. B, Biointerfaces* **2012**, *92*, 98–105.

82. Zhang, H. Z.; Li, X. M.; Gao, F. P.; Liu, L. R.; Zhou, Z. M.; Zhang, Q. Q., Preparation of folate-modified pullulan acetate nanoparticles for tumor-targeted drug delivery. *Drug Deliv* **2010**, *17* (1), 48–57.

83. Khalil, N. M.; do Nascimento, T. C.; Casa, D. M.; Dalmolin, L. F.; de Mattos, A. C.; Hoss, I.; Romano, M. A.; Mainardes, R. M., Pharmacokinetics of curcumin-loaded PLGA and PLGA-PEG blend nanoparticles after oral administration in rats. *Colloids and Surfaces. B, Biointerfaces* **2013**, *101*, 353–60.

84. Yallapu, M. M.; Dobberpuhl, M. R.; Maher, D. M.; Jaggi, M.; Chauhan, S. C., Design of curcumin loaded cellulose nanoparticles for prostate cancer. *Curr Drug Metab* **2012**, *13* (1), 120–8.

85. Mukerjee, A.; Vishwanatha, J. K., Formulation, characterization and evaluation of curcumin-loaded PLGA nanospheres for cancer therapy. *Anticancer Res* **2009**, *29* (10), 3867–75.

86. Anselmo, A. C.; Mitragotri, S., Nanoparticles in the clinic: An update. *Bioeng Transl Med* **2019**, *4* (3), e10143.

87. Li, Z.; Tan, S.; Li, S.; Shen, Q.; Wang, K., Cancer drug delivery in the nano era: An overview and perspectives (review). *Oncol Rep* **2017**, *38* (2), 611–624.

88. Lee, R. C.; Feinbaum, R. L.; Ambros, V., The C. elegans heterochronic gene lin-4 encodes small RNAs with antisense complementarity to lin-14. *Cell* **1993**, *75* (5), 843–54.

89. Ghildiyal, M.; Zamore, P. D., Small silencing RNAs: An expanding universe. *Nat Rev Genet* **2009**, *10* (2), 94–108.

90. Rao, D. D.; Vorhies, J. S.; Senzer, N.; Nemunaitis, J., siRNA vs. shRNA: Similarities and differences. *Adv Drug Deliv Rev* **2009**, *61* (9), 746–59.

91. Tokatlian, T.; Segura, T., siRNA applications in nanomedicine. *Wiley Interdisciplinary Reviews. Nanomedicine And Nanobiotechnology* **2010**, *2* (3), 305–15.

92. Ozpolat, B.; Sood, A. K.; Lopez-Berestein, G., Nanomedicine based approaches for the delivery of siRNA in cancer. *J Intern Med* **2010**, *267* (1), 44–53.

93. Hammond, S. M., MicroRNAs as oncogenes. *Curr Opin Genet Dev* **2006**, *16* (1), 4–9.

94. Hamilton, A. J.; Baulcombe, D. C., A species of small antisense RNA in posttranscriptional gene silencing in plants. *Science* **1999**, *286* (5441), 950–2.

95. Woodrow, K. A.; Cu, Y.; Booth, C. J.; Saucier-Sawyer, J. K.; Wood, M. J.; Saltzman, W. M., Intravaginal gene silencing using biodegradable polymer nanoparticles densely loaded with small-interfering RNA. *Nat Mater* **2009**, *8* (6), 526–33.

96. Zeng, P.; Xu, Y.; Zeng, C.; Ren, H.; Peng, M., Chitosan-modified poly(D,L-lactide-co-glycolide) nanospheres for plasmid DNA delivery and HBV gene-silencing. *International Journal of Pharmaceutics* **2011**, *415* (1 & 2), 259–66.

97. Du, J.; Shi, Q. S.; Sun, Y.; Liu, P. F.; Zhu, M. J.; Du, L. F.; Duan, Y. R., Enhanced delivery of monomethoxypoly(ethylene glycol)-poly (lactic-*co*-glycolic acid)-poly l-lysine nanoparticles loading platelet-derived growth factor BB small interfering RNA by ultrasound and/or microbubbles to rat retinal pigment epithelium cells. *J Gene Med* **2011**, *13* (6), 312–23.

98. Du, J.; Sun, Y.; Shi, Q. S.; Liu, P. F.; Zhu, M. J.; Wang, C. H.; Du, L. F.; Duan, Y. R., Biodegradable nanoparticles of mPEG-PLGA-PLL

triblock copolymers as novel non-viral vectors for improving siRNA delivery and gene silencing. *International Journal of Molecular Sciences* **2012**, *13* (1), 516–33.

99. Li, Y. H.; Shi, Q. S.; Du, J.; Jin, L. F.; Du, L. F.; Liu, P. F.; Duan, Y. R., Targeted delivery of biodegradable nanoparticles with ultrasound-targeted microbubble destruction-mediated hVEGF-siRNA transfection in human PC-3 cells in vitro. *Int J Mol Med* **2013**, *31* (1), 163–71.

100. Lu, Z. X.; Liu, L. T.; Qi, X. R., Development of small interfering RNA delivery system using PEI-PEG-APRPG polymer for antiangiogenic vascular endothelial growth factor tumor-targeted therapy. *International Journal of Nanomedicine* **2011**, *6*, 1661–73.

101. Liu, X. Q.; Song, W. J.; Sun, T. M.; Zhang, P. Z.; Wang, J., Targeted delivery of antisense inhibitor of miRNA for antiangiogenesis therapy using cRGD-functionalized nanoparticles. *Molecular Pharmaceutics* **2011**, *8* (1), 250–9.

102. Dragomira Nikolova, D. T., RNA interference – Regulations and application in oncology. *Journal of Cancer Molecules* **2008**, *4* (3), 67–77.

103. Rao, D. D.; Vorhies, J. S.; Senzer, N.; Nemunaitis, J., siRNA vs. shRNA: Similarities and differences. *Adv Drug Deliv Rev* **2009**, *61* (9), 746–59.

104. Paddison, P. J.; Caudy, A. A.; Bernstein, E.; Hannon, G. J.; Conklin, D. S., Short hairpin RNAs (shRNAs) induce sequence-specific silencing in mammalian cells. *Genes Dev* **2002**, *16* (8), 948–58.

105. Zhang, C.; Wang, Y. S.; Wu, H.; Zhang, Z. X.; Cai, Y.; Hou, H. Y.; Zhao, W.; Yang, X. M.; Ma, J. X., Inhibitory efficacy of hypoxia-inducible factor 1alpha short hairpin RNA plasmid DNA-loaded poly (D, L-lactide-co-glycolide) nanoparticles on choroidal neovascularization in a laser-induced rat model. *Gene Therapy* **2010**, *17* (3), 338–51.

106. Qazi, Y.; Stagg, B.; Singh, N.; Singh, S.; Zhang, X.; Luo, L.; Simonis, J.; Kompella, U. B.; Ambati, B. K., Nanoparticle-mediated delivery of shRNA.VEGF-a plasmids regresses corneal neovascularization. *Invest Ophthalmol Vis Sci* **2012**, *53* (6), 2837–44.

107. Yang, J.; Xie, S. X.; Huang, Y.; Ling, M.; Liu, J.; Ran, Y.; Wang, Y.; Thrasher, J. B.; Berkland, C.; Li, B., Prostate-targeted biodegradable nanoparticles loaded with androgen receptor silencing constructs eradicate xenograft tumors in mice. *Nanomedicine* **2012**, *7* (9), 1297–309.

108. Ferrara, N., VEGF and the quest for tumour angiogenesis factors. *Nat Rev Cancer* **2002**, *2* (10), 795–803.

109. Kim, D. H.; Rossi, J. J., Strategies for silencing human disease using RNA interference. *Nat Rev Genet* **2007**, *8* (3), 173–84.

110. Whitehead, K. A.; Langer, R.; Anderson, D. G., Knocking down barriers: Advances in siRNA delivery. *Nat Rev Drug Discov* **2009**, *8* (2), 129–38.

111. Huang, Z.; Dong, L.; Chen, J.; Gao, F.; Zhang, Z.; Zhang, J., Low-molecular weight chitosan/vascular endothelial growth factor short hairpin RNA for the treatment of hepatocellular carcinoma. *Life Sci* **2012**, *91* (23 & 24), 1207–15.

112. Yin, Q.; Gao, Y.; Zhang, Z.; Zhang, P.; Li, Y., Bioreducible poly (beta-amino esters)/shRNA complex nanoparticles for efficient RNA delivery. *Journal of Controlled Release: Official Journal of the Controlled Release Society* **2011**, *151* (1), 35–44.

113. Bartel, D. P., MicroRNAs: Target recognition and regulatory functions. *Cell* **2009**, *136* (2), 215–33.

114. Bartel, D. P., MicroRNAs: Genomics, biogenesis, mechanism, and function. *Cell* **2004**, *116* (2), 281–97.

115. Bhardwaj, A.; Singh, S.; Singh, A. P., MicroRNA-based cancer therapeutics: Big hope from small RNAs. *Mol Cell Pharmacol* **2010**, *2* (5), 213–219.

116. Cimmino, A.; Calin, G. A.; Fabbri, M.; Iorio, M. V.; Ferracin, M.; Shimizu, M.; Wojcik, S. E.; Aqeilan, R. I.; Zupo, S.; Dono, M.; Rassenti, L.; Alder, H.; Volinia, S.; Liu, C. G.; Kipps, T. J.; Negrini, M.; Croce, C. M., miR-15 and miR-16 induce apoptosis by targeting BCL2. *Proc Natl Acad Sci USA* **2005**, *102* (39), 13944–9.

117. Elbashir, S. M.; Harborth, J.; Lendeckel, W.; Yalcin, A.; Weber, K.; Tuschl, T., Duplexes of 21-nucleotide RNAs mediate RNA interference in cultured mammalian cells. *Nature* **2001**, *411* (6836), 494–8.

118. Hammond, S. M.; Bernstein, E.; Beach, D.; Hannon, G. J., An RNA-directed nuclease mediates post-transcriptional gene silencing in Drosophila cells. *Nature* **2000**, *404* (6775), 293–6.

119. Krutzfeldt, J.; Rajewsky, N.; Braich, R.; Rajeev, K. G.; Tuschl, T.; Manoharan, M.; Stoffel, M., Silencing of microRNAs in vivo with 'antagomirs'. *Nature* **2005**, *438* (7068), 685–9.

120. Baigude, H.; Rana, T. M., Interfering nanoparticles for silencing microRNAs. *Methods in Enzymology* **2012**, *509*, 339–53.

121. Mattes, J.; Yang, M.; Foster, P. S., Regulation of microRNA by antagomirs: A new class of pharmacological antagonists for the

specific regulation of gene function? *Am J Respir Cell Mol Biol* **2007**, *36* (1), 8–12.

122. Babar, I. A.; Cheng, C. J.; Booth, C. J.; Liang, X.; Weidhaas, J. B.; Saltzman, W. M.; Slack, F. J., Nanoparticle-based therapy in an in vivo microRNA-155 (miR-155)-dependent mouse model of lymphoma. *Proc Natl Acad Sci USA* **2012**, *109* (26), E1695–704.

123. Jin, H.; Yu, Y.; Chrisler, W. B.; Xiong, Y.; Hu, D.; Lei, C., Delivery of MicroRNA-10b with polylysine nanoparticles for inhibition of breast cancer cell wound healing. *Breast Cancer (Auckl)* **2012**, *6*, 9–19.

124. Ren, Y.; Kang, C. S.; Yuan, X. B.; Zhou, X.; Xu, P.; Han, L.; Wang, G. X.; Jia, Z.; Zhong, Y.; Yu, S.; Sheng, J.; Pu, P. Y., Co-delivery of as-miR-21 and 5-FU by poly(amidoamine) dendrimer attenuates human glioma cell growth in vitro. *J Biomater Sci Polym Ed* **2010**, *21* (3), 303–14.

125. Liang, G. F.; Zhu, Y. L.; Sun, B.; Hu, F. H.; Tian, T.; Li, S. C.; Xiao, Z. D., PLGA-based gene delivering nanoparticle enhance suppression effect of miRNA in HePG2 cells. *Nanoscale Res Lett* **2011**, *6*, 447.

126. Jin Zhang, C. Q. L.; Michael, Post; Benoit, Simard; Yves, Deslandes; Tsung, Han Hsieh, Design of nanoparticles as drug carriers for cancer therapy. *Cancer Genomics & Proteomics* **2006**, *3*, 147–157.

127. Brewer, E.; Coleman, J.; Lowman, A., Emerging technologies of polymeric nanoparticles in cancer drug delivery. *Journal of Nanomaterials* **2011**, *2011*.

128. Peer, D.; Karp, J. M.; Hong, S.; Farokhzad, O. C.; Margalit, R.; Langer, R., Nanocarriers as an emerging platform for cancer therapy. *Nat Nanotechnol* **2007**, *2* (12), 751–60.

129. Acharya, S.; Sahoo, S. K., PLGA nanoparticles containing various anticancer agents and tumour delivery by EPR effect. *Advanced Drug Delivery Reviews* **2011**, *63* (3), 170–183.

130. Hochwald, S. N., Molecular-targeted therapy for cancer and nanotechnology. *Methods in Molecular Biology* **2010**, *624*, 11–23.

131. Bae, Y. H.; Park, K., Targeted drug delivery to tumors: Myths, reality and possibility. *Journal of Controlled Release: Official Journal of the Controlled Release Society* **2011**, *153* (3), 198–205.

132. Yildirimer, L.; Thanh, N. T.; Loizidou, M.; Seifalian, A. M., Toxicology and clinical potential of nanoparticles. *Nano Today* **2011**, *6* (6), 585–607.

133. Serrano-Sevilla, I.; Artiga, A.; Mitchell, S. G.; De Matteis, L.; de la Fuente, J. M., Natural polysaccharides for siRNA delivery: Nanocarriers based on chitosan, hyaluronic acid, and their derivatives. *Molecules* **2019**, *24* (14).

134. Mahmoodi Chalbatani, G.; Dana, H.; Gharagouzloo, E.; Grijalvo, S.; Eritja, R.; Logsdon, C. D.; Memari, F.; Miri, S. R.; Rad, M. R.; Marmari, V., Small interfering RNAs (siRNAs) in cancer therapy: A nano-based approach. *Int J Nanomedicine* **2019**, *14*, 3111–3128.

135. Bonneau, E.; Neveu, B.; Kostantin, E.; Tsongalis, G. J.; De Guire, V., How close are miRNAs from clinical practice? A perspective on the diagnostic and therapeutic market. *EJIFCC* **2019**, *30* (2), 114–127.

136. Devulapally, R.; Lee, T.; Barghava-Shah, A.; Sekar, T. V.; Foygel, K.; Bachawal, S. V.; Willmann, J. K.; Paulmurugan, R., Ultrasound-guided delivery of thymidine kinase-nitroreductase dual therapeutic genes by PEGylated-PLGA/PIE nanoparticles for enhanced triple negative breast cancer therapy. *Nanomedicine (Lond)* **2018**, *13* (9), 1051–1066.

137. Mullick Chowdhury, S.; Wang, T. Y.; Bachawal, S.; Devulapally, R.; Choe, J. W.; Abou Elkacem, L.; Yakub, B. K.; Wang, D. S.; Tian, L.; Paulmurugan, R.; Willmann, J. K., Ultrasound-guided therapeutic modulation of hepatocellular carcinoma using complementary microRNAs. *J Control Release* **2016**, *238*, 272–280.

138. Wang, T. Y.; Choe, J. W.; Pu, K.; Devulapally, R.; Bachawal, S.; Machtaler, S.; Chowdhury, S. M.; Luong, R.; Tian, L.; Khuri-Yakub, B.; Rao, J.; Paulmurugan, R.; Willmann, J. K., Ultrasound-guided delivery of microRNA loaded nanoparticles into cancer. *J Control Release* **2015**, *203*, 99–108.

Chapter 14

Theranostics: A New Holistic Approach in Nanomedicine

Ankit Rochani[a] and Sreejith Raveendran[b]

[a]*Jefferson College of Pharmacy, Thomas Jefferson University, Philadelphia, Pennsylvania 19107, USA*
[b]*Nanoscale Research Facility, Indian Institute of Technology Delhi, Hauz Khas, New Delhi 110016, India*

anrochani@gmail.com, sreejithr84@gmail.com

14.1 Introduction

Nanotechnology is the science that comprises nanomedicine, nanomaterial, and the latest technology of nanorobotics. Nanoscience focuses mostly on the development of the nanoparticles (NPs) or nanomaterials used in the development of medicines, therapies, diagnostics, communication sensors, and other related tools. In the world of medicine, nanotechnology has played a significant role in developing treatment of various age-related conditions such as tumor or non-tumor forming cancers (Doxil®, first anticancer liposomal medicine), diabetes, tissue degeneration, and infectious conditions (AmBiosome®, first liposomal antibiotic) [1]. Following are some basic concepts that are most commonly explored for developing better nanoformulations or nanotheranostic designs:

Bionanotechnology in Cancer: Diagnosis and Therapy
Edited by D. Sakthi Kumar and Aswathy Ravindran Girija

ISBN 978-981-4800-30-3 (Hardcover), 978-0-429-42291-1 (eBook)
www.jennystanford.com

- Targeted delivery of active pharmaceutical ingredients
- Stimulated controlled delivery systems
- NPs with multiple modes of treatments
- Nanomaterial implants for localized therapies

The NP-based delivery systems have been evolving over time and have led to the development of novel nanomaterials with hybrid properties or improved characteristics compared to their parent material. For example, the development of solid lipid NPs, polymeric metallic NPs, antibody-tagged polymeric or metallic NPs, and alloy NPs are some examples that were developed to provide better therapeutic profiles [2–4]. The conventional passive- or active-targeted nanoformulation therapies do not provide real-time analysis of their primary therapeutic goals. The problem of visualizing tissue distribution of the NPs using imaging techniques while performing therapy prompted researchers to develop technologies can combine the therapy and diagnostic abilities of nanoformulations. Independent and historic scientific breakthroughs have led to the foundation of therapeutic-diagnostic treatment approaches that can be applied to various disease conditions.

Early studies conducted with radioactive isotopic seed for cancer therapy provided an early foundation for the development of theranostics [5]. It all started after Enrico Fermi developed the first isolated ^{128}I in 1934, which led to the development of ^{130}I and ^{131}I for the treatment of thyrotoxicosis in 1941 and for thyroid cancer in 1943. After World War II, this material was used for diagnostic testing and therapy. Radioactive isotopes of rare or noble earth metals were some of the first few materials (in the early 1950s) to be explored for possible theranostic medicine [6, 7]. These inorganic or metallic particles have been commonly imaged using techniques like positron emission tomography (PET), X-rays, or radioactive ray mappings with scintillation counters. On the other hand, certain organic materials (doxorubicin (DOX), curcumin, ellipticine, and other coumarin drugs) act as fluorescent materials [8–11]. In recent past, various important concepts were developed for image-guided cancer targeted therapies, like self-assembled polymeric nano or microparticles, or conjugational hybrid fluorescent materials that use a simple fluorescent imaging technique. As a result, around 2002, a new concept was introduced to perform therapy and real-time diagnosis of the disease condition like solid tumors. This technology was named theranostics or theragnostics [12].

The development of nanotheranostics mostly focuses on (i) efficient synthesis strategy (for new particle designs with therapeutic potential), and (ii) high-resolution imaging technique. In terms of synthesis, self-assembly has been the primary focus of most medicinal NP synthesis; for instance, self-assembly of polymeric-drug or fluorophore-conjugated system or the polymer coating of metallic or organo-metallic NPs for developing theranostic nanoparticles (TNPs). These NPs can be seen using imaging technologies like photoacoustic imaging (PAI), fluorescence imaging, magnetic resonance imaging (MRI), electron microscopy, and mass spectrometry imaging [12]. In particular, mass spectrometry imaging for NPs is relatively new and still under development [13], whereas other techniques have been validated and used in multiple preclinical and clinical investigations. Further, some studies have also suggested that molecular theranostics is also one of the emerging technologies, wherein researchers have developed fluorescence tagged aptamers or antibodies for the targeted therapies along with diagnostic capabilities [14, 15]. This knowledge has been extensively explored for developing target-specific nanotheranostic therapies.

The science of theranostics is exponentially growing (Figure 14.1). Our data analysis from Scopus publication records shows that theranostics covers only a small literature space (~13%) of the NP research database (shown in Figure 14.1). Currently, more studies are focused on researching fundamentals of newly developed hybrid NPs (with potential as theranostic agents), which could shine or contrast in the presence of light (for diagnosis) and simultaneously provide a therapeutic response.

The world of theranostics is constantly changing and developing new concepts. As per data from the Scopus database (Figure 14.1 a–c), the number of publications has consistently

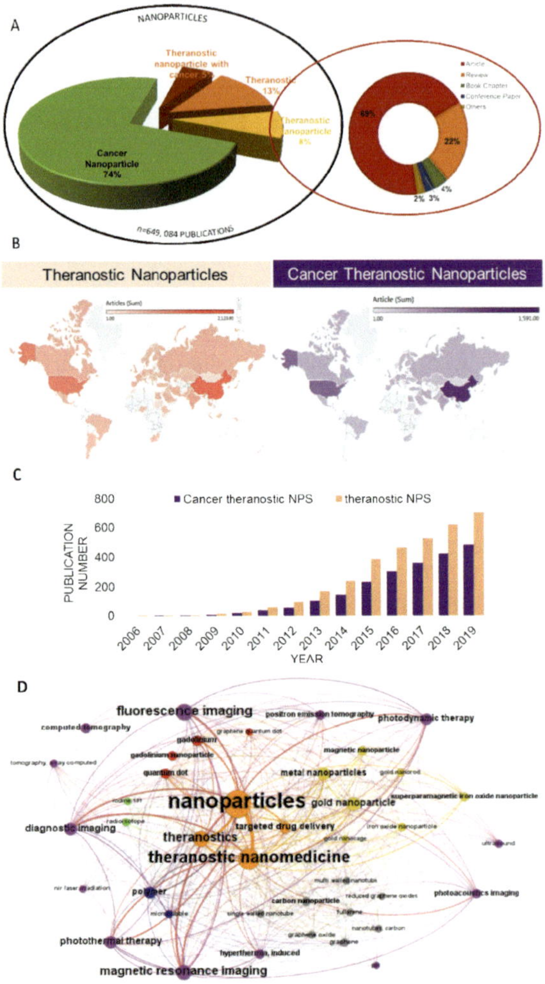

Figure 14.1 (A) Distribution of research work for theranostics in the publication landscape of NPs. (B) Distribution of publication data for TNPs with and without application in cancer by various countries in the world with respective heat maps. (C) Trend of publication numbers over period of time (for TNPs with and without application in cancer therapy) across the world. All the data was taken from Scopus and analyzed for understanding the publication landscape of TNPs. (D) Shows the network trend between important keywords in the landscape of TNPs. The lines indicate weightage of co-occurrence and relationship of some important keywords derived (n = 5515) from Scopus database. This clearly shows the transition of traditional nanomedicine toward theranostics.

increased since 2006. Nearly 69% of the publications are research articles, and <4% are books, conference materials, or other articles about theranostics. The analysis shows that China and the US, followed by India, produce most of the knowledge in NP research in the development of TNPs with or without application in cancer therapy.

In the last decade (according to PubMed), there were around 40 published records that discuss the use of theranostic systems in Phase I and Phase II clinical trials, which align with data provided by clinicaltrials.gov, with around 39 registered clinical trials [16]. These included trials in which metallic NPs and formulation designs based on the conjugation of radioisotopes labeled with proteins or oligonucleotides (ON) were tested. It took nearly six decades to bring these therapeutic designs (with the potential of being theranostic systems) into clinical studies.

In this chapter, we have discussed various intriguing TNP designs and strategies that were developed using polymeric or organo-metallic systems. These include the use of natural (like ON, proteins, and polypeptide), semi-synthetic or synthetic self-assembled TNPs, or bioconjugated materials for theranostic application [17, 18]. Most of the classical molecular and nanotheranostic systems developed were using either conjugation or non-conjugated, synthetic, or self-assembled NPs. Hence, we have also discussed the chemical nature of these small organic and organo-metallic self-assembled nanostructures and their fluorescence potential in their native and assembled polymeric NP states. The present work provides a new holistic overview of the nanotheranostic delivery systems.

14.2 Bioconjugation Technology for Theranostic Materials

Deliveries of gene or polypeptide-material-tagged fluorescent labels have been extensively explored using molecular cell biology techniques to map the therapeutic targets. Most conjugation chemistry systems follow the classical ethyl (dimethylaminopropyl) carbodiimide (EDC)/N-hydroxysuccinimide (NHS) reaction between the free carboxyl and primary amines to create amide linkages. For instance, the development of fluorescein isothiocyanate (FITC)-conjugated biomolecular systems (antibody, polymer, and ONs) is the most commonly explored for theranostic applications [19–21]. The use of FITC in conjugated state or particle-encapsulated state has become a common practice for cancer biologists for a variety of diagnostic applications.

Bioconjugation chemistries between dyes have been extensively explored for developing novel molecular structures with potential theranostic applications. Godbey et al. developed polyethylenimine (PEI)-DNA labeled with Oregon Green 488 and rhodamine. Upon examination with confocal laser microscopy, these systems can provide a yellow signal from green PEI and DNA materials. They even studied PEI and DNA individually by using green and red signals [22]. Although this technique may have some limitations, this material could provide an important concept for being explored as a potential theranostic tool [12].

Most bioconjugation concepts depend on technologies like fluorescence resonance energy transfer or froster resonance energy transfer (FRET), fluorescence correlation spectroscopy (FCS), or Fluorescence fluctuation spectroscopy (FFS). Lucas et al. developed polymeric pegylated poly[2-(dimethylamino) ethyl] methacrylate-co-aminoethyl methacrylate ((PEG-PDMAEMA-co-AEMA))-Cy5-labeled poly(L-Lysine) and rhodamine-labeled-ON complex. They used FFS for tracking the nanocomposite in cell lines [23]. Due to the high resolution of FFS, they were able to map the dissociation of the polymeric part from labeled-ON within the cytoplasm. Bioconjugational chemistry is an important tool for synthesis of novel materials that can be used for developing TNPs for advanced drug delivery applications [24–27]. Fluorescence imaging (a type of photoluminescence (PL) imaging) has been used to develop image-guided surgeries and other medicinal applications. These conjugation techniques and PL imaging methods are the basic principle behind the development of more complex and advanced NP-based theranostic systems. The classical conjugation chemistry works better with molecules with free proton-donating and proton-accepting moieties to create covalent conjugation. However, the conjugation between fluorescing drug molecules and polymeric biomaterials (protein, ON, or polymers) can be difficult to execute in certain cases. Hence, self-assembled particles can come in handy in these situations to create theranostic or imaging particles.

14.3 Polymeric Nano and Microstructures (Natural & Synthetic)

Apart from chemically conjugated systems, nanocomposites have also been explored for use as a novel theranostic-based delivery platform for cancer therapy. These systems follow two main design principles: (i) fluorescent-material-encapsulated self-assembly NPs, and (ii) use of bioconjugated fluorescent materials (discussed above) for NP development. In this section, we have highlighted the use of some fluorescent materials that are commonly explored on experimental grounds to develop better theranostic nanomedicine. We have also discussed the use of polymers as delivery systems for metallic NP theranostics.

Some classical fluorescent drug molecules like carotenoids (lutein, lutein epoxide, violaxanthin, and others), coumarins, DOX, and curcumin serve as important fluorescent probes that are used extensively as nanocomposites for anticancer theranostic applications [28 31]. These materials have been used as a combination of metallic NPs or quantum dots (QDs) for developing both active and passive targeted theranostic systems. Table 14.1 lists all the potential fluorescent and FRET probes.

Coumarin is commonly explored in the development of the conjugated delivery system for the theranostic application. Examples of this type of formulation are drug–coumarin–protein bioconjugates. Sukhendu et al. developed Gemcitabine–Coumarin–Biotin conjugates for anticancer theranostic applications against A549 and WI38 cell lines [28]. In another similar design, xanthene–coumarin–chlorambucil conjugates (prodrug) were used for developing a theranostic system using FRET imaging [33]. These concepts of conjugation can be used for the development of better nanotheranostic systems, such as photo-responsive multifunctional magnetic nanoparticles (MNPs) tagged with coumarin moiety as a photo trigger for the release of chlorambucil. This complex NP design was used for cellular luminescence imaging and had a strong fluorescence signal along with a controlled release of the drug molecule, indicating that it could function as a theranostic system [34].

Table 14.1 List of molecules with potential as fluorophores (with excitation and emission wavelengths) and FRET probes with their potentials as theranostic agents [32]

Fluorophore	Excitation wavelength range (nm)	Emission wavelength range (nm)	Excitation wavelength (nm)	Emission maximum (nm)	Peak extinction coefficient ($L \cdot mol^{-1} \cdot cm^{-1}$)
DiO	350–550	475–650	484	501	154,000
DiI	400–605d	530–700	549	565	148,000
DiD	495–700	600–800	644	665	264,000
DiR	500–830	720–900	750	779	270,000
Cy3	402–610	499–740	555	570	150,000
Cy5	400–700	600–800	646	662	250,000
Cy5.5	510–750	635–800	673	707	209,000
Cy7	500–850	700–850	750	773	199,000
Cy7.5	546–900	750–890	788	808	223,000
Alexa Fluor 488	400–600	475–680	490	525	72,000
Alexa Fluor 546	360–674	530–700	556	573	112,000
BODIPY-FL	400–550	470–640	503	509	80,000
BODIPY-TR	448–600	515–700	589	616	69,000
FITC	300–540	480–675	490	525	78,000
Coumarin 6	350–505	450–750	467	502	NA
Rhodamine B	440–585	520–760	555	580	NA

Molecules like DOX and curcumin tend to fluoresce and can be visualized in the cellular system using confocal microscopy, and they are also an important anticancer molecule. DOX has been explored by developing various hybrid (Fe or AuNPs) or non-hybrid polymeric NPs systems with synthetic and natural polymers for targeted and untargeted anticancer theranostic applications [35]. One such simple design is anti-nucleolin (AS1411)-targeted DOX-loaded PLGA NPs for theranostic application.

This was tested and proven effective against the C26 (colon) cancer cell line under in vitro and in vivo conditions (as shown in Figure 14.2) [36]. In recent past, it was shown that transferrin (Tf)-templated copper nanocluster (Tf-Cu) was also synthesized and formulated into spherical Tf-Cu DOX NPs due to electrostatic interaction between DOX and Tf-Cu. It acts as a theranostic system when analyzed using the FRET imaging technique [37]. Like DOX, curcumin is also extensively explored for its theranostic potential in the form of polymeric (PLGA, MPEG-PCL, β-cyclodextrin, poly(butyl cyanoacrylate, polyethylene glycol (PEG)-cholesterol, and others) NP systems [9]. Curcumin is a generic material that has been extensively explored for its use in various infectious and non-infectious diseases. This natural molecule has been considered a model system in developing various types of interesting TNP designs by combining different types of metallic or organic drug molecules.

Polymeric biomaterials have also been used to develop self-assembled nanotheranostic agents that could be diagnosed using ultrasound imaging. Ultrasound provides a relatively cheaper option for bioimaging and provides means for controlled delivery of microbubbles. Typical ultrasound machines work in the frequency range of 2–60 MHz, and conventional instruments operate in the range of 2–15 MHz. An increase in the frequency helps in improving resolution and imaging depth. The 50 MHz ultrasound can provide imaging of 8–9 mm deep [38]. Some commonly used ultrasound frequencies for conventional imaging applications are provided in Table 14.2.

Table 14.2 List of frequencies used for specific imaging applications [39, 40]

Frequency	Imaging application
2.5 MHz	Obstetric, gynecological, and deep abdomen imaging
3.5 MHz	Obstetric, general abdomen, and gynecological imaging
5 MHz	Vascular, breast, and general abdomen imaging
7.5 MHz	Breast and thyroid
10 MHz	Thyroid, breast, superficial masses and veins, and musculoskeletal imaging
15 MHz	Musculoskeletal and superficial structures imaging

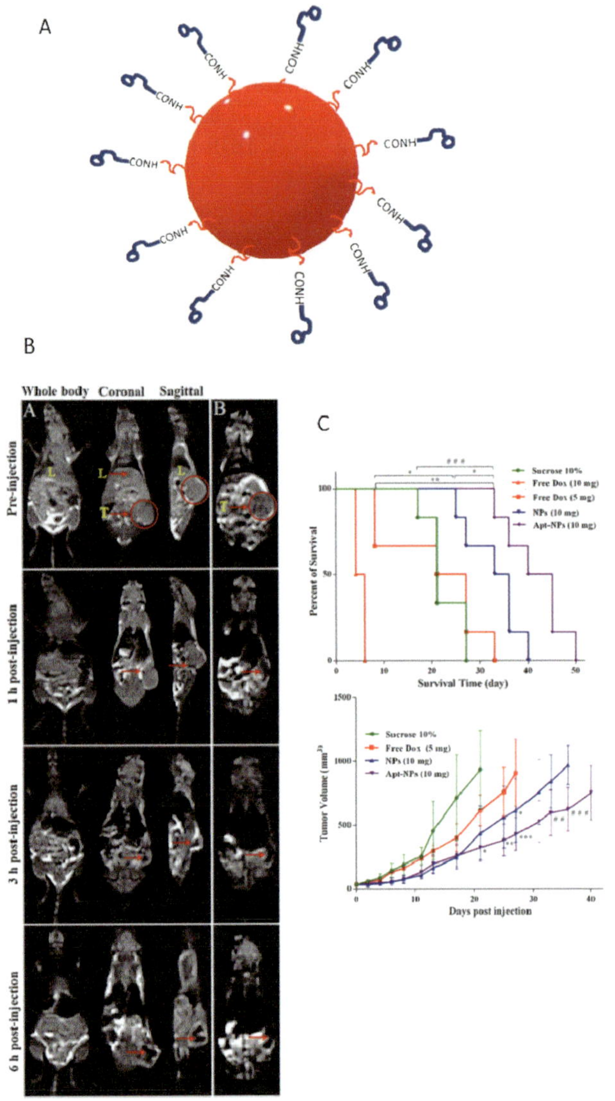

Figure 14.2 (A) NP design for DOX and Fe_3O_4 loaded nucleolin–aptamer-tagged PEG–PLGA magnetic NPs of size around 165 nm. (B) MRI images of female BALB/c mice bearing C26 colon cancer. The arrow shows before and after injection of NP with an increase in contrast from time 1 h to 6 h post-injection of NPs. (C) Increased survival of treatment group with aptamer-NPs (test) compared to controls free DOX (+ve control), NPs (-ve control), sucrose 10% (untreated -ve control). Reproduced with permission from Ref. [36].

Ultrasound technology, with an increase in frequency, also helps generate localized heating potential. For instance, 3 MHz ultrasound may provide heating at 0.5 cm deeper in the tissue. The temperature of 40 °C was generated within 4 min of ultrasound exposure in n = 18 healthy volunteers [41]. This suggests that the technique could be used to generate a theranostic controlled delivery platform. Typically, microbubbles are used as ultrasound-based theranostic agents. These microbubbles are synthesized using proteins, polysaccharides, polymers, lipids, and surfactants. As the name suggests, microbubbles consist of biomaterial shells filled with an inner core containing air, nitrogen, sulfur hexafluoride (SF6), and perfluorocarbons (PFCs) [42]. Microbubbles expand and relax when exposed to ultrasound frequencies. These vibrations increase at a higher frequency. This makes these systems better for imaging abnormalities rather than the normal tissues. The vibrations are the harmonic signals, which ultrasound can listen to and create enhanced greyscale images [43]. The first commercialized air-filed microbubble preparation, called albunex, was used for diagnostic applications. Today, we have both conventional microbubbles (beyond the scope of discussion) and nanobubble systems that are being explored for theranostic applications [42].

Pharmacological inspiration for the synthesis of nanobubbles comes from the fact that most cancer tumors have leaky vasculature of pore size in the range of 400–600 nm. This makes it easy for nanobubbles to get internalized in the tumors and can be used for diagnostic or therapeutic applications. Various microbubble designs (using different types of biomaterials) provided the necessary background to develop nanobubbles. For instance, Xu et al. developed colon cancer-targeted PLGA nanobubbles for the dual mode of targeting cancer imaging using optical and ultrasound applications [44]. In another simple design, a 0.25% albumin solution was used to prepare nanobubbles filed with PFC gas instead of air that was used for the synthesis of albunex (the first commercially available microbubble) [45, 46]. Recently, albumin-stabilized nanobubbles were compared with lipid nanobubbles for their efficiency in gene transfer using handheld ultrasound scanners at 1 MHz and 2.3 MHz frequencies under in vitro and in vivo conditions, respectively.

It was observed that lipid nanobubbles could work better than albumin nanobubbles for gene delivery during both in vitro and in vivo experiments [47]. Nanobubbles have been extensively researched and synthesized for siRNA deliveries, MNPs, drug molecules, and other means of delivery using ultrasound [48–51] (Figure 14.3). Some studies also show theranostic potential via dual modes of treatment by ultrasound and magnetic targeting. Overall these studies clearly show that nanobubbles could be used for controlled theranostic application depending upon the payloads they are carrying.

14.4 Radio-Isotopic Nanomaterials

Radioactive isotopes or seeds have been extensively explored as implants for post-surgical removal of tumors. These radioactive seeds generate necessary gamma radiation in the resected tumor cavity and prevent the recurrence of the tumor by killing the left-over cancer cells. These radioactive seeds also work as potential contrast agents and can be visualized using PET-CT or MRI scans [52].

Most radioactive materials emit gamma rays or positrons that are used for imaging applications. For instance, materials like technetium-99m (99mTc) or iodine-123 (^{123}I) use a gamma camera or single photo emission computed tomography (SPECT) for an imaging application. It is also suggested that high resolution can be achieved using PET imaging. Historically, radium was the first to be used for the treatment of prostate cancer. The first synthetic isotope was ^{198}Au, which was used for implantation in prostate cancer. Later, ^{125}I and ^{103}Pd were used for the treatment. There are several reports of the development of implantable radioactive seeds.

However, the use of radio nanoseeds can provide better theranostics agents. For instance, ^{103}Pd@Au nanoseeds were synthesized for the treatment of prostate tumors in mice models. The study showed that five weeks of single-dose treatment helped in greater than 80% reduction of tumor burden, and more than 95% of seeds were retained in the tumor, observed by SPECT due to gamma radiation emission by ^{103}Pd [53]. Radioactive therapy with gold nanoparticles (AuNPs) (such as ^{177}Lu-AuNPs, Au-dithiolated diethylenetriamine pentaacetic acid

(DTDPA), or ^{103}Pd-AuNPs) has been extensively evaluated as a potential therapeutic agent [54–56]. These materials have been gaining attention regarding their potential for treating of tumors, but there exist other safe and green nanomaterial designs being used for therapeutic purposes.

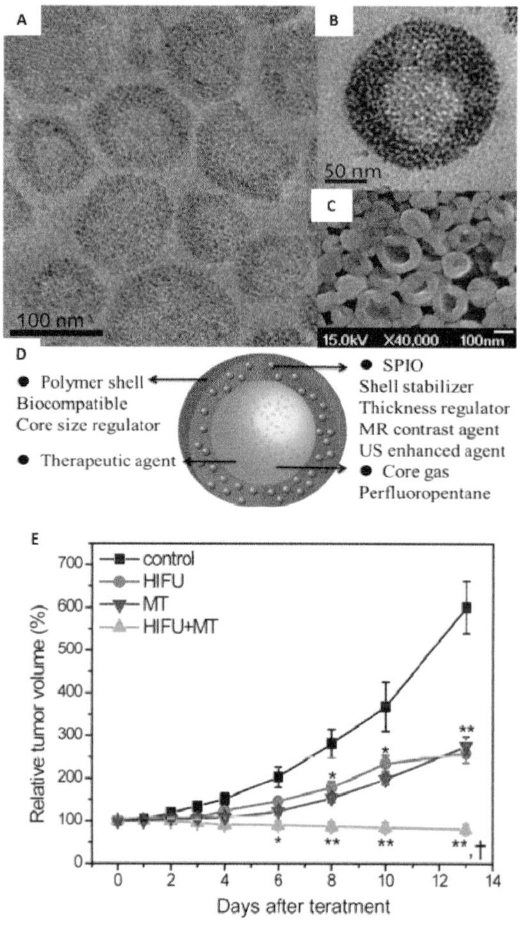

Figure 14.3 (A), (B) TEM images of nanobubbles under different magnification. (C) SEM images of collapsed nanobubbles under vacuum caused by instrument. (D) Schematic of nanobubble consists of core as gas (perfluoropentane) with a therapeutic agent, a polymeric shell consisting of MNPs. (E) Analysis of tumor growth in vivo rat model in the presence of high intensity frequency ultrasound (HIFU) and magnetic targeting. It shows reduction in tumor under dual therapy of HIFU+MT. Reproduced with permission from Ref. [50].

14.5 Nano Carbon Structures

Other than optical imaging (having penetration depth of ~1 mm in tissue), PAI is another imaging technology that is commonly explored for the deep tissue diagnostic application. This technology helps in converting optical energy to ultrasound waves in biological tissues. By principle, the contrast (also known as a chromophore) agents present in the tissue absorb laser light, leading to thermal contraction and relaxation process, which causes the formation of ultrasound (also known as photoacoustic) waves in tissue. This technique provides imaging depth of ~4 cm (using hemoglobin as a PAI agent at 800 nm wavelength) [57, 58] and ~11.6 cm (with phosphorus phthalocyanine (P-Pc) as a PAI agent at 1046 nm wavelength) [59]. The laser used for PAI has a wavelength in the ultraviolet (UV (190–700 nm)) or near-infrared (NIR (700–2500 nm)) region.

Various photoacoustic contrast agents (hemoglobin, melanin, and lipids) that are used in biomedical applications use PAI for diagnostic application. These are classified as endogenous contrast agents. Table 14.3 provides a list of photoacoustic contrast (organic or inorganic) nanomaterials, including a list of MNPs, carbon NPs, and dyes [60]. The present section provides important carbon nanostructure designs that could be potentially used as a theranostic agent. Many of these materials (nano or bioconjugated systems) could be used for photoacoustic deep tissue imaging or conventional fluorescence imaging for theranostic applications [59, 61], whereas others could be visualized using fluorescence imaging or contrast imaging techniques.

Other than PAI, nanomaterials like gold and other carbon nanostructures can be used for optical NIR imaging applications, due to their property of getting excited in NIR light. Further, some pieces of evidence suggest that carbon nanomaterials can also be used in FRET-based bioimaging applications. In summary, various carbon nanostructures have the theranostic potential to be used for developing nanotheranostic designs using various imaging and therapeutic strategies, but other materials exist as well.

Table 14.3 Materials that are commonly used as contrast agents for photoacoustic imaging [60]

Nature of material	Types of NP
Metallic nanomaterial	Au rods, nanostars, nanocages, nanoshell, nanovesicles; Ag nanoplates, palladium nanoplates; antimony NPs
Carbon-based nanomaterial	Carbon nanotube, graphene, and carbon dots
TMC	CuS, WS_2, MoS_2, FeS, Bi_2S_3, CuSe, CooSeg, Bi_2Se_3
Dye	Porphyrin, cyanine-based dyes, e.g., ICG, IR780, IR825, etc.
Polymer	Polypyrrole, polyaniline, polydopamine, semiconducting polymers

14.5.1 Graphene as Theranostic Agent

Graphene was introduced to the scientific community in 2004 as a thin carbon film by Novoselov et al. [62]. It is a 2D material with a monoatomic layer of sp2 hybridized carbon atoms organized in the form of a honeycomb lattice. New synthetic procedures have allowed it to have reduced graphene oxide (rGO) and oxidized (GO) form of graphene. Due to the presence of free carboxyl function, GO has shown potential for various functionalizations in developing conjugated graphene structures for drug delivery applications for anticancer therapy [63, 64]. To date, various drugs like DOX, camptothecin, lucathone, paclitaxel, and others have been evaluated using GO/rGO or GO derivatives for drug delivery applications [65, 66]. Apart from these drug carrying properties, GO and rGO possess optical properties like PL and increased absorbance in the NIR region, which make them excellent photo-responsive material that can be used for photothermal and photodynamic therapies for cancer treatment. rGO and its derivatives have a relatively high response to photothermal therapy through absorbance in the NIR region [67]. Studies have shown that due to absorption in the NIR region, the material (rGO) has important potential for being used as a contrast agent for PAI [68]. Further, some studies also suggest that various ribbon-shaped rGO could also prove to be effective as PAI and thermoacoustic imaging agents with possible bioimaging applications [69].

Other than these natural abilities of graphene to act as a theranostic agent, GO can show blue PL and rGO can show red PL [70]. Further, PL is usually categorized as fluorescence and phosphorescence. GO naturally acts as a fluorophore, and this property can be varied by changing the chemical composition, size, and other parameters. GO shows an excitation-based fluorescent effect called Giant Red-Effect in the presence of polar solvents like water. Some interesting studies have shown that an excitation wavelength of 440 nm gives a broad emission peak around 668 nm. With an increase to ~pH 12.7, the emission peak disappears. At the optimum ~pH 7–7.8, there is a broad, intense peak of ~668 nm; beyond this, the peak starts to drop, and doublet appears around 500 nm. This indicates that the graphene material exhibits fluorescence in polar solvents (including aqueous phase) and can work as an important imaging molecule using fluorescence imaging techniques [71–73]. Another study showed that PL imaging uses graphene oxide QDs (GOQDs) under in vivo conditions [74]. It was also suggested that graphene QDs could exhibit 3.5 times more fluorescence than GOQDs. A study showed the use of graphene QD as potential fluorescence material for tracking of normal cells using confocal microscopy [75], unlike other dyes and fluorescent materials like QDs that emit relatively high fluorescence irrespective of the excitation wavelengths. Hence, the use of fluorescent dyes or QDs is relatively more common for developing fluorescent-based diagnostic imaging.

However, the use of graphene or its QDs as dyes for developing diagnostic material requires significant technological enhancement to enable them to be viable material for diagnostic applications. Moreover, graphene has a high biocompatibility and low toxicity profile compared to other dyes and QDs [76]. Further, GO also exhibits FRET donor and acceptor, in the presence of certain materials, as shown in Table 14.4 [77]. However, the technology required for the clinical translation is still under development.

Table 14.4 List of graphene-based FRET assays [77]

Donor/acceptor	Concentration
Fluorescence GO/AuNP	200 nM, 10^5 pfu mL^{-1}, 0.5 μg/L, 0.3 μg/L
Fluorescence GO-NHBu/AgNP	1 nM, 0.1 pM, 20 pM
Fluorescence GO-PVA/Au^{+3}	275 ppb
GQD/AuNP	1 nM
GQD/CNT	0.4 nM
Dual-color GQD/CNT	3.6 nM (21 bases), 4.2 nM (24 bases)
GQD/GO	75 pM
GQD/4-NP	9 ng/mL
GQD/TNP	0.091 μmoL^{-1}
FITC/GO	6.25 pM
FAM/GO	40 pM
QD/GO	100 fg/μL
GQD/GO	75 pM
UCNP/GO	5 pM
RB/GO	10 nM
FAM-peptide/GO	0.1 μM
FAM-aptamer/GO	25 cells/mL

GO, Graphene oxide; PVA, polyvinyl alcohol; GQD, Graphene QD; CNT, Carbon nanotube; 4-NP, paranitrophenol; TNP, 2,4,6-trinitrophenol; UCNP, Up-conversion NPs; RB, Rhodamine B.

Currently, the most commonly explored designs for developing bioimaging application using graphene include bioconjugation of florescent dyes to GO or chemical modification of graphene to enhance the fluorescent emissions [78, 79]. Graphene has also been used for bioimaging applications using single-photon excitation, wavelength-dependent PL (using low intensity, 5–10 mW/cm^2 laser) for bioimaging, combined with exposure to NIR light (@ 980 nm) to create a dual-mode of photodynamic and photothermal destruction of solid tumors. The in vivo experiments suggested that GO-PEG-folate design in the same study was found to be more effective compared to native nanomaterial photothermal therapy (NaPTT) at 808 nm light irradiation for the treatment of melanoma tumors. The study confirmed that compared to native GO, GO-PEG-folate is an effective theranostic material for further clinical investigations [80]. Figure 14.4 shows a classic example of a theranostic application of cancer-targeted anticancer therapy. A similar design of GO-PEG-sinoporphyrin sodium (DVDMS) was investigated for theranostic efficiency by fluorescence-guided photodynamic therapy. Nearly 100% in vivo tumor elimination was observed by intravenous injection of GO-PEG-sinoporphyrin at 2 mg/kg bw of DVDMS, with no recurrent loss of body weight or any other noticeable toxicity [81].

Radioactive isotope tagged graphene (known as radio-graphene) is one of the interesting designs being explored for theranostic applications. ^{131}I and ^{125}I labeled PEG-coated GOs were used for the development of nuclear imaging-guided radio and photothermal therapy [82]. Further, ^{105}TRC tagged PEG-GO was developed, which was later tagged with ^{64}Cu using 1, 4, 7-triazacyclononane-1, 4, 7-triacetic acid ((NOTA), a known chelator for ^{64}Cu labeling) [83]. Various radio-graphene materials have been developed to date and shown to be effective against various forms of cancer conditions [84]. However, the safe removal of the radioactive material from the body or its metabolism remains unanswered. These products help increase targeting to cancer tissue and therapeutic effect via localized heating due to gamma radiation. PET scan or gamma imaging is the most commonly used imaging technique for the radio-graphene theranostic system.

14.5.2 Nanodiamonds as Theranostic Agents

Nanodiamond (ND) is a type of carbon nanomaterial that is commonly used for therapeutic and diagnostic applications. It was discovered and confirmed by a group of researchers in 1963 and 1984. It was an accidental discovery while studying the diamond synthesis by shock compression of non-diamond carbon materials in blast chambers [85]. The particle size typically ranges from 1 nm to 150 nm, and it exists in various sizes, depending on the manufacturing method.

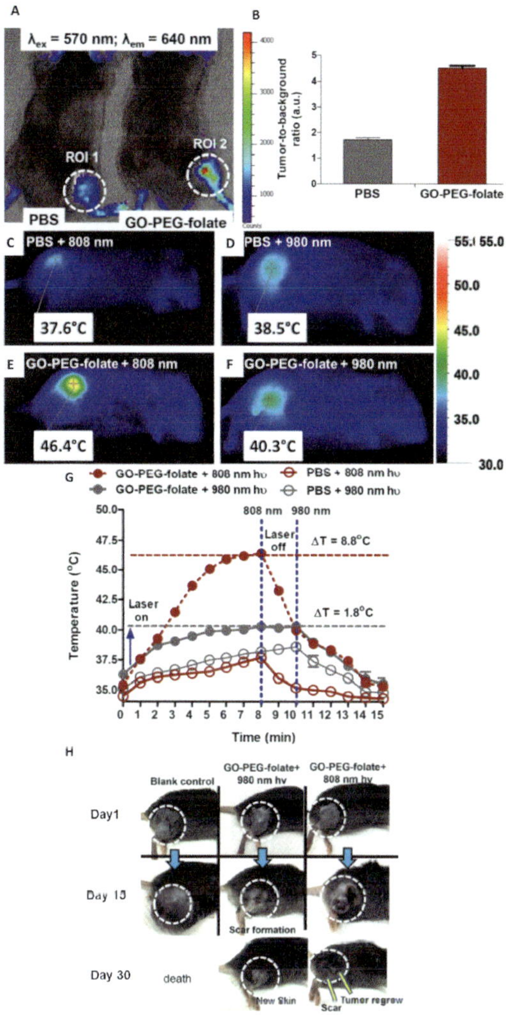

Figure 14.4 Shows theranostic application of GO for targeted cancer therapy: (A) In vivo fluorescence of PBS and GO-PEG-folate in mice after 24 h injection; (B) Tumor-to-background ratio shows increased uptake of GO-PEG-folate compared to blank PBS; (C–F) Photothermal images of mice injected with PBS and GO-PEG-FA exposed to NIR laser at 808 nm and 980 nm wavelength; (G) NIR exposure of formulation at two wavelengths causes rise in temperature to 47.5 °C (@808 nm) and 40 °C (@980 nm); and (E) In vivo anticancer effect shows almost complete recovery of mice injected with GO-PEG-Folate @980 nm wavelength exposure after day 30, whereas exposure @808 nm led to increased scar tissue and tumor relapse. Reproduced with permission from Ref. [80].

The most commonly used method for the synthesis of NDs is detonation, high pressure, and high temperature (HPHT). Detonation synthesized ND (DND) is sp3 hybridized with a partial shell having sp2 hybridized carbon. DNDs have a particle size of 5 nm with round and oval shapes. They have high impurity and defects; whereas, NDs produced from HPHT are the purest form of diamonds with sp3 hybridized carbons. These NDs have a broad range of particle sizes. They have been explored for developing photothermal and photodynamic therapies. Recent progress shows that certain ND materials could also be used for bioimaging purpose [86]. NDs have been modified by various means to get functionalized NDs for possible theranostic applications.

NDs have tendencies to absorb NIR 700–1100 nm and this can help ablate cancer cells. Using phase change material (PCM) consisting of chlorine e6 (Ce6) and NDs, researchers developed on-off mechanism-based photodynamic and photothermal ablation-based NPs. The temperature of PCM NPs containing NDs increased to 45 °C after exposure to laser for 5 min. The singlet oxygen generation by PCM NPs containing Ce^6 was gradually increased with respect to the laser exposure time. The study showed that laser ablation causes localized heating of cells and reduction in tumor volume, due to photodynamic and photothermal ablations, as shown in Figure 14.5 [87]. A conjugated system of ND-PEG-GLY-DOX (NPGD) was developed as a theranostic platform for cancer therapy. This NPGD works as off-on mediated theranostic response. The NPs, when excited at 480 nm, show fluorescence emission around 600 nm [88]. Similarly, in another design, ND-PEG-HYD (hydrazine hydrate)-FA/DOX was developed, and it was evaluated for its fluorescence-based theranostic potential [89]. Further, carboxylated NDs were peptide coupled to carboxylated gadolinium III (Gd III) to create ND-G conjugates, and DOX was electrostatically adsorbed on to ND-G (5:1 ND-G: Dox ratio) to get ND-G-Dox NPs. MRI imaging was used as a diagnostic platform for the theranostic system [90]. Research on developing better theranostic agents using NDs has been evolving. Although native NDs are effective in absorbing light in the NIR region and can help thermal ablation of cancer cells, they can also be used as theranostic tools by conjugating them with fluorescent organic material.

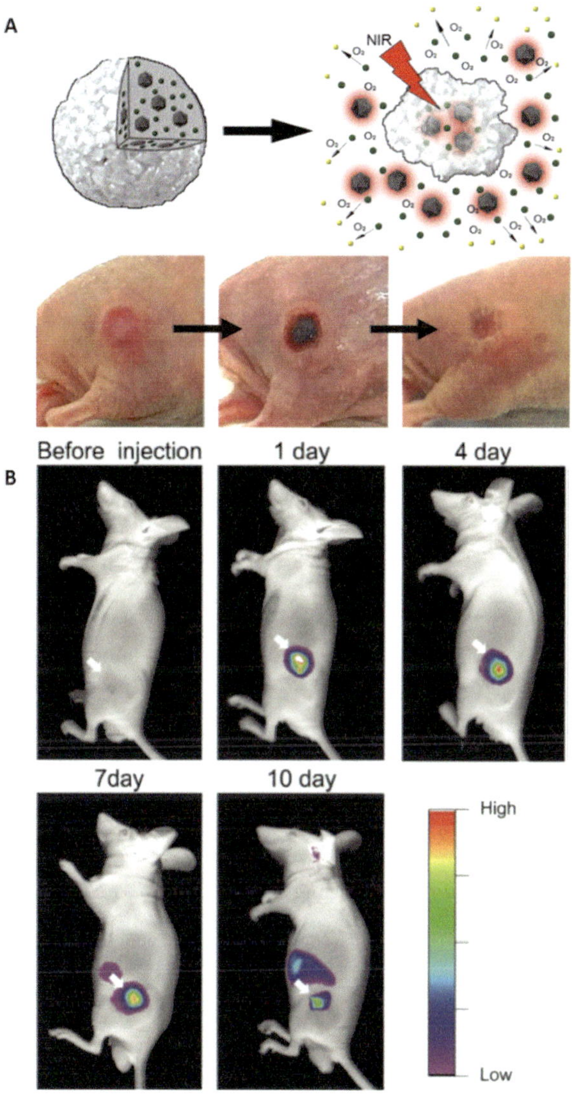

Figure 14.5 (A) Choline 6 (C6), ND-loaded PCM NPs for dual-mode (photodynamic and photothermal) therapeutic and fluorescent imaging nanotheranostic material. Mice injected with NPs were exposed to He-Ne laser (650 nm wavelength, 2 W/cm^2 power), which raise the temperature to around 40–45 °C at 0.5 and 1 mg/mL concentration. (B) Shows mice injected with NDs (0.1 mL, ND 5 mg/kg body weight) and fluorescence recorded for a period of 10 days after injection. Reproduced with permission from Ref. [87].

Researchers have been trying to modify the inherent chemical structures of NDs to create fluorescent nanodiamonds (FNDs) for better fluorescent signals and bioimaging purposes.

FNDs normally have nitrogen-vacancy defects that can emit fluorescence at 600 nm and 800 nm (NIR region) after irradiation, which can be used for a deep tissue imaging application. The excitation wavelength is usually between 510 nm and 560 nm, which is usually absorbed by the organic molecules that cause interference in the desired emission signals used for carrying out bioimaging work. Hence, the best solution to this problem is integration with NPs that can absorb the NIR light and convert it into wavelengths, which are useful for FND excitation. Some studies show that Au or up-conversion NaYF4: Yb, Er NPs help in increasing the fluorescence signals [86, 91]. Further, FND was developed by incorporation of imperfections like silicon vacancy defects, N3 vacancy defects, H3 defects, and EU3 defects in NDs. Each of these different types of FNDs was created with different color centers due to the varied types of defects [92–95]. The use of FNDs for developing theranostics is still under development.

14.5.3 Fullerene and Carbon Nanotubes

Buckminsterfullerene (C60) or fullerene is a carbon allotrope that was discovered in 1985. It has sp2-hybridized zero-dimensional carbon nanomaterials [96]. Fullerenes have interesting photophysical properties and can cause the generation of reactive oxygen species (ROSs) that could help in the creation of photodynamic therapies. Fullerence has a hollow cage structure that provides interesting possibilities to alter it and turn the molecule into metallofullerene or functionalized fullerene. These functionalized fullerenes (Gd III diethyltriaminepentaacetic acid (Gd(III)-DTPA) and Gd-tetraazacyclododecanetetraacetic acid (Gd(III)-DOTA)) have been currently marketed as Omniscan® and ProHance® as contrast agents for MRI scanning. The major limitation is the release of free metal ions upon the metabolism of Gd^{+3} chelate complexes, leading to toxicities. Hence, Gd is encapsulated in the fullerene cages to preserve the metallic properties and prevent the release of the metal ion that can provide less toxicity. The two most commonly explored structures are Gd@C*n* (*n* = 60 and 82) (shown in Figure 14.6) and $Gd_3N@C80$ for the contrast imaging applications [97]. Further, by ion implantation, researchers developed a new material, endohedral iron-fullerene [98]. Another study shows that iron carbonyl complexes with C60 and C70 by irradiation of $Fe(CO)_5$ to a thin sheet of C60 and C70 [99]. Apart from being used as imaging material, fullerenes like hydroxyl (Figure 14.6) and carboxyl have demonstrated hyperthermia properties when exposed to low-intensity lasers [100]. Probably these new materials could be an important key to the development of structures like metal (Gd)-encapsulated or tagged fullerene. Although the structure of fullerenes gives limited access for surface functionalization, researchers used conjugation chemistry to develop C60@Au-PEG/DOX for theranostic application. The system provided a four-in-one photodynamic, hyperthermia, and chemotherapeutic effects, where materials like Au and DOX can provide fluorescence and metallic contrast for X-ray imaging-based diagnostic capabilities [101, 102]. Due to limitations for conjugation of C60 (methoxy polyethylene glycol) mPEG-GO was used to create mPEG-GO-C60 hydride for developing photodynamic and photothermal therapy. This system could be an interesting theranostic system [103].

Free carboxyl groups and conjugational chemistry for surface modification are the important biochemical tools for developing theranostic systems. Soon after the discovery of fullerenes in 1985, carboxyl-group-terminated fullerenes were introduced [104]. The synthesis of these materials consists of two important steps: (i) cyclopropanation, followed by (ii) hydrogenolysis of fullerene ester. For the use of carboxylated fullerene to be used for biological applications, it is important to be biodegradable and safe for human consumption. Being a new synthetic material, limited data is available about the aqueous solubility of the carboxylated fullerenes as well as biosafety [104]. For instance, $C60[C(COOH)_2]_3$ was found to have antioxidant and prooxidant activity. It was shown that human serum albumin (HSA) can work as a transport protein for this molecule and can exhibit dose-dependent hemolytic activity. However, the material is non-genotoxic [105], and has significant potential for being explored for theranostic application. However, to our knowledge, no data is available for the application of these materials for their use in a theranostic application using classical conjugational chemistries, as shown by other carbon structures like graphene or CNTs.

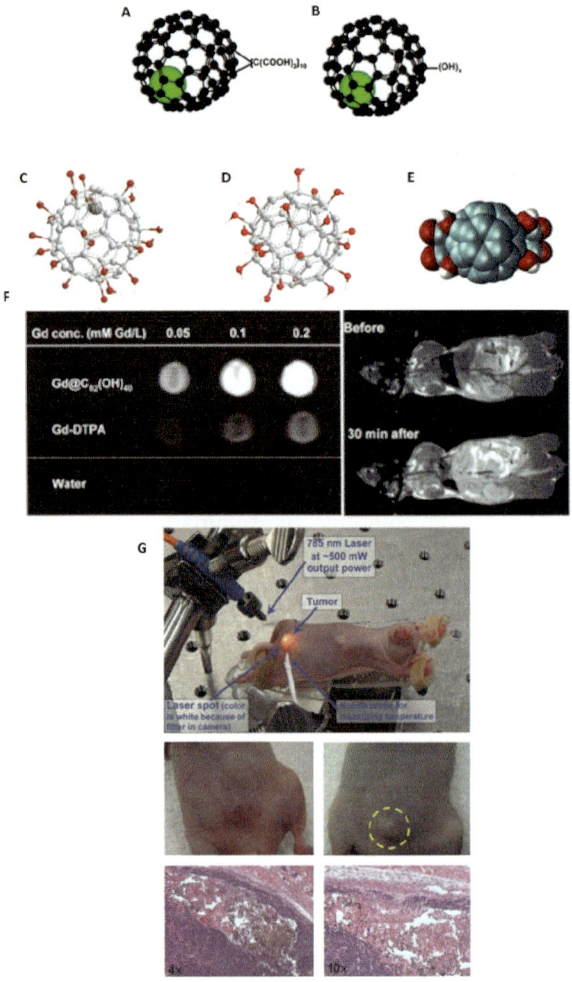

Figure 14.6 (A) Gd@C60[C(COOH)$_2$]$_{10}$; (B) Gd@C60(OH)x; (C) Gd@C82(OH)$_{22}$; (D) C60(OH)x; (E) C60[C(COOH)$_2$]$_2$; (F) MRI scans of Gd@C$_{82}$(OH)$_{40}$ and Gd-DTPA at various concentrations. It also shows MRI scans of mice injected with Gd@C82(OH)$_{40}$ through the tail vein, before and after 30 min of injection. (G) Hyperthermia using photothermal ablation of tumor treated with fullerene shows blistered area after laser treatment that is confirmed by necrotic tissue. Reproduced with permission from Ref. [97].

Table 14.5 Data for use of CNTs under various hyperthermia-based therapy [106]

Cancer type	Experimental model	Material (effective dose)	Laser (wavelength; radiant exposure)
Brain (human)	In vitro; adherent primary glioblastoma cells	SWCNT (2.5 μg/well of 24 well plate)	808 nm; 600 J/cm^2
	In vitro; non-adherent U251 glioblastoma cells	SWCNT (10 μg/ml)	808 nm; 600 J/cm^2
	In vitro; non-adherent MDA-MB-231 cells	Gold-coated SWCNTs (estimated 0.5–1×10^4 SWCNTs per cell)	850 nm; 0.5 J/cm^2
	In vivo murine flank tumor; primary glioblastoma cells	MWCNT (10 μg, ex vivo)	808 nm; 600 J/cm^2
Breast (human)	In vitro; transformed human mammary epithelial cells	MWCNT (50 μg/mL)	1064 nm; 90 J/cm^2
	In vitro; adherent BT-474 cells	SWCNT (5–10 μg/ml)	808 nm; 5130 J/cm^2
	In vitro; adherent SK-BR-3 cells	SWCNT (4 μg/ml)	808 nm; 900 J/cm^2
	In vitro; non-adherent BT474 cells	SWCNT (100 μg/ml)	800 nm;12 J/cm^2
	In vitro; non-adherent MCF-7 cells	SWCNT (not reported)	808 nm; 144 J/cm^2
	In vivo murine flank tumor; MDA-MB-231 cells	MWCNT (100 μg IT)	1064 nm; 90 J/cm^2
	In vivo murine flank tumor; transformed human mammary epithelial cells	MWCNT (100 μg IT)	1064 nm; 90 J/cm^2
Breast (murine)	In vitro; adherent EMT6 cells	SWCNT (3.5 μg/ml)	980 nm; 120 J/cm^2
	In vitro; adherent EMT6 cells	SWCNT (50 μg/ml)	980 nm; 60–150 J/cm^2
	In vivo murine flank tumor; 4T1 cells	SWCNT (70 μg IV)	808 nm; 180 J/cm^2
	In vivo murine flank tumor; 4T1 cells	SWCNT (100 μg IV)	808 nm; 300 J/cm^2

Table 14.5 (*Continued*)

Cancer type	Experimental model	Material (effective dose)	Laser (wavelength; radiant exposure)
	In vivo murine flank tumor; EMT6 cells	SWCNT (20–25 μg IT)	980 nm; 300 J/cm^2
	In vivo murine flank tumor; EMT6 cells	SWCNT (1 mg IT)	980 nm; 450 J/cm^2
Cervical (human)	In vitro; adherent Hela cells	MWCNT (10 μg/ml)	1064 nm; 128 J/cm^2
	In vitro; non-adherent HeLa cells	SWCNT (25 μg/ml)	808 nm; 168 J/cm^2
Erlich ascites (murine)	In vitro; non-adherent Erlich ascitic carcinoma cells	MWCNT (100 μg/ml)	780–1400 nm; 315 J/cm^2
Kidney (human)	In vitro; adherent CRL 1932 cells	Nitrogen-doped MWCNT (approx. 83 μg/ml)	1064 nm; 720 J/cm^2
Kidney (murine)	In vitro; non-adherent RENCA cells	MWCNT (100 μg/ml)	1064 nm; 135 J/cm^2
	In vitro; non-adherent RENCA cells	MWCNT (100 μg/ml)	1064 nm; 4590 J/cm^2
	In vivo murine flank tumor; RENCA cells	MWCNT (100 μg IT)	1064 nm; 90 J/cm^2
Liver (human)	In vitro; adherent HepG2 or CRL 4020 cells	MWCNT (1–50 μg/ml)	808 nm; 7680 J/cm^2
	In vitro; adherent SK-BR-3 cells	SWCNT (20 μg/ml)	1064 nm; 4 J/cm^2
Lymph (human)	In vitro; non-adherent Daudi Burkitt's lymphoma cells	SWCNT (90 μg/ml)	808 nm; 2100 J/cm^2
Mouth (human)	In vivo murine flank tumor; KB epidermoid mouth carcinoma cells	SWCNT (12 μg IT)	808 nm; 684 J/cm^2
Neuroend ocrine (human)	In vitro; adherent stNB-V1 neuroblastoma cells	MWCNT (5–10 μg/ml)	808 nm; 3000 J/cm^2
Prostate (human)	In vitro; non-adherent PC-3 cells	MWCNT (100 μg/ml)	1064 nm; 4590 J/cm^2
	In vivo murine flank tumor; PC-3 cells	MWCNT (50 μg IT)	1064 nm; 175 J/cm^2
Skin (murine)	In vivo murine flank tumor; SCCVII squamous carcinoma cells	SWCNT (60–100 μg IT)	785 nm; 120 J/cm^2

CNTs are one-dimensional sp2 hybridized nanomaterials [96]. They exist in the form of single-walled carbon nanotubes (SWCNTs) or multi-walled carbon nanotubes (MWCNTs). Like graphene, CNTs also absorb in the NIR region and can be used for hyperthermia-based treatment when exposed to UV light in the NIR region. Studies have shown that SWCNTs and MWCNTs work at various wavelengths and laser, as shown in Table 14.5 [106]. Some studies have shown that tumor tissue exposure to SWCNT in the presence of low power (200 mW/cm^2) NIR for 10 min can raise temperature to 55 °C.

SWCNTs are purified version of MWCNTs, with less metallic impurities. Studies with SWCNTs showed differences in the density of state at Fermi level and in the semiconducting nature of the metallic SWCNTs. This shows that metallic SWCNTs can absorb and convert more NIR into heating. This concept was also observed by other studies, which found that MWCNTs show relatively 10–100 fold more conversion of NIR to heat. The in vivo studies with MWCNTs show long-term (>6 months) survival of mice treated with MWCNTs when exposed to NIR (3 W/cm^2; 30 s) [107]. The energy of laser required for treatment with MWCNTs was comparatively less than what was required with SWCNTs. Designs like pegylated-CNTs were also developed for improved hyperthermia therapy. Attachment of PEG helps increase the hydrophilicity and tumor uptake. The surface of CNTs was also modified with antibodies targeting Her2, CD133, CD44, and insulin-like growth factor 1 (IGF1R) for improved targeted anticancer therapy through hyperthermia [106, 108–110]. Since CNTs also show emissions in the NIR region, a few studies show that this property can also be explored for imaging purposes. For instance, CNTs show the emissions in NIR II region of 950–1400 nm. These bright CNTs can be used for in vivo bioimaging applications [111–113]. Further, SWCNTs were used for imaging applications by generating PL in the spectral window of 1300–1400 nm [114]. This study clearly shows the significance of CNTs in the development of an efficient theranostic system mediated by hyperthermia and PL in NIR region. Here, use of NIR detectors becomes important for the use of SWCNTs in imaging applications. But studies have also shown the use of other detection methods like MRI (using radionucleotide-tagged

SWCNTs), photoacoustic tomography, and Raman spectroscopy being used for bioimaging applications [114–116]. Most carbon materials (NDs, graphene, fullerenes, and CNTs) require external stimulus or excitation for the emission of desired PL or contrasting effect for bioimaging applications. These materials offer a relatively safe and new technological concept for developing an efficient therapeutic application. However, other simpler designs are more commonly explored currently for theranostic applications.

14.6 Quantum Dots

While working on semiconductor nanocrystals made from CdS, CdSe, CuCl, and CuBr, two researchers, Alexei Ekimov and Alexander Efros, found that these nanocrystals absorb light at lower wavelength than expected [117]. They demonstrated that the optical properties of these semiconductor nanocrystals were dominated by their size. Around the same time, Louis Brus developed a method to control the size of CdS nanocrystals at around 4.5 nm and showed that these nanocrystals exhibited substantial blue shift compared to bulk CdS, which was consistent with the theory of quantum confinement of electron and holes previously stated [117]. Over the last few decades, the technology of QDs has been commercially used for enhancing the visual qualities of product displays by companies like Amazon and Samsung. Yet QDs in their native states are toxic. Hence, polymeric-encapsulated QDs have been developed for bioimaging applications. Various studies have been published about the development of natural, polymeric-material-based QDs for bioimaging applications. Some notable examples include PEG (CdS, CdSe, and CdTe) QDs for imaging of breast cancer, and gelatin- encapsulated (CdTe/CdSe) QDs for imaging of cancer and non-cancer cell lines [118, 119]. Various core/shell-based QD designs (like NIR luminescent QDs, silica-coated-antibody-tagged QDs, gold-coated QDs, or iron-coated QDs) have been developed to improve the toxicity profile for bioimaging and potential use as theranostic agents [120–122].

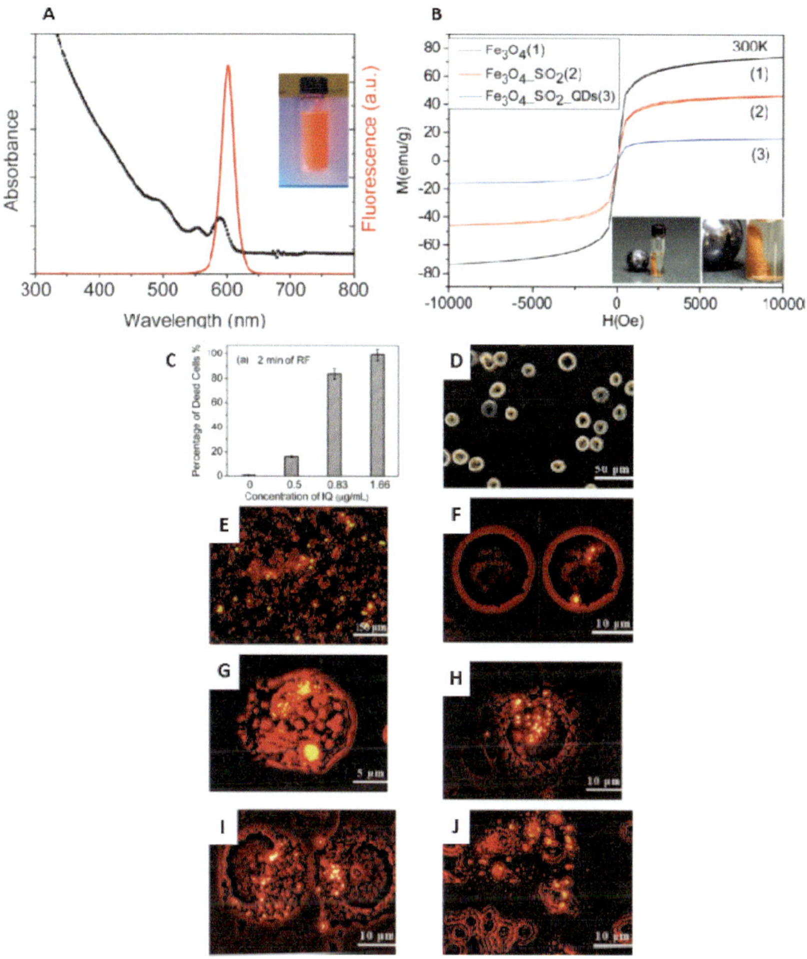

Figure 14.7 (A) Fluorescence spectra for Fe_3O_4/SiO_2 (core/shell)-CdSe/ZnS magnetic QDs. (B) Strength of magnetic QDs. (C) Anticancer effect of magnetic QDs in the presence of applied RF and % dead cells after 2 min exposure. (D) Normal panc-1 cells. (E) Incubation of panc-1 cells in the presence of magnetic QDs for 24 h. (F) Magnetic QDs in the cytoplasm of the panc-1 cell. (G–J) Cell shrinkage, membrane blebbing, and disintegration of membrane and apoptotic bodies as they resulted from exposure to RF 350 kHz. Reproduced with permission from Ref. [127].

The most commonly explored design for theranostic application includes the creation of polymer encapsulation of QDs and chemotherapeutic agents for anticancer therapy.

To increase the selectivity toward cancer cells, the antibody or ON is tagged to a polymeric shell mostly by EDC/NHS conjugation chemistry. CdSe/ZnS (core/shell) QDs were tagged with aptamer (Apt) and DOX to create QD-Apt-DOX. Once the DOX intercalated with covalently tagged QD-aptamer system and slowly released the drug to targeted prostate cancer cells over time, fluorescence imaging was used for theranostic application [123]. Similarly, other designs like QD-mucin 1 aptamer-DOX conjugates, trioctylphosphine oxide (TOPO)-coated DOX-loaded green-emitting CdSe/ZnS hybrid vesicles, 3-mercaptopropionic acid-capped daunorubicin CdTe QDs, and others have been developed for the potential of theranostic application. Most of these designs involve polymeric co-encapsulation strategies. The most commonly explored biomaterials used for the co-encapsulation strategies include PEG, albumin, PEG-poly-lactic acid (PLA), and PEG-TPGS [124]. Other than targeted drug-delivery-based chemotherapy, some reports suggest that QDs can also be used for photodynamic cancer therapy by the generation of ROSs [125, 126]. Further, Fe_3O_4/SiO_2-SH-CdSe/ZnS nanocomposite (SiO_2 forms covering for with -SH functionality for Fe_3O_4 core and QDs are tagged through this thiol functionality) was developed with the aim of combining bioimaging due to QDs along with a radio frequency (RF)-mediated self-heating mechanism for anticancer therapy (shown in Figure 14.7) [127]. Researchers have been developing better biocompatible and less toxic options by developing complex and unique delivery designs for QDs with less photobleaching.

14.7 Gold Nanostructures

After colloidal gold was produced in 1857 by Faraday, gold chloride was reduced by phosphorus and stabilization, using carbon sulfide, to form AuNPs. Later, in 1951, it was reported that gold particles could be formed using trisodium citrate to reduce tetrachloroauric acid (gold (III) chloride [HAuCl4]) in water, and later this was improved for the formation of AuNPs with various sizes and shapes. The physicochemical properties of AuNPs can be altered with different shapes of AuNPs. To date, various forms of AuNPs have been synthesized that include nanospheres (2–100 nm), nanorods (10–100 nm), core-shell nanospheres (100 nm), and hollow nanocubes (40–50 nm). Change in the shape helps tune an optical property called surface plasmon resonance (SPR), a phenomenon at which the amplitude of the oscillating free-electron reaches its maximum at a specific frequency. A strong absorption band is observed at SPR and can be visualized using a UV-visible spectrometer. SPR of noble metals like silver and gold is higher than other metals. AuNP can act as a contrast agent in the NIR region (630–680 nm) in bioimaging applications. A study showed the use of AuNP conjugates and detection using confocal microscopy in reflectance mode using laser excitation at 647 nm. Only the wavelength of light corresponding to SPR of AuNPs conjugates was clearly imaged with AuNPs in black contrast for imaging applications. Due to their property of providing a strong absorption peak in the NIR window 600–1350 nm, these materials are an ideal choice for carrying out PAI. For instance, Au nanospheres and cages have been reported for imaging of human breast cancer tumor xenografts and for detecting B16 melanomas at 532 nm and 778 nm [128], respectively.

A few studies have shown that PEG-Au nanoprisms (PEG-AuNPrs) have been used as signal amplifiers for optoacoustic tomography imaging to visualize gastrointestinal cancers. Staged-Au nanorods (AuNRs) have also been used for cancer detection with an absorption peak at 800 nm [128]. AuNPs have been extensively explored for photoacoustic-based imaging of the brain, lymph nodes, circulating tumor cells, atherosclerotic plaques, and other deep tissue applications [128]. AuNPs can also be used as CT imaging agents due to their excellent X-ray attenuation properties (atomic number Z = 79; k-edge value = 80.7 keV). Unlike other molecular contrast agents that get cleared off quickly, AuNPs help in improving the pharmacokinetic profile on pegylation and can be useful for cancer imaging. AuNPs can also help in enhancing Raman scattering signal of adjacent molecules, hence surface-enhanced Raman spectroscopy (SERS) imaging-guided AuNPs have been widely used for the detection of virus and cancer cells [129]. Like other systems, AuNPs can be surface-functionalized with relevant antibodies or targeting moieties like folates, aptamers, and other ONs for increasing the cancer selectivity for NPs. For instance, a study showed that gold nanorod tagged with a peptide gets degraded with matrix metalloproteinase (MMP) enzyme, affects the fluorescence in presence of NIR light, and can be imaged in biological matrix (as shown in Figure 14.8). The system works as a therapeutic agent by increasing localized temperature, owing to the hypothermia due to exposure to NIR light [130].

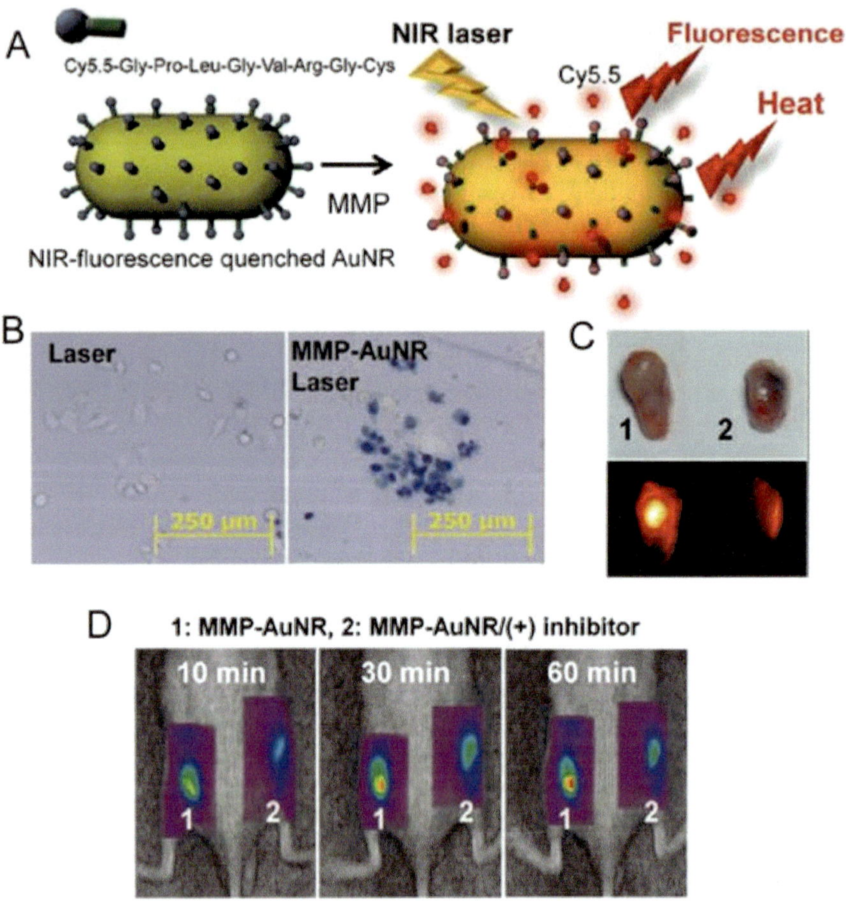

Figure 14.8 (A) Schematic peptide-linker-tagged AuNP that gets degraded in the presence of MMP for simultaneous imaging and photothermal therapy. (B) HeLa cells in trypan blue after treatment. (C) Optical and fluorescent images (after exposing to NIR light) of tumors treated without and with MMP inhibitor. (D) Mice treated with AuNPs without and with MMP inhibitor. Reproduced with permission from Ref. [130].

One interesting property of AuNPs is their affinity toward thiol functionality. As a result, thiolated AuNP (AuNP-SH) helps in functionalization with amine moieties like peptides and proteins using EDC/NHS reaction for developing cancer-targeted therapies. As stated earlier, exposure to NIR laser at wavelengths of 532–808 nm at power density of 0.4–2.45 W/cm^2 causes hyperthermia by increasing temperature to >50 °C within a few minutes of exposure [131, 132]. AuNPs also act as photosensitizers when exposed to light in NIR region by producing ROSs that help the cytotoxic effect. This makes AuNPs in the form of their native states (with different shapes) polymeric nanocomposites or targeted nanoconjugates that work as ideal theranostic agents for anticancer therapy.

14.8 Magnetic Nanoparticles

Unlike other noble metals like silver and gold, MNPs have been extensively explored as contrast agents for MRI imaging techniques, including development of fluorescent superparamagnetic iron oxide nanoparticle (SPION), as shown in Figure 14.9, with excellent imaging capabilities [133]. These particles have an important property of causing hyperthermic shock (Figure 14.9) and temperature rises when exposed to an alternating magnetic field [134]. This unique property of MNP makes it an intriguing material for drug delivery and theranostic applications. The most commonly explored MNPs include Fe_3O_4 and Fe_2O_3. Most MNPs can convert magnetic energy into heat through hysteresis loss or the Neel/Brownian relaxation process in the presence of an alternating magnetic field.

MNPs are the most commonly explored in developing novel passive and active targeted nanodrug delivery strategies with the potential of being used as theranostic agents. One of the most commonly explored strategies is to develop multiple-modal (hyperthermia and chemotherapy) anticancer therapies. The most commonly explored nano delivery strategy is the co-encapsulation of MNPs in synthetic, semi-synthetic, and natural polymers (PLGA, PEG, PEG-PLGA, chitosan, gelatin, and others) with or without anticancer drug molecules (DOX, gemcitabine, 5FU, and others) [135–138].

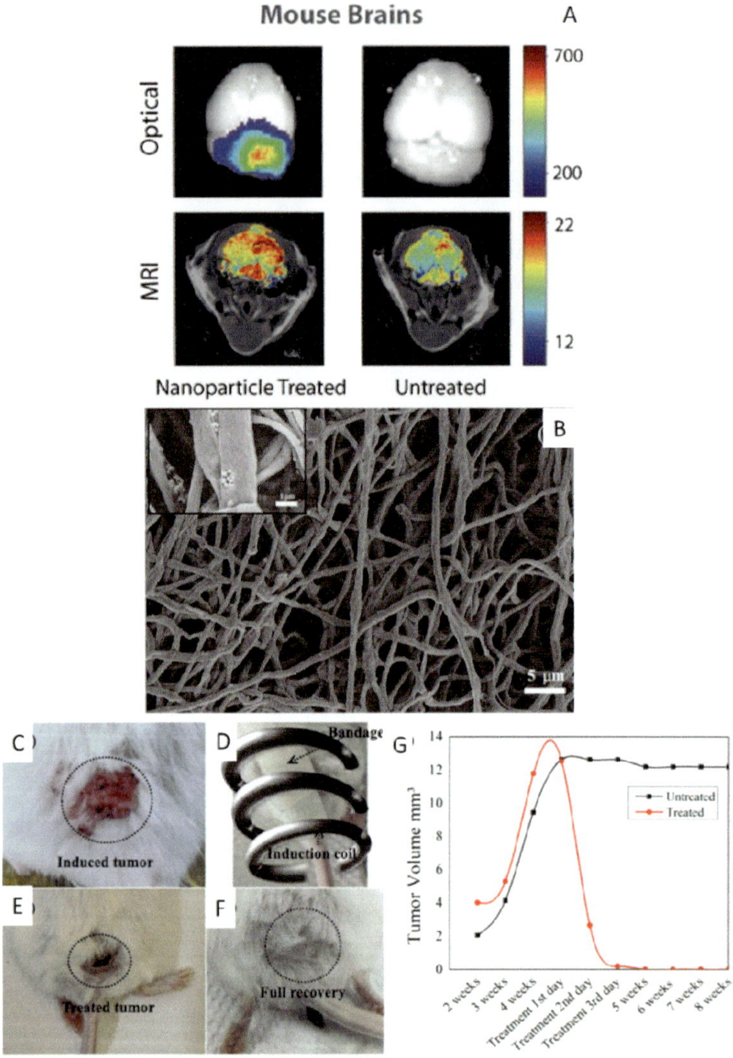

Figure 14.9 (A) Shows ex vivo optical image and in vivo MRI image of brain tumor treated (left) with Cy5.5-labeled MNPs and untreated animal (right). (B) Poly (ε-caprolactone) (PCl)-Fe_3O_4 system nanofibers. (C–G) Exposure of induced tumor treated with PCl-Fe_3O_4 nanofiber bandage exposed to alternating magnetic field causes tumor reduction compared to untreated conditions. Reproduced with permission from Ref. [133, 145].

All these co-encapsulated designs show the potential of MNPs for theranostics application. The drug loading on the polymeric MNPs has also been carried out by classical conjugation of drug molecules to the polymers coating on MNPs. For instance, the carboxyl group of DOX was conjugated on amine-functionalized MNPs (obtained by treatment with (3-Aminopropyl) triethoxysilane (APTES)) by forming imine linkage. Drug-conjugated MNPs were separated using magnets. The drug-conjugated MNP systems retained the magnetic behavior that could be used as dual-targeting carriers (drug and MNPs) for anticancer therapy. Likewise, various conjugational strategies could be adopted to design efficient drug delivery system using surface-modified MNPs such as: (i) amine-MNPs, (ii) carboxyl-MNPs, (iii) hydroxyl-MNPs, (iv) epoxy-MNPs, (v) thiol-MNPs, (vi) azide-MNPs, (vii) aldehyde-MNPs, and (viii) hydrazide-MNPs. To our knowledge, other than DOX, drugs like methotrexate, paclitaxel, and gemcitabine have been used to covalently conjugate to MNPs by covalent or non-covalent pathways [139].

Other conjugational chemistries, metallic alloys (FePt or FeAu) [140, 141], and core/shell nanostructures have also been developed to tap into the properties of the individual metallic system, such as the development of gold-coated MNPs with modified plasmonic properties [142, 143]. Another interesting design is Janus "nanobullets" for magnetic targeting of liver cancer therapy. Here, the structure is made of mesoporous silica consisting of DOX drug with MNP head. This bullet can be targeted using an applied magnetic field to a specific site for anticancer therapy [144]. These NPs are mostly developed with the idea of preventing corrosion of iron particles in the biological matrix by covering them with a layer of noble metals while retaining the magnetic properties that could be necessary for theranostic applications. Most of these studies with metallic or organo-metallic therapies answer questions related to increasing site or tissue-specific therapy. This is the next step in the field of nanotechnology, formulation, and theranostic science.

14.9 Clinical Translations of Theranostic Materials

Translation of nanoformulations has been evident in the ongoing clinical trials and nanoformulations that are available in the field for the treatment of chronic conditions like cancer. However, the translation is severely hindered due to problems such as: (i) deciding route of administration, (ii) model variability (animal to human) in absorption, distribution, metabolism, and elimination (ADME) profile, (iii) large-scale manufacturing, (iv) nanoformulations design based EPR-mediated accumulation, (vi) unknown molecular mechanisms for protein-targeted nanoformulations, and (vii) determining stability, impurity profile, and other biosafety parameters set by the Food and Drug Administration (FDA).

Non-scientific problem that hamper the translation include intellectual property (IP) rights and government regulations (FDA, Pharmacopeias, and others). To our knowledge, the government regulations are limited about the use of specific biomaterial for developing clinically acceptable nanoformulations, unlike regulations for other formulations. The first-generation of nanomedicine (like Doxil) got approval for testing by meeting the basic criteria of being applicable as medical materials. This criterion is generally regarded as safe (GRAS) for use, but can be tricky to extrapolate for nanomedicine applications due to the complex nature of nanomedical designs [146].

To date, many nanomedicines products have entered the market and are routinely used for the treatment of various chronic conditions [146]. New nanomedicine designs under clinical investigations are the key to the development of nanotheranostics. For instance, around 443 clinical trials are registered regarding NPs in clinicaltrials.gov. These include three NP trials registered for the treatment of COVID-19 patients. Although the number looks big, it occupies only a fraction of the clinical translation landscape. From an optimistic perception, it is important to notice that many of the classical NP designs have managed to get translated to human testing under various categories of trials for the treatment of various chronic and infectious diseases.

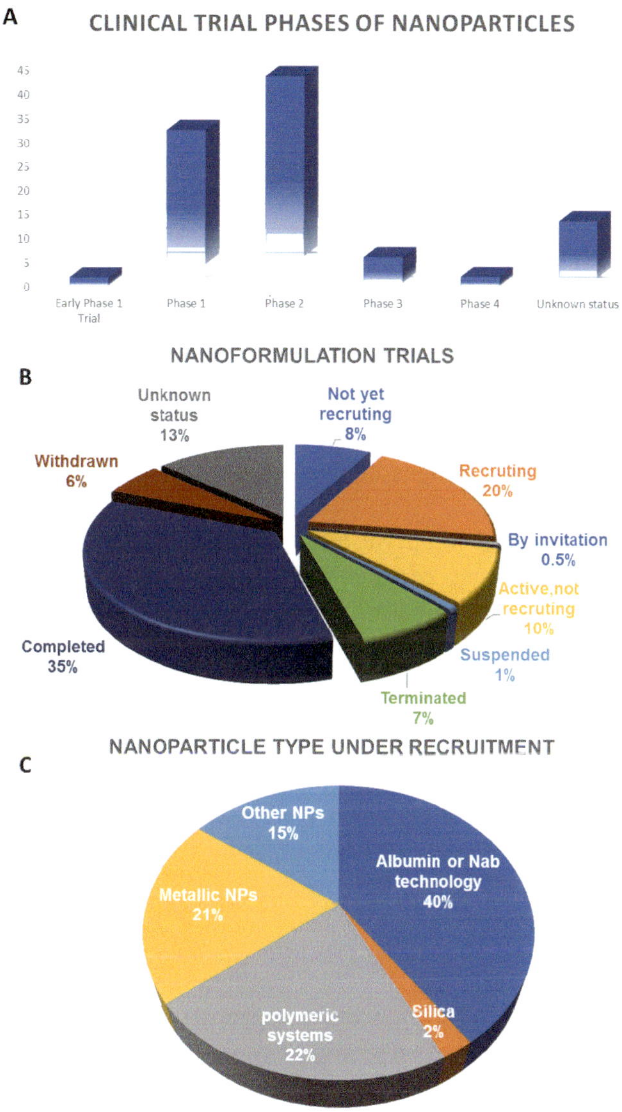

Figure 14.10 (A) Clinical trial phase distribution of NPs. (B) Trials under various stages of completions. (C) Percentage distribution of various types of NPs explored in clinical trials.

As of 2020, nearly 32% and 44% of the NP trials are under Phase I and Phase II, respectively. Further, we also observed that 8% of the total NP trials had been translated to advance trials

(Phases III and IV), with a translation rate of 6%, which is significantly lower than Phase I to Phase II translation (~100%). This shows that although the nanomaterial toxicology profile seems to be under control (usually studied in Phases I and II trials), researchers may be facing challenges in addressing primary clinical outcomes like controlling disease conditions under multiple scientific or non-scientific variables (mentioned above). There are around 10 trials in Phase IV that are either completed or recruiting, which shows the use of nab-paclitaxel and MNPs in (breast and pancreatic) cancer therapy and MRI-based imaging (pancreatic cancer), respectively. Currently, nearly 35% of trials are completed, and around 30% of the total trials are either actively recruiting or active but non-recruiting. This shows the enormous efforts being made for the translation of NP systems.

Of all the registered NP trials (from clinicaltrials.gov), 40% are currently being conducted on drug-loaded albumin NPs. Other polymeric (including liposomes) and metallic NP systems are explored in 22% and 21% of trials, respectively (shown in Figure 14.10). Moreover, concepts of NIR plasmonic nanotherapy with AuNPs have also been explored in clinical trials for dissolving atherosclerotic plaques, but their use for cancer treatment in humans remains unexplored. A pilot Proof of Principle clinical study for the use of gold-coated silica (shell-core design) was used to perform photothermal ablation (using NIR laser) of prostate tumors [147]. AuNPs as a delivery vector for gene and peptide delivery are explored in Phase I trials for glioblastoma and type-1 diabetes therapy (NCT03020017 and NCT02837094), respectively. Other than these magnetic and gold metallic systems being investigated as potential theranostic systems, Gd-based NPs have also been studied in human trials for their use in chemoradiation and brachytherapy. This metallic system is known for developing QDs for bioimaging applications, as discussed previously. All these pieces of evidence show that the translation of NPs to human trials has been slow and steady. Moreover, this indirect evidence suggests that the material discussed in this chapter has a chance of being explored as a theranostic application. Currently, there is no direct human trial that is being carried out for translation of TNPs, but some basic radioactive isotropic theranostic materials like ^{68}Gd-DOTA, ^{177}Lu-DOTA, ^{90}Y-DOTA, ^{67}Cu-SARTATE^{TM}, and others have been explored for theranostic purposes.

14.10 Conclusion and Future Perspectives

Studies discussed in this chapter show the importance of polymer-encapsulated, conjugated, surface-coated, and composite nanosystems for the treatment and diagnosis of chronic conditions like cancer. These designs have been adopted with the fundamental attempt of combining the knowledge of various bioimaging techniques (PL, MRI, FRET, ultrasound, PET, PET-CT, and others) along with therapeutic (hyperthermia, chemotherapy, radiotherapy, photodynamic, and others) potential of organic or metallic particles for theranostic applications. Novel nanodesigns, like a rtificial molecular machines or nanorobots that work on the energy supplied by simple catalytic reactions [148], are still under development, and very little is known about their functioning as well as their application in nanomedicine or theranostics. Given the success of fundamental NPs, theranostic nanosystems will provide the relatively better real-time option of imaging-guided tactical therapeutic options to reduce the unnecessary side effects caused due to classical therapies.

Conflict of Interest

Authors of this chapter have no conflict of interest.

Acknowledgement

The authors would like to thank Ms. Vandana Tibbani for her help in creating the bibliometric analysis for this work and for creating a network analysis of keyword occurrence in published work in the field of TNPs from the Scopus database. The authors would like to thank Ms. Pamela Walter (Office for Professional Writing, Publishing, and Communication), Thomas Jefferson University, for carefully reading and editing this manuscript.

References

1. Weissig V, Pettinger TK, Murdock N. Nanopharmaceuticals (part 1): Products on the market. *Int J Nanomedicine*. 2014; 9:4357–73.
2. Ferreira Soares DC, Domingues SC, Viana DB, Tebaldi ML. Polymer-hybrid nanoparticles: Current advances in biomedical applications. *Biomed Pharmacother*. 2020; 131:110695.
3. Ye X, Shi H, He X, Yu Y, He D, Tang J, et al. Cu-Au alloy nanostructures coated with aptamers: A simple, stable and highly effective platform for in vivo cancer theranostics. *Nanoscale*. 2016; 8(4):2260–7.
4. Juan A, Cimas FJ, Bravo I, Pandiella A, Ocana A, Alonso-Moreno C. An overview of antibody conjugated polymeric nanoparticles for breast cancer therapy. *Pharmaceutics*. 2020; 12(9).
5. Silberstein EB. Radioiodine: The classic theranostic agent. *Semin Nucl Med*. 2012; 42(3):164–70.
6. Francois PE, Mallard JR, McKinnell A. Seeds of pure beta-ray emitter (yttrium-90) for radiation hypophysectomy. *Nature*. 1956; 178(4544):1240–1.
7. James AG, Henschke UK, Myers WG. The clinical use of radioactive gold (Au198) seeds. *Cancer*. 1953; 6(5):1034–9.
8. Etrych T, Lucas H, Janouskova O, Chytil P, Mueller T, Mader K. Fluorescence optical imaging in anticancer drug delivery. *J Control Release*. 2016; 226:168–81.
9. Yallapu MM, Jaggi M, Chauhan SC. Curcumin nanomedicine: A road to cancer therapeutics. *Curr Pharm Des*. 2013; 19(11):1994–2010.
10. Shah S, Chandra A, Kaur A, Sabnis N, Lacko A, Gryczynski Z, et al. Fluorescence properties of doxorubicin in PBS buffer and PVA films. *J Photochem Photobiol B*. 2017; 170:65–9.
11. Ali Z, Saleem M, Atta BM, Khan SS, Hammad G. Determination of curcuminoid content in turmeric using fluorescence spectroscopy. *Spectrochim Acta A Mol Biomol Spectrosc*. 2019; 213:192–8.
12. Kelkar SS, Reineke TM. Theranostics: Combining imaging and therapy. *Bioconjug Chem*. 2011; 22(10):1879–903.
13. Zhang J, Du Q, Song X, Gao S, Pang X, Li Y, et al. Evaluation of the tumor-targeting efficiency and intratumor heterogeneity of anticancer drugs using quantitative mass spectrometry imaging. *Theranostics*. 2020; 10(6):2621–30.

14. Li Q, Maier SH, Li P, Peterhansl J, Belka C, Mayerle J, et al. Aptamers: A novel targeted theranostic platform for pancreatic ductal adenocarcinoma. *Radiat Oncol.* 2020; 15(1):189.

15. Wu X, Chen J, Wu M, Zhao JX. Aptamers: Active targeting ligands for cancer diagnosis and therapy. *Theranostics.* 2015; 5(4):322–44.

16. Clinicaltrials.gov. Theranostic 2020. Available from: https://clinicaltrials.gov/ct2/results?cond=&term=theranostic&cntry=&state=&city=&dist=.

17. Raveendran S, Rochani AK, Maekawa T, Kumar DS. Smart carriers and nanohealers: A nanomedical insight on natural polymers. *Materials (Basel).* 2017; 10(8).

18. Ankit Rochani SR, D. Sakthi Kumar. Sustainable green polymeric nanoconstructs for active and passive cancer therapeutics. In: Torchilin V (ed.). Handbook of Materials for Nanomedicine Polymeric Nanomaterials. 1st Ed. New York: Taylor and Francis Group. 2020; 69.

19. Zheng Y, Ji S, Czerwinski A, Valenzuela F, Pennington M, Liu S. FITC-conjugated cyclic RGD peptides as fluorescent probes for staining integrin alphavbeta3/alphavbeta5 in tumor tissues. *Bioconjug Chem.* 2014; 25(11):1925–41.

20. Kozuch SD, Cultrara CN, Beck AE, Heller CJ, Shah S, Patel MR, et al. Enhanced cancer theranostics with self-assembled, multilabeled siRNAs. *ACS Omega.* 2018; 3(10):12975–84.

21. Thermoscientific. Anti-FITC Antibodies 2020. Available from: https://www.thermofisher.com/us/en/home/life-science/antibodies/primary-antibodies/epitope-tag-antibodies/anti-fitc antibodies.html.

22. Godbey WT, Wu KK, Mikos AG. Tracking the intracellular path of poly(ethylenimine)/DNA complexes for gene delivery. *Proc Natl Acad Sci USA.* 1999; 96(9):5177–81.

23. Lucas B, Remaut K, Sanders NN, Braeckmans K, De Smedt SC, Demeester J. Studying the intracellular dissociation of polymer-oligonucleotide complexes by dual color fluorescence fluctuation spectroscopy and confocal imaging. *Biochemistry.* 2005; 44(29):9905–12.

24. Kaneo Y, Tanaka T, Nakano T, Yamaguchi Y. Evidence for receptor-mediated hepatic uptake of pullulan in rats. J *Control Release.* 2001; 70(3):365–73.

25. Jiang L, Li X, Liu L, Zhang Q. Cellular uptake mechanism and intracellular fate of hydrophobically modified pullulan nanoparticles. *Int J Nanomedicine*. 2013; 8:1825–34.
26. Qaqish R, Amiji M. Synthesis of a fluorescent chitosan derivative and its application for the study of chitosan–mucin interactions. *Carbohydrate Polymers*. 1999; 38(2):99–107.
27. Chopra A. Cy5.5-conjugated glycol chitosan-5beta-cholanic acid nanoparticles. Molecular Imaging and Contrast Agent Database (MICAD). Bethesda (MD). 2004.
28. Maiti S, Park N, Han JH, Jeon HM, Lee JH, Bhuniya S, et al. Gemcitabine-coumarin-biotin conjugates: A target specific theranostic anticancer prodrug. *J Am Chem Soc*. 2013; 135(11):4567–72.
29. Wolf FT, Stevens MV. The fluorescence of carotenoids. *Photochemistry and Photobiology*. 1967; 6(8):597–9.
30. Fu Y, Finney NS. Small-molecule fluorescent probes and their design. *RSC Advances*. 2018; 8(51):29051–61.
31. Abdalkader R, Kawakami S, Unga J, Suzuki R, Maruyama K, Yamashita F, et al. Evaluation of the potential of doxorubicin loaded microbubbles as a theranostic modality using a murine tumor model. *Acta Biomater*. 2015; 19:112–8.
32. Chen T, He B, Tao J, He Y, Deng H, Wang X, et al. Application of forster resonance energy transfer (FRET) technique to elucidate intracellular and in vivo biofate of nanomedicines. *Adv Drug Deliv Rev*. 2019; 143:177–205.
33. Gangopadhyay M, Mengji R, Paul A, Venkatesh Y, Vangala V, Jana A, et al. Redox-responsive xanthene–coumarin–chlorambucil-based FRET-guided theranostics for "activatable" combination therapy with real-time monitoring. *Chemical Communications*. 2017; 53(65):9109–12.
34. Karthik S, Puvvada N, Kumar BN, Rajput S, Pathak A, Mandal M, et al. Photoresponsive coumarin-tethered multifunctional magnetic nanoparticles for release of anticancer drug. *ACS Appl Mater Interfaces*. 2013; 5(11):5232–8.
35. Gautier J, Allard-Vannier E, Munnier E, Souce M, Chourpa I. Recent advances in theranostic nanocarriers of doxorubicin based on iron oxide and gold nanoparticles. *J Control Release*. 2013; 169(1–2):48–61.
36. Mosafer J, Abnous K, Tafaghodi M, Mokhtarzadeh A, Ramezani M. In vitro and in vivo evaluation of anti-nucleolin-targeted magnetic

PLGA nanoparticles loaded with doxorubicin as a theranostic agent for enhanced targeted cancer imaging and therapy. *Eur J Pharm Biopharm.* 2017; 113:60–74.

37. Goswami U, Dutta A, Raza A, Kandimalla R, Kalita S, Ghosh SS, et al. Transferrin-copper nanocluster-doxorubicin nanoparticles as targeted theranostic cancer nanodrug. *ACS Appl Mater Interfaces.* 2018; 10(4):3282–94.
38. Shung KK. High frequency ultrasonic imaging. *J Med Ultrasound.* 2009; 17(1):25–30.
39. Szabo TL, Lewin PA. Ultrasound transducer selection in clinical imaging practice. *J Ultrasound Med.* 2013; 32(4):573–82.
40. Sindoni A, Fama F, Rosano A, Scisca C, Dionigi G, Koch CA, et al. Thyroid nodules coexisting with either cystic or solid breast nodules: A new clue for this association between nodules coming from ultrasonography. *Gland Surg.* 2017; 6(6):630–7.
41. Hayes BT, Merrick MA, Sandrey MA, Cordova ML. Three-MHz ultrasound heats deeper into the tissues than originally theorized. *J Athl Train.* 2004; 39(3):230–4.
42. Fu L, Ke HT. Nanomaterials incorporated ultrasound contrast agents for cancer theranostics. *Cancer Biol Med.* 2016; 13(3):313–24.
43. Blomley MJ, Cooke JC, Unger EC, Monaghan MJ, Cosgrove DO. Microbubble contrast agents: A new era in ultrasound. *BMJ.* 2001; 322(7296):1222–5.
44. Xu JS, Huang J, Qin R, Hinkle GH, Povoski SP, Martin EW, et al. Synthesizing and binding dual-mode poly (lactic-co-glycolic acid) (PLGA) nanobubbles for cancer targeting and imaging. *Biomaterials.* 2010; 31(7):1716–22.
45. Watanabe A, Sheng H, Endo H, Feril LB, Irie Y, Ogawa K, et al. Echographic and physical characterization of albumin-stabilized nanobubbles. *Heliyon.* 2019; 5(6):e01907.
46. Lafond M, Watanabe A, Yoshizawa S, Umemura SI, Tachibana K. Cavitation-threshold determination and rheological-parameters estimation of albumin-stabilized nanobubbles. *Sci Rep.* 2018; 8(1):7472.
47. Kida H, Nishimura K, Ogawa K, Watanabe A, Feril LB, Irie Y, et al. Nanobubble mediated gene delivery in conjunction with a hand-held ultrasound scanner. *Front Pharmacol.* 2020; 11:363.
48. Xie X, Lin W, Li M, Yang Y, Deng J, Liu H, et al. Efficient siRNA delivery using novel cell-penetrating peptide-sirna conjugate-

loaded nanobubbles and ultrasound. *Ultrasound Med Biol.* 2016; 42(6):1362–74.

49. Zullino S, Argenziano M, Ansari S, Ciprian R, Nasi L, Albertini F, et al. Superparamagnetic oxygen-loaded nanobubbles to enhance tumor oxygenation during hyperthermia. *Front Pharmacol.* 2019; 10:1001.
50. Huang HY, Hu SH, Hung SY, Chiang CS, Liu HL, Chiu TL, et al. SPIO nanoparticle-stabilized PAA-F127 thermosensitive nanobubbles with MR/US dual-modality imaging and HIFU-triggered drug release for magnetically guided in vivo tumor therapy. *J Control Release.* 2013; 172(1):118–27.
51. Khan MS, Hwang J, Lee K, Choi Y, Seo Y, Jeon H, et al. Anti-tumor drug-loaded oxygen nanobubbles for the degradation of HIF-1alpha and the upregulation of reactive oxygen species in tumor cells. *Cancers (Basel).* 2019; 11(10).
52. Yordanova A, Eppard E, Kurpig S, Bundschuh RA, Schonberger S, Gonzalez-Carmona M, et al. Theranostics in nuclear medicine practice. *Onco Targets Ther.* 2017; 10:4821–8.
53. Moeendarbari S, Tekade R, Mulgaonkar A, Christensen P, Ramezani S, Hassan G, et al. Theranostic nanoseeds for efficacious internal radiation therapy of unresectable solid tumors. *Sci Rep.* 2016; 6:20614.
54. Butterworth KT, Nicol JR, Ghita M, Rosa S, Chaudhary P, McGarry CK, et al. Preclinical evaluation of gold-DTDTPA nanoparticles as theranostic agents in prostate cancer radiotherapy. *Nanomedicine (Lond).* 2016; 11(16):2035–47.
55. Yook S, Cai Z, Lu Y, Winnik MA, Pignol JP, Reilly RM. Intratumorally injected 177Lu-labeled gold nanoparticles: Gold nanoseed brachytherapy with application for neoadjuvant treatment of locally advanced breast cancer. *J Nucl Med.* 2016; 57(6):936–42.
56. Laprise-Pelletier M, Lagueux J, Cote MF, LaGrange T, Fortin MA. Low-dose prostate cancer brachytherapy with radioactive palladium-gold nanoparticles. *Adv Healthc Mater.* 2017; 6(4).
57. Beard P. Biomedical photoacoustic imaging. *Interface Focus.* 2011; 1(4):602–31.
58. Kruger RA, Lam RB, Reinecke DR, Del Rio SP, Doyle RP. Photoacoustic angiography of the breast. *Med Phys.* 2010; 37(11):6096–100.

59. Upputuri PK, Pramanik M. Recent advances in photoacoustic contrast agents for in vivo imaging. Wiley *Interdiscip Rev Nanomed Nanobiotechnol*. 2020; 12(4):e1618.

60. Wang S, Lin J, Wang T, Chen X, Huang P. Recent advances in photoacoustic imaging for deep-tissue biomedical applications. *Theranostics*. 2016; 6(13):2394–413.

61. Wang L, Yang PP, Zhao XX, Wang H. Self-assembled nanomaterials for photoacoustic imaging. *Nanoscale*. 2016; 8(5):2488–509.

62. Novoselov KS, Geim AK, Morozov SV, Jiang D, Zhang Y, Dubonos SV, et al. Electric field effect in atomically thin carbon films. *Science*. 2004; 306(5696):666.

63. Liu Z, Robinson JT, Sun X, Dai H. PEGylated nanographene oxide for delivery of water-insoluble cancer drugs. *J Am Chem Soc*. 2008; 130(33):10876–7.

64. Kim H, Lee D, Kim J, Kim TI, Kim WJ. Photothermally triggered cytosolic drug delivery via endosome disruption using a functionalized reduced graphene oxide. *ACS Nano*. 2013; 7(8):6735–46.

65. Patel SC, Lee S, Lalwani G, Suhrland C, Chowdhury SM, Sitharaman B. Graphene-based platforms for cancer therapeutics. *Ther Deliv*. 2016; 7(2):101–16.

66. Pei X, Zhu Z, Gan Z, Chen J, Zhang X, Cheng X, et al. PEGylated nano-graphene oxide as a nanocarrier for delivering mixed anticancer drugs to improve anticancer activity. *Sci Rep*. 2020; 10(1):2717.

67. Robinson JT, Tabakman SM, Liang Y, Wang H, Casalongue HS, Vinh D, et al. Ultrasmall reduced graphene oxide with high near-infrared absorbance for photothermal therapy. *J Am Chem Soc*. 2011; 133(17):6825–31.

68. Sheng Z, Song L, Zheng J, Hu D, He M, Zheng M, et al. Protein-assisted fabrication of nano-reduced graphene oxide for combined in vivo photoacoustic imaging and photothermal therapy. *Biomaterials*. 2013; 34(21):5236–43.

69. Lalwani G, Cai X, Nie L, Wang LV, Sitharaman B. Graphene-based contrast agents for photoacoustic and thermoacoustic tomography. *Photoacoustics*. 2013; 1(3–4):62–7.

70. Mei Q, Liu B, Han G, Liu R, Han M-Y, Zhang Z. Graphene oxide: From tunable structures to diverse luminescence behaviors. *Advanced Science*. 2019; 6(14):1900855.

71. Zheng P, Wu N. Fluorescence and sensing applications of graphene oxide and graphene quantum dots: A review. *Chem Asian J.* 2017; 12(18):2343–53.

72. Shang J, Ma L, Li J, Ai W, Yu T, Gurzadyan GG. The origin of fluorescence from graphene oxide. *Sci Rep.* 2012; 2:792.

73. Cushing SK, Li M, Huang F, Wu N. Origin of strong excitation wavelength dependent fluorescence of graphene oxide. *ACS Nano.* 2014; 8(1):1002–13.

74. Yan C, Hu X, Guan P, Hou T, Chen P, Wan D, et al. Highly biocompatible graphene quantum dots: green synthesis, toxicity comparison and fluorescence imaging. *Journal of Materials Science.* 2020; 55(3):1198–215.

75. Kumawat MK, Thakur M, Gurung RB, Srivastava R. Graphene quantum dots for cell proliferation, nucleus imaging, and photoluminescent sensing applications. *Sci Rep.* 2017; 7(1):15858.

76. Syama S, Mohanan PV. Comprehensive application of graphene: Emphasis on biomedical concerns. *Nano-Micro Letters.* 2019; 11(1):6.

77. Tian F, Lyu J, Shi J, Yang M. Graphene and graphene-like two-denominational materials based fluorescence resonance energy transfer (FRET) assays for biological applications. *Biosens Bioelectron.* 2017; 89(Pt 1):123–35.

78. Lei L, Ma H, Yang M, Qin Y, Ma Y, Wang T, et al. Fluorophore-functionalized graphene oxide with application in cell imaging. *New Journal of Chemistry.* 2017; 41(21):12375–9.

79. Senapati S, Patel DK, Ray B, Maiti P. Fluorescent-functionalized graphene oxide for selective labeling of tumor cells. *J Biomed Mater Res A.* 2019; 107(9):1917–24.

80. Kalluru P, Vankayala R, Chiang CS, Hwang KC. Nano-graphene oxide-mediated In vivo fluorescence imaging and bimodal photodynamic and photothermal destruction of tumors. *Biomaterials.* 2016; 95:1–10.

81. Yan X, Niu G, Lin J, Jin AJ, Hu H, Tang Y, et al. Enhanced fluorescence imaging guided photodynamic therapy of sinoporphyrin sodium loaded graphene oxide. *Biomaterials.* 2015; 42:94–102.

82. Chen L, Zhong X, Yi X, Huang M, Ning P, Liu T, et al. Radionuclide (131) labeled reduced graphene oxide for nuclear imaging guided combined radio and photothermal therapy of cancer. *Biomaterials.* 2015; 66:21–8.

83. Hong H, Yang K, Zhang Y, Engle JW, Feng L, Yang Y, et al. In vivo targeting and imaging of tumor vasculature with radiolabeled, antibody-conjugated nanographene. *ACS Nano*. 2012; 6(3):2361–70.

84. Hwang DW. Radio-graphene in theranostic perspectives. *Nucl Med Mol Imaging*. 2017; 51(1):17–21.

85. Danilenko VV. On the history of the discovery of nanodiamond synthesis. *Physics of the Solid State*. 2004; 46(4):595–9.

86. Gao G, Guo Q, Zhi J. Nanodiamond-based theranostic platform for drug delivery and bioimaging. *Small*. 2019; 15(48):e1902238.

87. Ryu TK, Baek SW, Kang RH, Jeong KY, Jun DR, Choi SW. Photodynamic and photothermal tumor therapy using phase-change material nanoparticles containing chlorin e6 and nanodiamonds. *J Control Release*. 2018; 270:237–45.

88. Wei S, Li L, Du X, Li Y. OFF–ON nanodiamond drug platform for targeted cancer imaging and therapy. *Journal of Materials Chemistry B*. 2019; 7(21):3390–402.

89. Du X, Li L, Wei S, Wang S, Li Y. A tumor-targeted, intracellular activatable and theranostic nanodiamond drug platform for strongly enhanced in vivo antitumor therapy. *J Mater Chem B*. 2020; 8(8):1660–71.

90. Rammohan N, Filicko A, MacRenaris K, Ho D, Meade TJ. Theranostic nanodiamond-gadolinium (III)-doxorubicin conjugates improve chemotherapy of cancer cells that can be monitored by magnetic resonance imaging. *International Journal of Radiation Oncology, Biology, Physics*. 2016; 96(2):E589.

91. Zhang K, Zhao Q, Qin S, Fu Y, Liu R, Zhi J, et al. Nanodiamonds conjugated upconversion nanoparticles for bio-imaging and drug delivery. *J Colloid Interface Sci*. 2019;537:316–24.

92. Wee T-L, Mau Y-W, Fang C-Y, Hsu H-L, Han C-C, Chang H-C. Preparation and characterization of green fluorescent nanodiamonds for biological applications. *Diamond and Related Materials*. 2009; 18(2):567–73.

93. Merson TD, Castelletto S, Aharonovich I, Turbic A, Kilpatrick TJ, Turnley AM. Nanodiamonds with silicon vacancy defects for nontoxic photostable fluorescent labeling of neural precursor cells. *Opt Lett*. 2013; 38(20):4170–3.

94. Magyar A, Hu W, Shanley T, Flatté ME, Hu E, Aharonovich I. Synthesis of luminescent europium defects in diamond. *Nature Communications*. 2014; 5(1):3523.

95. Torelli MD, Nunn NA, Shenderova OA. A perspective on fluorescent nanodiamond bioimaging. *Small.* 2019; 15(48):e1902151.

96. Liu Z, Liang XJ. Nano-carbons as theranostics. *Theranostics.* 2012; 2(3):235–7.

97. Chen Z, Ma L, Liu Y, Chen C. Applications of functionalized fullerenes in tumor theranostics. *Theranostics.* 2012; 2(3):238–50.

98. Minezaki H, Ishihara S, Uchida T, Muramatsu M, Racz R, Asaji T, et al. Synthesis of endohedral iron-fullerenes by ion implantation. *Rev Sci Instrum.* 2014; 85(2):02A945.

99. Garcia-hernandez DA, Cataldo F, Manchado A. About the iron carbonyl complex with C60 and C70 fullerene: [Fe(CO)4(η2C60)] and [Fe(CO)4(η2C70)]. *Fullerenes, Nanotubes and Carbon Nanostructures.* 2016; 24(3):225–33.

100. Krishna V, Stevens N, Koopman B, Moudgil B. Optical heating and rapid transformation of functionalized fullerenes. *Nat Nanotechnol.* 2010; 5(5):330–4.

101. Shi J, Chen Z, Wang L, Wang B, Xu L, Hou L, et al. A tumor-specific cleavable nanosystem of PEG-modified C60@Au hybrid aggregates for radio frequency-controlled release, hyperthermia, photodynamic therapy and X-ray imaging. *Acta Biomater.* 2016; 29:282–97.

102. Hamblin MR. Fullerenes as photosensitizers in photodynamic therapy: Pros and cons. *Photochem Photobiol Sci.* 2018; 17(11):1515–33.

103. Li Q, Hong L, Li H, Liu C. Graphene oxide-fullerene C60 (GO-C60) hybrid for photodynamic and photothermal therapy triggered by near-infrared light. *Biosens Bioelectron.* 2017; 89(Pt 1):477–82.

104. Semenov KN, Andrusenko EV, Charykov NA, Litasova EV, Panova GG, Penkova AV, et al. Carboxylated fullerenes: Physico-chemical properties and potential applications. *Progress in Solid State Chemistry.* 2017; 47–48:19–36.

105. Gaponenko IN, Ageev SV, Iurev GO, Shemchuk OS, Meshcheriakov AA, Petrov AV, et al. Biological evaluation and molecular dynamics simulation of water-soluble fullerene derivative C60[C(COOH)2]3. *Toxicol In Vitro.* 2020; 62:104683.

106. Singh R, Torti SV. Carbon nanotubes in hyperthermia therapy. *Adv Drug Deliv Rev.* 2013; 65(15):2045–60.

107. Burke A, Ding X, Singh R, Kraft RA, Levi-Polyachenko N, Rylander MN, et al. Long-term survival following a single treatment of kidney tumors with multiwalled carbon nanotubes and near-infrared radiation. *Proc Natl Acad Sci USA.* 2009; 106(31):12897–902.

108. Wang CH, Chiou SH, Chou CP, Chen YC, Huang YJ, Peng CA. Photothermolysis of glioblastoma stem-like cells targeted by carbon nanotubes conjugated with CD133 monoclonal antibody. *Nanomedicine.* 2011; 7(1):69–79.

109. Shao N, Lu S, Wickstrom E, Panchapakesan B. Integrated molecular targeting of IGF1R and HER2 surface receptors and destruction of breast cancer cells using single wall carbon nanotubes. *Nanotechnology.* 2007; 18(31):315101.

110. Galanzha EI, Kim JW, Zharov VP. Nanotechnology-based molecular photoacoustic and photothermal flow cytometry platform for in-vivo detection and killing of circulating cancer stem cells. *J Biophotonics.* 2009; 2(12):725–35.

111. O'Connell MJ, Bachilo SM, Huffman CB, Moore VC, Strano MS, Haroz EH, et al. Band gap fluorescence from individual single-walled carbon nanotubes. *Science.* 2002; 297(5581):593–6.

112. Smith AM, Mancini MC, Nie S. Bioimaging: Second window for in vivo imaging. *Nat Nanotechnol.* 2009; 4(11):710–1.

113. Ceppi L, Bardhan NM, Na Y, Siegel A, Rajan N, Fruscio R, et al. Real-time single-walled carbon nanotube-based fluorescence imaging improves survival after debulking surgery in an ovarian cancer model. *ACS Nano.* 2019; 13(5):5356–65.

114. Hong G, Diao S, Chang J, Antaris AL, Chen C, Zhang B, et al. Through-skull fluorescence imaging of the brain in a new near-infrared window. *Nat Photonics.* 2014; 8(9):723–30.

115. Hong H, Gao T, Cai W. Molecular imaging with single-walled carbon nanotubes. *Nano Today.* 2009; 4(3):252–61.

116. Gong H, Peng R, Liu Z. Carbon nanotubes for biomedical imaging: The recent advances. *Adv Drug Deliv Rev.* 2013; 65(15):1951–63.

117. Nanocrystals in their prime. *Nature Nanotechnology.* 2014; 9(5):325.

118. Girija Aswathy R, Sivakumar B, Brahatheeshwaran D, Ukai T, Yoshida Y, Maekawa T, et al. Biocompatible fluorescent jelly quantum dots for bioimaging. *Materials Express.* 2011; 1(4):291–8.

119. Cheruvathoor Poulose A, Veeranarayanan S, Mohamed MS, Raveendran S, Nagaoka Y, Yoshida Y, et al. PEG coated biocompatible cadmium chalcogenide quantum dots for targeted imaging of cancer cells. *Journal of Fluorescence.* 2012; 22(3):931–44.

120. Jin Y, Gao X. Plasmonic fluorescent quantum dots. *Nature Nanotechnology.* 2009; 4(9):571–6.

121. Veeranarayanan S, Poulose AC, Mohamed MS, Nagaoka Y, Iwai S, Nakagame Y, et al. Synthesis and application of luminescent single CdS quantum dot encapsulated silica nanoparticles directed for precision optical bioimaging. *Int J Nanomedicine.* 2012; 7:3769–86.

122. Aswathy RG, Yoshida Y, Maekawa T, Kumar DS. Near-infrared quantum dots for deep tissue imaging. *Anal Bioanal Chem.* 2010; 397(4):1417–35.

123. Bagalkot V, Zhang L, Levy-Nissenbaum E, Jon S, Kantoff PW, Langer R, et al. Quantum dot-aptamer conjugates for synchronous cancer imaging, therapy, and sensing of drug delivery based on bi-fluorescence resonance energy transfer. *Nano Lett.* 2007; 7(10):3065–70.

124. Tripathi SK, Kaur G, Khurana RK, Kapoor S, Singh B. Quantum dots and their potential role in cancer theranostics. *Crit Rev Ther Drug Carrier Syst.* 2015; 32(6):461–502.

125. Juzenas P, Chen W, Sun YP, Coelho MA, Generalov R, Generalova N, et al. Quantum dots and nanoparticles for photodynamic and radiation therapies of cancer. *Adv Drug Deliv Rev.* 2008; 60(15):1600–14.

126. Bakalova R, Ohba H, Zhelev Z, Ishikawa M, Baba Y. Quantum dots as photosensitizers? *Nat Biotechnol.* 2004; 22(11):1360–1.

127. Xu Y, Karmakar A, Wang D, Mahmood MW, Watanabe F, Zhang Y, et al. Multifunctional Fe_3O_4 cored magnetic-quantum dot fluorescent nanocomposites for RF nanohyperthermia of cancer cells. *Journal of Physical Chemistry C.* 2010; 114(11):5020–6.

128. Li W, Chen X. Gold nanoparticles for photoacoustic imaging. *Nanomedicine (Lond).* 2015; 10(2):299–320.

129. Guo J, Rahme K, He Y, Li LL, Holmes JD, O'Driscoll CM. Gold nanoparticles enlighten the future of cancer theranostics. *Int J Nanomedicine.* 2017; 12:6131–52.

130. Zhang Z, Wang J, Chen C. Gold nanorods based platforms for light-mediated theranostics. *Theranostics.* 2013; 3(3):223–38.

131. Choi J, Yang J, Bang D, Park J, Suh JS, Huh YM, et al. Targetable gold nanorods for epithelial cancer therapy guided by near-IR absorption imaging. *Small.* 2012; 8(5):746–53.

132. Kim HS, Lee DY. Near-infrared-responsive cancer photothermal and photodynamic therapy using gold nanoparticles. *Polymers (Basel).* 2018; 10(9).

133. Stephen ZR, Kievit FM, Zhang M. Magnetite nanoparticles for medical MR imaging. *Mater Today (Kidlington)*. 2011; 14(7–8):330–8.

134. Rochani AK, Balasubramanian S, Ravindran Girija A, Raveendran S, Borah A, Nagaoka Y, et al. Dual mode of cancer cell destruction for pancreatic cancer therapy using Hsp90 inhibitor loaded polymeric nano magnetic formulation. *Int J Pharm*. 2016; 511(1):648–58.

135. Niu C, Wang Z, Lu G, Krupka TM, Sun Y, You Y, et al. Doxorubicin loaded superparamagnetic PLGA-iron oxide multifunctional microbubbles for dual-mode US/MR imaging and therapy of metastasis in lymph nodes. *Biomaterials*. 2013; 34(9):2307–17.

136. Liang C, Li N, Cai Z, Liang R, Zheng X, Deng L, et al. Co-encapsulation of magnetic Fe_3O_4 nanoparticles and doxorubicin into biocompatible PLGA-PEG nanocarriers for early detection and treatment of tumors. *Artif Cells Nanomed Biotechnol*. 2019; 47(1):4211–21.

137. Balasubramanian S, Girija AR, Nagaoka Y, Iwai S, Suzuki M, Kizhikkilot V, et al. Curcumin and 5-fluorouracil-loaded, folate- and transferrin-decorated polymeric magnetic nanoformulation: A synergistic cancer therapeutic approach, accelerated by magnetic hyperthermia. *Int J Nanomedicine*. 2014; 9:437–59.

138. Parsian M, Unsoy G, Mutlu P, Yalcin S, Tezcaner A, Gunduz U. Loading of gemcitabine on chitosan magnetic nanoparticles increases the anti-cancer efficacy of the drug. *Eur J Pharmacol*. 2016; 784:121–8.

139. El-Boubbou K. Magnetic iron oxide nanoparticles as drug carriers: Preparation, conjugation and delivery. *Nanomedicine (Lond)*. 2018; 13(8):929–52.

140. Taylor RM, Huber DL, Monson TC, Esch V, Sillerud LO. Structural and magnetic characterization of superparamagnetic iron platinum nanoparticle contrast agents for magnetic resonance imaging. *J Vac Sci Technol B Nanotechnol Microelectron*. 2012; 30(2):02C101.

141. Chiang IC, Chen DH. Synthesis of monodisperse FeAu nanoparticles with tunable magnetic and optical properties. *Advanced Functional Materials*. 2007; 17(8):1311–6.

142. Xu Z, Hou Y, Sun S. Magnetic core/shell Fe3O4/Au and Fe_3O_4/Au/Ag nanoparticles with tunable plasmonic properties. *Journal of the American Chemical Society*. 2007; 129(28):8698–9.

143. Chen M, Yamamuro S, Farrell D, Majetich SA. Gold-coated iron nanoparticles for biomedical applications. *Journal of Applied Physics*. 2003; 93(10):7551–3.

144. Shao D, Li J, Zheng X, Pan Y, Wang Z, Zhang M, et al. Janus "nano-bullets" for magnetic targeting liver cancer chemotherapy. *Biomaterials.* 2016; 100:118–33.

145. Suneet K, De T, Rangarajan A, Jain S. Magnetic nanofibers based bandage for skin cancer treatment: A non-invasive hyperthermia therapy. *Cancer Rep (Hoboken).* 2020; e1281.

146. Hua S, de Matos MBC, Metselaar JM, Storm G. Current trends and challenges in the clinical translation of nanoparticulate nanomedicines: Pathways for translational development and commercialization. *Front Pharmacol.* 2018; 9:790.

147. Rastinehad AR, Anastos H, Wajswol E, Winoker JS, Sfakianos JP, Doppalapudi SK, et al. Gold nanoshell-localized photothermal ablation of prostate tumors in a clinical pilot device study. *Proc Natl Acad Sci USA.* 2019; 116(37):18590–6.

148. Yu G, Yung BC, Zhou Z, Mao Z, Chen X. Artificial molecular machines in nanotheranostics. *ACS Nano.* 2018; 12(1):7–12.

Glossary

ABC transporters: ATP binding cassette (ABC) transporters

ABC transporters are the proteins used to transport ions, nutrients, biomolecules, etc. across the plasma membrane into the cells, and toxins and wastes out of the cells. These transporters use the binding and hydrolysis of ATP to power the transportation of these micro- and macro-molecules across the plasma membrane.

AChE: Acetylcholinesterase

An enzyme that catalyzes the breakdown of acetylcholine and some other choline esters, and functions as neurotransmitters.

Afb molecules: Affibody molecules

Affibody molecules are small, robust proteins engineered to have the affinity to a large number of target proteins or peptides with high affinity and selectivity.

APCs: Antigen presenting cells

Antigen presenting cells are group of immune cells that after processing antigens present T cells for the recognition of the same. APCs include dendritic cells, Macrophages, B cells, etc.

Aptamers:

Aptamers are oligonucleotides or peptides having specific affinity to specific target molecules.

Apo: Apolipoproteins

Apolipoproteins are proteins that bind to lipids to form lipoproteins, whose main function is to transport lipids.

Atherosclerosis or atheromata:

Atherosclerosis or atheromata is an abnormal fatty deposition or plaque inside the artery, the blood vessel that carries oxygen rich blood from heart to other parts of the body. This fatty deposition can be based on macrophage cells, cholesterol, debris containing calcium, lipids, or connective tissues. This plaque will narrow down the artery and will cause the blockade for the blood flow.

ATP: Adenosine triphosphate

An energy carrying molecule, which is found in the cells, carries chemical energy obtained from the breakdown of food molecules and releases it to support other cellular process.

Attenuation coefficient

Attenuation coefficient is described as the fraction of the beam (e.g., X-rays, gamma rays, etc.) absorbed or scattered by unit thickness of the material that passes through.

BBB: Blood brain barrier

Blood brain barrier is a highly selective semi-permeable membrane separating brain from circulating blood and extracellular fluid in the central nerve system. This barrier protects brain by selectively allowing the movements of ions, molecules, and cells between blood and brain.

Blood half-life:

Blood half-life is the time taken for a drug or a metabolite to lose half of its pharmacological or physiological activities while circulating in the blood.

BPCA: Blood pool contrast agent

Blood pool contrast agents (BPCAs) or blood pool agents (BPAs) or intravascular contrast agents (ICAs) are the contrast agents used for arterial and venous imaging. This material usually has high molecular weight and relaxivity (if it is a magnetic material). Due to its high molecular weight, it will not diffuse through vascular epithelium to leak into interstitial space; hence, it will stay in the vascular system itself comparatively for longer time, which will enable to get a good image of vascular system or circular system in a human body.

BSA: Bovine serum albumin

Bovine serum albumin is a globular protein derived from cow and used in numerous biochemical applications.

CLSM: Confocal laser scanning microscopy

CLSM is based on a conventional microscope and it replaces the normal light source with laser. In conventional microscope, light travels through the specimen as far as it can penetrate, while a confocal microscope focuses a smaller beam of light at one narrow depth level at a time. This will help us to get a good high resolution image in 3D, which may not be possible to obtain by conventional microscope.

Compton effect:

Scattering of photon by a charged particle results in decreasing of photon's energy. A part of the energy will be transferred to charged particle.

CPPs: Cell penetrating peptides

Cell penetrating peptides (CPPs) also known as protein transduction domains are peptides with 5–30 amino acids. CPPs are divided into anionic, amphipathic and hydrophobic peptides. These peptides can carry small molecules, plasmid DNA, siRNA, proteins, viruses, imaging moieties, and nanoparticles across the cellular membranes, resulting in delivery of these cargos intact in cells.

CSC: Cancer stem cells

Cancer stem cells are rare immortal cells within a tumor that can both self-renew by dividing and give rise to many cell types that constitute to the formation of tumor. Even after the treatment, sometimes these cells may survive, which will result in relapse. Such cells have been found in various types of human tumors and might be attractive target for cancer treatment too.

Cytokines

Cytokines are signaling molecules that aid in communicating cells to cells in immune responses and stimulate cells to move toward the site of infection, inflammation, etc. Though the name of cytokine is general, there are different classes in this, such as lymphokine (cytokine produced by lymphocytes), monokine (cytokine produced by monocytes), chemokine (cytokine developed by chemotactic activities), and interleukin (cytokine developed by leukocytes and acting on other leukocytes).

DCs: Dendritic cells

Dendritic cells are antigen presenting cells. Their main function is to process the antigen and present the surface of the antigen infected cells to the T cells of the immune system to take necessary steps to promote immunity against the antigens.

Dendrimers

Dendrimers are highly branched, tree like polymers. The world dendrimer is derived from the Greek word dendron meaning tree.

DNA: Deoxyribonucleic acid

A molecule that carries genetic instructions used in the growth, development, functioning, and reproduction of all living organisms.

ECM: Extra cellular matrix

Extra cellular matrix is a non-cellular component presents in all tissues and organs that provides physical structural support as well as bio chemical support to surrounding cells. Most animal cells release materials into extra cellular space, creating a complex mesh work of proteins and carbohydrates. This physical scaffolding recognizes first the biochemical and biomechanical cues, which are required for tissue morphogenesis, differentiation, and homeostasis.

EGFR: Epidermal growth factor receptor

Epidermal growth factor receptor is a cell surface protein that binds to the epidermal growth factor. Its one end is situated inside cell membrane and the other side projects outside the cell. Epidermal growth factor stimulates cell growth and differentiates by binding to its receptor. When the EGF binds to the EGFR then the cells divide, and the EGFR will be abnormally high on the surface of cancer cells.

ELIP: Echogenic immuno liposomes

Echogenic immuno liposomes are the liposome formulations conjugated to antibodies that can target the cells, which can be used to reflect the ultra sound waves for echo imaging.

Embolism

Embolism is the lodging of an embolus, a blockage-causing piece of material, inside a blood vessel. The embolus may be a blood clot (thrombus), a fat globule (fat embolism), a bubble of air or other gas (gas embolism), or foreign material.

Endocytosis

Endocytosis is a process in which cells are taking up material (food, etc.) inside them. The material to be internalized is surrounded by an area of plasma membrane, which then buds off inside the cell to form a vesicle containing the ingested material.

EPR: Enhanced permeability and retention

Enhanced permeability (permeation) and retention is an effect in which, a drug of certain size accumulates more into the cancer cells compared to normal cells. Due to the fast division of cancer cells, these cells compromise the cell integrity and structure, etc. that make the cells to become leaky and having poor lymphatic drainage, which will result in permeation and retention of the drug inside the cells compared to normal cells.

FDA: Food and Drug Administration

FDA is a federal agency of the Department of Health and Human Services of the United States of America (USA), which is responsible for protecting and promoting the public health through the control and supervision of food safety, dietary supplements, medicines, etc.

FRET: Forster (fluorescence) resonance energy transfer

Fluorescence resonance energy transfer is the mechanism of the energy transfer between two nearby light sensitive molecules. The efficiency of the energy transfer is determined by their distance and it is inversely proportional to the sixth power the distance between them. This technique can be used to determine the proximity of two molecules.

HCD: Hypercholesterolemic diet

Hypercholesterolemic diet is the diet used to reduce high cholesterol level in the blood.

Heparin

Heparin is a drug used as an anticoagulant (blood thinner) to inhibit the clotting of the blood and to prevent thrombosis.

HSP: Heat shock protein

Heat shock protein is the protein developed in the cell based on the stressful condition that experienced from heat, cold, UV light, during wound, etc. This helps new proteins in the cell to fold in a correct way or refold that were damaged under stress.

IC50

IC50 is the half maximal inhibitory concentration, which is a measure of the effectiveness of a substance inhibiting a specific biological or biochemical function to 50%.

IFN: Interferon

Interferons are the signaling proteins released by the cells of the immune system, such as white blood cells, natural killer cells, fibroblast, etc. There are three classes of interferons such as alpha, beta, and gamma.

Ig: Immunoglobulin

It is a kind of antibody, whose main function is immune defense, and it plays a pivotal role in response to allergens.

ILs: Interleukins

It is a type of cytokines, developed by leukocytes and used to communicate between the leukocytes, plays a very important role in nearly all aspects of inflammation and immunity.

K-edge energy

Binding energy of K shell electron of atoms.

Kupffer cells

Kupffer cells are resident liver macrophages and play a critical role in maintaining liver functions. Under physiological conditions, they are the first innate immune cells and protect the liver from bacterial infections.

LD_{50}

LD_{50} is lethal dose 50% or median lethal dose. It is the amount of the substance required (usually per body weight) to kill 50% of the test population or test samples.

Leukocytes

Leukocytes or white blood cells circulate in the blood and body fluids to counter attack the invasion of foreign bodies into the body.

LSPR: Localized surface plasmon resonance

Localized surface plasmon arises due to the confinement of plasmons inside the nanoparticles of the size less than the wavelength used to excite the plasmons. In this case, the electric field near to the surface is greatly enhanced when it excited using the resonance frequency. In noble metals like gold, the resonance frequency occurs at visible wavelength region.

Liposome

Liposome is a spherical vesicle having at least one lipid bilayer. This can be used to transport drugs inside the body.

Macrophages

Macrophages are big eating cells that are a type of white blood cells of the immune system, which detect and digest cellular debris, foreign substances, microbes, cancer cells, bacteria, or any other harmful organisms.

MDR: Multiple drug resistance

Multiple drug resistance is the cellular mechanism that resists the drug intake by cancer cells by fluxing out the drug from the cells. This generally executes by the cell membrane anchored efflux transporter proteins.

Micelles

Micelles are aggregate surfactant molecules dispersed in a liquid colloid. Normally these molecules will have both hydrophilic and hydrophobic heads. These heads arrange in the liquid according to its nature.

MMP: Matrix metalloproteinase

A group of enzymes that is responsible for the degradation of most extracellular matrix proteins during organogenesis, growth, and normal tissue turnover.

Ms: Saturation of magnetization

In some magnetic materials, increase in the external magnetic field will not increase the magnetization of the materials further after some stage. This stage is considered as the saturation magnetization of those materials.

MSCs: Mesenchymal stem cells

Mesenchymal stem cells are multipotent adult stem cells that can differentiate into different cell types such as osteoblasts (bone cells), chondrocytes (cartilage cells), myocytes (muscle cells), and adipocytes (fat cells).

MTT assay

MTT assay is a calorimetric assay that determines the cell metabolic activity. This assay can provide the information about dead and live cells.

NF-kB: Nuclear factor – kappa B

A protein complex that is involved in cellular responses to various stimuli, such as stress, cytokines, free radicals, heavy metals, and bacterial or viral antigens.

NIR biological window

NIR biological window is known as optical window or therapeutic window in a range of wavelengths of 650–1350 nm. It comprises two windows: the first biological window (generally used) of wavelengths ranging 650–900 nm, and the second biological window (rarely used nowadays, however, a lot of research work is going on to use this window effectively) of wavelengths ranging 1100–1350 nm. In these ranges, the absorption and scattering, due to biological components such as hemoglobin, water, etc., will be minimum that will help deep penetration into body tissue.

NLS: Nuclear localization signal/sequence

Nuclear localization sequence is an amino acid sequence that can be tagged to a protein or to a cargo material, which can be transported into cell nucleus.

NMR: Nuclear magnetic resonance

It is technique that exploits the magnetic properties of certain atomic nuclei. This type of spectroscopy determines the physical properties of atoms or molecules.

PCR: Polymerase chain reaction

Polymerase chain reaction is a lab technique used to make many fold copies of DNA from a single DNA or a few DNAs.

PEGylation

PEGylation (pegylation) is the process in which poly(ethylene glycol) (PEG) molecules are covalently or non-covalently attached to molecules (usually drug) or micro structures. Coating of PEG on the surface of materials also known as pegylation and the process renders biocompatibility.

P-gp: P-glycoprotein

P-gp is a protein pump, an ABC transporter, which is capable of extruding toxins, drugs, and xenobiotics out of the cells across plasma membrane and also plays significant role in absorption and disposition of various drugs and nutrients. It extrudes even needed several chemotherapeutic drugs that results in reduced drug target interaction.

Photoelectric effect

Photoelectric effect is a phenomenon in which electrically charged particles are released from or within a material when it absorbs electromagnetic radiation such as light. The effect is often defined as the ejection of electrons from a metal plate when light falls on it.

Photon attenuation

Photon attenuation is the gradual loss of the flux intensity of photon when it interacts with tissues, materials, etc.

Pinocytosis

Pinocytosis, also known as liquid endocytosis, is the ingestion of liquid into a cell by budding of small vesicles from the cell membrane.

Plasma half-life

Plasma half-life is the time taken for the concentration or amount of drugs in human body to become one-half. A drug plasma half-life is depending on how quickly it eliminates from the body.

PLGA - Poly(lactic-co-glycolic acid):

PLGA is a copolymer of poly lactic acid (PLA) and poly glycolic acid (PGA). PLGA are a family of FDA-approved biodegradable polymers that are highly biocompatible and have been extensively studied as delivery vehicles for drugs, proteins and various other macromolecules such as DNA, RNA and peptides

qRT-PCR: Quantitative reverse transcription polymerase chain reaction

It is a laboratory technique of molecular biology on the polymerase chain reaction (PCR). It monitors the amplification of a targeted DNA molecule during the PCR.

RES: Reticuloendothelial system

Reticuloendothelial system or macrophage system is a part of immune system that consists of phagocytic cells located in reticular connective tissues such as blood, spleen, liver, lungs, bone marrow, and lymph nodes.

RNA: Ribonucleic acid

It is a polymeric molecule that is essential in various biological roles in coding, decoding, regulation, and expression of genes.

ROSs: Reactive oxygen species

ROSs are used to describe a number of reactive molecules and free radicals derived from molecular oxygen. They are known to cause the death of the cells.

SCF: Super critical fluid

The super critical fluid is a stage at which any substance at a temperature and pressure above its critical point, where its distinct liquid and gas phases do not exist. It can effuse through solids like a gas, and dissolve materials like a liquid.

Sentinel lymph nodes

The sentinel lymph node is the hypothetical lymph node or a group of nodes to which the metastasis of tumor reaches first from its primary point. From this sentinel lymph node, the cancer cells metastasis to other areas in the body. Surgeons check the sentinel lymph nodes while doing surgery and if they found that the cancer is not reached there, and then they will not remove any other lymph nodes, since there is no spread of tumor.

SOD: Superoxide dismutase

SOD is an enzyme that repairs cells and reduces the damage done to them by superoxide, which is the most common free radical in the body.

Super paramagentism

Super paramagnetism is a state of magnetism appears in ferromagnetic or ferrimagnetic nanoparticles. The advantage of this state is that the material having this state will show absolute zero magnetism in the absence of external field, but under external magnetic field, it will show magnetism.

$t_{1/2}$: Half-life

The period of time required for radiative material mass to be reduced to one-half of the material, or the concentration or amount of drug in the body to be reduced by one-half.

TF: Tissue factor

Tissue factor is also known as thromboplastin, which is the principal initiator of the coagulation cascade.

Tf: Transferrin

It is a plasma protein that transports iron through the blood to the liver, spleen, and bone marrow, which controls the level of free iron (Fe) in biological fluids.

TK: Tyrosine kinase

Tyrosine kinase is an enzyme that can transfer a phosphate group from ATP to a protein in a cell. It functions as an "on" or "off" switch in many cellular functions.

TLR: Toll like receptor

Toll like receptors are a class of proteins that play a key role in the innate immune system. They are single, membrane-spanning, non-catalytic receptors usually expressed on sentinel cells, such as macrophages and dendritic cells, which recognize structurally conserved molecules derived from microbes.

TNF: Tumor necrosis factor

TNF is a cell signaling protein (cytokine) involved in systemic inflammation and it is one of the cytokines that makes up the acute phase reaction. The primary role of TNF is regulation of immune cells.

VEGF: Vascular endothelial growth factor

VEGF is a signal protein produced by cells that stimulates vasculogenesis and angiogenesis.

XRD- Xray diffraction:

X-ray powder diffraction (XRD) is a rapid analytical technique primarily used for phase identification of a crystalline material and can provide information on unit cell dimensions.

Zeta potential:

Zeta potential is the potential difference existing between the surface of a solid particle immersed in a conducting liquid and the bulk of the liquid.

Index

acoustically reflective nanoparticles for tumor diagnosis 143–158
alternating magnetic field (AMF) 92, 344–345, 350, 354–356, 605–606
AMF, *see* alternating magnetic field
androgen receptor (AR) 551
angiogenesis 16–17, 20, 26, 90, 145, 147–148, 176, 638
antibodies 31, 78, 92, 118, 128, 146–147, 202–203, 210, 217, 221, 227, 237, 240, 272, 301–302, 356, 381, 383, 406, 465, 470, 496, 501, 504, 511, 540, 575, 578, 599, 602–603, 629, 631
anticancer activity 302
anticancer agents 303, 544, 546
anticancer drug loading 273, 295
anticancer drugs 23, 151, 228, 272, 297, 544, 546
  nanoparticle-loaded 546
anticancer therapy 16, 24, 33, 544–546, 587, 601–602, 605, 607
  conventional 22–23, 25
antigen delivery 514
antigen presentation 500–502, 512
antigen presenting cells (APCs) 496, 500, 502, 510, 512–515, 625, 628
  artificial 513–515, 518
antigens 236–237, 496, 506–507, 512, 514–515, 625, 628
antitumor immunity 358, 496, 517
APCs, *see* antigen presenting cells
apolipoprotein 183, 277, 625
apoptosis 6, 21, 26–29, 31, 54, 81, 297, 305, 316, 339–341, 351, 355, 375, 384–385, 411, 456, 458–460, 473, 475, 552, 554
aptamers 146, 202, 205, 227, 299, 302, 349, 354, 356, 406, 418, 511, 602–603, 625
AR, *see* androgen receptor
atherosclerosis 117–118, 146–147, 626
AuNPs/GNPs, *see* gold nanoparticles
AuNRs, *see* gold nanorods
AuNSts, *see* gold nanostars
autofluorescence 197–198, 204, 220, 234, 237
autophagy 15, 27–28, 456, 459–460

biocompatibility 29, 128, 149, 180–181, 183, 215, 226, 229, 240, 291, 300, 305–306, 349, 352, 408,

464–465, 469, 471–472, 531, 539, 635
biocompatible SWCNTs 221
bioimaging 204, 226, 236, 238, 581, 589, 592, 594, 600, 602
bioimaging applications 97, 198, 219, 588–589, 600, 603, 610
biological fluids 50, 145, 150, 271
biological function 287, 530–531
biomaterials, polymeric 578, 581
biomolecules 32, 55–56, 62, 202–203, 234, 236, 286, 348, 392, 625
bionano-based advanced anticancer strategies 1–34
bladder cancer 9, 24, 396, 506
  risk of 9
block copolymers 205, 213, 294–296
blood 2, 5, 50, 52, 78, 91, 111, 116, 123, 144, 150, 156, 173, 178, 181, 216, 280, 285, 287, 350, 358, 471, 512, 555, 626, 630–632, 636–637
blood-brain barrier 97, 343, 626
blood clots 149, 629
blood pool contrast agents (BPCAs) 173–174, 627
blood vessels 5, 13–14, 16, 22, 49, 90, 92, 115, 144, 147, 173, 176, 181, 206, 208, 211, 218, 276, 303, 341, 626, 629
bloodstream 128, 275–278, 280, 287, 289–291, 294–297, 308, 318
body cavity 339, 455
body tissues 157, 216, 220, 530, 634
bovine serum albumin (BSA) 232, 355, 385, 627
BPCAs, *see* blood pool contrast agents
brachytherapy 446, 448–449, 455–456, 610
breast cancer cells 204, 217–218, 227, 303, 305, 461, 539, 545
breast cancer metastases 207, 554
breast cancer patient 497
breast cancer radiation therapy 448
BSA, *see* bovine serum albumin

CAFs, *see* cancer-associated fibroblasts
CAM, *see* cell adhesion molecules
camptothecin (CPT) 300–301, 587
cancer 1–28, 30–34, 45–46, 59–60, 63, 75, 77, 83, 85, 88, 109, 182, 195–196, 226, 228, 271–272, 289, 291–292, 303, 307–309, 311, 316, 318, 337–340, 342, 344, 346, 348, 350, 352–354, 356–358, 373–374, 378, 380, 445–448, 453–454, 495, 497–498, 502–503, 505–507, 509–510, 517–518, 527, 529, 535, 537–538, 540, 544–547, 551–552, 554
  breast 5, 9, 11, 22, 118, 148, 176, 205, 233, 309, 311–312, 315, 317, 381, 448, 462, 497, 502, 559, 600

colon 278, 557, 582
conventional therapies 1–2, 4, 6, 8, 10, 12, 14, 16, 18, 20, 22, 24, 26, 28, 30, 32, 34
metastasis 95, 211
metastatic 92, 339
neck 89, 396, 449–450, 452, 502, 504
primary 5, 503
secondary 462
skin 10–11, 396
solid 16, 497
squamous cell 12
stomach 2
cancer-associated fibroblasts (CAFs) 13, 17–18
cancer cell imaging 228
cancer cell proliferation 19, 303
cancer cells 23–24, 292, 342, 347, 355
cultured 288
darkfield imaging of 383
diploid polyclonal 7
metastasis 637
prostate 205
cancer chemotherapy 308, 318
cancer detection 603
cancer development 1, 6, 10–11, 552
cancer diagnosis 2, 29–31, 75–76, 78, 80, 82, 84, 86, 88, 90, 92, 94, 96, 109–110, 112, 114, 116, 118, 120, 122, 124, 126, 128, 342, 475
cancer diagnostics 229, 342
cancer drugs, aromatic 304
cancer immunoediting 498–499
cancer immunology 495, 497
cancer immunosurveillance 498
cancer immunotherapy 495–506, 508, 510–512, 514, 516–518
blockade-based 495, 503
role of nanoparticles in 495–518
cancer nanomedicine 274, 277, 318
cancer nanotechnology 46, 52
cancer radiotherapy 452
cancer stem cells 453, 628
cancer theranostics 474
cancer therapeutics 271, 296, 377, 423
cancer therapy 46, 272, 275, 338, 340, 342, 346, 459, 528, 533, 542, 544, 547–548, 550, 556, 559, 574, 576–577, 579, 592, 610
photodynamic 602
radiation-based 445, 459, 461
cancer treatment 11, 22–24, 26, 32, 45–46, 289, 303, 338, 340, 342, 346–347, 352, 374–375, 381, 396, 403, 420, 445–446, 455, 460, 463, 472, 495, 497, 500, 503, 505, 510, 517–519, 587, 610, 628
conventional 555
cancer tumors 30, 583
human breast 92
cancer vaccines 506
carbon 83, 240, 389–390, 395, 418, 452, 537–538
carbon-based nanomaterial 377, 587
carbon nanomaterials 390–391, 538, 542, 586, 590
carbon nanoparticles 538, 542
carbon nanostructures 303–305, 586
carbon nanotubes (CNTs) 90–91, 218–220, 223, 274, 303–305, 377, 390–392,

395, 407, 418, 538, 542, 587, 589, 594, 596–597, 599–600
carcinogenesis 3, 7–8, 20
carcinogens, chemical 3
carcinoma cells 598
cardiotoxicity 309
cationic lipids 286–287, 531–532, 535, 557
caveola-mediated nanoparticle delivery 48
cell activation 498–499, 503, 510, 514
cell adhesion molecules (CAM) 210, 410
cell cycle 9, 447, 456, 552
cell death 26–28, 297, 340–341, 352, 375, 385, 389, 392, 447, 456–461, 469
cell death mechanisms 457, 459
cell membranes 12, 47, 49–50, 57, 237, 279, 288–289, 293, 341, 398, 402, 418, 528, 533, 535, 538, 629, 635
cell-penetrating peptides (CPPs) 49–50, 289, 473, 554, 627
cells
  apoptotic 27, 81, 458
  cancerous 1, 341, 378
  cultured 292, 297, 304
  damaged 22, 27
  diseased 114, 395, 400
  inflammatory 13, 17–18
  irradiated 457, 460
  labeled 123
  liver 54
  malignant 4, 13, 92
  mammalian 48, 514, 548
  muscle 48, 633
  panc-1 601
  phagocytic 636
  stromal 279
cervical cancer 2, 12, 215, 456, 504
cervical cancer cells 302, 305, 473
  detection 215
chemical vapor deposition (CVD) 219, 222
chemiluminescence 215, 232
chemokines 509, 628
chemoradiation 505, 610
chemotherapeutic agents 13, 33, 307, 339, 355, 601
chemotherapeutic drugs 33, 635
  platinum-based 468
chemotherapeutics 29, 299, 341, 355, 464–465
chemotherapy 3, 22–23, 45, 111, 147, 271–272, 274, 276–278, 280, 282, 284, 286, 288, 290, 292, 294, 296, 298, 300, 302, 304, 306–308, 310, 312, 314, 316, 318, 338–341, 356, 373–374, 384–385, 390, 446, 458–461, 497, 504, 510, 512, 517, 605, 611
  anticancer 545
  nanotechnology in 308
  targeted drug-delivery-based 602
chemotherapy drugs 504
  platinum-based 468
chitosan 227–228, 295, 354, 407–408, 410, 543, 556, 605
chitosan nanoparticles 413
chlorins 304, 410, 417–418
CHPP, *see* continuous hyperthermic peritoneal perfusion
chromophores 157, 399, 401, 415, 586
cisplatin 272, 297, 301, 307, 468, 544

CLSM, *see* confocal laser scanning microscopy
CNTs, *see* carbon nanotubes
cobalt nanoparticles 386
colon cancer cells 307, 422, 470
colorectal cancer 6, 9, 210, 504
combination therapies 318, 518
computed tomography 96, 123, 171, 381, 471, 584
computerized tomography (CT) 76, 94–95, 123, 171–173, 175, 179, 181–184, 243, 381, 536
confocal laser scanning microscopy (CLSM) 228, 627
continuous hyperthermic peritoneal perfusion (CHPP) 339
contrast agents 76, 93, 112, 114–117, 119, 122–124, 145, 148–149, 156, 158–159, 175–176, 178–179, 186, 222, 229, 236, 470, 536–537, 587, 594, 603, 605, 627
  blood pool 173, 627
  iodinated 176–177, 179–180
  MRI 128–129
  for tumor diagnosis 171–172, 174, 176, 178, 180, 182, 184, 186
CPPs, *see* cell-penetrating peptides
CPT, *see* camptothecin
CT, *see* computerized tomography
CVD, *see* chemical vapor deposition
cyclodextrin 227–228, 293, 299, 350, 407–408
cytarabine 308–310, 318, 546
cytokines 502, 509, 514, 628, 631, 634, 638

DDR, *see* DNA damage response
DDSs, *see* drug delivery systems
dendrimers 30, 83, 86, 173, 176, 213, 274, 300–301, 308, 407, 517–518, 542, 628
dendritic cells 496, 625, 628, 638
detonation synthesized ND (DNDs) 592
diseases 1, 9, 12, 19, 46, 53, 57, 61–62, 145–147, 171, 177, 195–197, 272, 274, 337, 396, 400–401, 409, 466, 497, 527, 529, 544, 548, 556–557
DLS, *see* dynamic light scattering
DMCA, *see* dual-mode contrast agents
DNA 3, 7, 12, 19, 23, 25, 204, 223, 240–241, 272, 286, 293, 343, 400, 447, 467–469, 531, 578, 629, 634, 636
DNA damage 18, 23, 452, 456–457, 460–462, 468–469
DNA damage response (DDR) 461
DNDs, *see* detonation synthesized ND
DOX, *see* doxorubicin
DOX-loaded micelles 296–297, 299
doxorubicin (DOX) 92, 151, 153–154, 272, 276, 286, 290, 296–297, 299–301, 304, 306–307, 309–310,

318, 384–385, 390, 410–411, 545–546, 574, 579–582, 587, 592, 595, 602, 605, 607
drug delivery 33, 46, 60, 92, 111, 152–153, 229, 272–273, 350, 357, 378, 393, 402, 412–413, 539, 543, 554, 556, 605
  anticancer 544
  controlled 61, 352, 542
  hyperthermia-based controlled 350
  nanoparticle-mediated 536
  targeted 21, 29, 31, 46, 228, 276, 291, 346
  targeted anti-cancer 31
drug delivery systems (DDSs) 52, 271–281, 283, 285, 287, 293, 295, 304, 308, 318, 410, 555
  lipid-based 284–285, 287, 289, 291
  polymer-based 293–295, 297, 299, 301
drug loading 286, 296–298, 304, 306, 382, 409, 607
drugs
  anti-cancer 31–33, 293, 404
  anticancer 300
  encapsulated 290–292
  hydrophilic 286, 291, 404, 543
  macromolecular 274, 296
  nucleic acid 286, 289
  small molecule 271–272, 278, 294, 296, 531, 537, 540, 550, 559
dual-mode contrast agents (DMCA) 119–120
dyes 204, 229, 236, 409, 578, 586–588
dynamic light scattering (DLS) 541, 553
EBRT, *see* external beam radiation therapy
echogenic immuno liposomes 629
ECM, *see* extracellular matrix
external beam radiation therapy (EBRT) 448, 456
extracellular matrix (ECM) 14–15, 17, 19–20, 47, 226, 406, 629
FA, *see* folic acid
FI, *see* fluorescent imaging
fluorescence 121, 197, 199–200, 207–210, 215, 232, 243, 306, 381, 395, 415, 577, 588, 593, 595, 604, 630
fluorescence resonance energy transfer (FRET) 204, 237, 415, 417, 578, 611, 630
fluorescent imaging (FI) 122, 181–185, 575, 593
fluorescent nanodiamonds (FNDs) 594
fluorophores 122, 159, 197–198, 415, 466, 580, 588
FNDs, *see* fluorescent nanodiamonds
folate receptors 93, 288, 305, 422
folic acid (FA) 93, 118, 238, 288, 302, 306, 350, 382, 385, 387, 392–394, 538
free radicals 10, 12, 446, 634, 636
FRET, *see* fluorescence resonance energy transfer
fullerenes 90, 303–306, 377, 390–391, 395, 415, 417–418, 594–596, 600
  carboxylated 595

gadolinium 115–116, 124, 184, 242, 292, 465, 470
gastric cancer 10, 310, 504
  risk of 10
gastric cancer cells 384
gene delivery 60, 549, 554, 584
gene silencing 59, 237, 547–548
genes 3, 6–8, 10, 17, 19, 54–55, 81, 272, 316, 392, 527, 538, 540, 549, 557, 577, 610, 636
  full-length 527–529, 533
GFP, *see* green fluorescent protein
glioblastoma cancer cells 223
glioma cells 469–470, 473–474
gold 84, 122, 174–175, 178, 186, 213–214, 217, 302, 349, 377–378, 380–382, 388–389, 407–408, 465, 475, 537, 539, 545, 586, 602–603, 605, 632
gold-based nano contrast agents 173, 175, 177
gold nanocages 212, 214, 377, 388
gold nanomaterials 123, 377–379
gold nanoparticles (AuNPs/GNPs) 29, 31, 33, 54–55, 58, 175, 178, 212–218, 464, 512, 514
  colloidal 378
  heparin-coated 175
  spherical 465
gold nanorods (AuNRs) 157, 214, 303, 377, 379, 382–385, 389, 603–604
  synthesis of 214
gold nanoshells 214, 216, 377, 379
gold nanospheres 212–213, 215
  hollow 387
gold nanostars (AuNSts) 377, 379, 389–390
gold nanostructures 382, 602–603
graphene oxide 304, 393, 537–538, 589
green fluorescent protein (GFP) 549–550
gross tumor volume (GTV) 453
ground state (GS) 113, 197, 235, 375, 397–399
GS, *see* ground state
GTV, *see* gross tumor volume

heat generating sources for hyperthermia 341
HeLa cells 237, 387, 390, 411–414, 473, 544–545
HePG2 cells, miRNA in 554
Her2-negative cells 152
Her2-positive cells 152
HIFU, *see* high intensity focused ultrasound
high intensity focused ultrasound (HIFU) 152, 585
Hipco SWCNTs 223
homeostasis 15, 17, 19–20, 55, 629
HSA, *see* human serum albumin
human breast cancer tumor xenografts 603
human ovarian cancer SKOV3 cells 546
human pancreatic cancer xenografts 86
human serum albumin (HSA) 79, 233, 382, 411, 545, 595
human umbilical vein endothelial cells (HUVECs) 148, 158, 533, 554

HUVECs, *see* human umbilical vein endothelial cells
hydrophilic surfactants 203
hydrophobicity 52, 402, 409–410
hyperthermia 111, 220, 289, 337–350, 352–358, 595–596, 599, 605, 611
 local 339
 nanoparticle-mediated 303
 regional 339
 treatment 339, 341, 344, 348–349, 357
hypoxia 14, 16–17

IBRT, *see* internal beam radiation therapy
IEL, *see* internal energy loss
IGRT, *see* image-guided radiation therapy
ILCs, *see* innate lymphoid cells
image-guided radiation therapy (IGRT) 450–452, 475
imaging
 biological 219, 221
 biomedical 90, 234, 238, 243
 blood pool 116, 178, 181, 185–186
 bright field 123
 cancer 84, 583, 603
 cellular 199, 393
 chemiluminescence 232
 contrast 179, 532
 deep tissue 94, 120, 123, 184–185, 204, 216, 219, 234, 389, 586
 dual-modal 120, 126–127
 fluorescence 181, 197, 205, 219, 232, 382, 390, 575, 578, 586, 602
 liver 180–181
 mass spectrometry 575
 multiplexed Raman 224
 nuclear 75, 78, 87
 quantitative 96–97
 real-time 124, 159, 237
 tomographic 82
 tumor 88, 182–183, 215–216, 450
 tumor tissue 216
 tumor vasculature 91
 vascular 237
immune surveillance 18, 498, 500, 507–508
immunoinhibitory receptors 495, 503, 510
immunosuppressive cells 517–518
immunotherapy 22, 24, 340, 353, 374, 495, 503, 510–511
 active 24, 501–502
 nano 511, 513, 515
 passive 24, 501–502
 tumor 21, 510
IMRT, *see* intensity-modulated radiotherapy
inertness, chemical 468, 472
inhibitory receptors 504, 508
innate lymphoid cells (ILCs) 509
integrins 90, 92, 210
intensity-modulated radiotherapy (IMRT) 449–450, 452
internal beam radiation therapy (IBRT) 448
internal energy loss (IEL) 235
iodine 124, 174, 176, 178–179, 218, 465
ionizing radiation 12, 23, 124, 179, 308, 341, 459–460, 462, 473, 497
iron oxide nanoparticles 29, 92, 292, 356, 387, 473–474

irradiation 216, 287, 290, 303–304, 341, 380–381, 384–385, 387, 389–390, 394, 396, 403, 411, 419, 457, 465–470, 474, 594–595
  laser 381, 384–385, 387, 389, 391, 393–394, 410
  light 303, 307, 402, 412, 590

kidney 91, 156–157, 177–181, 232, 276, 379, 470, 531, 535, 543, 598
kinesin spindle protein (KSP) 533
KSP, *see* kinesin spindle protein

latent membrane protein (LMP) 507
layered double hydroxide (LDH) 239–243
LDH, *see* layered double hydroxide
ligand exchange 128, 202
ligands 20–21, 29, 49, 61, 110, 128, 146, 148, 196, 202–203, 227–228, 236–237, 279, 288, 347, 392, 406, 458, 468, 503–504, 508–509, 514, 518, 555
  targeted 532, 555
light scattering 197–198, 215, 375, 379, 381, 388, 415, 418
lipid nanoparticles 274, 310
lipid–polymer hybrid nanoparticles (LPNs) 291
liposomes 285
  echogenic 149–150
  hybrid 292
  magnetic cationic 353
  magnetite cationic 353
  thermoresponsive 289–290
Listeria monocytogenes Listeriolysin O (LLO) 58
liver 2, 5, 11, 23, 29, 88, 91, 95, 111, 121–122, 144, 156, 174–178, 180–182, 206–207, 211, 232, 317, 340, 379, 381, 476, 543–544, 552, 598, 632, 636–637
liver cancer 210, 448, 504
liver cancer therapy 607
liver cancer treatment 448
liver tumor tissues 551
LLO, *see* Listeria monocytogenes Listeriolysin O
LMP, *see* latent membrane protein
localized surface plasmon resonance (LSPR) 378, 383, 388, 422, 632
LPNs, *see* lipid–polymer hybrid nanoparticles
LSPR, *see* localized surface plasmon resonance
lung cancer 9, 11, 16, 85, 175, 396
lung cancer cells 475
lymph nodes 95, 111, 121, 178, 180, 206, 208–209, 211, 448–449, 603, 636–637
  sentinel 96, 127, 208–209, 233, 637
lysolipids 290
lysosomes 15, 28, 55–57, 154, 241, 473

macromolecules 16, 636
macrophages, tumor-associated 13
magnetic cores 350, 413
magnetic hyperthermia 92, 110, 227–228, 348–349, 352, 355–357, 472
magnetic iron oxide nanoparticles 345
magnetic nanomaterials 122, 342, 346
  for hyperthermia 346–347
magnetic nanoparticles 109, 227, 342, 605, 607
  for cancer diagnosis 109–110, 112, 114, 116, 118, 120, 122, 124, 126, 128
  for hyperthermia 337–338, 340, 342, 344, 346, 348, 350, 352, 354, 356, 358
  photo-responsive multifunctional 579
magnetic particle imaging (MPI) 124–126, 129
magnetic resonance (MR) 183–184, 195, 381–382, 536
magnetic resonance imaging (MRI) 76, 94–96, 109, 112–114, 117–124, 126–127, 154–157, 181–185, 242–243, 348–349, 395, 470–471, 532, 575, 599, 611
magnetism 114, 222, 637
magnetite 347, 413
magnetization 113, 348, 633
matrix metalloproteinases 17, 19, 284, 604, 633
MB, *see* methylene blue
MCF-7/AZ breast cancer cells 421
MCF-7 breast cancer cells 475
MCF-7 cells 307, 473
MCF7 breast cancer cells 549
MDA-MB-231 breast cancer cells 554
MDSCs, *see* myeloid-derived suppressive cells
melanoma 12, 316, 339, 353–354, 387, 453, 505, 513
melanoma tumors 590
mesenchymal stem cells (MSCs) 153, 633
mesoporous silica nanoparticles (MSNs) 152–153, 412, 542
methylene blue (MB) 144–148, 152, 154, 156–159, 410–412, 549
micelles, polymeric 30, 274, 279, 294–297, 318, 542
miRNA delivery
  nanoparticle-based 557
  nanoparticle-mediated 558
  polymer nanoparticles for 552
MPI, *see* magnetic particle imaging
MR, *see* magnetic resonance
MRI, *see* magnetic resonance imaging
MSCs, *see* mesenchymal stem cells
MSNs, *see* mesoporous silica nanoparticles
multi-walled carbon nanotubes (MWCNTs) 218–220, 391–393, 597–599
multifunctional nanoparticles 126, 342–343, 348
  lipid-based 532
MWCNTs, *see* multi-walled carbon nanotubes

myeloid-derived suppressive cells (MDSCs) 509, 516–518

nano carbon structures 586–587, 589, 591, 593, 595, 597, 599
nanobubbles 144, 148, 151, 583–585
  polymer-based 149, 151
nanocages 212, 218, 388–389, 587
nanocarriers, polymeric 61–62, 294
nanocomposites 90, 350, 578–579
nanodrugs 46, 406
nanoformulations 46, 233, 308, 534, 545, 550, 554, 573–574, 608
nanomaterials
  in chemotherapy 271–318
  inorganic 179, 200, 230, 240, 302
nanomedicine, theranostic 579
nanoparticles
  biodegradable 408
  carbon-based 391
  cationic 49, 535
  ceramic 303, 404
  chitosan-based 411
  engineered 46
  hybrid 291, 384, 535, 537
  magnetic Fe_3O_4 413
  metallic 111–112, 117, 212, 347, 378–379, 413, 574, 577, 579
  non-biodegradable 408, 412
  organic 535, 537, 542
  passive 406–408
  photosensitizer 406, 414–415
  polymer-based 543, 546
  polymer-based biodegradable 406, 408
  polymeric 291
  reflective 144, 146, 148, 150, 152, 154, 156, 158
  scintillation 419
  self-lighting 406, 414, 419
  silver 54, 464, 542
  spherical 379, 383
  synthetic 536
nanoscale metal-organic frameworks 306–307
nanoshells 212, 217, 379–381, 387, 587
nanospheres 212, 217–218, 603
nanotubes 84, 221–224, 538
NCT, *see* neutron capture therapy
ND-PEG-GLY-DOX (NPGD) 592
near infra-red (NIR) 61, 92, 123, 184, 198, 208, 211, 220–223, 233, 238, 303, 374, 383, 386, 389, 392, 394–395, 469, 545, 586, 592, 599, 634
neutron capture therapy (NCT) 470
NIH-3T3 cells 221
NIR, *see* near infra-red
NIR laser 381, 421, 591, 605, 610
NIR QDs 95, 204, 206–207, 210, 223
NMR, *see* nuclear magnetic resonance
non-small cell lung cancer (NSCLC) 175, 309, 311, 317, 454, 504–505
normalization therapy 501
NPGD, *see* ND-PEG-GLY-DOX

NSCLC, *see* non-small cell lung cancer
nuclear magnetic resonance (NMR) 86, 112, 634
nucleic acids 29, 240, 286, 293–294, 296, 341, 418, 530–537, 540, 543, 556
  charged 532, 534, 538, 540, 556
  loaded 533–535
nucleotides 240, 547, 550, 552

OCT, *see* optical coherence tomography
OI, *see* optical imaging
oncogenes 3, 6–8
  activation of 4–6
optical coherence tomography (OCT) 197, 216, 379
optical imaging (OI) 76, 96, 121, 123, 159, 171, 183, 195–197, 199, 201, 203, 205, 207, 209, 211, 213–217, 219, 221, 223–225, 227, 229–233, 235–239, 241, 349, 532, 536, 539, 586
  contrast agents 159
  non-invasive 229, 233
optical nanoprobes for diagnosis 195–244
ovarian cancer 274, 309, 311, 419, 448, 533
oxygen, molecular 396, 398, 636

PA, *see* photoacoustic imaging
paclitaxel 227, 272, 296, 309–311, 313–314, 516, 544, 587, 607
pancreatic cancer 85, 278, 307, 311, 317, 547, 610
pancreatic cancer cells 349, 394
pancreatic cancer therapy 227
pancreatic cancer treatment 228, 394
PARP, *see* poly-ADP-ribose polymerase
PAT, *see* photoacoustic tomography
PC, *see* protein corona
PCR, *see* polymerase chain reaction
PET, *see* positron emission tomography
photoacoustic imaging (PA) 123, 127, 154–155, 157, 220, 379, 395, 495, 575, 587
photoacoustic tomography (PAT) 216–217, 600
photodynamic therapy 22, 24, 233, 290, 395, 397, 399, 401, 537, 587, 592, 594
photons 25, 77, 184, 196–198, 216, 289, 397, 446, 627, 635
  high-energy 235
  low-energy 235
photosensitivity 400, 402–404
photosensitizer (PS) 395–402, 404, 408, 415–417, 419–420
  delivery capability of 402–403
  early release of 403–404
  first-generation 399, 402
  hydrophobic 402, 404, 410
  second-generation 399, 402–403
  third-generation 400, 403
photosensitizers 25, 59, 303, 381, 395, 397–398,

403–404, 407, 410–411, 413–419, 421, 605
phototherapy 373–374, 376, 378, 380, 382, 384, 386, 388, 390, 392, 394, 396, 398, 400, 402, 404, 406, 408, 410, 412, 414, 416, 418, 420, 422
photothermal therapy (PTT) 220, 303, 374–375, 377–387, 389–391, 393–395, 422, 537, 587, 590, 595, 604
of cancer cells 376
imaging-guided 394–395
photothermal treatments 378–379
phthalocyanine 230, 233, 303, 409, 413
piRNAs, *see* piwi-interacting RNAs
piwi-interacting RNAs (piRNAs) 528–529, 547
PLA, *see* poly-lactic acid
poly-ADP-ribose polymerase (PARP) 27
poly-lactic acid (PLA) 151, 227, 407–409, 543, 602, 636
poly-vinyl chloride (PVC) 226
polycations 295–296
polyesters 226–227, 295, 301, 350
polyethers 295–296
polymer micelles 294–296
polymer nanoparticles 404, 537, 543–546, 551, 554–555
biodegradable 412, 548, 555
polymerase chain reaction (PCR) 634, 636
polymeric nanoparticles 226, 291, 407–408, 542–543
lipid-coated 291
polymers
biodegradable 181, 298, 546, 559
cationic 58, 534
CD-embedded 300
hybrid 540–541, 556
linear 300–301
natural 229, 555–556, 580, 605
polymersomes 227, 294, 298
polypeptides 227, 295–296, 315, 577
polysaccharides 227–229, 293, 295, 349, 583
porphyrin 233, 240, 381, 399, 402, 409–411, 413, 419, 470, 587
positron emission tomography (PET) 75–77, 79–86, 90–91, 94–96, 120, 154–155, 157, 181–183, 185, 221, 449, 574, 611
positrons 76, 79–80, 90, 584
prolonged blood half-life 277, 315
prostate cancer 10–11, 22, 148, 283, 379, 382, 449–450, 452, 456, 551, 584
advanced 314
prostate cancer cells 9, 602
prostate cancer treatment 273
protein corona (PC) 50–52, 209
proteins 6, 12, 15, 24, 26, 32, 47, 50–52, 57–58, 62, 85, 90, 118, 202–203, 227, 229, 272, 286, 288, 293–294, 296, 299, 305, 312, 340–341, 343, 375, 400, 418, 447, 456, 459, 510, 531, 537, 542, 552, 556,

577, 579, 583, 605, 625, 628–629, 631, 633–634, 636, 638
heat shock 85, 339, 357, 631
proton sponge effect 57–58, 61
protons 58, 112–113, 398, 446, 452–453
PS, *see* photosensitizer
PS drugs 405, 423
PS-loaded nanoparticles 405, 423
PTT, *see* photothermal therapy
PVC, *see* poly-vinyl chloride

QDs, *see* quantum dots
quantum dots (QDs) 95, 122–123, 198, 200, 202–206, 208–211, 219, 229, 234, 236, 303, 393, 395, 415–417, 537, 539, 579, 588, 600–602
for imaging of breast cancer 600
for imaging of cancer 600
magnetic 601
semiconductor 84, 200–201, 234, 415
quantum yield (QY) 202–203, 222, 395, 397, 399, 403, 411, 415
QY, *see* quantum yield

radiation, electromagnetic 374–375, 378, 635
radiation delivery 450–451
radiation-induced cancer cells 460
radiation induced cell death mechanisms 456–457, 459
radiation-induced secondary malignancy (RISM) 462
radiation oncologists 446, 449, 451–452
radiation therapy (RT) 22–23, 338, 445–449, 451, 453, 455, 462, 466–467, 471–473, 475, 497
internal 455
nanotechnology-mediated 445–476
primary 462
stereotactic body 448, 453
radio-isotopic nanomaterials 584–585
radioisotopes 75–79, 85, 95, 121, 182, 241, 474–475, 577
radiolabeled nanoparticles (RNPs) 83–85, 87, 89–93, 95, 97
radiolabeled nanoparticles
for cancer diagnosis 75–76, 78, 80, 82, 84, 86, 88, 90, 92, 94, 96
for nuclear imaging 87
radiolabeling 92–93, 95
radionuclide 78, 80, 83–84, 86–88, 97, 474
radiopharmaceuticals 75, 79, 83
radiosensitivity 453, 467, 470–471, 474
radiosensitizers 464–467, 470, 473–475
radiotherapy 3, 338, 340–341, 373–374, 446–447, 453–454, 458–461, 465, 471–472, 474–476, 497, 611

conventional 461
intensity-modulated 449
particle 452–453
Raman imaging of SWCNTs in live cells 223
RBE, *see* relative biological effectiveness
reactive oxygen species (ROSs) 25, 54, 284, 297, 355, 384, 399, 401, 417, 459, 473, 594, 602, 636
relative biological effectiveness (RBE) 453
RENCA cells, non-adherent 598
rhodamine 122, 159, 198, 578, 580, 589
RISM, *see* radiation-induced secondary malignancy
RNA-based therapeutics 548
RNA interference (RNAi) 289, 534, 547
RNAi, *see* RNA interference
RNAi-based therapeutics 551
RNAs 205, 272, 286, 400, 528, 530–531, 535, 543, 547–548, 636
RNPs, *see* radiolabeled nanoparticles
ROSs, *see* reactive oxygen species
RT, *see* radiation therapy

SAR, *see* specific absorption rate
saturation magnetization 115–117, 633
SBRT, *see* stereotactic body radiation therapy
scintillation nanoparticles 419
SCNPs, *see* scintillation nanoparticles
senescent cells 460–461
SERS, *see* surface enhanced Raman spectroscopy
shRNA 528, 540, 542, 547, 550–551, 559
shRNA delivery, polymer nanoparticles for 550
silica nanoparticles 412, 542
mesoporous 152, 412
single photon emission computed tomography (SPECT) 75–79, 81–84, 94–96, 120, 154–157, 183, 475, 584
single-walled carbon nanotubes (SWCNTs) 90–91, 218–224, 226, 391–392, 597–600
in imaging applications 599
metallic 599
siRNA 59, 211, 237, 241, 528, 531, 533–535, 537–542, 547–550, 556–557, 559, 628
loaded 533, 538–539, 542
siRNA delivery 538, 548–549, 556, 584
nanoparticle-mediated 557
polymer nanoparticles for 548
SLNs, *see* solid lipid nanoparticles
small RNA 528–530, 533, 535, 537–539, 542, 547, 549–551, 555, 557, 559
small RNA delivery 528, 531, 533, 535, 537–539, 541–543, 545, 547, 549, 551, 553, 556
carbon nanoparticles for 537, 542
in clinical trials 556–557

nanoparticle-mediated 527–558
organic nanoparticles for 542
structured nanoparticles for 535, 537, 539, 541, 543, 545, 547, 549, 551, 553
SNALPs, *see* stable nucleic acid lipid particles
solid lipid nanoparticles (SLNs) 29, 85
specific absorption rate (SAR) 344, 352
SPECT, *see* single photon emission computed tomography
SPIONs, *see* super paramagnetic iron oxide nanoparticles
spleen 88, 91, 95, 111, 156, 175–176, 178, 180–181, 206, 211, 379, 636–637
SPR, *see* surface plasmon resonance
stable nucleic acid lipid particles 534
stable nucleic acid lipid particles (SNALPs) 534–535
stem cells 154
human mesenchymal 209
stereotactic body radiation therapy (SBRT) 448, 452–455
superparamagnetic iron oxide 114–115
superparamagnetic iron oxide nanoparticles (SPIONs) 61, 121, 124–125, 127–128, 316–317, 346–350, 353–354, 464, 472, 474, 605
surface enhanced Raman spectroscopy (SERS) 124, 204, 215–216, 218, 224, 384, 603
surface plasmon resonance (SPR) 214–215, 302, 376, 378–379, 389, 603
surgery, image-guided 243, 578
SWCNTs, *see* single-walled carbon nanotubes

T-cell receptor (TCR) 496, 514
T-cells 312, 499, 515
TAAs, *see* tumor associated antigens
TAMs, *see* tumor-associated macrophages
target tissues 398–400, 403, 466, 530, 535, 559
targeted delivery 30, 228, 312–313, 348, 407, 413, 532, 535, 540, 549, 574
TAT-functionalized NPs 473
TCR, *see* T-cell receptor
theranostic agents 93, 408, 575, 580, 584, 586–588, 590, 592, 600, 605
theranostic materials 577, 608–609
theranostic nanoparticles 575–577, 589, 610–611
theranostics 123, 152, 159, 240, 357, 573–584, 586, 588, 590, 592, 594, 596, 598, 600, 602, 604, 606, 608, 610–612
therapeutic agents 45, 120, 175, 273, 296, 299, 307, 384–385, 390, 413, 416, 418, 469, 471, 537, 585, 604

therapeutic anticancer drug delivery 544
  polymer nanoparticles for 544
TLRs, *see* toll-like receptors
TME, *see* tumor microenvironment
TNF, *see* tumor necrosis factor
tobacco 2, 10–11
toll-like receptors (TLRs) 496, 502, 638
toxicity 31, 33, 61, 117, 129, 174–175, 180, 184, 205, 242, 348, 352, 408, 446, 461, 469, 475, 531, 535, 539, 546, 590, 594
transferrin 86, 206, 211, 227–228, 288, 354–355, 581, 637
tumor ablation 216, 356
tumor angiogenesis 13–14, 146, 148, 303
tumor associated antigens (TAAs) 499, 510, 512–513
tumor-associated macrophages (TAMs) 13, 509, 516–518
tumor blood vessels 95, 176, 224, 278–279
tumor cell proliferation 291, 411
tumor-cell receptors 21
tumor cells 2, 4–5, 19–21, 24, 32, 60, 83, 90, 118, 210, 228, 311–313, 344, 374, 403, 405, 411, 413–414, 420, 448, 468, 471, 498–499, 501–502, 507–510, 554–555
  fibrosarcoma 414
  nascent 498
  stomach 292
tumor development 13, 18, 551
  recurring 342
tumor diagnosis 143–158, 171–186, 395
tumor growth 5, 13, 15–17, 19–20, 23, 211, 289, 297, 394, 470, 500, 552, 585
tumor imaging systems 84
tumor-immunity cycle 507
tumor immunosurveillance 498, 510
tumor initiation 3
tumor localization 86, 454
tumor metastasis 2–4
tumor microenvironment (TME) 13, 18, 32, 60, 500, 509–510, 516–517
tumor models 93, 297, 518
  human ovarian cancer xenograft 148
tumor necrosis factor (TNF) 302, 458, 638
tumor oxygenation 469
tumor penetration 30, 295
tumor progression 3–4, 18–20, 159
tumor regression 92, 340, 353
tumor resection 22, 31
tumor sites 284, 291, 512
tumor suppressor genes 3–6, 8
tumor vasculature 116, 148, 223, 278–279
tumor vessels 33, 209, 225, 406
tumor volume 305, 381, 592
tumors
  cancerous 358
  hepatic 119
  human 299, 467, 628
  hypo-vascularized 277–278
  malignant 5, 80
  mammary carcinoma xenograft 218
  prostate 382, 584, 610
  responsive 509

solid 15–16, 209, 308, 408, 457, 462, 509, 555, 557, 575, 589
unresponsive 509
xenograft 151

ultra-small particles of iron oxide (USPIOs) 111, 156
ultrasound (US) 76, 120, 124, 126–127, 143–149, 151–157, 159, 195, 217, 243, 284, 289, 308, 341, 375, 451, 549, 577, 581, 583–584, 586, 611
contrast agents 144–145, 152
dual-modal imaging 126–127
enhanced 146–147, 154
upconversion luminescence 96, 235
upconversion luminescence imaging 96
upconversion nanoparticles 230, 406, 414, 420
US, *see* ultrasound
USPIOs, *see* ultra-small particles of iron oxide
UV radiation 11–12

vascular cell adhesion protein 1 (VCAM-1) 118, 146–147
VCAM-1, *see* vascular cell adhesion protein 1
virosomes 288, 293, 316

water 111–112, 114, 154, 156, 178, 197–198, 213, 216, 220, 240, 283–284, 298, 305, 346, 351, 375, 415–417, 550, 588, 602, 634
WBH, *see* whole body hyperthermia
white blood cells 631–632
whole body hyperthermia (WBH) 339–340

X-ray 123, 143, 172, 179, 243, 419, 446, 448, 452, 471, 497, 574, 595, 626, 638
X-ray absorption 175, 182
X-ray beams 172, 184
X-ray CT 180–181, 218
X-Ray CT 171–172, 181, 183, 185, 195, 216, 218
X-ray diffraction (XRD) 242, 638
X-ray imaging 171–172, 178, 183, 536
XRD, *see* X-ray diffraction

For Product Safety Concerns and Information please contact our EU representative GPSR@taylorandfrancis.com Taylor & Francis Verlag GmbH, Kaufingerstraße 24, 80331 München, Germany